Additive Manufacturing:

Breakthroughs in Research and Practice

Information Resources Management Association
USA

IGI Global
DISSEMINATOR OF KNOWLEDGE

Published in the United States of America by
IGI Global
Engineering Science Reference (an imprint of IGI Global)
701 E. Chocolate Avenue
Hershey PA, USA 17033
Tel: 717-533-8845
Fax: 717-533-8661
E-mail: cust@igi-global.com
Web site: http://www.igi-global.com

Library of Congress Cataloging-in-Publication Data

Names: Information Resources Management Association, editor.
Title: Additive manufacturing : breakthroughs in research and practice / Information Resources Management Association, editor.
Other titles: Additive manufacturing (Engineering Science Reference (Publisher))
Description: Hershey, PA : Engineering Science Reference, an imprint of IGI Global, [2020] | Includes bibliographical references and index.
Identifiers: LCCN 2019006766| ISBN 9781522596240 (hardcover) | ISBN 9781522596257 (ebook)
Subjects: LCSH: Three-dimensional printing. | Manufacturing processes.
Classification: LCC TS171.95 .A438 2020 | DDC 621.9/88--dc23 LC record available at https://lccn.loc.gov/2019006766

British Cataloguing in Publication Data
A Cataloguing in Publication record for this book is available from the British Library.

For electronic access to this publication, please contact: eresources@igi-global.com.

List of Contributors

Addo-Tenkorang, Richard / *Aalborg University, Denmark* 116
Akinlabi, Esther T. / *University of Johannesburg, South Africa* 220, 245
Akinlabi, Esther Titilayo / *University of Johannesburg, South Africa* 271
Alabi, Micheal Omotayo / *University of the Witwatersrand, South Africa* 503
Amuda, Muhammed Olawale Hakeem / *University of Lagos, Nigeria & University of Johannesburg, South Africa* 271
Burmaoglu, Serhat / *Izmir Katip Celebi University, Turkey* 478
Canning, Samuel / *Griffith University, Australia* 86
Chiavenna, Andrea / *Institute of Industrial Technologies and Automation, Italy* 394
Dhaimini, Karim / *Alma Mater Studiorum University of Bologna, Italy* 52
Dinon, Tito / *Institute of Industrial Technologies and Automation, Italy* 394
Donnici, Giampiero / *Alma Mater Studiorum University of Bologna, Italy* 52
Doukas, Michael / *University of Patras, Greece* 446
Erinosho, Mutiu F. / *University of Johannesburg, South Africa* 245
Frizziero, Leonardo / *Alma Mater Studiorum University of Bologna, Italy* 52
Garfella-Rubio, José T. / *Universitat Jaume I. Avinguda Sos Baynat s/n, Spain* 163
Gaur, Loveleen / *Amity University, India* 332
Gual-Ortí, Jaume / *Universitat Jaume I. Avinguda Sos Baynat s/n, Spain* 163
Gwangwava, Norman / *Botswana International University of Science and Technology, Botswana* 116
Hartmann, Timo / *Maastricht University, The Netherlands* 136
Jenab, Kouroush / *Society of Reliability Engineering – Ottawa Chapter, Canada* 370
Kumar, Kaushik / *Birla Institute of Technology, India* 321, 353
Liverani, Alfredo / *Alma Mater Studiorum University of Bologna, Italy* 52
Loy, Jennifer / *University of Technology Sydney, Australia* 1, 86
Mahamood, Rasheedat M. / *University of Johannesburg, South Africa & University of Ilorin, Nigeria* 220
Malosio, Matteo / *Institute of Industrial Technologies and Automation, Italy* 394
Máñez-Pitarch, Jesús / *Universitat Jaume I. Avinguda Sos Baynat s/n, Spain* 163
Martínez-Moya, Joaquín A. / *Universitat Jaume I. Avinguda Sos Baynat s/n, Spain* 163
Mohan, Kayalvizhi / *Agni College of Technology, India* 424
Mourtzis, Dimitris / *University of Patras, Greece* 446
Novak, James I. / *Griffith University, Australia* 1
Ogunmuyiwa, Enock / *Botswana International University of Science and Technology, Botswana* 116

Ozdemir, Basak / *Independent Researcher, Turkey* 478
Pathak, Vimal Kumar / *Malaviya National Institute of Technology, India* 207
Pityana, Sisa / *National Laser Centre, South Africa* 220, 245
Prakash, Chander / *Lovely Professional University, India* 296
Prini, Alessio / *Institute of Industrial Technologies and Automation, Italy* 394
Ramakrishnan, Ravi / *Amity University, India* 332
Ruttico, Pierpaolo / *Politecnico di Milano, Italy* 25
Scano, Alessandro / *Institute of Industrial Technologies and Automation, Italy* 394
Shukla, Mukul / *University of Johannesburg, South Africa & MNNIT Allahabad, India* 220
Singh, Amit Kumar / *Malaviya National Institute of Technology, India* 207
Singh, Sunpreet / *Lovely Professional University, India* 296
Spagnuolo, Giulio / *Institute of Industrial Technologies and Automation, Italy* 394
Tosatti, Lorenzo Molinari / *Institute of Industrial Technologies and Automation, Italy* 394
Ude, Albert U / *Botswana International University of Science and Technology, Botswana* 116
Uthayakumar, M. / *Kalasalingam University, India* 296
Vanpoucke, Evelyne / *Université Libre de Bruxelles, Belgium* 136
Weinsier, Philip D. / *Bowling Green State University – Firelands, USA* 370
Yalcin, Haydar / *Izmir Katip Celebi University, Turkey* 478
Yayla, Kemal / *Ege University, Turkey* 478
Zindani, Divya / *National Institute of Technology Silchar, India* 321, 353

Table of Contents

Preface x

Section 1
Design and Production

Chapter 1
Digital Technologies and 4D Customized Design: Challenging Conventions With Responsive Design 1
James I. Novak, Griffith University, Australia
Jennifer Loy, University of Technology Sydney, Australia

Chapter 2
Innovative Construction Systems: New Design Aesthetics Through Competitive Realization Processes 25
Pierpaolo Ruttico, Politecnico di Milano, Italy

Chapter 3
Design for Additive Manufacturing and Advanced Development Methods Applied to an Innovative Multifunctional Fan 52
Leonardo Frizziero, Alma Mater Studiorum University of Bologna, Italy
Giampiero Donnici, Alma Mater Studiorum University of Bologna, Italy
Alfredo Liverani, Alma Mater Studiorum University of Bologna, Italy
Karim Dhaimini, Alma Mater Studiorum University of Bologna, Italy

Chapter 4
Clash of Cultures: Fashion, Engineering, and 3D Printing 86
Jennifer Loy, Griffith University, Australia
Samuel Canning, Griffith University, Australia

Section 2
Modeling

Chapter 5
Cloud Based 3D Printing Business Modeling in the Digital Economy ... 116
Norman Gwangwava, Botswana International University of Science and Technology, Botswana
Albert U Ude, Botswana International University of Science and Technology, Botswana
Enock Ogunmuyiwa, Botswana International University of Science and Technology, Botswana
Richard Addo-Tenkorang, Aalborg University, Denmark

Chapter 6
User Acceptance of Technologies in Their Infancy: The Case of 3D Printing Business Models ... 136
Timo Hartmann, Maastricht University, The Netherlands
Evelyne Vanpoucke, Université Libre de Bruxelles, Belgium

Chapter 7
Study on Different Graphic Representations in Architectural Heritage: Digital and Physical Modelling ... 163
José T. Garfella-Rubio, Universitat Jaume I. Avinguda Sos Baynat s/n, Spain
Jesús Máñez-Pitarch, Universitat Jaume I. Avinguda Sos Baynat s/n, Spain
Joaquín A. Martínez-Moya, Universitat Jaume I. Avinguda Sos Baynat s/n, Spain
Jaume Gual-Ortí, Universitat Jaume I. Avinguda Sos Baynat s/n, Spain

Section 3
Processes and Techniques

Chapter 8
A Particle Swarm Optimization Approach for Minimizing GD&T Error in Additive Manufactured Parts: PSO Based GD&T Minimization ... 207
Vimal Kumar Pathak, Malaviya National Institute of Technology, India
Amit Kumar Singh, Malaviya National Institute of Technology, India

Chapter 9
Improving Surface Integrity Using Laser Metal Deposition Process ... 220
Rasheedat M. Mahamood, University of Johannesburg, South Africa & University of Ilorin, Nigeria
Esther T. Akinlabi, University of Johannesburg, South Africa
Mukul Shukla, University of Johannesburg, South Africa & MNNIT Allahabad, India
Sisa Pityana, National Laser Centre, South Africa

Chapter 10
Enhancement of Surface Integrity of Titanium Alloy With Copper by Means of Laser Metal Deposition Process 245
Mutiu F. Erinosho, University of Johannesburg, South Africa
Esther T. Akinlabi, University of Johannesburg, South Africa
Sisa Pityana, National Laser Centre, South Africa

Chapter 11
Trend and Development in Laser Surface Modification for Enhanced Materials Properties 271
Muhammed Olawale Hakeem Amuda, University of Lagos, Nigeria & University of Johannesburg, South Africa
Esther Titilayo Akinlabi, University of Johannesburg, South Africa

Chapter 12
Recent Advancements in Customized Investment Castings Through Additive Manufacturing: Implication of Additive Manufacturing in Investment Casting 296
Sunpreet Singh, Lovely Professional University, India
Chander Prakash, Lovely Professional University, India
M. Uthayakumar, Kalasalingam University, India

Section 4
Prototyping

Chapter 13
Integrated Manufacturing System for Complex Geometries: Towards Zero Waste in Additive Manufacturing 321
Divya Zindani, National Institute of Technology Silchar, India
Kaushik Kumar, Birla Institute of Technology, India

Chapter 14
Innovation in Product Design: IoT Objects Driven New Product Innovation and Prototyping Using 3D Printers 332
Ravi Ramakrishnan, Amity University, India
Loveleen Gaur, Amity University, India

Chapter 15
Selection of Prototyping Process and Part Orientation for Virtually Manufactured Gears 353
Divya Zindani, NIT Silchar, India
Kaushik Kumar, Birla Institute of Technology, India

Chapter 16
Complexity Analysis in Additive Manufacturing for the Production of Tissue Engineering Constructs 370
Kouroush Jenab, Society of Reliability Engineering – Ottawa Chapter, Canada
Philip D. Weinsier, Bowling Green State University – Firelands, USA

Chapter 17
The "Arm" Line of Devices for Neurological Rehabilitation 394
Alessandro Scano, Institute of Industrial Technologies and Automation, Italy
Andrea Chiavenna, Institute of Industrial Technologies and Automation, Italy
Tito Dinon, Institute of Industrial Technologies and Automation, Italy
Alessio Prini, Institute of Industrial Technologies and Automation, Italy
Giulio Spagnuolo, Institute of Industrial Technologies and Automation, Italy
Matteo Malosio, Institute of Industrial Technologies and Automation, Italy
Lorenzo Molinari Tosatti, Institute of Industrial Technologies and Automation, Italy

Chapter 18
Recent Trends in 3D Printing of Dental Models: Rapid Prototyping in Dental Implants 424
Kayalvizhi Mohan, Agni College of Technology, India

Section 5
Trends and Future Directions

Chapter 19
The Evolution of Manufacturing Systems: From Craftsmanship to the Era of Customisation 446
Dimitris Mourtzis, University of Patras, Greece
Michael Doukas, University of Patras, Greece

Chapter 20
3D Printing Technology Diffusion: A Revolution or an Illusion? 478
Kemal Yayla, Ege University, Turkey
Basak Ozdemir, Independent Researcher, Turkey
Serhat Burmaoglu, Izmir Katip Celebi University, Turkey
Haydar Yalcin, Izmir Katip Celebi University, Turkey

Chapter 21
Big Data, 3D Printing Technology, and Industry of the Future 503
Micheal Omotayo Alabi, University of the Witwatersrand, South Africa

Index 521

Preface

Additive manufacturing (AM) is the process of building 3D objects by adding layer upon layer of materials, which can include plastic, metal, concrete, and other common building substances, until the item is complete. This seemingly simplistic process is continually advancing and can be applied in a variety of fields and industries including business, manufacturing, and medicine. As this emerging technology grows, new materials are explored in the construction of objects, such as human tissues, which lead to revolutionary advancements. In its current state, AM equipment uses computer-aided design (CAD) to add layers of the required material to ultimately create the desired 3D object.

Although additive manufacturing is the umbrella term used to describe the layering process, the term espouses numerous sub-manufacturing technologies such as 3D printing, rapid prototyping (RP), direct digital manufacturing (DDM), layered manufacturing, and additive fabrication. Benefits of this approach include speed, greater range of producible shapes, stronger materials, less wasted materials, and more applications across various disciplines and industries. Moving forward, medical industries aim to further utilize 3D printing and additive manufacturing to produce artificial organs and tissue to aid in patient care and recovery at a quicker pace than previously imagined.

While additive manufacturing can have its drawbacks, such as slow production time and additional effort and time in the application design phase, businesses are rapidly leaning towards this process for their production needs. Despite these drawbacks, additive manufacturing requires no assembly and creates little product waste while ensuring any changes to the final design are easily and quickly made, creating a production environment that will outweigh most cost or build disadvantages.

The everchanging landscape surrounding the diverse applications of different scientific areas can make it very challenging to stay on the forefront of innovative research trends. That is why IGI Global is pleased to offer this one-volume comprehensive reference that will empower engineers, researchers, developers, manufacturers, factory managers, students, researchers, practitioners, technicians, engineering designers, and academicians with a stronger understanding of additive manufacturing.

This compilation is designed to act as a single reference source on conceptual, methodological, and technical aspects, and will provide insight into emerging topics including but not limited to 3D and 4D design and printing, laser modification, enhanced materials, medical technology, and print manufacturing systems. The chapters within this publication are sure to provide readers with the tools necessary for further research and discovery in their respective industries and/or fields.

Additive Manufacturing: Breakthroughs in Research and Practice is organized into five sections that provide comprehensive coverage of important topics. The sections are:

1. Design and Production;
2. Modeling;

3. Processes and Techniques;
4. Prototyping; and
5. Trends and Future Directions .

The following paragraphs provide a summary of what to expect from this invaluable reference source:

Section 1, "Design and Production," opens this extensive reference source by highlighting the latest trends in digital design and innovative printing systems. The first chapter in this section, "Digital Technologies and 4D Customized Design: Challenging Conventions With Responsive Design," by authors Prof. James I. Novak from Griffith University, Australia and Prof. Jennifer Loy from the University of Technology Sydney, Australia, highlights how "4D products" respond to the vision laid out three decades ago for ubiquitous computing and how they have the potential to enhance human experiences by creating more seamless human-centered relationships with technology. In the second chapter, "Innovative Construction Systems: New Design Aesthetics Through Competitive Realization Processes," Prof. Pierpalo Ruttico from Politecnico di Milano, Italy, analyzes adaptive and flexible systems that facilitate the design and production processes of mass-customized building components through digital fabrication and assembly. In the following chapter, "Design for Additive Manufacturing and Advanced Development Methods Applied to an Innovative Multifunctional Fan," the authors, Prof. Leonardo Frizziero, Prof. Giampiero Donnici, Prof. Alfredo Liverani, and Prof. Karim Dhaimini from Alma Mater Studiorum University of Bologna, Italy, study the entire product development process taking advantage of the most modern models and technologies for the final realization of a case study that involves the design and prototyping of an innovative multifunctional fan (lamp, aroma diffuser, and fan) through the multi-jet fusion of HP. In the final chapter of this section, "Clash of Cultures: Fashion, Engineering, and 3D Printing," Prof. Jennifer Loy and Prof. Samuel Canning of Griffith University, Australia provide an example of how fashion design has become a creative stimulus for the development of the technology by illustrating how disruptive creativity has the potential to advance scientific research, with the two worlds of engineering and fashion coming together through a collaboration with industrial design.

Section 2, "Modeling," includes chapters on emerging innovations in cloud-based printing models and new technologies. In the first chapter in this section, "Cloud Based 3D Printing Business Modeling in the Digital Economy," Prof. Noman Gwangwava, Prof. Albert U. Ude, Prof. Enock Ogunmuyiwa from Botswana International University of Science and Technology, Botswana and Prof. Richard Addo-Tenkorang from Aalborg University, Denmark present a business case for a cloud-based approach in consumer product niches such as 3D printing. In the following chapter, "User Acceptance of Technologies in Their Infancy: The Case of 3D Printing Business Models," the authors, Prof. Timo Hartmann from Maastricht University, The Netherlands and Prof. Evelyn Vanpoucke from Université Libre de Bruxelles, Belgium, study user acceptance of early stage technology, in particular 3D printing, for home users. Within the final chapter of this section, "Study on Different Graphic Representations in Architectural Heritage: Digital and Physical Modelling," the authors, Prof. José T. Garfella-Rubio, Prof. Jesús Máñez-Pitarch, Prof. Joaquín A. Martínez-Moya, and Prof. Jaume Gual-Ortí from Universitat Jaume I, Spain, analyze the advantages and disadvantages of additive printing, 3D scanners, and digital representations in architectural heritage studies.

Section 3, "Processes and Techniques," presents coverage on novel strategies to further laser metal productivity and customization in modern manufacturing. The first chapter in the section, "A Particle Swarm Optimization Approach for Minimizing GD&T Error in Additive Manufactured Parts: PSO Based GD&T Minimization," by authors Prof. Vimal Kumar Pathak and Prof. Amit Kumar Singh from

Malaviya National Institute of Technology, India, presents a particle swarm optimization (PSO) approach to improve the geometrical accuracy of additive manufacturing (AM) parts by minimizing geometrical dimensioning and tolerancing (GD&T) error. In the following chapter, "Improving Surface Integrity Using Laser Metal Deposition Process," the authors, Prof. Rasheedat M. Mahamood from University of Johannesburg, South Africa & University of Ilorin, Nigeria; Prof. Esther T. Akinlabi from University of Johannesburg, South Africa; Prof. Mukul Shukla from University of Johannesburg, South Africa & MNNIT Allahabad, India; and Prof. Sisa Pityana from the National Laser Centre, South Africa, establish the LMD in the production of complex and one of its kind parts, its ability to improve surface properties, repair high-valued parts, and reduce the buy-to-fly ratio in the production of aerospace parts, and highlights the use of non-traditional finishing techniques on laser deposited parts to further improve the surface integrity of components. Within the third chapter of this section, "Enhancement of Surface Integrity of Titanium Alloy with Copper by Means of Laser Metal Deposition Process," the authors, Prof. Mutiu F. Erinosho from the University of Johannesburg, South Africa; Prof. Esther T. Akinlabi from University of Johannesburg, South Africa; and Prof. Sisa Pityana from the National Laser Centre, South Africa, provide the surface modification of titanium alloy with the addition of percentage range of Cu within its lattices as well as the results obtained from the characterizations conducted on the laser deposited Ti6Al4V/Cu alloys. In one of the concluding chapters of this section, "Trend and Development in Laser Surface Modification for Enhanced Materials Properties," the authors, Prof. Muhammed Olawale Hakeem Amuda from University of Lagos, Nigeria & University of Johannesburg, South Africa and Prof. Esther Titilayo Akinlabi from the University of Johannesburg, South Africa, present a process review of the commonly available laser surface modification techniques for surface property enhancement. The final chapter, "Recent Advancements in Customized Investment Castings Through Additive Manufacturing: Implication of Additive Manufacturing in Investment Casting," by authors Prof. Sunpreet Singh from Lovely Professional University, India; Prof. Chander Prakash from Lovely Professional University, India; and Prof. M. Uthayakumar from Kalasalingam University, India, discusses additive manufacturing as it assists in conventional investment casting.

Section 4, "Prototyping," discusses coverage and research perspectives on innovations in product design, complex manufacturing systems, and virtual constructs. The first chapter in the section, "Integrated Manufacturing System for Complex Geometries: Towards Zero Waste in Additive Manufacturing," by Prof. Divya Zindani from the National Institute of Technology Silchar, India and Prof. Kaushik Kumar from Birla Institute of Technology, India, proposes an integrated manufacturing system consisting of three main components: digital prototyping, physical prototyping, and lost core technology. The following chapter, "Innovation in Product Design: IoT Objects Driven New Product Innovation and Prototyping Using 3D Printers," by Prof. Ravi Ramakrishnan and Prof. Loveleen Gaur from Amity University, India, aims to understand how the introduction of IoT sensors embedded in customer appliances or wearable's sending real time customer information coupled with rapid prototyping using remotely located 3D printers can help address design considerations for new products, and provides an overview of how using IoT data and 3D printers for new product development and prototyping as an early stage activity can be done without using human imagination of restricted market survey data. In a noteworthy chapter in this section, "Selection of Prototyping Process and Part Orientation for Virtually Manufactured Gears," the authors, Prof. Divya Zindani from the National Institute of Technology Silchar, India and Prof. Kaushik Kumar from Birla Institute of Technology, India, explore the transformation of a virtually produced gear to a physical product using rapid prototyping technique. Within the following chapter, "Complexity Analysis in Additive Manufacturing for the Production of Tissue Engineering Constructs," the authors,

Prof. Kouroush Jenab from the Society of Reliability Engineering – Ottawa Chapter, Canada and Prof. Philip D. Weinsier from Bowling Green State University – Firelands, USA, report an analytical model, which is able to rank AM processes based on their relative complexities, for evaluating process complexity that takes into account uncertain situations and additive manufacturing process technologies. In another noteworthy chapter, "The 'Arm' Line of Devices for Neurological Rehabilitation," the authors, Prof. Alessandro Scano, Prof. Andrea Chiavenna, Prof. Tito Dinon, Prof. Alessio Prini, Prof. Giulio Spagnuolo, Prof. Matteo Malosio, and Prof. Lorenzo Molinari Tosatti from the Institute of Industrial Technologies and Automation, Italy, explore the benefits of "the Arm," a line of prototypical devices for rehabilitating the upper limb. In the concluding chapter of this section, "Recent Trends in 3D Printing of Dental Models: Rapid Prototyping in Dental Implants," Prof. Kayalvizhi Mohan from Agni College of Technology, India discusses the advantages and disadvantages in 3D printing in dentistry.

Section 5, "Trends and Future Directions," discusses emerging areas in manufacturing systems and forthcoming trends in 3D printing technology. The first chapter in this section, "The Evolution of Manufacturing Systems: From Craftsmanship to the Era of Customisation," by authors Prof. Dimitris Mourtzis and Prof. Michael Doukas from the University of Patras, Greece, specifies the evolution of manufacturing paradigms, their basic principles, and the links among them. In the following chapter, "3D Printing Technology Diffusion: A Revolution or an Illusion?" the authors, Prof. Kemal Yayla from Ege University, Turkey; Prof. Basak Ozdemir, an Independent Researcher from Turkey; Prof. Serhat Burmaoglu from Izmir Katip Celebi University, Turkey; and Prof. Haydar Yalcin from Izmir Katip Celebi University, Turkey, review 3D printing technology to understand its diffusion among manufacturing businesses. In the final chapter of this section, "Big Data, 3D Printing Technology, and Industry of the Future," Prof. Michael Omotayo Alabi from the University of the Witwatersrand, South Africa presents the recent trends in 3D printing technologies, big data and Industry 4.0, including the benefits and the application areas of these technologies, as well as emerging and near future application areas of 3D printing, and possible future research areas in 3D printing and big data technologies as relating to Industry 4.0.

Although the primary organization of the contents in this work is based on its five sections, offering a progression of coverage of the important concepts, methodologies, technologies, applications, social issues, and emerging trends, the reader can also identify specific contents by utilizing the extensive indexing system listed at the end.

Section 1
Design and Production

Chapter 1
Digital Technologies and 4D Customized Design:
Challenging Conventions With Responsive Design

James I. Novak
Griffith University, Australia

Jennifer Loy
University of Technology Sydney, Australia

ABSTRACT

Digital design tools are rapidly changing and blurring the boundaries between design disciplines. By extension, the relationship between humans and products is also changing, to the point where opportunities are emerging for products that can co-evolve with their human users over time. This chapter highlights how these '4D products' respond to the vision laid out three decades ago for ubiquitous computing, and have the potential to enhance human experiences by creating more seamless human-centered relationships with technology. These developments are examined in context with broader shifts in sociocultural and environmental concerns, as well as similar developments being researched in Responsive Architecture, 4D printing and systems designed to empower individuals during the design process through interactive, parametric model platforms. Technology is fundamentally changing the way designers create physical products, and new understandings are needed to positively guide these changes.

INTRODUCTION

Human-centered product design, where the end users of a product remain at the forefront of all design decisions, is evolving rapidly as the opportunities of the digital age become better understood. The boundaries between design disciplines are blurring as digital technologies break down traditional practices, creating products that challenge design conventions. This chapter discusses the significance of changes to product design and design systems enabled by digital technologies in relation to human

DOI: 10.4018/978-1-5225-9624-0.ch001

development. In particular it considers how the relationship between people and products could change with a shift in thinking from static, resolved outcomes to digitally enabled products, in particular ones that are capable of changing over time—4D products—and the implications of that change. This affects the designer-product relationship, and the customer-product relationship as well as the systems within which they operate.

Examples of practice-led design research in the latter part of this chapter introduce different approaches to the development of 4D products. These provide starting points for a practical and theoretical framework for design research and education in the changing digital design environment from user experience, through interaction, to a digitally enabled product service system. Designs based on ubiquitous computing allow for the development of products that change the relationships and experiences in human-centered product design. Based on additive manufacturing processes, 4D printing allows for products that morph under certain conditions, as do examples from the field of responsive architecture (Meagher, 2014). These approaches challenge designers to think less about static, final outcomes, and more about the opportunities for objects to evolve into different states through their lifespan. As such, designers are called on to develop systems and parameters for product permutations, rather than fixed product outcomes, and rethink their role and relationship to products and users, and the impact of their designs in a digitally connected world. This aligns with current theory on the relationship between human evolution and digital technologies.

HUMAN EVOLUTION IN A DIGITAL ERA

There is an argument that the biological mechanisms that have governed human evolution for 3.5 million years have been disrupted by the development of human cognition and cultural behaviors, overwhelming natural systems, and resulting in what is termed "Human Evolutionary Stasis" (Powell, 2012). The suggestion is that humans have the ability to collectively circumvent the challenges they may otherwise face as individuals, and that this is impacting the biological evolution of the species as a whole.

The human organism is a paradigmatic case of ontogenetic adaptation: thanks to an enormously flexible cognitive and behavioral repertoire, including the ability to acquire and transmit cumulative (intergenerational) cultural adaptations, humans can survive and reproduce across a wide range of otherwise hostile developmental conditions. (Powell, 2012, p. 150)

Yet the impact of technological development on early learning and ontogenetic adaptation could be argued to be challenging the idea of a stalled evolution of the species. If human evolution is seen as referring to its adaptation to the complex systems in which humans operate, then human development in a technological age is evidenced by the ability of each successive generation to adapt more quickly to evolving digital systems:

Gen Y have grown up in a world of rapid technological advances affecting the way they learn, their approach to knowledge acquisition and the forms of interaction between themselves… as a result of their techno-dependency and the fact they are accustomed to using computers and Internet to perform any given task, Gen Y has formed unique characteristics and competences… (Petrova, 2014, p. 525)

As the pace of technological innovation increases, there is a tendency to assume that humans will continue to adapt cognitively to keep pace, yet, as suggested by Al Gore, humans will soon be in a situation where their cognitive abilities are surpassed by their own inventions:

The emergence of a planet-wide electronic communications grid connecting the thoughts and feelings of billions of people and linking them to rapidly expanding volumes of data, to a fast growing world of sensors being embedded ubiquitously throughout the world, and to increasingly intelligent devices, robots, and thinking machines, the smartest of which already exceed the capabilities of humans….and may soon surpass us in manifestations of intelligence we have always assumed would remain the unique province of our species. (Gore, 2013, xiv)

For designers operating in a world where technological innovation could assume an intelligence that drives change itself—beyond human cognition—it is tempting to focus on the potential of technology to create new object mitigated interactions and complex support systems without taking into account the implications for humanity. This is a pitfall into which the commercial world has already fallen: "Could the people who began the Industrial Revolution foresee the ecological effects and loss of life caused by the rise of factory systems, chemical manufacturing, machine tools, coal burning, and mining?" (St Clair, 2016, p. 42) The publication of *Wealth of Nations* through to the development of the moving assembly line by Ford (Sparke, 1987) and the establishment of the Ford Motor Co in 1903 could hardly have been expected to lead to the economic problems for Detroit in the 1950s. Understanding economic and social implications in a period of significant change are consistently problematic. Understanding the implications for the development of the human psyche, already linked to cognitive behavior changes, brought on by technological immersion at an early age of subsequent generations in a rapidly evolving digital age, are nigh on impossible. Who could have predicted the overwhelming initial response to the launch of the augmented reality Nintendo game of Pokémon Go (*Pokemon Go*, 2016) in 2016 and the unexpected consequences of players, such as being injured through inattention to their surroundings and collaborating with strangers, all in the pursuit of Pokémon?

The challenge for all people, from designers creating new products to the end users openly embracing them, is that "we are morphing so fast that our ability to invent new things outpaces the rate we can civilize them. These days it takes us a decade after a technology appears to develop a social consensus on what it means" (Kelly, 2010, p.3). Similarly Gore argues that there have never before been "so many revolutionary changes unfolding simultaneously and converging with one another" and that there is a "clear consensus that the future now emerging will be extremely different from anything we have ever known in the past. It is a difference not of degree but of kind" (2013, xv). His reasoning is that global digital connectivity, the expansion of information collection and data mining techniques are fostering the emergence of a collective "global mind" that requires a re-imagining of human interaction and organization (Gore, 2013).

One of the inevitabilities widely assumed over the last 30 years has been cultural homogenization through globalization. A consequence of the digital connectivity described by Gore is that country borders become porous, and after an unprecedented era of relative stability across the globe allowing for increased trade and co-operation, the predictions have been for a linear pattern of growth in the amalgamation of individual countries into larger trade groups, such as in the European Union. Yet the signs in 2016 are that globalization is facing a crisis. The fallout from the Brexit vote, the crisis in France under the onslaught of ISIS, and the concerns in Germany over their open borders policy during the Syrian refugee crisis provide contrary indications that suggest the possibility of a trend reversal.

Within product design, there is evidence of a corresponding drive for diversity over homogenization, with cultural difference and individuality paradoxically supported by the very digital technology predicted to alienate people from production. While Aldersey-Williams' (2011) predictions of a return to a pre-industrial era of individual craftsmanship may have sounded farfetched in 2011, the exponential growth of the digitally enabled maker society highlighted by Anderson (2012) suggests there could be a change in paradigm, where people become significantly more engaged in the design and development of the products around them.

This shift in thinking is already resulting in a new breed of digitally enabled product designer, who uses generative modeling techniques as part of their process, only possible because new Computer-Aided Design (CAD) software and 3D printing (additive manufacturing) allows for individual objects to be built without tooling. Designers such as Lionel Dean (Dean, 2016) and Fung Kwok Pan ("Fung Kwok Pan: Fluid Vase," 2010) evidence this trend where the designer and consumer both contribute to decision making on the form of the product, communicating through digital platforms. Also in both these cases, the designers are exploring the interaction itself and the power of digital communication facilitated by additive manufacturing. Recognizing the human-centered aspects to the development of products in a digital age remains at the core of the role of the designer. Yet the nature of that human-centered focus is changing as the digital age matures, and designers will have to renegotiate their role and responsibilities alongside it, while also keeping pace with broader design theory including the sustainability imperative and systems thinking.

PEOPLE, PRODUCTS AND SYSTEMS

Over the last 20 years, design theory has moved the focus of design from discrete products to the design of systems (Hawken, Lovins & Lovins, 2000). This strategy began as a response to economic drivers with practices such as "Lean Manufacturing" and "Just in Time" extending out to the re-evaluation of the supply chain to a value chain approach, but it gained momentum with the sustainability imperative introduced to design and manufacturing in response to the Brundtland commission report in 1987. Technology and the digital age in which humans now live is allowing for a more holistic understanding of the lifecycle of designs not previously possible, with quantifiable measurement and standardization at all stages making designers increasingly accountable for their decisions. It is only in the current era that this complex connectivity is becoming overt, changing the relationship between designers, products and consumers. However, these rapid changes in information gathering and technical possibility need to take into account the corresponding changes to the collective human psyche mentioned in the previous section, as well as the specific experiences of individuals. The scope, pace and scale of change currently being experienced have not previously been part of human evolutionary history. Therefore the impact of current thinking in technology as it informs the design of products and its subsequent impact on humans requires sufficient attention, and the responsibilities designers face for affecting ontogenetic adaptation need to be addressed.

It is essential for humans at this time in our evolution to begin to identify what is being lost through our utter dependence on technology and to determine how it may affect humanity long term. Furthermore, we must not think only of the long-term effects: we must also ask ourselves about the intentions of our actions. (St Clair, 2016, p. 43)

The impact on human development and the realities of the human response to the covert introduction of technology into everyday living is not linear and predictable. There are unexpected consequences, just as there were with the impact of manufacturing on the health of the planet, and the human response can be equally unexpected, even perverse, and often emotional—as illustrated by the recent resurgence in "low-tech" vinyl record sales, for example, despite the proliferation of high quality digital music. One of the issues is the technology promise tends to focus on ideas of simplifying the human experience, and reducing choice, rather than focusing on user experience and the individual needs of people to be unique and valued for their differences. The growing move to cloud-based services and intangible goods can clash with a human need to be connected to the tangible world. Focusing on the direct marketing aspects in transactions with online travel companies, clothing shops and bookshops, ignores the very human need for the hunt—for meeting the curiosity cognitive driver identified in the updated Maslow's hierarchy (McLeod, 2016).

The problem of information technology (IT) driven change is that the user-centered experiences are rarely understood or catered for, let alone central to the product discourse. Digitally enabled user-experience products have typically evolved from user-interface, screen-based product design. IT professionals are not traditionally trained to focus on the user experience beyond the boundaries of the screen in the way that product designers are, but, equally, product design disciplines do not usually contain the technical understanding of electronics and coding needed to maximize the possibilities for digitally enabled design. Yet, that human understanding and focus are vital for design—both in a business sense and in a broader, social and human development point of view.

As technology evolves, designers and IT specialists alike have the opportunity to integrate more functionality, data capture, information sorting and communication into everyday products, but the challenge is how to keep a level head to maximize integrated digital technologies for actual benefit to the users. Evgeny Morozov describes this modern dilemma as "solutionism": when a designer "presumes rather than investigates the problems that [the product] is trying to solve" (2013, p. 6) and too readily adopts new technologies without thoroughly understanding the human-centered problem—if there is even a problem at all. With the rapid pace of technological adoption, the opportunity to slow down to understand all facets of a problem as suggested by Morozov can be a challenge. If one of the fundamental characteristics predicted for the post digital revolution era is that technology will evolve at an ever more rapid pace, then creating products that center the human experience within that change become even more important. Helping individuals to define and support their identities should be a driver for design in an era of homogenization. Equally, supporting the confidence of the individual to navigate in a technological era should be vital for designers, rather than losing all decision making power to machines, as emphasized by St Clair (2016).

Developing users who are effective lifelong learners matters in a time of technological development, as does designing products that are fully expected to evolve as situations and needs change. Developing lifelong learners depends on empowering the individual to take control of their own learning going forward. In her work on student learning in higher education, Weimer (2002) discusses the importance of shifting the balance of power in the learning experience from the lecturer to the student. She argues that this enables the student to build resilience in learning terms to facing new challenges that are initially difficult. If this approach is applied to design, then for designers to create products that allow for an evolution of use or interaction, they must fundamentally empower the user during use, rather than disempowering the individual in service of the organizational system. This making of the individual experience subservient to support the efficiencies in management practices is short sighted and could well lead to a very human rejection of technologically based products.

Collaboration between disciplines in developing technology-driven products is becoming essential in the current sociocultural and environmental sustainability context (Loy, Canning, & Haskell, 2016). The goal is arguably to reinforce identity and sense of self, supporting diversity in a global community, by using technology to collaborate to enhance the human experience and sense of engagement and responsibility, rather than to make interaction more efficient, or organizationally more effective.

DESIGNING IN CONTEXT

For the last 20 years, designers have been striving to design more "lightly"; that is, to design to reduce the impact of the manufacture, use, and disposal of their products on the environment. This has extended to include the social impact of their products for their full lifecycle right from the impact on communities of sourcing raw materials (e.g. foreign-owned mono-culture plantations on local communities dependent on natural, diverse forestation, or the mining of bauxite for aluminum smelting on aboriginal communities and farming communities). In *Design Activism: Beautiful Strangeness in a Sustainable World*, Fuad-Luke (2009) looked at the discourses and context informing design strategies at a point in time. Hindsight highlights the unexpected consequences of design decisions, suggesting that the information available at a point in time is context specific, and that there could be layers of impact initially unknown that could eventually become clear. The impact on human development is one of those layers. So, just as sustainable design theory has shifted the focus of product design from a "cradle to grave" approach to a "cradle to cradle" approach (Braungart & McDonough, 2002), then correspondingly an approach to design taking into account human development needs to be considered equally "lightly". "In considering 'value' during product definition, then, we should expand this usual definition to include intangibles such as personal aspirations and model codes of behavior" (Rowland, Goodman, Charlier, Light, & Lui, 2015, p. 187).

This moral imperative suggests designers need to consider their work even more holistically, as part of a complex team that includes sociologists and psychologists. However, with the bourgeoning digital technology environment, IT specialists need to be integrated into that team more effectively than in current practice. Where user-interface design is morphing into user-experience design, the body of knowledge in product design is lost or diluted. A new approach is needed, perhaps one whereby designers are able to adopt new technologies to take greater control of—or at least have a greater understanding of—the design solution in its entirety. Without a vantage point to see both the problem and solution in entirety, it is difficult for a designer to appreciate the impact of their designs on the users, and on stakeholders indirectly affected during the product's lifecycle.

Designer-maker John Makepeace (1995) described good design as being informed by the knowledge and understandings available at a particular point in time. He also argued that good design embodies the aspirations for society of the period, expressing its values and ideals. Rowland et al. suggest that ideas of value and aspirations are linked to what people care about: "If your business model involves helping people feel closer to specific dreams, it's worth taking time to clearly articulate what those dreams are" (Rowland et al., 2015, p.187). Whether individuals currently care sufficiently about the impact of the digital age on human development in order for it to be overtly mandated as a design imperative is open to debate, but for the design discipline, leading the development of products post the digital revolution, these issues need to be articulated and discussed. Designers working to Makepeace's ideal should not ignore the potential of their work to influence human development and, equally, should recognize that where humans are concerned, the impacts are impossible to predict and they should tread lightly.

The main learning from the pace of technological development is that change is perpetual, and the current rate of change should be a key consideration for the development of products. Aligned with the recognition that agile and evolving products are part of a product service systems approach is the planned "graceful degradation" of products (products that decline to obsolescence with as little harm as possible) and the "cradle to cradle" product lifecycle ideal that addresses the sustainability imperative (economic, environmental and sociological). Extending the essential criteria of this approach to a new era of data generation, collection and analysis, and technologically enhanced functionality, designers need to focus on the fundamentals of really designing for people and for change, not impose values and behaviors on users for the sake of efficiencies and technical enhancements. Just as sustainability theory has been extended to include socio-cultural sustainability in the face of challenges to cultural diversity brought about by digital connectivity and globalization (Loy et al., 2016), so design theory needs to respond to current thinking around supporting the development of the human psyche, and human learning after the digital revolution. The following sections of this chapter describe ideas on the development of responsive products as the role and responsibilities of the designer change again.

THE CHANGING ROLE OF DESIGNERS

A key theme of this chapter is that technological development is becoming symbiotic with humanity. Indeed, Kelly, cofounder of *Wired* magazine, argues that technology can be viewed as a living force:

The technium [the broader interconnected system of technologies] is now as great a force in our world as nature, and our response to the technium should be similar to our response to nature. We can't demand that technology obey us any more than we can demand that life obey us. Sometimes we should surrender to its lead and bask in its abundance, and sometimes we should try to bend its natural course to meet our own. (Kelly, 2010, p. 17)

In Kelly's view, if technology is considered as a living force then the next logical step forward is for "living products" or products that act independently, directed by technologies referred to as ubiquitous computing. The term ubiquitous computing can be used interchangeably with the terms pervasive computing and physical computing (Greenfield, 2006, p. 1), and is credited to the seminal paper by Mark Weiser in 1991 titled "The Computer for the 21st Century." In this paper Weiser describes the most profound technologies as "…those that disappear. They weave themselves into the fabric of everyday life until they are indistinguishable from it" (1991, p.1). These go beyond conspicuous computers, such as smart phones and tablets, to more rudimentary products that have sensors and computing power in them, as seamlessly gathering and processing data, and reacting to this information.

The Internet of Things (IoT) is an extension of ubiquitous computing, with the key difference described by Adrian McEwen and Hakim Cassimally as the element of "the Internet" (2013, p. 10). Assa Ashuach's AI light (*Assa Ashuach Studio*, 2016) is an example of ubiquitous computing because it responds to movement in a room autonomously, without needing to be connected to the Internet for control. In contrast, home LED lighting solutions that can be controlled via mobile phone applications, such as the Philips Hue (*Philips Hue*, 2016) are examples of Internet of Things solutions where the product relies on being able to connect to other devices via the Internet in order to function.

While it is clear that many elements of theorist Weiser's ubiquitous computing vision from the early 1990s have come to pass, in particular the concepts of "tabs [mobile phones], pads [tablet computing] and boards [interactive displays]" (Weiser, 1991, p. 9), the examples of research-led design described in the following section take this vision further. The concepts explore the creation of products embedded with computing technology to add value to users' experiences while remaining relatively inconspicuous during daily life. As with many digital technologies, ubiquitous computing has been a research focus for IT and computer science disciplines, with innovation largely a response to the opportunities provided by new technologies and reduced size of processors and sensors in line with Moore's Law. User-driven design has taken a back seat. However, Adam Greenfield (2006, p. 3) and Peter Nowak (2015, p. 7) argue that the majority of technical hurdles to ubiquitous computing in products have now been overcome, suggesting it is time for designers to determine how ubiquitous computing can enhance the experiences of consumers. As Greenfield asserts, "All the necessary pieces of the puzzle are sitting there on the tabletop, waiting for us to pick them up and put them together" (2006, p. 92).

One of the challenges for designers in creating ubiquitous computing products is the quantity and variation of skills involved in working in this field. In addition to the traditional understanding of form, function, materials and sustainability familiar to industrial designers, there is an added necessity to work with electrical hardware, computer programs, interfaces and Internet protocols. Splitting the product into separate components and responsibilities, as is historically the case in product development where industrial designers, engineers, IT, graphic designers, manufacturers and numerous other parties work on aspects of the same product in relative isolation, is unlikely to result in the high value output described by visions of ubiquitous computing. The results can be feature creep or solutionism, rather than the genuine engagement with improving user experience needed for a paradigm shift. This is because, by definition, a product designed to seamlessly integrate into the complex daily lives of humans is made up of numerous elements, systems and sub-systems that must operate as a unified whole, or else expose themselves through glitches and errors that break the illusion of being at one with the surrounding environment. The fractured growth of the field has yet to provide a unified, informed approach sufficient to meet the challenges involved.

Bruce Sterling describes current offerings of ubiquitous computing products as low-level gizmos—products that "have enough functionality to actively nag people. Their deployment demands extensive, sustained interaction: upgrades, grooming, plug-ins, plug-outs, unsought messages, security threats and so forth" (Sterling, 2005, p. 11). An example would be the mobile phone, a so-called "smart" device that demands attention with ring tones, flashing lights, vibrations and messages. While alerts may relate to a human-centered social contact, the phone itself may send alerts to demand an update to an app, its system, or security settings. Most smart objects and humans are still essentially discrete entities, and while Kevin Kelly describes humans as "coevolving with our technology" (2010, p. 37), the technology is not yet a reflection of humanity. Sterling describes this when he says:

It's mentally easier to divide humans and objects than to understand them as a comprehensive and interdependent system: people are alive, objects are inert, people can think, objects just lie there. But this taxonomical division blinds us to the ways and means by which objects do change, and it obscures the areas of intervention where design can reshape things. (Sterling, 2005, pp. 8-9)

Perhaps the struggle for designers in developing innovative ubiquitous computing products lies in historical baggage, including the term "industrial" design. In Stuart Walker's book *Sustainable by Design: Explorations in Theory and Practice*, this concept is explored from a sustainability perspective, but applies equally to the struggles in adopting new digital technologies. Walker argues:

The very term 'industrial design' brings with it a great deal of baggage. It defines an approach, carrying with it a set of expectations, and it represents a particular set of knowledge and skills. Industrial design is closely linked, as a term and as a discipline, with the development of mass production and consumer goods in the early years of the 20th century. It is generally understood as the design of products for large-scale distribution and it is commonly described as a service profession to the manufacturing industry. (2006, p. 35)

However, design is no longer just industrial in nature, nor is it necessarily focused on mass production or even the creation of physical objects. The digital revolution provides a paradigm shift away from focusing on the atoms of products to the bits of technology. Gore cites an example of this shift in his book *The Future*, through the company Narrative Science, which uses computer algorithms to write entirely new newspaper and magazine articles on topics ranging from sports to finance. Gore observes, "The few human writers who work for the company have become 'meta-journalists' who design the templates, frames, and angles into which the algorithm inserts data" (2013, p. 8). The meta-journalists, or rather, their algorithms, can now write millions of stories at a time rather than just one, with huge implications for the future employment opportunities for journalists. Should this trend to rethink entire operational approaches to conventional disciplines spread as is predicted by Gore, all industries, including industrial design, must adapt or risk becoming obsolete to computers.

Adapting is made all the more challenging because of the rapid progress of technologies. No sooner has the latest software upgrade to a smart phone been delivered than a new model is created with yet more features and power. This is also the case with software. The digital tools used by industrial designers, including optimized software to keep pace with technological developments such as 3D printing (additive manufacturing), 3D scanning, virtual reality, and the Internet of Things challenge designers in staying current and relevant in their practice. An example to illustrate how industrial design is changing as a profession is found in the evolution of bicycle design. Bicycles are industrial products, with their materials, construction, and form a result of the machine age. While new materials have been applied to bicycle frames over the decades, and there are multiple configurations of tube cross-sections and frame size, bicycle design has remained relatively unchanged over the last hundred years. Under the mass production paradigm, bicycle manufacturers need to sell a large number of the same bicycle to offset the expense of tooling its individual components. To reduce costs, models are produced in a limited range of sizes. Customers choose the closest-fitting size for their anthropometry, but frames are rarely perfect for the rider because of different body shapes, so adjustments are made through components such as the saddle and handlebars, to create as close to fit as possible. This longstanding model of practice is poised for significant change—in part because of the opportunities provided by additive manufacturing (3D printing) technology, and in part because of the advances in 3D CAD tools. For example, topological optimization software utilizes mathematics and computer simulation to calculate "the 'logical place' for material—normally using iterative steps and finite element analysis" (Renishaw, 2014). The design of a product—from both a functional and an aesthetic point of view—can theoretically be calculated by computer, and because of the opportunities to use additive manufacturing to bring complexity into the real world, a topologically optimized seat post by Renishaw can be designed 44% lighter than before (Renishaw, 2014). From the digital design perspective, if software using topology optimization is creating complex lightweight forms for the bicycle frame previously impossible to manufacture, then consider the following:

1. If the computer can automatically design a better product from an engineering perspective than a human being, does this reduce the need for designers?
2. If additive manufacturing allows for every design to be different, will designers be left to micro-manage every variation of a design rather than creating a single static product and moving on to something else?

This shift could be viewed as confrontational for the design discipline. However, depending on the response of the industrial design discipline, it is possible that the designer's role could evolve positively alongside emerging digital tools. To research this, the customization of products through data manipulation has been investigated, looking at projects such as "Fab Form" (Shugrina, Shamir, & Matusik, 2015), "Project Shapeshifter" (AutoDesk, 2016), and "Parametric Parts" (*Parametric Parts*, 2016). These projects use parametric relationships as constraints while limiting objects to a range of values known to produce manufacturable results. This means a novice user can, within limits, customize the CAD data for a pre-defined product within a simplified user interface. A shift in the industrial designer's role is required because these systems mean the designer is no longer creating a single 3D model of a final product, but is responsible for creating the system whereby a product concept can be manipulated into forms not necessarily envisaged by the designer. This is referred to as "Meta-Design" (Bezirtzis, Lewis, & Christeson, 2007; Fischer & Scharff, 2000). A second shift is that the customer is a participant in the design process, as discussed by Tseng, Jiao and Wang (2010, p. 177) and Jack Hu (2013, p. 6)—no longer being limited in what they must purchase by designers, but actively involved in creating it themselves. This shift from consumer to "prosumer" (Abel, Evers, Klaassen, & Troxler, 2011; Ahluwalia & Miller, 2014) radically transforms the relationship between people buying products and those creating products to sell. It is also part of a response to the imperative to create proactive users, who are invested in the products they use, with potential positive ramifications for the sustainability imperative.

However, the changes confronting modern designers go deeper: "Over the last two decades designers have mastered algorithms alongside visual thinking, and today we can witness the first generation of computational designers mature. The coming generation of designers sees code as a kind of material just as a potter sees clay" (Klanten, Ehmann, & Hanschke, 2011, p. 5). Algorithms and coding have developed in order to create software or control complex electronics. Intuitive visual programming, while not new, is only now becoming readily accessible, with programs such as Scratch (MIT Media Lab, 2016) and App Inventor (MIT, 2015) providing a visual method of coding where blocks of code can be graphically linked together in-tune with designers' visual understanding of the world. Furthermore, Rhino, popular CAD program amongst industrial designers and architects, has a range of plug-in packages including Grasshopper and Firefly, allowing the form of a product to be developed alongside its electronics or mobile application, within the same piece of software. These capabilities are also being newly embedded within PTC's Creo 3D design software through their addition of the "Designworx" Internet of Things platform. In both cases, Arduino and other electronic devices can be directly prototyped and controlled in the real world utilizing plug-ins, allowing the designer to manage additional aspects of the design relating to digital technologies, and rapidly iterate this integrated approach during design development more efficiently than previously. These examples signal a breaking down of barriers between formerly distinct roles, such as Electrical Engineer, Software Developer, User Interface Designer and Industrial Designer, with a single creator able to design, prototype, test and even manufacture complex ubiquitous computing based products. A new breed of industrial designer, the "Digital Technologist" (Loy, Canning, & Little, 2015), is emerging as education responds to these changes.

While designers might have access to a rapidly transforming range of new tools and technologies, it is crucial that human-centered design remains the focus of all design endeavors. How would humans react if simple tasks such as running a bath or making a phone call became part of autonomous systems, and what would be the level of frustration when those autonomous systems break down or misinterpret the sensor data designed to predict a person's needs? No doubt most people with a smart phone have experimented with the voice-recognition features, whether it is Siri on the Apple system, Google Now on Android, or Kortana from Microsoft, and experienced the frustrations of having to repeat the query or speak in an unnatural way in order for the system to "hear" what was being said. These systems highlight how challenging a ubiquitous system could be if it struggled to interpret user needs. Adam Greenfield calls this a needless application of technology complexification, and warns "whatever marginal 'improvement' is enacted by overlaying daily life with digital mediation has to be balanced against the risk of screwing up something that already works, however gracelessly or inelegantly" (2006, p. 125). The design of ubiquitous systems is a delicate balancing act that designers are still working to understand and implement with the kind of finesse imagined by Mark Weiser.

Emerging digital design tools, such as those in Rhino or Creo, will give designers the ability to take a holistic control of product development combined with sensors and data collection tools. However, designers will face a seemingly contradictory challenge—to design "invisible" products that enhance user experiences from the background, while being obvious in function so that they can be understood. As products move towards being truly ubiquitous, consumers will be faced with the challenge of not necessarily knowing if a product they encounter is "smart" or not, or to what degree it might be capable of responding to their needs. Greenfield explains "we should get used to the idea that there will henceforth be little correlation between the appearance of an artifact and its capabilities—no obvious indications as to how to invoke basic functionality nor that the artifact is capable of doing anything at all" (2006, p. 134). Traditionally designers include indicators of how a product functions with a visual language intended to transcend written language and culture. An example would be a handle that indicates where a product should be held, a few raised edges on a plastic enclosure to show how it can be opened, a variation in color or material to suggest where a lid can be removed. These cues replace an instruction manual, as the intention is that the user can unconsciously recognize them through their life experiences. However, how does a designer integrate visual cues describing to someone that in order to have a custom multi-vitamin 3D printed each day, as is under development by researchers at the Nestle Institute of Health Sciences (Boyle, 2014)? They may need to use a variety of inter-connected devices, such as a toilet that can analyze urine, a physical activity tracker, a fridge monitoring the foods consumed, and a pill that will only print in response to a retinal eye scan. Such complex, networked systems are as yet more science fiction than reality, but the individual technologies are possible, and demonstrate the sort of complex interactions that designers will need to consider if ubiquitous computing is embraced by mainstream consumers.

INTRODUCING 4D PRODUCT DESIGN

The authors propose an emerging type of product to fulfill the vision of Mark Weiser's ubiquitous computing—the 4D product. A 4D product may be described as one having "four dimensions, typically the three dimensions of space (length, breadth, and depth) plus time" ("four-dimensional," 2010), with time being the critical factor in this category. Time is not related to the longevity of the product, nor the time to produce it, but the ability for the product to physically evolve over time to suit changes in user needs. The emergence of 4D products is strengthened by the growing interest in 4D printing by digital revolution authors such as Eujin Pei (2014) and Dan Headrick (2015). It also relates to the field of responsive architecture at the "intersection of architecture and computer science... and implies the capacity of the building to respond dynamically to changing stimuli" (Meagher, 2014, p. 95). With relevant literature emerging relatively recently, the relationship between time and products specifically is yet to find a clear voice. With their capacity to seamlessly integrate into existing human experiences, 4D products not only meet the goals of ubiquitous computing, but they are only made possible by the concurrent developments in digital tools for designers.

The origins of 4D products are found in nature: there are numerous living examples of creatures capable of transforming or adapting to their environment in real time, such as color-changing chameleons and the blue-ringed octopus. There is also the example of the Mutable Rain Frog, only discovered in 2006, capable of transforming its skin texture from soft and smooth to sharp and spiny. Even humans are capable of a similar, if smaller, change, as cold, stress, or even euphoria can cause small bumps, known as goosebumps (or goose pimples), to be raised on the skin. This is thought to be left-over reflex from when sapiens were covered in thick hair, and when confronted by a potential predator, their goosebumps would cause the hair to stand out, making the person look more intimidating. This reaction can still be seen in animals such as porcupines and cats. If technology is a natural evolutionary force, and humans have spent millennia surrounded by living organisms capable of rapid and ongoing change, then 4D products could be considered a form of biomimicry inevitable because of technological evolution. To illustrate what a 4D product does and why it represents a shift in product design thinking, the following section provides product examples considered in the context of developments such as responsive architecture, 4D printing, and robotics.

4D DESIGN EXAMPLES

The MIT Media Lab has developed "bioLogic" (Yao et al., 2015), a form of nano-actuator powered by bacteria cells. When exposed to moisture, the cells expand to move thin biofilm materials, working like a hinge to open small vents in clothing. The product operates autonomously without the user needing to control the vents. The effect is similar to the reflexive goosebumps mentioned earlier. Although in the early stages of development, critically this work embodies the concept of being 4D in terms of its human-object relationship changing form continuously over time without specific instructions from the user or even an interface to interact with. Another company experimenting with moving forms is BMW, with both their GINA and more recent Vision Next 100 concept cars. While both cars employ a range of technologies in response to the future of driving and automation, most relevant is their capacity to change their exterior shape for functional purposes. In the case of GINA, the frame is hydraulically controlled to reveal things such as headlights or the engine, or change the shape of the rear of the car to increase traction at high speeds. As with MIT's bioLogic, these form changes are automatically controlled by the car's onboard sensors to improve the driving experience and safety on the road. The Vision Next 100 car adopts the same capabilities of "physically altering the car itself on the go and anticipating and even improving the driver's performance" (Szondy, 2016), with the addition of moving fender skirts to expand the car body as the wheels turn, reducing the drag coefficient to reportedly 0.18 C_d. This feature also improves visibility on the road, revealing a glowing red sub-structure to alert oncoming drivers the car is turning. While such a dynamic car body is yet to be seen on the mainstream market, elements of this transformation can be seen in Formula 1 racing cars. These vehicles have movable rear wing flaps to modify drag for overtaking in certain sections of a race (known as the Drag Reduction System or DRS). Consumer-level cars, such as the Audi TT, are also capable of automatically increasing the rear spoiler angle at high speeds to increase traction and safety. This suggests these life-like movements are slowly finding their way into the mainstream.

Inspired by these conceptual visions of the future, "Dynaero" by James Novak is a research project focusing on the development of a bicycle helmet capable of autonomously adjusting its aerodynamic and thermal regulation properties. The intention is to enhance the performance of the wearer through the use of integrated electronics, sensors and actuators. Research shows that "an aerodynamically efficient helmet can reduce the riders drag by up to 8% of the total aerodynamic drag that a cyclist usually generates" (Alam et al., 2010). During time trial events, cyclists typically wear aerodynamic race helmets with no ventilation, sacrificing thermal regulation for speed over short distances; however over long distances this becomes dangerous for the rider's health so they must sacrifice aerodynamics for helmets with large vents to maintain their body temperature. New helmets from Bell (Star Pro) and KASK (Infinity) have recently emerged allowing riders to manually adjust ventilation with sliders while riding. However, Dynaero draws on recent computing developments in cycling, where sensors relay data on the bodily responses of the cyclist and their bicycle during performance to the support team. Dynaero uses this data to create a responsive helmet that automatically adjusts its vents based on the riders' data, and independent of that rider or even the support team within a predetermined system. This ensures accuracy in using the data patterns to adjust the vents, rather than relying on rider awareness in the heat of competition (Figure 1).

The current working prototype of Dynaero (Figure 2) operates from a custom-designed mobile application and uses the built-in mobile phone sensors to calculate the rider's movements, sending a signal to the helmet via Bluetooth technology to control the opening angle of the large vents. The full design of this system, including both hardware and software, has been created by Novak through the use of graphical tools such as MIT App Inventor and Rhino with Grasshopper discussed previously, exemplifying the way in which designers are becoming empowered to control a greater breadth of the design process through new digital tools. While the prototype is linked to a mobile phone, the final product would be driven by its own sensors and expanded to engage with a variety of common sensors already used by cyclists and transmitting and recording using wireless signals, such as Bluetooth. As such, this project is an example of a product that upholds the ubiquitous computing vision—the helmet responds autonomously to changes in the riders' needs, with its own sensors, and those already installed on a bicycle, communicating simultaneously with the helmet to determine the second-by-second optimum position of the openings. In this way, just like the previous examples from MIT and BMW, the athlete can be immersed in the experience of riding without distraction, yet their performance and safety is improved. Such an "awareable" (i.e., a wearable technology that is itself aware of its surroundings or use) device aims to reduce the attentive demands of the user by raising the "awareness" of the products.

Figure 1. James Novak, Key stages of elite cycling races with relevant speed and power information indicating specific combinations that can be used to control air vents

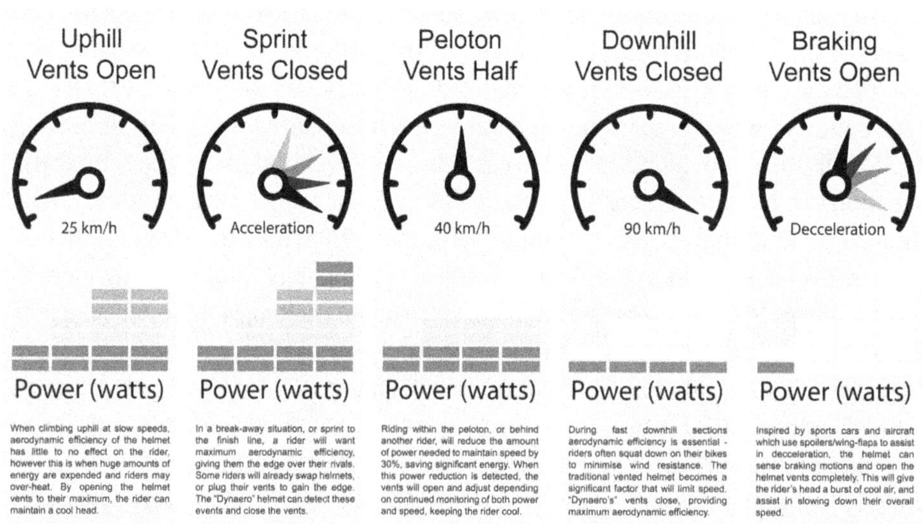

Figure 2. James Novak, Dynaero 2.0 with prototype mobile application 2016, polyamide 3D printed

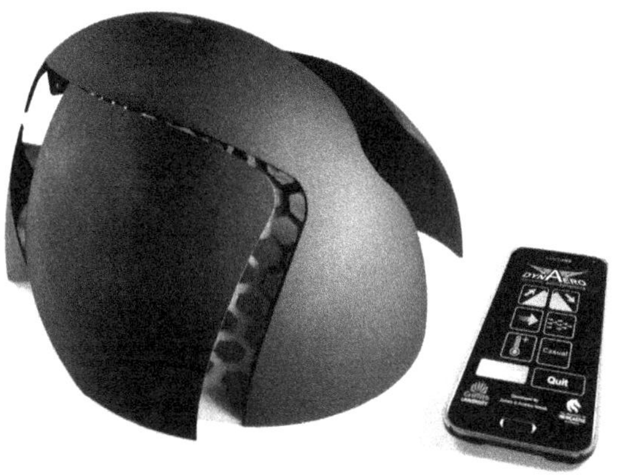

There are key criteria for the design of 4D products (as defined in this chapter). Fundamentally, a 4D product must be human-centered in this digitally enabled era. It must physically adapt over time—this emerging field is not concerned with software responses alone, such as notifications or other methods of alerting a user as currently utilized by gizmos such as smart phones. The product must autonomously analyze its situation based on the information from sensors, and reflexively respond through actuators. As such, 4D products are inherently tangible and tactile, and act symbiotically with the human user. Adam Greenfield suggests that the word "user" is inadequate to describe these new relationships with the ubiquitous computing objects; "the whole point of designing ubiquitous systems was that they would be ambient, peripheral, and not focally attended to in the way that something actively "used" must be" (Greenfield, 2006, p. 70). In their most basic sense, 4D products are driven by functional and emotional requirements, to enhance the human experience in some way that relates to performance, safety, comfort, ergonomics etc. While ideally these will be engaging aesthetic transformations, the examples discussed here are all driven from a need to improve the functional characteristics of the object with measurable results. However, this is the starting point for the development of 4D products rather than their conclusion.

THE 4D LANDSCAPE

Products that are 4D relate to the field of responsive architecture that emerged in environmentally responsive building practices (Meagher, 2014). In Jean Nouvel's Institut du Monde Arabe, the building's facade includes hundreds of mechanical diaphragms that open and close in response to the changing climate to regulate building temperature and lighting. The Al Bahar Towers in Abu Dhabi, designed by Aedas Architects, features a similar mechanically responsive climate control façade. Designed to suit the harsh desert weather, the façade elements open and close as the sun tracks around them, shutting the building off from direct sun to keep them cool, and then opening as the sun passes to let in natural light without the heat. It is estimated that such a system could reduce solar gain by up to 50% (Cilento, 2012). Both examples use sensors and weather simulation to control the facades, operating as ubiquitous systems with no need for human control. They are also designed to improve the experience of users inside through temperature control and exposure to natural light.

The development of 4D products can also be related to the emerging field of 4D printing, a term coined by the head of Massachusetts Institute of Technology (MIT)'s Self-Assembly Lab, Skylar Tibbits. In his own words:

4D printing... entails multi-material prints with the capability to transform over time, or a customized material system that can change from one shape to another, directly off the print bed. This technique offers a streamlined path from idea to reality with performance-driven functionality built directly into the materials. The fourth dimension is described here as the transformation over time, emphasising that printed structures are no longer simply static, dead objects; rather, they are programmably active and can transform independently. (Tibbits, 2014, p. 119)

4D printing uses 3D printing technology to create structures capable of transforming shape over time under certain stimuli, including liquids or electrical current. This means the object created by the printer will not initially look like the object 15 minutes later, or a week later, or a year later. The object is "programmed" to morph into various states under specific conditions, with much in common to the bioLogic garment discussed earlier. In fact, it is possible to imagine bioLogic as a 4D print as it too uses multiple materials in order to create the opening effect of the panels. However, it currently uses screen-printing techniques to layer materials rather than a 3D printer. The development and application of 4D printing can be considered as related to the broader development of 4D products, and has potential to be a valid method of manufacturing a 4D product similar to the bioLogic garment. Tibbits acknowledges this idea, suggesting that it is the broader ideas around 4D printing that are of significance. The opportunities—and challenges—of programming materials and what this might mean for designers and consumers (Headrick, 2015, p. 8) add to the body of knowledge for the design discipline in the evolving design landscape.

An example of a 4D printed product is the "Minimal Shoe" by Christophe Guberan and Carlo Clopath from MIT's Self-Assembly Lab. The shoe starts out as a shallow 3D print created with Fused Deposition Modeling (FDM) printed over a stretched textile. When removed from the printer, the material automatically folds itself into a three-dimensional shoe, and will further morph to suit the wearer's foot over time. This can be combined with other materials for the shoe sole for durability as a hybrid manufacturing approach, allowing for greater customization. This type of product has movements free of electronics, whereas 4D products, such as the Dynaero helmet or BMW concept cars, rely on actuators, power and programming and change the relationship between users and products. For example, although fountain pens were often valued objects, disposable pens have generally superseded these invested objects. However, if a person's relationship to the use of a pen were regarded as a system, then advances in digital technology create a different approach to design. "MyPen", published in the proceedings for the 2015 Drawing International Brisbane conference (Novak, 2015), embeds sensors in a pen that detect the user's grip and how the pen moves through space. This information can be used to 3D print a customized pen that can be embedded with sensors that monitor the evolution of a person's grip over time—short term, for different uses or from childhood to adulthood. Each evolution of the print refines the grip, responding to changes of writing style or weaknesses or strengths in the hand. Further, this system allows for change in consultation with a clinician if the user experienced difficulties, for example, following a medical incident, such as a stroke. This is arguably a form of evolutionary robotics, or alternatively part of what A. E. Eiben terms the Evolution of Things (Eiben, 2014), and figure 3 shows four different pens customized to the grip and aesthetic preference of four different test subjects in Novak's study.

Just as MyPen builds a body of knowledge on the evolving requirements of an individual, product designer Tom Dixon used digital technology to build up a body of cultural knowledge. He embedded the cultural history in a product, designed to evolve over time when he relaunched second-hand Artek stools on the market with the story of where they had previously been used contained in an RFID chip and accessed by smart phone (Loy, 2008). In *Sustainable by Design* (2006), Walker discusses building such opportunities for emotional connections between people and objects to invest objects with additional meaning to keep the product in use for longer, spreading the embodied energy invested in its manufacture. Digitally enabled 4D products designed for socio-cultural sustainability embed more than

Figure 3. James Novak, MyPen customized by four test subjects 2015, ABS 3D Prints

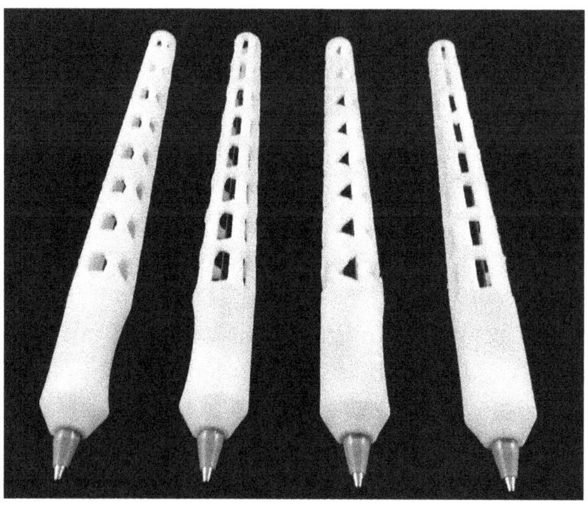

technology; they embed cultural information, such as storytelling, images, sounds and other sensory information specific to an individual or a group. This approach builds shared memory into product systems that supports the evolution of the group, and retains and enhances community. Based on this approach, conventional ideas of digital customization through 3D printing and parametric modeling have to change, with products customized through engagement, building value through use because of the embedded cultural connections. A 4D product may therefore itself change over time, or may spawn a new iteration of itself when it senses it is no longer performing optimally, or it may change as information tagged to that product or provided as data to a system is updated. This begins to describe what Bruce Sterling calls SPIMES, objects that:

...begin and end as data. They are designed on screens, fabricated by digital means, and precisely tracked through space and time throughout their earthly sojourn. SPIMES are sustainable, enhanceable, uniquely identifiable, and made of substances that can and will be folded back into the production stream of future SPIMES. (Sterling, 2005, p. 11)

With the digital revolution, and complex sustainability imperatives, including socio-cultural sustainability as well as environmental, and the aspirations for a society that responds to inclusion and diversity there is a shift towards the fourth dimension that will impact the man-made world, from the macro to the micro. This challenges what it means to be a designer and raises the question as to what degree this will affect human development in an increasingly digital age. Responsive architecture, 4D products, and 4D printing together suggest how designers are embracing new technologies with an understanding of the flux attached to rapid technological development, choosing to embed their designs with the capability to change and adapt to a changing world.

EMPOWERING PRODUCTS

It has been shown how the relationship between designers and products inherently changes as the designer's role moves from creating single products for mass production, to more of a systems approach where products change over time. The relationships between consumers and 4D products are likely to be different to those already in existence and need to be studied. This is suggested by the emotional responses noted in "The Never Hungry Caterpillar" project. At a functional level, this is an extension cable for the home designed by Matthias Laschke, Marc Hassenzahl, and Sarah Diefenbach (2011). However rather than blending into the background, this device will squirm, twist and turn as if in pain when it detects devices wasting power in standby mode, using "the metaphor of a caterpillar [to] touch upon (at least some) people's tendency to help and take care of living things" (Laschke et al., 2011). Trial users in the study could not help but unplug devices in standby mode to stop the perceived "pain" of the device. Of course "technically, the Caterpillar could just detect connected devices in stand-by and simply switch them off automatically. This would be practical, it would save energy, but it would not engage in any argument or dialogue" (Laschke et al., 2011).

This type of product is 4D in nature, having the capacity to autonomously transform in order to affect the habits of the user. What is particularly important to consider in 4D products, especially those that may be as life-like as The Never Hungry Caterpillar, is the psychological and emotional connection between users and such products, who may begin to see them as living entities to be cared for like a pet, rather than an inanimate object with little value as would be the case of a standard extension cord. This is highlighted in the work of Moyle et al. (2013) on the use of robots in the care of Alzheimer patients, where the emotional relationship between the robot and the user was explored through studies on robotic animals and toys. What happens when these "living" products stop working? Will users be more likely to try and repair these products than throw them away, much like a beloved family pet, and could this be a response to more sustainable design? With 4D products only newly emerging through research it is difficult to say, however it is conceivable based on research in robotics that this will be the likely scenario. Matthias Scheutz explores these unidirectional relationships, explaining that robots:

... affect humans in very much the same way that animals (e.g., pets) or even other people affect humans. In particular, the rule-governed mobility of social robots allows for, and ultimately prompts, humans to ascribe intentions to social robots in order to be able to make sense of their behaviors (e.g., the robot did not clean in the corner because it thought it could not get there). The claim is that the autonomy of social robots is among the critical properties that cause people to view robots differently from other artefacts such as computers or cars. (Scheutz, 2011, p. 207)

Scheutz uses numerous examples to evidence this trend, from soldiers becoming attached to robots used to detect and detonate improvised explosive devices in Iraq, to the popular robotic Roomba vacuum cleaner and the personality's people ascribe to them. As such, new scenarios and systems become important long-term considerations for 4D products, including the necessity to provide better means for product repair and maintenance, ethical regulation to prevent products influencing humans to do things inappropriate or outside their normal behavior, psychological and counseling services to help people move on after a product 'dies', control over product memory so that traits or sensitive data can be removed when a product is sold second hand. As technology moves closer to nature, these considerations will transform the relationship between consumers and products, and the industries built around them.

In responding to the ideas expressed in this chapter about human empowerment in an increasingly technological age, it might be assumed that ambient intelligence would be at odds with ideas of enhancing the human experience based on empowerment. However, this would be an oversimplification. Ubiquitous computing might be a utopian ideal, alongside modernist ideas of houses as "machines for living", but the development of ambient intelligence just needs to be developed with an underlying recognition of the importance of empowerment in the same way environments must focus on human emotion, social interaction and behavior rather than being completely machine-like. Light bulbs that wirelessly transmit their status and trigger a replacement order, as shown at the 2013 *The Future Is Here* exhibition in the London Design Museum walk a fine line between enhancing the human experience through the development of freedom from day to day minutia, and alienating people from their environment. Complete freedom from responsibility can equate to feelings of a loss of control. This is not aligned with current aspirations for society, and alternative strategies, such as delegation, should be considered in the development of the product service system.

There's a tendency in connected products and service design to assume that having machines take over human work is a good in itself....however, there are good reasons to treat automation as a choice, not a default. Instead of talking about automation, philosophers of technology often talk about 'delegation'—that is, the distribution of responsibility and effort among different kinds of humans and machines. (Rowland et al., 2015, p. 211)

However, there is an opposing argument that computing provides humans with the opportunity to extend their faculties and connection to the real world, becoming more sensitive to the nuances of their surroundings and better informed about the systems that operate within it. For example, MIT have embedded sensors into marshland (Tidmarsh, Plymouth, Massachusetts) and even their Media Lab building. This allows users to virtually explore the landscape, honing in on details such as pockets of noise, or temperature change that would otherwise be impossible for individuals to be aware of. This has the potential to create people with a very different sense of the world, in the same way that television and the Internet has changed collective understandings of the world by different populations (depending on the way it is presented). That shift in perception, triggered by an increase in information, causing a transformational change in understanding is illustrated by, for example, the reported impact of the first image of the earth taken as the Apollo rounded the moon in 1968.

Although the signs have been there with augmented reality products such as Google Glass and QR codes, Pokémon Go has heralded in a democratization of augmented reality technology that has the potential to change the human perception of the world and interactions with their environment and each other. Kuniavsky makes the comment that to have computing embedded in a forest would make no sense "...in an office it is not surprising to have wires running under the floor, but forest trees with embedded electronics seem wrong" (Kuniavsky, 2010, p. 286), yet if that technology allowed the individual to walk through the forest with the option of technologically heightened senses, they could zoom in on the movement of insects in the forest floor, hear chicks hatch high above their heads, feel the direction and strength of the wind at the top of the trees. Biohackers such as Stelarc and Neil Harbisson suggest that information sent to the brain in any form can be learned and interpreted by the brain to increase the information received through traditional senses. Augmented experience and ambient intelligence could have the potential to support the human development in a digital age if the focus is on the human experience.

Even so, the inevitability of unexpected consequences could have an equally negative impact on development. Will real life become less real? As the opportunities provided by ubiquitous computing create temptations for designers to strive for all encompassing systems, the ability to opt out at any point must be provided. The most amazing technology in the world "...will mean little if we don't, as individuals, have genuine power to evaluate its merits on our own terms, and make decisions accordingly...a paradise without choice is no paradise at all" (Greenfield, 2006, p. 247).

CONCLUSION

While there is a cautionary note in working with the power of digital technology in human-centered products, the ontogenetic adaptation capabilities of humans should allow for the evolution of the collective to rise to the opportunities—and challenges—digital technology provides. The essential criteria, though, involve designing in context, and not get swept along with what is technologically possible, but rather what responds to design theory, aspirations and values for now, for positive human development. That is, for designs to align with current thinking about sustainability, connectivity, experience-centered design, design for change, and to support inclusion and diversity.

As defined here, 4D products are about responding to the drivers behind design now, and working with time as an essential ingredient of that design approach. Going beyond the design of customizable, parametric models for individual 3D printing, and beyond the design of products embedded with sensors that send information to a management system, and going beyond augmented reality, 4D products are capable of changing over time. Ideas evolve, people's needs and desires change. Sustainability drivers support the idea of designing for that change. Digitally enabled 4D products are an emerging class of products that work towards these ideals, in sync with the natural flux of human behaviors, needs and learning.

This shift from static, resolved 3D products that make up today's retail market to 4D products is only beginning to emerge. It has the potential to encompass the next state beyond cradle to cradle, with an iterative development lifecycle, rather than a rehabilitated but still linear lifeline. The proposal is that they can work with the reality of human behaviors and interactions. The assumption will be that they require repair and are therefore designed for that inevitability, and be flexible enough to have their functionality revised and updated over time. They will be responsive, agile, and embedded with information on how to use, how to repair, and how to alter for new needs—data is their life-blood.

What does that mean for designers? What does it mean in terms of the use of technology in products? What does it mean in relation to the psychology of those products? If an essential driver for selling products is for economic growth, then there will be significant changes in how economic growth is sustained. In the past, products made for manufacturing utilized as many common components as possible, and costs of tooling constrained ideas of iterative change. Digitally enabled design has the potential to alter understandings of systems of products and production, not only for practical reasons, but also in response to the normal, human desire for change. The need now is to expand this approach to a more fluid, flexible, empowering approach. Design can become more iterative, allowing for bespoke development and evolution of products. This ability for customization in a technological era responds to the need for designers to keep referring back to the human-centered core in their design work, highlighted by Greenfield (2006, p.13): "It would be wise for us all to remember that, while our information technology may be digital in nature, the human beings interacting with it will always be infuriatingly and delightfully analogue".

The difficulty of designing in this way is that even basic designs are creating complex problems for teams across disciplines. If designers are to respond effectively, creating positive product experiences that respond to the aspirations for society at this time, they need to be basing their work on updated understandings of the human psyche and ecological interconnection as well as the technological opportunities and potential pitfalls. A new approach to the education and development of designers and teams working in this space needs to be devised, and, fundamentally, a paradigm shift in thinking and research is needed to keep pace with designing for the human experience in a digital age.

REFERENCES

Abel, B. V., Evers, L., Klaassen, R., & Troxler, P. (2011). *Open design now: Why design cannot remain exclusive*. Amsterdam, Netherlands: BIS.

Ahluwalia, P., & Miller, T. (2014). The prosumer. *Social Identities*, *20*(4-5), 259–261. doi:10.1080/13504630.2015.1004830

Alam, F., Chowdhury, H., Elmir, Z., Sayogo, A., Love, J., & Subic, A. (2010). An experimental study of thermal comfort and aerodynamic efficiency of recreational and racing bicycle helmets. *Procedia Engineering*, *2*(2), 2413–2418. doi:10.1016/j.proeng.2010.04.008

Aldersey-Williams, H. (2011). *The new tin ear: Manufacturing, materials and the rise of the user-maker.* London: RSA Design Projects.

Anderson, C. (2014). *Makers: The new industrial revolution.* New York, NY: Crown Business.

Assa Ashuach Studio. (2016). Retrieved from http://assaashuach.com/#!/portfolio/assa-studio-micro-3d-design/0

AutoDesk. (2016). *Project Shapeshifter.* Retrieved from www.shapeshifter.io

Bezirtzis, B. G., Lewis, M., & Christeson, C. (2007). Interactive evolution for industrial design. In *Proceedings of the 6th ACM SIGCHI conference on Creativity & Cognition* (pp. 183-192). New York, NY: ACM. 10.1145/1254960.1254986

Boyle, M. (2014, June 23). *Nestle aiming to develop a Nespresso of nutrients.* Retrieved from http://www.bloomberg.com/news/articles/2014-06-22/nestle-aiming-to-develop-a-nespresso-of-nutrients?_ga=1.134810162.210344514.1402054238

Braungart, M., & McDonough, W. (2002). *Cradle to cradle: Remaking the way we make things.* New York, NY: North Point Press.

Cilento, K. (2012, September 5). Al Bahar Towers responsive facade / Aedas. *Archdaily.* Retrieved from http://www.archdaily.com/270592/al-bahar-towers-responsive-facade-aedas

Dean, L. T. (2016). *Future factories.* Retrieved from www.futurefactories.com

Eiben, A. E. (2014). *In vivo veritas: Towards the evolution of things.* Paper presented at the 13th International Conference on Parallel Problem Solving from Nature, Ljubljana, Slovenia. 10.1007/978-3-319-10762-2_3

Fischer, G., & Scharff, E. (2000). Meta-design: design for designers. *Proceedings of the 3rd conference on Designing interactive systems: Processes, practices, methods, and techniques* (pp. 396-405). New York, NY: ACM.

Four-dimensional. (2010). *Oxford Dictionary of English* (3rd ed.). Oxford University Press.

Fuad-Luke, A. (2009). *Design activism: Beautiful strangeness for a sustainable world.* London: Earthscan.

Fung Kwok Pan. Fluid vase. (2010). *Design Boom Newsletter 440071.* Retrieved from http://www.designboom.com/technology/fung-kwok-pan-fluid-vase/

Gore, A. (2013). *The future.* New York, NY: Random House.

Greenfield, A. (2006). *Everyware: The dawning age of ubiquitous computing.* Berkeley, CA: New Riders.

Hawken, P., Lovins, A., & Lovins, L. H. (2000). *Natural capitalism: The next industrial revolution.* London: Earthscan.

Headrick, D. (2015). 4D printing transforms product design. *Research-Technology Management, 58*(2), 7–8.

Hu, S. J. (2013). Evolving paradigms of manufacturing: From mass production to mass customization and personalization. *Procedia CIRP*, *7*, 3–8. doi:10.1016/j.procir.2013.05.002

Kelly, K. (2010). *What technology wants*. New York, NY: Penguin Group.

Klanten, R., Ehmann, S., & Hanschke, V. (2011). *A touch of code: Interactive installations and experiences*. Berlin, Germany: Gestalten.

Kuniavsky, M. (2010). *Smart things: Ubiquitous computing user experience design*. Burlington, MA: Morgan Haufmann.

Laschke, M., Hassenzahl, M., & Diefenbach, S. (2011). *Things with attitude: Transformational products*. Presented at the Create 11 Conference, Shoreditch House, London, UK. Retrieved from http://static1.squarespace.com/static/52722face4b02a66778dc9d0/t/53318711e4b09837d5a328f8/1395754769468/Laschke%2C+Hassenzahl%2C+Diefenbach+-+Things+with+attitude+Transformational+Products+-+2011.pdf

Loy, J. (2008). Strategies for teaching sustainable design practice with product design students. *Proceedings of E&PDE 2008, the 10th International Engineering and Product Design Education Conference* (pp. 491-496). Design Society.

Loy, J., Canning, S., & Haskell, N. (2016). 3D printing sociocultural sustainability. In S. S. Muthu & M. M. Savalani (Eds.), *Handbook of sustainability in additive manufacturing, environmental footprints and eco-design of products and processes* (pp. 145–168). New York, NY: Springer. doi:10.1007/978-981-10-0549-7_7

Loy, J., Canning, S., & Little, C. (2015, June-July). *Digital industrial design*. Paper presented at the International Design Technology Conference, DesTech2015, Geelong, Australia.

Makepeace, J. (1995). *Makepeace: Spirit of adventure in craft and design*. London: Octopus.

Massachusetts Institute of Technology. (2015). *MIT App Inventor*. Retrieved from http://appinventor.mit.edu/explore/

McEwen, A., & Cassimally, H. (2013). *Designing the Internet of Things*. Oxford, UK: John Wiley & Sons.

McLeod, S. A. (2016). *Maslow's hierarchy of needs*. Retrieved from www.simplypsychology.org/maslow.html

Meagher, M. (2014). Responsive architecture and the problem of obsolescence. *International Journal of Architectural Research*, *8*(3), 95–104.

MIT Media Lab. (2016). *Scratch*. Retrieved from https://scratch.mit.edu/

Moyle, W., Beattie, E., Jones, C., Klein, B., Cook, G., & Gray, C. (2013). Exploring the effect of companion robots on emotional expression. *Journal of Gerontological Nursing*, *39*(5), 46–53. doi:10.3928/00989134-20130313-03 PMID:23506125

Novak, J. (2015). *Drawing the pen: From physical to digital and back again.* Presented at the Drawing International Brisbane Conference, Brisbane, Australia. Retrieved from http://static1.squarespace.com/static/55779bbce4b004acf1e1479d/t/56aef970b6aa60cdf1c253d6/1454307702608/JAMES+XXXX_DRAWING+THE+PEN_DIB2015.pdf

Nowak, P. (2015). *Humans 3.0: the upgrading of the species.* London: Harper Collins.

Parametric Parts. (2016). Retrieved from http://www.parametricparts.com/

Pei, E. (2014). 4D printing: Revolution or fad? *Assembly Automation, 34*(2), 123–127. doi:10.1108/AA-02-2014-014

Petrova, M. (2014). Educating designers from Generation Y: Challenges and alternatives. In *DS 78: Proceedings of the 16th International conference on Engineering and Product Design Education (E&PDE14), Design Education and Human Technology Relations* (pp. 524-529). New York, NY: ACM.

Philips Hue. (n.d.). Retrieved from http://www2.meethue.com/en-au/

Pokémon Go. (n.d.). Retrieved from http://www.pokemongo.com/

Powell, R. (2012). The future of human evolution. *The British Journal for the Philosophy of Science, 63*(1), 145–175. doi:10.1093/bjps/axr027

Renishaw. (2014, February 2). *First metal 3D printed bicycle frame manufactured by Renishaw for Empire Cycles.* Retrieved from http://www.renishaw.com/en/first-metal-3d-printed-bicycle-frame-manufactured-by-renishaw-for-empire-cycles--24154

Rowland, C., Goodman, E., Charlier, M., Light, A., & Lui, A. (2015). *Designing connected products: UX for the consumer internet of things.* Sebastopol, CA: O'Reilly.

Scheutz, M. (2011). *The inherent dangers of unidirectional emotional bonds between humans and social robots. In Robot ethics: The ethical and social implications of robotics* (pp. 205–221). Cambridge, MA: The MIT Press.

Shugrina, M., Shamir, A., & Matusik, W. (2015). *Fab forms: Customizable objects for fabrication with validity and geometry caching.* Paper presented at the ACM Siggraph 2015, Los Angeles, CA. Retrieved from http://cfg.mit.edu/content/fab-forms-customizable-objects-fabrication-validity-and-geometry-caching

Sparke, P. (1987). *Design in context.* London: Quarto Books.

St Clair, A. (2016). Critical thinking, wisdom and paying homage to the human experience. *Voices, 42*(1/2), 41-43.

Sterling, B. (2005). *Shaping things.* Cambridge, MA: MIT Press.

Szondy, D. (2016, March 9). BMW looks to the future with shape-shifting. *Vision Next 100 Concept.* Retrieved from http://www.gizmag.com/bmw-group-vision--next-100-concept/42194/

Tibbits, S. (2014). 4D printing: Multi-material shape change. *Architectural Design*, *84*(1), 116–121. doi:10.1002/ad.1710

Tseng, M. M., Jiao, R. J., & Wang, C. (2010). Design for mass personalization. *CIRP Annals: Manufacturing Technology*, *59*(1), 175–178. doi:10.1016/j.cirp.2010.03.097

Walker, S. (2006). *Sustainable by design: Explorations in theory and practice*. London: Earthscan.

Weimer, M. (2002). *Learner-centered teaching: Five key changes to practice*. San Francisco, CA: Jossey Bass.

Weiser, M. (1991). The computer for the 21st century. *Scientific American*, *265*(3), 94–110. doi:10.10 38cientificamerican0991-94 PMID:1675486

Yao, L., Ou, J., Cheng, C.-Y., Steiner, H., Wang, W., Wang, G., & Ishii, H. (2015). bioLogic: Natto cells as nanoactuators for shape changing interfaces. In *Proceedings of the 33rd Annual ACM Conference on Human Factors in Computing Systems* (pp. 1-10). New York, NY: ACM. 10.1145/2702123.2702611

KEY TERMS AND DEFINITIONS

Additive Manufacturing (3D Printing): A process of producing three-dimensional forms directly from 3D computer data by depositing material layer by layer. This is in contrast to subtractive manufacturing, whereby material is removed from a solid piece of material to reveal the final object.

Generation Y (Gen Y): People born between approximately 1980 and 1995.

Human-Centred Design: An approach to design where the user remains at the forefront of all design decisions, resulting in solutions that are more in tune with user needs.

Industrial Design: A profession responsible for developing the form and function of physical products, traditionally for the intention of mass production.

Internet of Things (IoT): Extends the concept of ubiquitous computing by adding the element of the Internet, allowing objects to communicate with each other or the cloud.

Ontogenetic: The entire sequence of events influencing the development of an organism.

Stasis: A state in which there is no action or progress.

Ubiquitous Computing: The concept of embedding computing capabilities into any ordinary object, in any location. May also be known as "pervasive computing".

This research was previously published in the Handbook of Research on Human Development in the Digital Age edited by Valerie C. Bryan, Ann T. Musgrove, and Jillian R. Powers, pages 403-426, copyright year 2018 by Information Science Reference (an imprint of IGI Global).

Chapter 2
Innovative Construction Systems:
New Design Aesthetics Through Competitive Realization Processes

Pierpaolo Ruttico
Politecnico di Milano, Italy

ABSTRACT

With the emergence of a new materialism in architecture and as a result of the ongoing advancements in digital technologies, computer numerically controlled (CNC) fabrication and 3D-printing are rapidly gaining popularity within the building construction industry. Robots, CNC machines, and algorithmic programming allow us to create new construction systems and innovative designs that otherwise might not be possible. This chapter analyzes adaptive and flexible systems that facilitate the design and the production processes of mass-customized building components through digital fabrication and assembly. The aspiration is to achieve morphological complexity and performance in material constituents: an engineering computational design process that envisions a sustainable built environment, with higher-level functionality and higher-level integration between material system and environment.

INTRODUCTION

This chapter describes the original contribution developed by the author with the help of his team (E.Gamba, M.Sangiorgio, D.Cucchi, M.Dragoljevic) regarding the digital design and production processes of three construction systems, specifically:

1. Robotic manufacturing system for mass-customized insulated wall panels;
2. Responsive louvers system for adaptive façades;
3. Topology optimization system for spatial structures.

DOI: 10.4018/978-1-5225-9624-0.ch002

Several models and prototypes have been built in order to validate the assumptions. The experiments draw attention to how digital technologies generate the efficiencies for mass customization of real world building systems.

This research demonstrates how digital fabrication can establish a loop between physical performance aspects and the development of a custom design system.

By integrating digital fabrication technologies into the design and production process, novel forms of architectural materialization can be explored.

The chapter begins with a description of the state of the art regarding contemporary design, manufacturing and assembly strategies; it is subsequently organized in three sectors, describing each system.

DESIGNING DESIGN, MANUFACTURING AND ASSEMBLY

Pattern and Transformation

Modernist architecture sought to establish consistency between parts and wholes through modular logics made possible by the processes of industrialization; these methods relied on regulating systems and planning grids to organize space.

Nowadays advances in digital technologies and algorithmic programming allow for a shift towards an automated deformation of the grid in order to morphologically adapt to the environment. The modernist model of the simple repetition of unchanging units has been replaced with a model of variable units repeated along variable trajectories (Reiser, J. 2006). The systems proposed in this chapter will show how it is possible to organize flexible part-to-whole relationships that can grow and differentiate to produce patterns of complex material-organizations.

The potential of digital design and fabrication processes can best be exploited where a large number of discrete elements are combined as part of a self-similar structure. The deployment in quantity and the difference in degree between the elements allow to perceive a sense of transformation of the whole. Essentially the whole becomes greater than the sum of its parts. The possibility of generating formal signs of "transformation" in the design of architecture is related to a mathematical model's ability to describe variation. In this respect, the notion of pattern has re-emerged within architectural design, specifically as a critical tool for managing large quantities of information throughout the design process (Picon, A. 2010).

The theme of transformation finds its foundations in the studies of the Master Albrecht Dürer (1471 – 1528), his work on human proportions and "deformation grids" (Dürer, A. 1528) proposes a relativist notion of beauty based on variety and could be considered a reference point for a contemporary architectural discourse. The biologist and mathematician D'Arcy Thompson (1860 – 1948) revisits Dürer's approach in order to illustrate and formalize shape differences between geometrical objects. Thompson described how the differences between the forms of related species can be represented geometrically through the transformation of the rectangular coordinate system into a curvilinear system that stretches the geometries of the analyzed species.

The definition of form as "a diagram of force" appeared in Thompson's book On Growth and Form (1917) and continues to instill the principle that the design of forms in the built environment should be driven by structural intuition in a dialogue among materials, performance, geometry and aesthetics.

Another inspiring contribution is the one embedded in the powerful paradigm of "Epigenetic Landscape", illustrated in a diagram as attracting valleys resulting from a complex regulatory network (Wad-

Figure 1. Modular Grid Subdivision for design and production, 60cm=6M. (Ciribini et al. 1965). The concept of serial repetition was the product of industrialization, this so-called pre-fabrication in architecture became possible through the design of the module within a geometric grid layout.

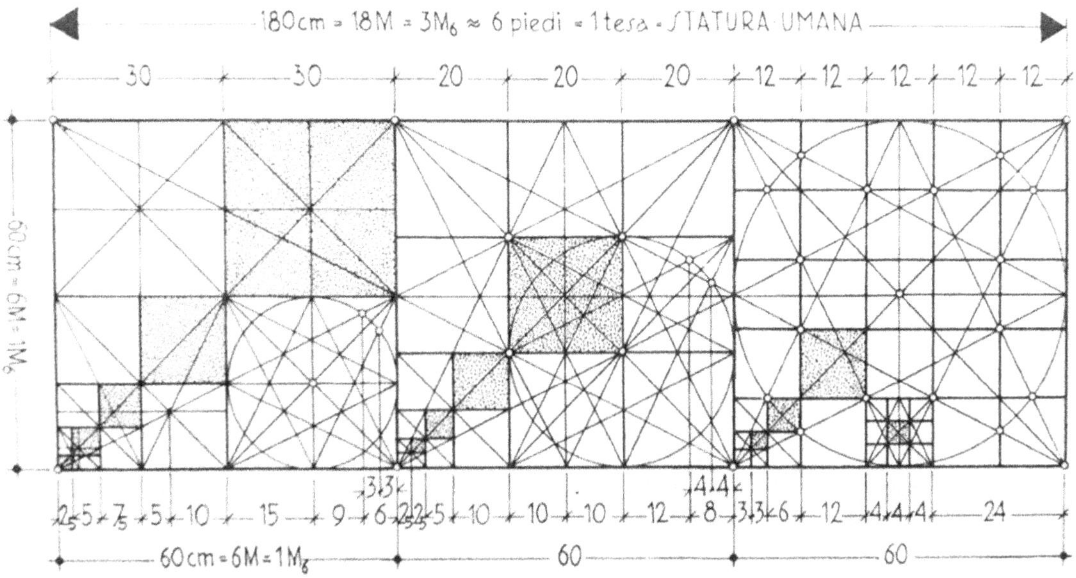

Figure 2. A pattern of repetition where threads are twisted in an endless fabric (Wachsmann, K. 1959). Konrad Wachsmann. Architecture D'Aujourd'Hui 91 Sep 1960: 185.

dinghton H 1957). In the article "Landscapes of Change", Sandford Kwinter describes how epigenetic landscapes were derived from Catastrophe theory. As defined by Kwinter: "Catastrophe theory is one method for describing the evolution of forms in nature" (p. 59). He suggests that even as some systems try to correct their behaviours, there are occurrences where a system can flip and give rise to a whole new state or form" (p. 60). A tendency towards variation and deviation from a condition of stability, regularity and constancy is at the very essence of our survival. The evolution of the pattern from one configuration to the next is what ensures the success of any system in space and time. In Kwinter's analogy of the trajectory of a ball describing epigenetic landscapes and its form, a possible interpretation is that one surface may appear chaotic, however, it is actually governed by an inherent algorithmic

Figure 3. Illustration taken from the Book "On Growth and Form", D'Arcy Thompson (1917). In order to give graphical expression to the underlying simplicity of the deformation pattern Thompson employed a Cartesian grid system in which the intersections of the grid lines were taken to represent corresponding points on both the starting and final forms of the species he analyzed.

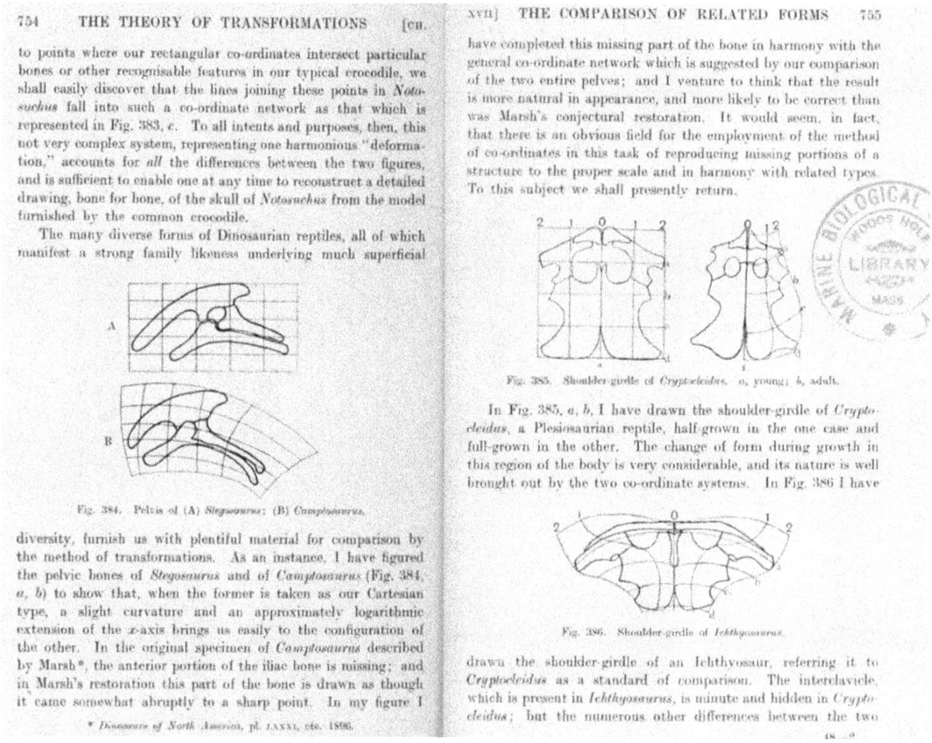

754 THE THEORY OF TRANSFORMATIONS [CH.

to points where our rectangular co-ordinates intersect particular bones or other recognisable features in our typical crocodile, we shall easily discover that the lines joining these points in *Notosuchus* fall into such a co-ordinate network as that which is represented in Fig. 383, *c*. To all intents and purposes, then, this not very complex system, representing one harmonious "deformation," accounts for *all* the differences between the two figures, and is sufficient to enable one at any time to reconstruct a detailed drawing, bone for bone, of the skull of *Notosuchus* from the model furnished by the common crocodile.

The many diverse forms of Dinosaurian reptiles, all of which manifest a strong family likeness underlying much superficial

Fig. 384. Pelvis of (A) *Stegosaurus*; (B) *Camptosaurus*.

diversity, furnish us with plentiful material for comparison by the method of transformations. As an instance, I have figured the pelvic bones of *Stegosaurus* and of *Camptosaurus* (Fig. 384, *a*, *b*) to show that, when the former is taken as our Cartesian type, a slight curvature and an approximately logarithmic extension of the *x*-axis brings us easily to the configuration of the other. In the original specimen of *Camptosaurus* described by Marsh*, the anterior portion of the iliac bone is missing; and in Marsh's restoration this part of the bone is drawn as though it came somewhat abruptly to a sharp point. In my figure I

* *Dinosaurs of North America*, pl. LXXXI, etc. 1896.

XVII] THE COMPARISON OF RELATED FORMS 755

have completed this missing part of the bone in harmony with the general co-ordinate network which is suggested by our comparison of the two entire pelves; and I venture to think that the result is more natural in appearance, and more likely to be correct than was Marsh's conjectural restoration. It would seem, in fact, that there is an obvious field for the employment of the method of co-ordinates in this task of reproducing missing portions of a structure to the proper scale and in harmony with related types. To this subject we shall presently return.

Fig. 385. Shoulder-girdle of *Cryptocleidus*. *a*, young; *b*, adult.

In Fig. 385, *a*, *b*, I have drawn the shoulder-girdle of *Cryptocleidus*, a Plesiosaurian reptile, half-grown in the one case and full-grown in the other. The change of form during growth in this region of the body is very considerable, and its nature is well brought out by the two co-ordinate systems. In Fig. 386 I have

Fig. 386. Shoulder-girdle of *Ichthyosaurus*.

drawn the shoulder-girdle of an Ichthyosaur, referring it to *Cryptocleidus* as a standard of comparison. The interclavicle, which is present in *Ichthyosaurus*, is minute and hidden in *Cryptocleidus*; but the numerous other differences between the two

Figure 4. Albrecht Dürer's studies of human proportion influenced Thompson's theory of coordinates. Illustration taken from the book Vier Bücher Von Menschlicher Proportionen, A. Dürer (1528).

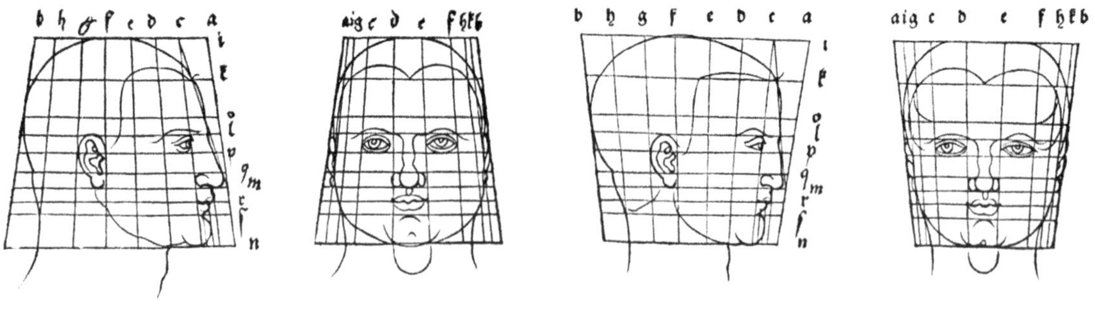

system. The concept of non-linearity of nature's systems, based on growth, evolution and feedback continues to inspire the discourse with regards to contemporary architecture. The pattern facilitates the search to find order and structure within the chaotic complexities of the world surrounding us. Through variation and differentiation in patterns, architecture becomes increasingly meaningful and expressive (Gramazio, F. 2008).

Algorithmic Workflows

The variation and differentiation affect different scales of design, more specifically, materials and building structures, spatial sequences and urban planning. Since the quantity of information, when dealing with repetition-with-variation, cannot be managed with conventional CAD-CAM processes, it is therefore necessary to shift towards algorithmic logics. The algorithm becomes a vital tool for invention that significantly expands the design capabilities of architects and engineers.

Rather than focusing on form and performance in an alternating sequence, the designer has the potential – through algorithmic logics - to integrate material complexities and performance analysis in the manner of a feedback loop, within the same design environment. The rules by which the space is originated are then defined by the designer, by means of variables and parameters, bounded by constraints, such as upper and lower limits. The variables allow for the exploration of possible design options; whilst parameters enable the setting of variation and differentiation; finally, constraints allow for the control of possible design options. Between Top-down strategies where the form is given a-priori – usually following aesthetic sensitivity, and Bottom-up strategies that optimize a system originating from local reciprocalities to address specific performance criteria, there is a wide spectrum of form-finding possibilities. The pertinent issue is how to structure a relationship between efficiency and creativity that balances their different strengths. An optimal design does not however necessarily equate to a good design. Moreover, a good design for one person could be perceived to be a sub-standard design by another person. Algorithms could help define design objectives and could allow design to be "democratized" (Kolarevic, B. 2008). Beyond efficiency and creativity we might use algorithms to examine values, the digital model could become a platform for further debate and discussion. For instance, the model could be tuned in different configurations according to different values. Different design permutations would correspond to different values. The model widens participation and enables the practice of trade-offs to become a feasible one. Throughout the negotiation process, architects, engineers, fabricators and citizens of all backgrounds could join the discussion and debate, consequently democratizing the design process (Brell-Cokcan, S. 2009, 2010).

Figure 5. Epigenetic Landscape, Waddinghton H 1957.

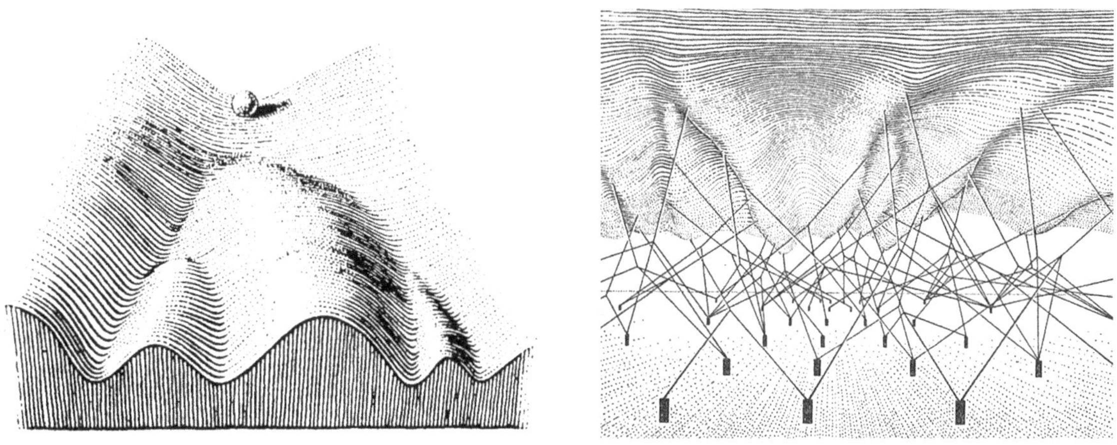

The very notion of the skills necessary to write custom programs are still a major deterrent for many designers embarking upon the process. Nevertheless, more accessible programming tools are being continually developed and taught to each subsequently new generation of students, architects and engineers.

Robotics

Current research in the field of digital fabrication in architecture is characterized by a shift from CNC (Computer Numerical Control) machinery towards more generic fabrication equipment such as industrial robots (Bartschi, M. et al 2010). CNC machines are typically tasks specific, whereas robotic arms are flexible and can perform any desired task such as welding, gripping, spinning, milling, cutting, bending, stacking, spraying, etc., depending on the application. Leading to an increasingly open design space robotic fabrication offers the opportunity for computational design to explore the design space for particularly promising areas. Cutting-edge research institutions and schools of architecture that embarked upon the challenge of robotic fabrication are creating new markets for the building construction industry and robots are gaining ever-more popularity.

ROBOTIC MANUFACTURING SYSTEM FOR MASS-CUSTOMIZED INSULATED WALL PANELS

This sector presents a system that allows to design and produce parametric wall components having highly expressive forms and inherent performative capacity. The design workflow and the manufacturing process are interwoven: geometric properties, material characteristics and fabrication techniques concur to meet the efficiency of the system. The form that is produced is the result of designing both the form and the production process. The system allows for intuitive control of parametric design configurations whilst providing an easy-to-use code for robotic fabrication. A series of patterning strategies lead to highly engineered surfaces that potentially optimize the wall from a structural, acoustic and thermal perspective. In order to validate the system, an initial spatial structure has been designed and realized

Figure 6. Algorithm editor Grasshopper, a Rhino Plug-in.

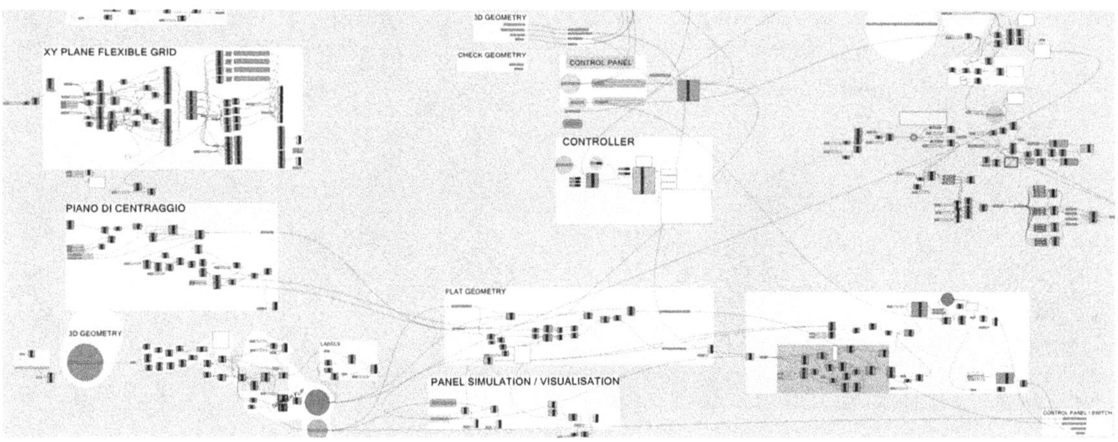

Figure 7. The map shows the current research institutions and companies that are focusing their research on robotics for architectural applications.

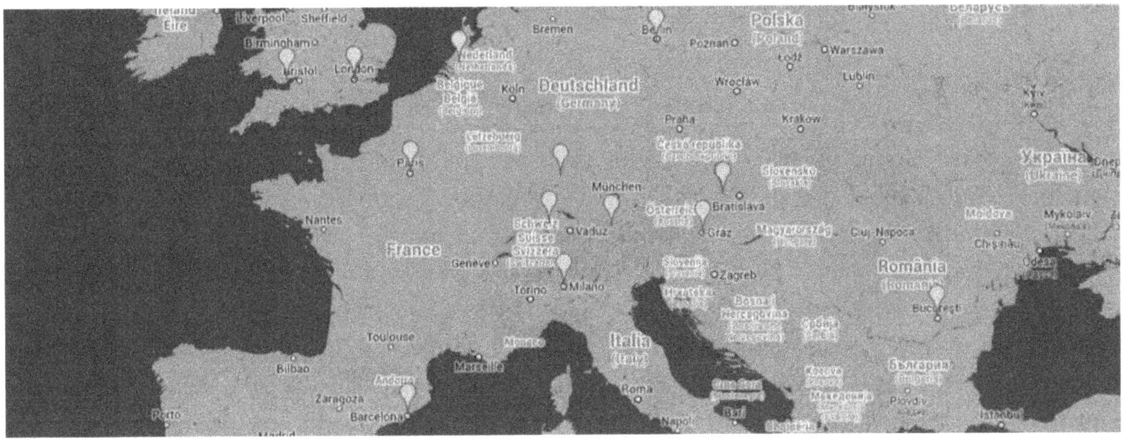

out of individual, stackable EPS (Expanded PolyStyrene) elements. The blocks are cut by a six-axis robot with a hot-wire end-effector. The continuously varying geometry is driven by a bitmap pattern that defines the aesthetics of the external wall surface and potentially optimizes the walls performance from a structural and energetic standpoint. Regarding the second prototype, the robotically manufactured EPS panels are coupled with thin metal sheets that adapt on the corresponding geometry. The application of metal sheets on mass-customized EPS ruled surfaces represents an innovative construction system that could be employed to insulate walls having complex geometries or refurbish older buildings improving their thermal performance. This system opens up new possibilities for architects to design more articulated and performative façades.

Background: Robotic Fabrication in Architecture

Robotic manufacturing has been deployed successfully in the automotive and aeronautics industries over the last 40 years, however the utilization of automated equipment for the fabrication and assembly of buildings has a limited history. Recently, this has been made possible through the current advancements in digital modelling and simulation. Since the architectural application differs from the original target industries in terms of scale, statics, aesthetics, to name a few, the investigation regarding how the design of a building system can be informed by robotic fabrication processes is now a dominant trend within architectural research (Werner, L. 2013).

The research on robotic fabrication, pioneered by Gramazio & Kohler, gains support from cutting-edge research institutions such as SCI-arc – Los Angeles, Taubman College – University of Michigan, Hyperbody – TU Delft, Institut für Computerbasiertes Entwerfen (ICD) – Universität Stuttgart, ITE – TU Graz, ROB Technologies AG – Zurich, ACIN – TU Wien, Iaac – Barcelona, EZCT – Ecole Nationale Supérieure d'Architecture Paris-Malaquais, Institute of Arts and Design – TU Wien, dFab – Carnegie Mellon University, Design Robotics Group – Harvard GSD, University of Technology – Sydney, American University of Sharjah, MIT. Contributions to the field of robotic fabrication come also from Robofold, Robots in Architecture association and the Smart Geometry group. Within all these institutions we find experiments on robotic foam cutting, but only speculative projects have been carried out. None of the experiments introduce the process of bit-map driven design and the manufacturing technique of combining EPS panels with sheet metal for wall design.

Project Description

The aim of this research project is to develop an adaptive and flexible design system that facilitates the designer in investigating a great variety of walls having complex geometry and helps the fabricator to automate the manufacturing process. The system allows intuitive control of parametric design configurations through bitmap-driven geometries and provides an easy-to-use code for immediate robotic fabrication.

A series of patterning strategies lead to highly engineered surfaces that potentially optimise the wall from a structural, acoustic and thermal perspective. In order to validate the system, a first spatial structure has been designed and realized out of individual, stackable EPS elements. The blocks are cut by a six-axis robot with a hot-wire end-effector. The continuously varying geometry is driven by a bit-map pattern that defines the aesthetics of the external wall surface and potentially optimizes the wall performances from a structural and thermal performance standpoint. On a second prototype, the robotically manufactured EPS panels are coupled with thin aluminium sheets. The application of metal sheets on mass-customized EPS ruled surfaces represents an innovative construction system that could be employed to insulate walls having complex geometries or refurbish older buildings improving their thermal performance. This system would open up new possibilities for architects to design more articulated and performative façades.

Digital Design Process

Due to its flexible open structure, the Grasshopper plug-in for Rhinoceros was selected as the algorithm editor to support the digital workflow. A custom-made script was developed in order to allow the user to simulate and visualize in real-time both the design and the production processes. The adopted parametric approach, compared to traditional CAD-CAM processes, is more flexible and adaptive. The designer can manipulate the initial CAD surface and simultaneously check the new block-layout configuration. The digital workflow is consequently fast and reliable.

Since a hot-wire machine cuts through a moving straight line, the double curved surface is approximated by developable surfaces. Specifically, the ruled surfaces are built upon XY section planes corresponding to the horizontal stack joints of the wall components. The amount of the uv subdivisions depends on the block dimension constraints, on the shifting layout pattern and on the curvature values.

In order to optimize the building system performances, it is possible to directly manipulate the geometric properties of the digital model on the basis of performative analysis. Specifically, daylight analysis, thermo-acoustic simulations, curvature analysis and structural simulations could be interpreted by various bitmap patterns.

The algorithm allows the user to modify the geometry of the outer face of the wall according to a bitmap image. The procedural texture is applied to the initial doubled curved surface. The grayscale values of that image are used to physically displace the geometry. Black represents the maximum displacement along the directrixnormal vector, whilst white represents the minimum displacement. The maximum value is determined with an iterative process that aims at maximizing the depth of the displacement while minimizing material waste, in accordance with the boundary conditions of the block units.

Figure 8. Pictures of the full-scale mock-up realization.

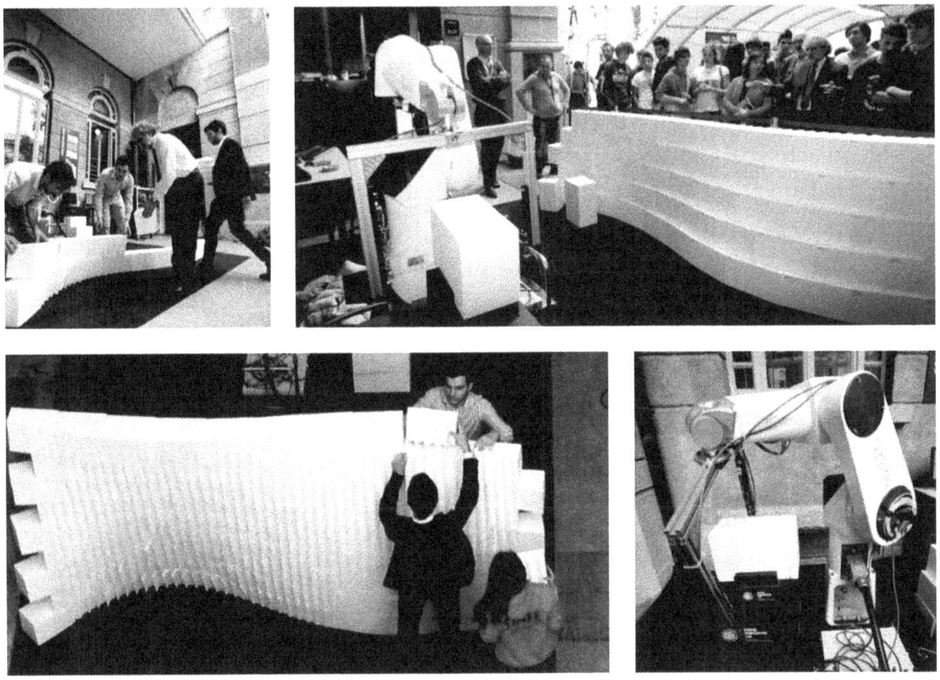

Cutting the EPS elements by utilization of the hot-wire cutter, implies the melting of the material where the wire passes. After experimenting with cutting speed and wire temperature on various densities of EPS, the GH definition can be easily set to take into consideration the thickness and consistency of the kerf. For a given wire temperature, the speed of the wire is arranged in order to avoid wire drag and poor surface finish, as well as kerf of varying thickness. The orientation of the end-effector is automatically calculated according to the façade geometry. The algorithm checks for collisions and optimizes dimensional parameters in a feedback loop way. The robot's joint parameters that provide the tool-path trajectories are automatically generated, through inverse kinematics, for each different block. According to the typology of the cutting tool, the system performs with the appropriate set of tolerances in terms of form, orientation and position of each block.

The experiment carried out at the Politecnico di Milano in march 2012 demonstrates the efficiency of the proposed system. A mono-material EPS wall, however, is too fragile for cladding real building facades without being coupled with hard finishes. On a second prototype, the robotically manufactured EPS panels are coupled with thin aluminum sheets. The application of metal sheets on mass-customized EPS ruled surfaces represents an innovative construction system that could be employed to insulate walls having complex geometries or refurbish older buildings improving their thermal performance. Further research has been carried out and several options investigated. In the following pages, several trials and mock-ups are illustrated.

EPS Sheet Metal Mass-Customized Wall Panels

Research on mass-customized insulated wall panels was conducted by following the intuition that developable ruled surface, defined by the straight line of the hot-wire cutter at specified trajectories, can be coupled with pre-cut and folded sheet metal. The process consists of generating a pattern through parametric design, discretizing and approximating the surface through ruled surface rationalization, and subsequently extrapolating the data for robotic hot-wire cutting and CNC laser cutting. Although 0.6mm aluminium sheet are malleable, a folding machine is needed to provide sharp edges. The folding process could be automated as well through Robotic machine tending. More experiments have been conducted to investigate the stack joints of the panels and the surface geometrical qualities, with different kind of curve typologies. In order to test adhesives, different samples have been realized. The geometric shapes should correspond to the worst-case scenarios in curvature degree.

Three type of adhesive have been applied to the samples and accelerated ageing tests have been initiated. The climatic chamber has been used by adopting the international standard EN/ISO 9142, cycle D2. The aging cycle for this standard consists of 4 hours at -20°C, 4 hours at -70°C 50%HR, 16 hours +40°C 100% HR).

After two weeks of cycles two out of three samples failed the test. Below a short description of the adhesive that have been utilised and tested with associated qualitative results.

The first type (the transparent-yellow one in the picture) is a very fast one-part polyurethane-based moisture curing adhesive. It is commonly used for bonding of hard insulating foams like extruded or expanded polystyrene and phenolic foam. The adhesive can be applied on one of the substrates to be bonded with a spray gun, we used a roller. One of the materials to be bonded must be porous or contain

Figure 9. EPS blocks combined with aluminium sheets through ruled surface engineering.

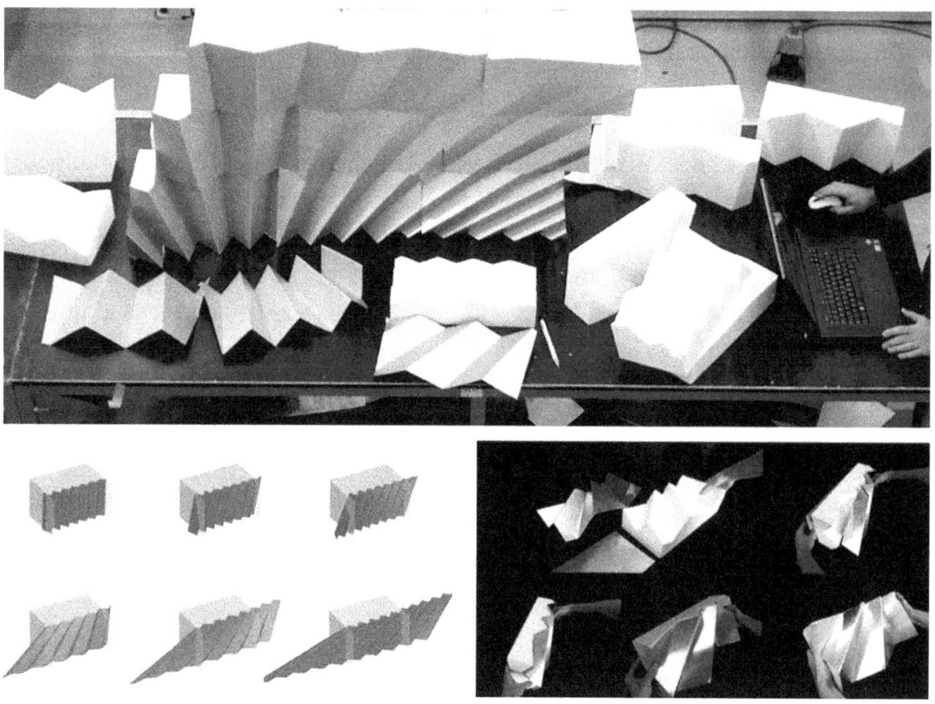

Figure 10. 6-axis robot with hot-wire cutter end-eff ector. Panel dimensions: mm500x1000x250. The hot-wire cutter tool is a rigid frame made out of aluminium standard profi les, with a wire stretched between the ends. A voltage is applied to the wire which heats it up allowing it to move throught the EPS. A spring is used to allow the wire to stretch due to thermal dilatations.

Figure 11. Samples and climatic chamber at 3M laboratories - Milan

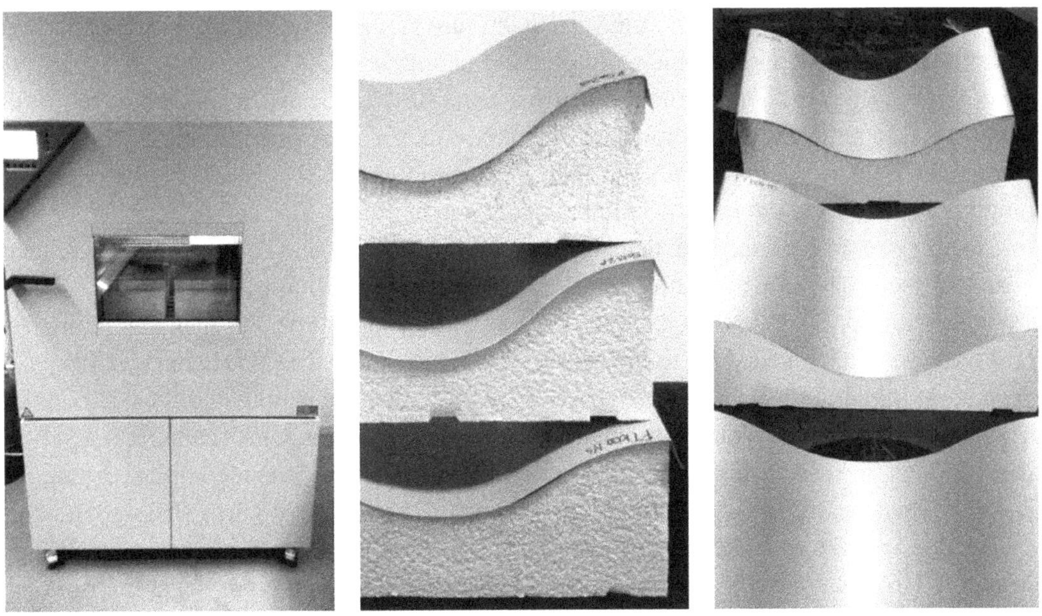

Figure 12. Adhesive deposition on one substrate of three samples [Top Left] 50132P Adhesive [Top Right] 75039 Adhesive [Bottom Left] FT1000 Adhesive [Bottom Right] Samples in the climatic chamber.

moisture. If both substrates are non-porous, it is recommended to spray a light water mist onto the adhesive film. For the experiment we used a roller to apply the adhesive and the complementary panel for applying the pressure needed. The Pressing time at 20°C is 10 minutes, the final setting time, 2 hours.

The sample with Adhesive type 50132P succesfully completed accelerating durability testing, providing data for further investigations (Fatigue, push-pull bond testing, etc.).

However, the process of applying the adhesive to the substrate must be carefully carried out, in order to avoid stains during the deposition and mold pressing phase.

The second type (the black one in the picture) is a water based adhesive, a single component elastomeric sealant based on silylated polymers curing under the effect of atmospheric humidity. Its temperature resistance ranges from -40°C to +90°C. The Skin formation time at 23°C and 50% HR takes from 20 to 30 minutes; this time is shorter at higher temperature and humidity. So, it is necessary to be sure that open time is not excessive. Otherwise, defective bonding could happen. The cure time - the length of time needed for the adhesive to cure fully - at 23°C and 50% HR is 24h for 3mm.

The sample with Adhesive type 75039 failed the accelerating aging test. The panel broke apart easily, some of the adhesive remained stuck to the metal, some to the EPS. The third adhesive (the blue one in the picture) is a high-performance water-based, one part adhesive formulated for fast bonding and long term heat resistant bonds. Adheres to many types of flexible foam, latex foams. It is fast bonding, and - depending on substrates - handling strength bonds can be made without complete drying. It is repositionable while adhesive is wet and aggressively tacky, this could help when dealing with complex shapes. The adhesive may activate sufficiently to permit making bonds within 30 seconds. Bonds may be made up to 1-10 minutes after application depending on ambient temperature and humidity condition.

The adhesive was applied with a spray gun and the pressure sufficient to assure the substrates was applied by means of the panel which has the complementary shape. The adhesive type FT1000 deteriorated during the test, but remained flexible and kept sticking on EPS and Metal.

CONCLUSION AND FUTURE RESEARCH

As robotic fabrication is gaining popularity in architecture, the need for easy interoperability between architectural design tools and digital fabrication techniques will increase. The methodology presented in this experiment could be a possible approach towards mass customization of full scale building systems and components. Besides the hotwire cutting investigations the authors are planning to experiment with cnc milling, laser, waterjet and diamond wire cutting applications on various materials, such as vibrate and lightweight products for siding masonry, ceramics, natural stone. Work on development of a production system that combines molds with customized robotic operations is being experimented upon for potential future application.

Automation technology for developing new formwork systems for free-form cast-on-site concrete structures is also being currently trialled. The design tool system integrates construction parameters directly into the design stage, informing the architect about physical manufacturing constraints, such as surface discretization typology, formwork size requirements and expected gap patterns. This can potentially lead to sustainable and efficient subsequent realization phases.

The system allows for the performance of different ranges of curvature, increasing time savings and cost effectiveness, in comparison with traditional cnc-milling processes.

RESPONSIVE LOUVERS SYSTEM FOR ADAPTIVE FAÇADES

This part of the chapter presents the digital process to generate an adaptive façade system, designed to optimize the energetic behavior of building envelopes. The geometry of the system is obtained through algorithmic modelling, whose aim is providing the best energetic and lighting conditions as a shad-

Figure 13. The Adhesive type FT1000 sample before [Left] and after [Right] the test.

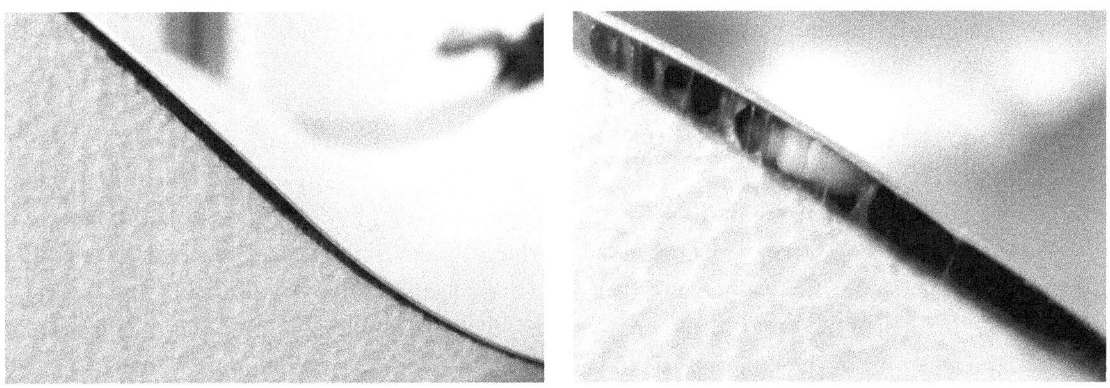

Figure 14. Double block mold and robotic manufacturing application for wall customization.

ing system. The digital process of form-finding is followed by the production of the elements, which consists in milling composite panels (aluminum and polyethylene) into elements that can be bended in their milled folds, orienting the units as the scheme obtained by the digital process. The potential of the system is twofold: once the target is achieved in terms of energetic behavior, the resulting shape can be completed in a single manufacturing process. Another peculiarity is the possibility to achieve an optimized result in terms of environmental control for the interiors of the building, considering a low number of geometric variables. Another quality of the system consists of its adaptability to many types of anchoring systems, which makes it possible to use it for refurbishment purposes or new buildings: the contours of the anchoring points are designed in the algorithmic form-finding phase - together with the global shape - and milled during the production phase. The workflow adopted in this digital fabrication process and the simple way the digital instruments are intertwined with, made it possible to produce two mock-ups of this architectural solution.

Background: Adaptive Façade in Architecture

The building envelope, being a medium between inside and outside, has great potential to make an enhancement of resident's comfort and reduce impact on the natural environment. Particularly, for building fenestration there are sophisticated demands relating to the users' comfort and health, and these have complex interactions between the requirements for good daylight quality and low energy consumption. As a result, the envelope of the building has to be very carefully designed, with a consideration of the

Figure 15. The formwork system consists of both standard parts and on-site robotically manufactured parts with different discretization patterns. The negative sides of EPS blocks are fixed on the erected formwork back-panel. After the rebar placement with spacers, the remaining EPS blocks are fixed to the formwork front-panel and appropriately positioned. The positive final shape is realized out of self-compacting concrete, pumping from the bottom of the formwork. In order to achieve smoother surfaces, thin rubber membranes are fixed to the EPS elements. Investigations regarding the application of textile reinforced mortar (TRM) materials to customized EPS panels have also been initiated. The TRM jacketing increases the shear resistance of the sandwich panels, opening up a wide range of promising possibilities for refurbishment / retrofitting operations, and new highly optimized building systems of complex architectural geometries.A fabrication methodology which integrates fabrication constraints within the architectural design process - through custom written algorithms - rises a whole new set of challenges for the architect and pushes the boundaries of the construction industry.

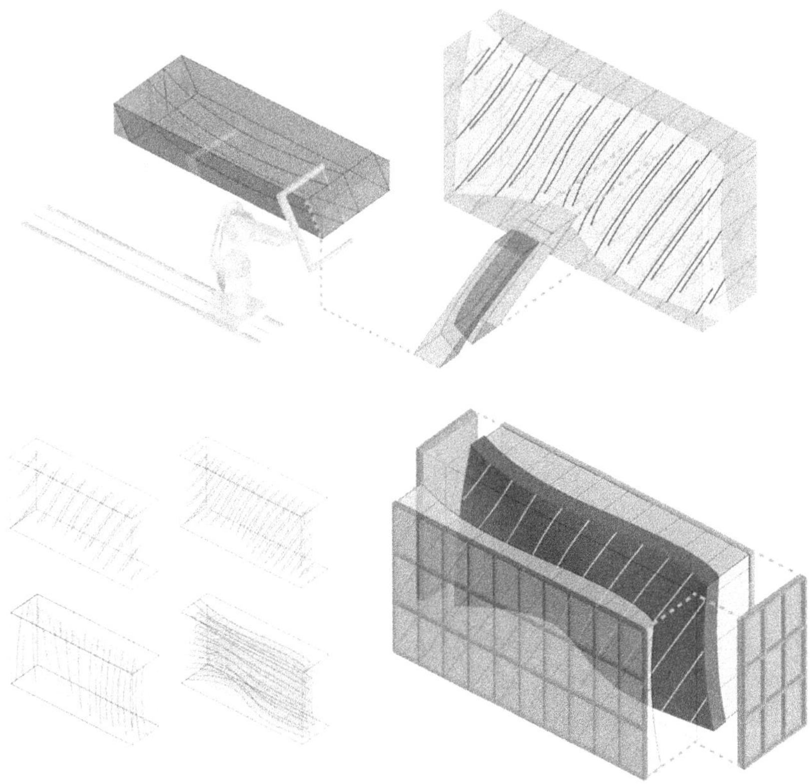

building performance. This research is aimed at proposing a façade system developed through algorithmic technique of form finding which directly depends on the specific material and manufacturing process chosen for the production. Because of this direct connection between modelling and realization, it is required that analysis, design and manufacture planning are highly integrated and developed simultaneously: complex dependencies between energetic performance geometry, and fabrication details is the key of this research.

The advances presented in this project include:

- The implementation of an adaptive façade design system through an algorithmic process aimed at obtaining the best daylight conditions in the indoor spaces;
- The design of the external skin of a residential building;
- The realization of a mock-up to validate the assumptions.

Related Work

The architecture machine, published in 1973, and the following books written by Nicholas Negroponte represent first significant attempts to define and produce responsive building systems. In the 1990s, as a result of the available technological advancements, numerous academic smart home projects were initiated. Recent projects by Patrick Teuffel and Tristan d'Estrée Sterk include structures that can monitor their shape and rigidity through the selective placements of sensors and actuators. Nowadays the Adaptive Building Initiative (ABI) is dedicated to designing a new generation of buildings that optimise their configuration in real time by responding to environmental changes. This design approach was previously pioneered by Jean Nouvel in his Institute du Mond Arabe, Paris, 1987, featuring an automated façade shading system, or Jean Prouvé's Square Mozart apartment building, Paris, 1953, with a manually controlled façade ventilation and shading system. Both undertook extensive architectural research on the uses of movable panels and endorsed the link between art and industry. A more recent example is the automated roof shading system of the City of Justice in Madrid, 2011.

INVESTIGATIONS

Material

This research presents a solution for an adaptive façade system, whose modelling phase is directly connected to its feasibility through a specific manufacturing process. For this reason, before considering the form-finding phase, it is essential to underline the potentialities and the limits of the chosen material and its production method.

The envelope of the building is considered as a set of panels made of two aluminum layers with a core of polyethylene; the final shape of the panel is reached by milling one of the two aluminum layers and bending the other together with the polymeric core. Considering this structure, this material is very resistant and can obtain a great variety of geometries in one single operation that is possible with a CNC milling machine which follows a CAD model. The panel can be bent in a specific range of angles on one side only.

Algorithmic Modelling

To summarize what is presented in the following paragraph as the workflow of the process, it can be divided into four steps: choice of the shape parameters for the envelope panels, both shading and cladding, choice of the target comfort conditions, algorithmic iterative process of form-finding to achieve the previously selected goal, tests to verify the level of comfort that is obtained.

Figure 16. Jean Prouvé's Square Mozart apartment building, Paris, 1953; Institute du Mond Arabe, Paris, 1987, City of Justice in Madrid, Foster + Partners 2011.

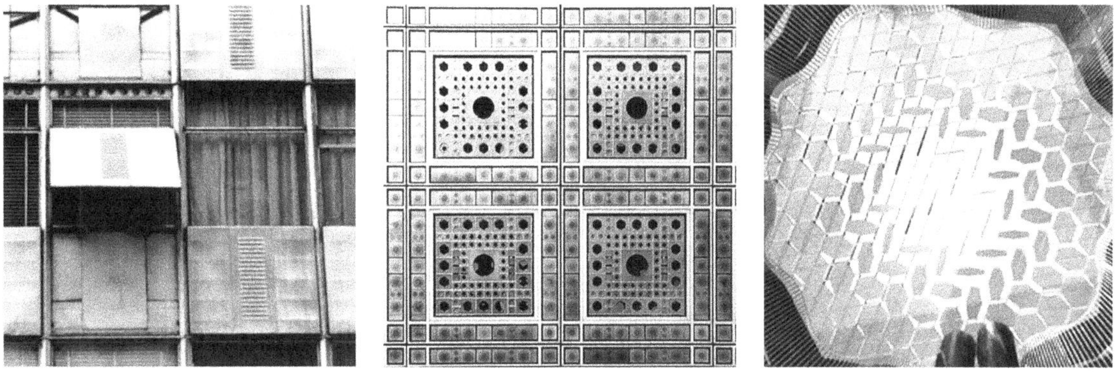

Due to its flexibility, the Grasshopper plug-in for Rhinoceros is chosen as the algorithm editor to support the digital modeling phases. The Grasshopper definition allows to define the shape of the system through a set of parameters, whose measure will be evaluated in the following performance-oriented steps. "The adopted parametric approach, compared to traditional CAD-CAM processes, is more flexible and adaptive: the designer can manipulate the initial CAD surface and simultaneously check for fabrication constraints. The digital workflow is consequently fast and reliable" (P. Ruttico, S. Arrighi, 2013).

The first step consists in choosing the parameters involved in the performance optimization process: the number of them must be a good compromise between a very complex system, which could highly increase processing time and data, and a system that do not exploit the potentiality of this material in the best way.

Subsequently, the process involves the choice of the building-physics' parameter which best outlines the visual comfort of indoor spaces. Some alternatives are initially considered, in the end the Illuminance is chosen as the most representative for residential buildings, which are the type of space the lighting analysis are performed in.

This phase is the real core of the process: this section results in the range of values the geometrical parameters previously chosen can obtain. It consists in performing an iterative simulation in which the geometry of the panels changes according to the parameters variations set at the beginning. The possible configurations are evaluated using "Galapagos", the optimizing function of Grasshopper, using as a target the best Illuminance value which is the output of a daylight analysis held in Autodesk Ecotect® through Geco. The Illuminance is calculated in pre-determined conditions, related to the weather, the place, the daytime and the period of the year: changing one of these would return a different result, which is the peculiarity of an adaptive system.

The algorithm does not identifies a single result as the best configurations, but provides a range of values which are considered eligible: the definitive shape for the envelope of the building is determined through a shifting between them that produces movement.

In the last phase of the process, the final shape of the envelope is defined. After adopting the parameters' values for the panels dimensions found in the previous section, which represent the best solution in terms of internal illuminance, the other lighting indicators are verified. This operation is managed

Figure 17. Dimension and tolerance of milling technique for aluminum-composite panels. Three types of milling are possible, depending on the bending angle required. Variable parameters for performance-oriented design of facade panels. Three parameters are chosen, in order to have enough dimensional combinations and a wide range of geometrical possibilities.

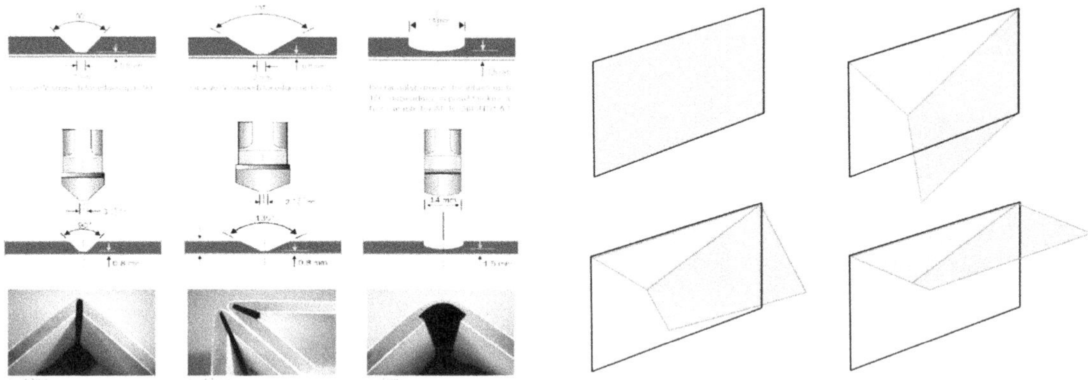

through Velux Daylight Visualizer software, which performs lighting analysis in pre-determined conditions. To validate the result, the Daylight Factor is evaluated, changing the chosen shape in case it does not satisfy the standard comfort conditions.

Case Studies

One case study and one mock-up are presented in this research; one as the theoretical and the other as the tangible result of the process previously described.

The process which was described in this project is adopted to design the envelope of a residential building: after finding the best orientation for the shading panels in correspondence of the curtain-wall, in order to optimize the lighting conditions indoor, the remaining panels are set manually to hide the slabs, achieving a sinuous global configuration.

The pavilions served as an opportunity to validate the digital and manufacturing techniques previously described. The pavilion is composed by 24 panels bolt onto a double-curvature waffle steel structure. Comparing to the standard process previously shown, in this example another geometric parameter is introduced: the panels present holes with different sizes, presenting the flexibility and the variability of this system both for aesthetic and performance purposes.

CONCLUSION AND FUTURE WORK

The presented process demonstrates to be very flexible, but simultaneously very simple in its development and real application. The simplicity directly depends both on the modelling process and on the production, which are the characterizing factors of Industry 4.0. The most relevant potentiality of the process which was developed in this research is its capability of being adopted for a wide range of diverse external conditions without being modified, this is possible because of the algorithmic nature of this procedure.

Figure 18. Examples of shading and cladding configurations analyzed in the simulation process, evaluating illuminance on a plane set at 1 m from the ground. The unit used for the analysis is a standard single bedroom: 3x3 m floor with a 1x3 m window.

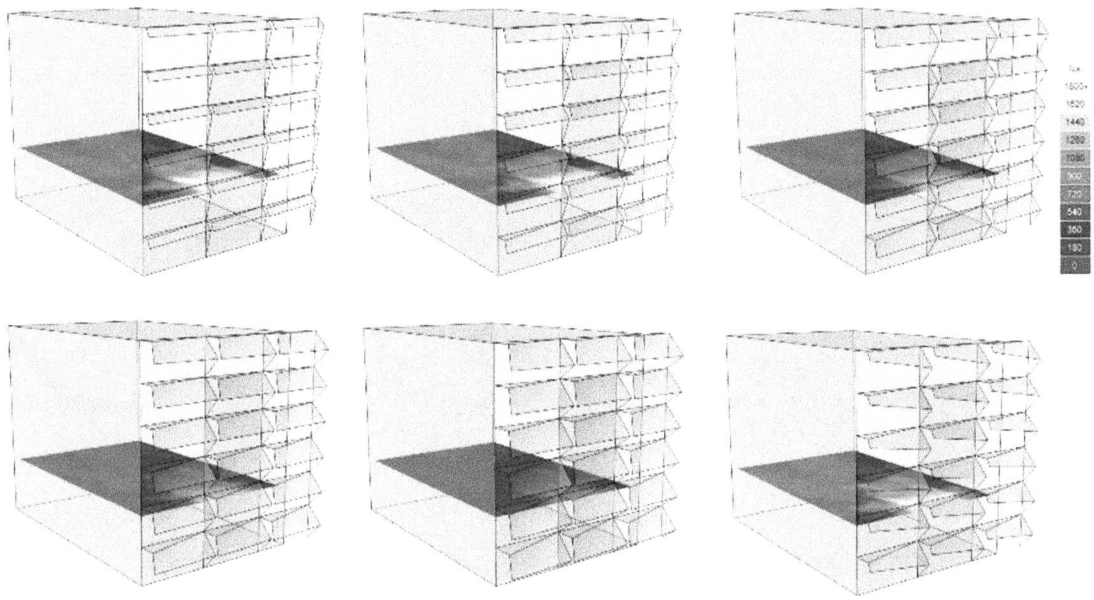

Figure 19. The image represents the final result of the implementation of the algorithmic performance-based process developed in this research, in this case adopted for a residential building.

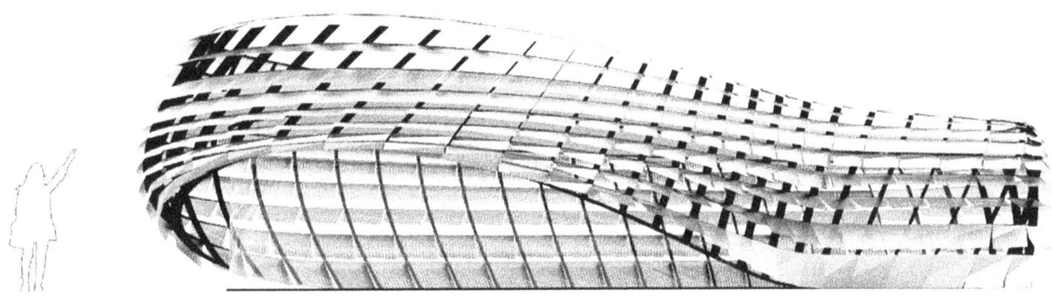

A further development of the proposed solution could be an integration with a moving digital-controlled structure. The research presented in this project provides the best shading configuration for average lighting condition over the year, unless providing a manual change of panels inclination. A self-changing solution, able to detect specific external condition and change one of the parameters of the panels (inclination), could represent a more performing system and could be more efficient in terms of natural light exploitation and in terms of shading, both aimed at energy saving. This represents an evolution of the current solution: the flexibility of the chosen material is suitable for this implementation and also the studied algorithm could be applied, considering the changing measured conditions as inputs.

Figure 20. The images represent a 3D model of the installation and the real mock-up.

TOPOLOGY OPTIMIZATION SYSTEM FOR SPATIAL STRUCTURES

This part of the chapter is aimed at presenting the design and realization process of tridimensional structural joints obtained through topology optimization, researching the potentialities of the additive manufacturing (AM) to produce optimized final shapes. Topology optimization is a method that enables designers to find a suitable structural layout for the required performances. The workflow starts from an existing structural element, which is a beam connection, whose shape is modified by utilizing a specific type of topology optimization process: Evolutionary Structural Optimization (ESO). ESO method is based on the simple concept of removing inefficient material from a structure so that the residual shape evolves towards the optimum. The obtained unconventional geometry is then produced through Additive Manufacturing, an innovative and efficient technique to produce such a complex object. Direct Metal Laser Sintering is chosen within the variety of AM techniques: it is a technology based on metal powder melted by a laser, layer by layer, in a closed and inert chamber. Subsequently, the research compares the result of the process with the initial objects for different parameters: structural resistance, mass and cost. Digital fabrication process allows to achieve both an optimization of the structure, as to the resulting geometry, and an optimization of the production process. This model is very flexible in terms of adaptability to different functions, it can be adopted for many architectural elements in addition to the one proposed in this research project as experimental example.

Introduction

"Structural optimization seeks to achieve the best performance for a structure while satisfying various constraints such as a given amount of material. Optimal structural design is becoming increasingly important due to the limited material resources, environmental impact and technological competition, all of which demand lightweight, low-cost and high-performance structures" (X. Huang, Y.M. Xie, 2010).

Additive manufacturing and topology optimization are two tightly connected concepts in specific and advanced fields of research: aerospace and automotive industry make a high use of those combined techniques, just to mention a few (Bendsoe, M.P. et al 2004). The building construction field is still researching potential applications. The aim of this research is to show how these design and production techniques can be adopted in architecture, presenting a process and a real application of it.

Contributions

This chapter presents the process to achieve an optimized geometry for a structural component, adopting the best manufacturing technology. After describing the procedure, the case study concerning the topology optimization of a node connecting four wooden beams in a gridshell structure is illustrated.

Background

The first example of Topology Optimization application in architecture for structural purposes through Additive Manufacturing was proposed by Salomé Galjaard's team at Arup in partnership with 3D Systems and Within Enhance by WithinLab (Galjaard, S. 2015). Their research focused on the development of a structural node whose aim was to connect the cables of a tensegrity-based bridge: the Kurilpa Bridge of Brisbane (AU). The intuition of using this technology is based on a peculiarity of this structure: every connection joint is one-of-a-kind, considering the different stress directions the component is subjected to.

INVESTIGATIONS

Evolutionary Structural Optimization

Topology Optimization can be defined as a process of form-finding whose aim is to develop a shape that is optimized for a specific reason: some examples are minimum mass, minimum surface, minimum volume. This project presents one method of T. O.: Evolutionary Structural Optimization introduced in 1993 by Y.M. Xie and G.P. Steven: according to P. Tanskanen (Tanskanen, P. 2005), ESO is based on the simple idea that the optimal structure (maximum stiffness, minimum weight) can be produced by gradually removing the ineffectively used material from the design domain. This process consists of an iterative sequence of Finite Element Analysis which consecutively reduces the mass of the initial model through a FE method where external loads and support conditions are applied to the element.

Additive Manufacturing

The result of the ESO does not have a shape suitable for traditional manufacturing techniques: using subtractive technologies would neutralize the material saving represented by Topology Optimization, while formative technologies would not be applicable at all. Additive manufacturing, because of its proper nature of depositing material only where expected (taking into account some modeling precautions), is the best manufacturing technology. Because of the structural requirements of the element object of this research, metal-based AM techniques are chosen.

Figure 21. Arup's research about Topology Optimization and Additive Manufacturing: 3D-printed model and Finite Element Analysis for initial and the optimized shape of the structural connection.

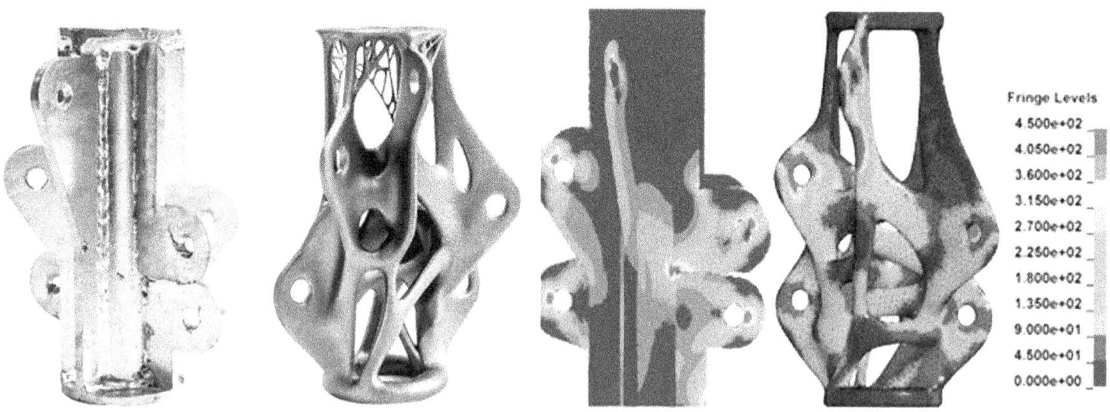

The process consists in the following phases:

1. Topology Optimization performed through iterative finite element analysis procedure;
2. Mechanical validation of the obtained geometry;
3. 3D-modeling optimization for Additive Manufacturing.

To achieve the optimization geometry from the structural component, a specific software is needed; this research was developed through Inspire by SolidThinking - Altair. What is needed to perform is an iterative process in which an initial 3D model and its material are set. The process starts from the definition of the area that is significant for the optimization to which loads (Ultimate State Limit) and constraints are applied. The following step consists in choosing the target of the optimization: it can be a determined percentage of the initial mass or a set factor of safety in terms of strains. After setting this parameters, the iterative process starts and provides a new geometry for the object.

In performing this process, you need indirectly to consider the technology you are using for the manufacture of the component; steel 3D-printed objects, in fact, do not demonstrate a perfect isotropy in load tests because in Z-direction they are composed of layers which are less cohesive along Z-axe than XY. The mechanical properties of the model (elastic modulus, tensile/compressive resistance) must be set according to the chosen technology and AM material.

This phase is aimed at validating through Finite Element Analysis the outcome of the previous section. For this research, the Inspire software tool for FEA was used. After verifying the geometry through this digital instrument, it could be necessary to validate it through further mechanical calculation regarding, for instance, instability in specific sections, depending on the final shape of the component. Geometry modifications are considered in this section. In this step, the specific geometrical requirements of the AM are taken into account. After choosing the best printing direction, considering a flat base and a vertical development of the global shape, the printed supports must be limited or eliminated at all. This can be done by avoiding hangovers with angle more than 45°/60° to the horizontal plane, depending on the specific machine. Minimizing supports represents material and printing-time savings.

Figure 22. Case study: global configuration of the gridshell the node is isolated from and construction detail of beams connection of the structure, before ESO process.

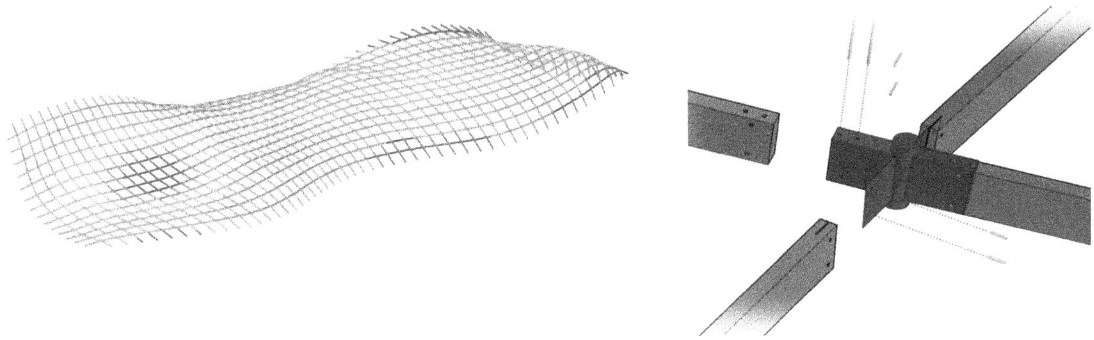

Case Study

For this research, the connection node of a double curvature gridshell structure is analyzed. To perform the form-finding phase, a preliminary FE Analysis of the global structure is needed, both to isolate a significant joint in terms of loads and to obtain the stress it submitted to.

Isolating the 3D-model to be considered for this process is the first step; after doing this, the boundary conditions must be applied: the beams loads are set on its extremities, considering two as hinges and two as joints, but they are not part of the model. Stainless Steel AISI346L is chosen and its properties set into Inspire software to perform the analysis.

After obtaining the output some modifications are necessary both in terms of mechanics, considering instability in the thinnest part, and in terms of printing technique, to reduce supports. The final result is 70% lighter than the initial node.

Figure 23. Case study: Finite Elements Analysis of stress-configuration and deformation at Ultimate Limit State for the node before the optimization process.

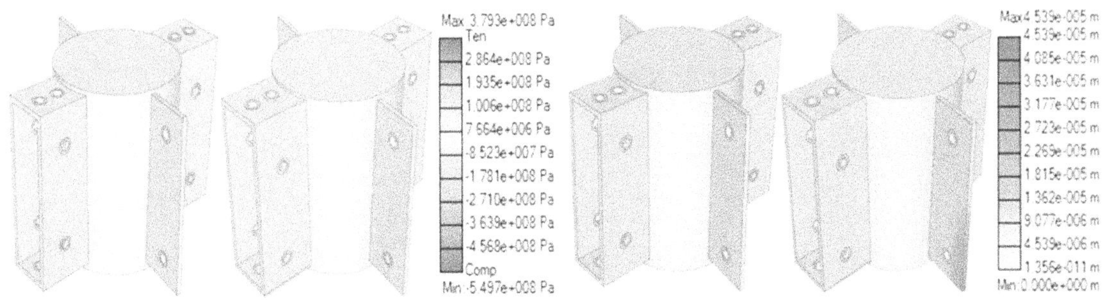

Figure 24. Case study: geometry of the node after the complete iterative ESO process (30% of the initial mass).

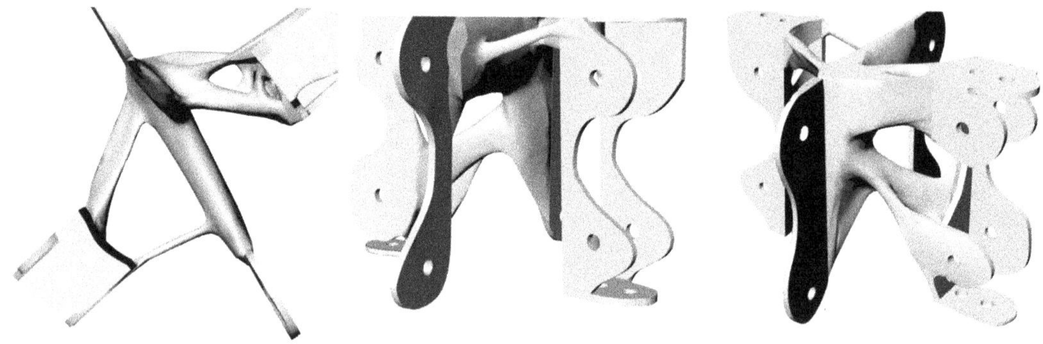

Figure 25. Case study: Finite Elements Analysis of stress-configuration and deformation at Ultimate Limit State for the node at 75% - 50% - 30% of the initial mass.

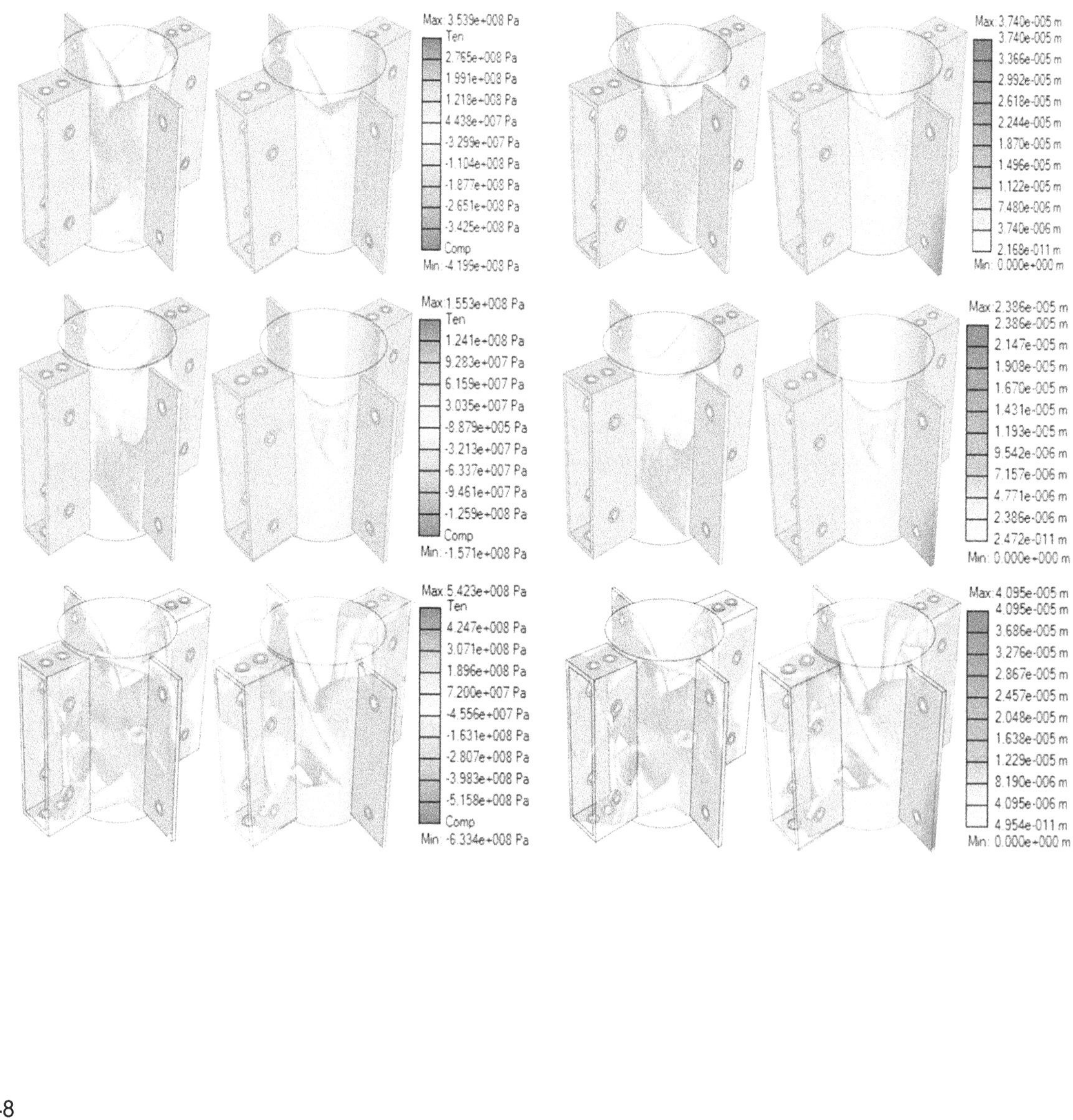

CONCLUSION AND FUTURE WORK

To validate the process explained in this research and the presented case study, it would be significant to produce the final result of the optimization through Direct Metal Laser Sintering and subject it to load tests; this would be interesting to compare these results to the ones obtained from the original complete part to verify the real bearing capacity of the outcomes, and evaluate the process both for its form-finding and production phases.

The direct consequence of this innovative technology is the expensiveness of this manufacturing technique, which is the reason why is has not taken hold in the construction field yet. Compared to other sectors, in fact, the economic advantage in terms of material saving and the structural benefits from the lower weight of the optimized component does not justify the unaffordable production costs. An improvement in efficiency in addition to a rising diffusion of this technology would certainly determine a decreasing of costs and would make it convenient to adopt it also in the construction field.

REFERENCES

Barazzetta, G. (2011). Progetto e cantiere, idea e costruzione. In Costruttori di Modernità. Milano: Academic Press.

Barazzetta, G. (2012). Costruzione. In *Architettura del Novecento, Teorie, Scuole, Eventi*. Electa Milano.

Bärtschi, M., Knauss, T., Bonwetsch, F., Gramazio, F., & Kohler, M. (2010). Wiggled Brick Bond. In Advances in Architectural Geometry 2010 (pp. 137-148). Vienna: Springer.

Bendsoe, M. P., & Sigmund, O. (2004). *Topology Optimization; Theory, methods and Applications*. Springer.

Brell-Cokcan, S., & Braumann, J. (2010). A New Parametric Design Tool for Robot Milling. *Proceedings of the 30th Annual Conference of the Association for Computer Aided Design in Architecture*, 357-363.

Brell-Cokcan, S., Reis, M., Schmiedhofer, H., & Braumann, J. (2009). Digital Design to Digital Production: Flank Milling with a 7-Axis Robot and Parametric Design. Computation: The New Realm of Architectural Design. *27th eCAADe Conference Proceedings*, 323-330.

Caneparo, L. (2014). *Digital Fabrication in Architecture*. New York, London: Engineering and Construction Springer. doi:10.1007/978-94-007-7137-6

Ciribini, G. (1958). *Architettura e Industria, Lineamenti di tecnica della produzione edilizia*. Libreria Editrice Politecnica Tamburini Milano.

Dürer, A. (1528). *Vier Bücher Von Menschlicher Proportionen, facsimile edition*. Zürich: J. Stocker-Schmid.

Frazer, J. (1995). *Themes VII: An Evolutionary Architecture*. London: Architectural Association.

Galjaard, S., Ren, S., & Hofman, S. (2015). *New Opportunities to Optimize Structural Designs in Metal by Using Additive Manufacturing*. Academic Press.

Giedion, S. (1948). *Mechanization Takes Command: A Contribution to Anonymous History*. New York: Oxford University Press.

Gottfried, A. (2014). *Produzione edilizia e automazione del cantiere*. Hoepli Milano.

Gramazio, F., & Kohler, M. (2008). *Digital Materiality in Architecture*. Baden: Lars Müller Publishers.

Hilberseimer, L. (1964). *Contemporary Architecture*. Chicago: Paul Theobald and Co.

Hoberman, C. (n.d.). *Transformable design*. Retrieved from www.hoberman.com

Iwamoto, L. (2009). *Digital Fabrication: Architectural and Material Techniques*. New York: Princeton Architectural Press.

Kolarevic, B. (2003). *Architecture in the Digital Age: Design and Manufacturing*. London: Spon Press.

Kolarevic, B., & Klinger, K. (2008). *Manufacturing Material effects, Rethinking Design and Making in Architecture*. New York: Routledge.

Kolarevic, B., Klinger, K. (2008). *Manufacturing Material Effects: Rethinking Design and Making in Architecture*. Academic Press.

Lynn, G. (1998). *Animate form*. New York: Princeton Architectural Press.

Meneguzzo, M., Morteo, E., Saibene, A. (2012). Programmare l'arte. In *Olivetti e le neoavanguardie cinetiche. Catalogo della mostra*. Johan & Levi editore.

Negroponte, N. (1970). *The architecture machine*. M.I.T. Press.

Pask, G. (1961). *An Approach to Cybernetics*. Harpen.

Pelli, C. (1999). *Observations for Young Architects*. Monacelli.

Petrignani, A., Blachère, G., & Ciribini. (1965). *Industrializzazione dell'edilizia*. Dedalo.

Pickering, A. (2009). *The Cybernetic Brain: Sketches of Another Future*. The University of Chicago Press.

Picon, A. (2010). *Digital culture in architecture: an introduction for the design professions*. Basel: Birkhauser.

Piller, F. (2004). Mass Customization: Refl ections on the State of the Concept. *International Journal of Flexible Manufacturing Systems*, *16*(4), 313–334. doi:10.100710696-005-5170-x

Reiser, J., & Umemoto, N. (2006). *Atlas of Novel Tectonics*. New York: Princeton Architectural Press.

Scheurer, F. (2010). *Materialising Complexity. Architectural Design 80*. Hoboken, NJ: Wiley and Sons.

Schumacher, P. (2011-2012). *The autopoiesis of architecture I: a new framework for architecture. II: a new agenda for architecture*. Wiley.

Tanskanen, P. (2005). The evolutionary structural optimization method: theoretical aspects. Computer Methods in Applied Mechanics and Engineering and Engineering, 191, 47–48.

Wachsmann, K. (1959). *Una svolta nelle costruzioni*. Mondadori Milano.

Waddington, C. H. (1957). *The Strategy of the Genes*. George Allen & Unwin.

Werner, L. (2013). *Encoding Architecture*. Pittsburgh, PA: Carnegie Mellon University publications.

This research was previously published in the Handbook of Research on Form and Morphogenesis in Modern Architectural Contexts edited by Domenico D'Uva, pages 354-379, copyright year 2018 by Information Science Reference (an imprint of IGI Global).

Chapter 3
Design for Additive Manufacturing and Advanced Development Methods Applied to an Innovative Multifunctional Fan

Leonardo Frizziero
Alma Mater Studiorum University of Bologna, Italy

Giampiero Donnici
https://orcid.org/0000-0002-6666-1309
Alma Mater Studiorum University of Bologna, Italy

Alfredo Liverani
Alma Mater Studiorum University of Bologna, Italy

Karim Dhaimini
https://orcid.org/0000-0003-2981-1168
Alma Mater Studiorum University of Bologna, Italy

ABSTRACT

In an increasingly competitive business world, the "time to market" of products has become a key factor for business success. There are different techniques that anticipate design mistakes and launch products on the market in less time. Among the most used methodologies in the design and definition of the requirements, quality function deployment (QFD) and design for Six Sigma (DFSS) can be used. In the prototyping phase, it is possible to address the emerging technology of additive manufacturing. Today, three-dimensional printing is already used as a rapid prototyping technique. However, the real challenge that industry is facing is the use of these machineries for large-scale production of parts, now possible with new HP multi-fusion. The aim of this article is to study the entire product development process taking advantage of the most modern models and technologies for the final realization of a case study that involves the design and prototyping of an innovative multifunctional fan (lamp, aroma diffuser and

DOI: 10.4018/978-1-5225-9624-0.ch003

fan) through the Multi Jet Fusion of HP. To begin with, issues related to the DFSS, the QFD and their application to identify the fan requirements are explored. Once the requirements have been defined, the modern CAD design systems and the CAE systems for the validation of the case study will be analyzed and applied. Finally, HP's Multi Jet Fusion methodology and design rules for additive manufacturing will be analyzed in detail, trying to exploit all the positive aspects it offers.

1. INTRODUCTION

New product development means new products for the market that have never appeared before. In some cases, these already exist on the market, but they are new to the company. A process of product development can also take place in terms of improving products made by the company and on the market, seeking to reduce costs. Generally speaking, innovation can be defined as market-pull innovation (starting from customer needs), technology-push innovation (made by scientific breakthrough) and design driven innovation.

The development and launch of new products in the market is one of the most frequent activities in the business environment and it is always a very risky action to take up. In many sectors, the ability to develop new products quickly and efficiently has become a fundamental and indispensable element for the success of the company. Therefore, the positive outcome of businesses depends on their ability to identify the needs of consumers, to design and manufacture products that meet these targets quickly and at low cost. Reaching these objectives is not just a commercial or design problem: it's closely linked to innovation and product development that involves all the company functions. For these reasons it is necessary to build and adopt modern models, processes and technologies to minimize risk and costs. In addition, nowadays businesses are marked by:

- Strong product customization
- High production mix
- Variability
- High quality
- Low Time to Market
- Prices

In order to fulfill these goals, businesses implemented different types of resources. For instance:

- Flexible manufacturing systems
- Product development techniques
- High product modularity
- Flexible Automation
- Integrated production systems and new technologies

For decades, economists and researchers have been focusing on the entire process of product development. A generic process involves different stages as shown above. The whole process can be compared

Figure 1. A generic product development process

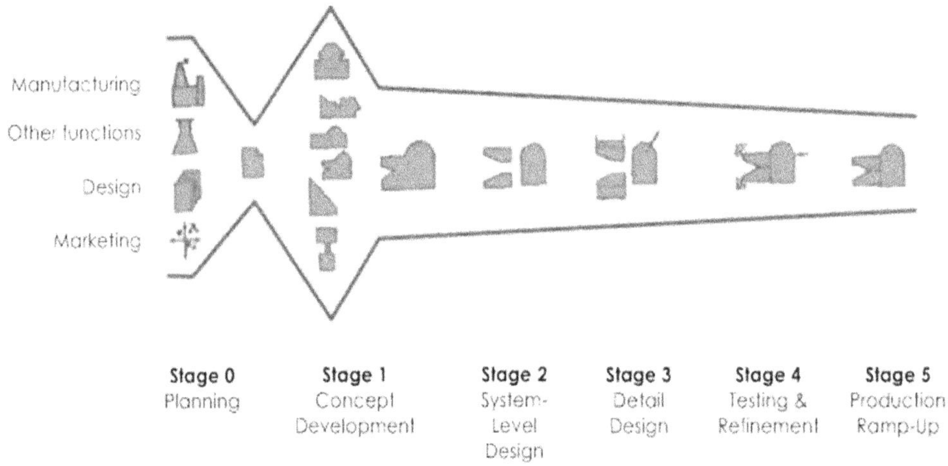

to a funnel (funnel model) that starts with a variety of ideas and a huge amount of information and ends with one idea in production. The process consists of different stages:

- **Planning:** Includes assessment of technological developments and market objectives. The result of the planning activity is the declaration of intent (mission statement) of the project. Beta tests, surveys and Crowdsourcing are usually accomplished during this phase;
- **Concept Development:** A concept is the description of the form, function and characteristics of a product. It is often accompanied by a set of specifications, an analysis of competitive products and a preliminary economic justification of the project. Techniques such as Conjoint Analysis, QFD and Design Failure Mode & Effects Analysis (DFMEA) are usually used;
- **System Level Design:** The layout of the product is the typical output of this stage;
- **Detail Design:** The result of this phase is the technical documentation with drawings and files that describe the components of the product and its specifications. Powerful tools such as CAD-CAM systems and CAE techniques are involved at this stage;
- **Testing and Refinement:** The initial prototypes are made by different processes such as the technology of Additive Manufacturing that will be discussed in the article;
- **Production and Ramp-Up:** The final stage is where costs rise up. It also contains the take-off phase and the timing of the product launch, essential for the final success. In the ramp-up phase the product is built using the planned production system. This last phase is indispensable to reduce the barriers that hinder or delay the adoption of innovation. Empirical evidence highlights a first phase of ramp-up of slow growth and finally a take-off phase with an increase in sales.

However, the main focus of this article is to discuss and present all modern techniques and technologies that support product development. As a matter of fact, in today's global market, competition for products and services has never been so resentful. Consumers have more choices for many similar products. Many manufacturing companies are continually striving for completely new products or entering new markets. Sometimes, the products meet the needs and expectations of consumers and sometimes not. Multiple redesign of a product is expensive. It would be much more beneficial if the product met the expectations of customers with a higher level of quality. It is necessary to understand what is really of value for the market and for this reason the article will focus on several techniques that anticipate design mistakes and reduced total costs of development. Many techniques and models such as Design For Six Sigma, CAD-CAM software and additive manufacturing techniques will be applied to a case study of a multifunctional fan, discussing all the advantages and disadvantages. To begin with, a series of techniques to support the design and development of new products will be introduced, starting from the DFSS which contains within it the most well-known techniques such as the QFD, CAD-CAM systems (Frizziero, 2018).

2. DESIGN FOR SIX SIGMA (DFSS)

The Design for Six Sigma (DFSS) is an alternative approach to developing new products or processes as it incorporates multiple methodologies that can be used. Design for Six Sigma aims to better meet customer demands and expectations by minimizing defects and contradictions found in products on the market and by focusing on the full understanding of customer needs, so as to obtain a new one that is more satisfying and innovative.

Considering the application of the procedure it is possible to distinguish between different approaches:

- **DMADV:** Define, Measure, Analyze, Design and Validate
- **CDDOV:** Concept, Define, Design, Optimization and Verification phases
- **CDOV:** Concept, Design, Optimize, Verify

With DFSS, the goal is to design products and processes by minimizing defects and root changes. It involves practices such as Triz, Design for X, QFD, Taguci, Computer-aided Design. The basic activities that are carried out in a Design for Six Sigma approach are the following: definition and investigation of customer requests and needs, analysis and classification of such requests, transformation of customer requests into engineering information, development of a design, validation of performances and implementation of processes strongly linked to the objective that needs to be achieved. The DFSS team must be cross-functional to ensure that all aspects of the product are taken into account, from the market research to the design phase, process implementation and product launch. The following paragraphs will discuss in detail the DMADV phases and techniques and technologies like QFD and Additive Manufacturing applied to the case study.

Figure 2. DFSS process

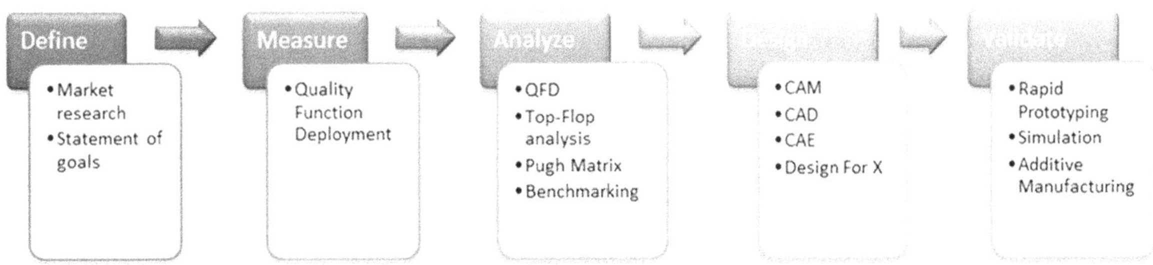

3. DEFINE, MEASURE, ANALYZE: STATEMENT OF GOALS, MARKET RESEARCH AND QFD

During these phases we collect information on what are the requests and needs of the customer and a detailed assessment is required. It is compulsory to understand what the customer really needs and what are its expectation. Moreover, it is essential to define the product and its functions. For this purpose, it is now required the development of the quality function deployment (QFD) (Caligiana et al., 2017). The latter consists of (Francia et al., 2018):

- Documentation of the environment by six questions: Who, What, Where, When, How, Why;
- Organization of a summary of customer needs;
- Assessment of need's importance;
- Technical response to customer's needs.

Basically, the QFD, conceived in Japan between 1960 and 1970, is a methodology based on the definition of the client requirement that may be implicit or explicitly stated. The customer's voice can be acquired through discussions, surveys and market studies. A market survey had been carried out for the case study. Generally speaking, QFD is considered a tool for product development that systematically translates VOC (Voice of customer) into guidelines for the design and production process. The QFD consists in the compilation of a series of matrices that translate customers' needs into design requirement. The customer is the first priority of the company, so one needs to understand exactly what he wants. In the next paragraph, QFD will be applied practically and numerically to the case study.

3.1. Environment Documentation by Six Questions

- **What?** The Innovation consists in compressing different functionalities into a single object. As a matter of fact, the fan provides an air recirculation in closed environments, it will have the function of a lamp to allow the visibility of the keyboard or to read a book. Furthermore, it will contain thermal resistances to allow heating. Finally, it could also contain a nebulizer or an aroma diffuser;
- **Who?** The Innovative fan will be mainly for people who work in the office, students or simply for people who need a multifunctional object. The market research itself will target a segment of the market that includes these categories and it is mainly addressed to an age group of less than fifty years;
- **Where?** The fan can be used in the office, on desks, in rooms, in the living room (as a furnishing object) and in all closed spaces;
- **When?** Thanks to the thermal resistance, the fan can be used during the whole calendar year. Furthermore, it can also be used exclusively as a lamp or as an aroma diffuser;
- **Why?** It will be used to refresh, illuminate, warm up and perfume interiors;
- **How?** The fan will use a DC motor powered by a transformer. The lighting will be guaranteed by a led strip and the heating function by thermal resistances.

Table 1. Environment documentation

What	Fan, Lamp, Heater and aroma diffuser
Who	Whoever works in an office or on a desk
Where	Office, desk, living room
When	During winter and summer
Why	Illuminates, refreshes and perfumes
How	Fan, thermal resistances and diffuser

3.2. Organization of a Summary of Customer Needs

In order to identify customer needs, it can be used a tree diagram which is a hierarchical structure of ideas. Tree diagrams are top-down methods. It is simply based on an intuitive decomposition of the various needs or technical requirements. Normally, it is identified by:

- Primary levels (the most abstract that measure the level of completeness and total correctness);
- Secondary levels;
- Tertiary levels.

Overall, the highest levels are used to explain the lower levels. The working group chooses a starting level to carry out its own analyzes. Below it is shown the tree diagram of the innovative fan.

3.3. Assessment of Customer Needs Importance: Statistical Analysis

Starting from the last level of the tree diagram, it is necessary to understand how these features are important. In order to discover customer needs and how much they are important, a market research has been carried out: an online questionnaire was given to a sample of 44 people, 22 women and 22 men (Figure 3).

Specifically, the Google Drive platform was used to provide pre-filled survey forms. In the questionnaire, the planned product was introduced to the customer and general questions were submitted. Finally, some questions concerning both the interest in the purchase and the importance of some product features were asked. The questionnaire is based on the Likert Scale [1-5] where:

1 = not important
5 = very important

The customer was asked to express his preference to characteristics regarding the innovative fan, for example weight, cost, reliability, quality, number of colours available. Data had been collected in a excel worksheet and then analysed. Below it is illustrated the distribution of preferences for the weight collected from the survey (Figure 4).

In addition, data regarding the customer had been collected to understand who the buyer is. Surely, market segmentation and the knowledge of the customer are extremely useful to design properly the

Figure 3. Fan Tree diagram

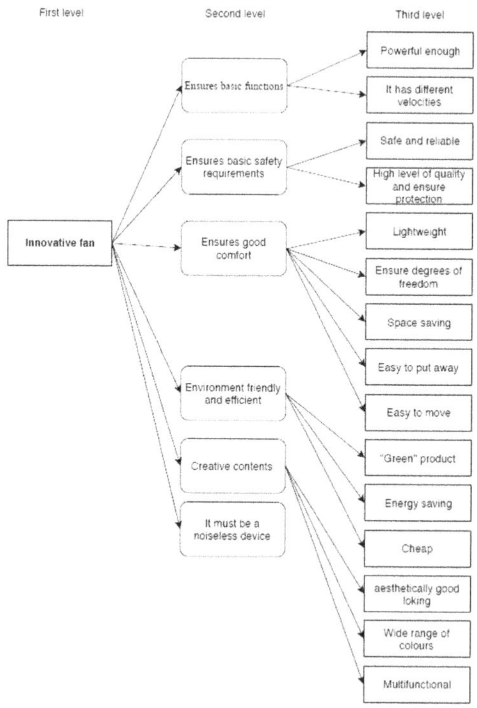

Figure 4. Example of data analysis

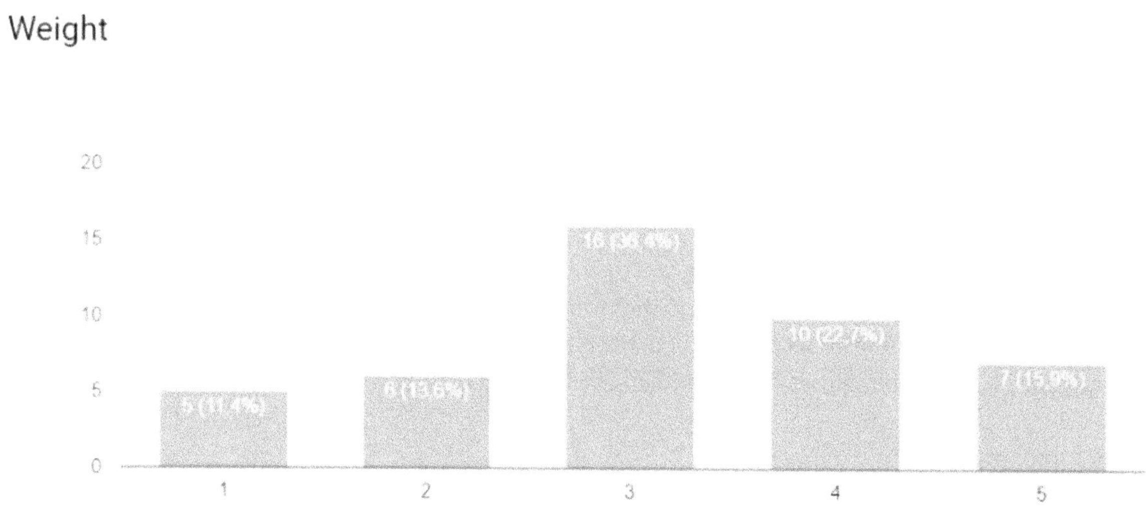

product and to develop specific features dedicated to the market segment. Below it is shown an example of data collected regarding the possible future client age of the innovative fan. It is easy to understand that students between 18 and 25 years old are chosen. Therefore, the product has to be cheap and compact to meet student needs (Figure 5).

Figure 5. Customer age

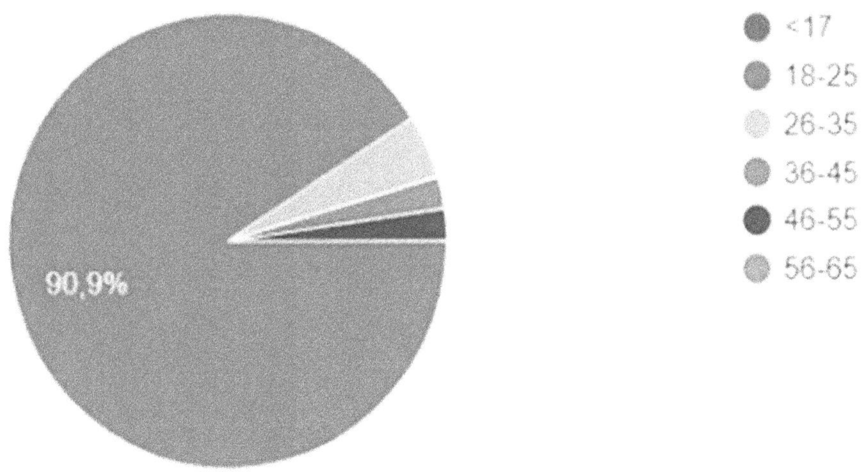

Once data are collected through the survey, it is essential to organize and structure them according to their importance. Therefore, data are organized into tables and frequencies, expected value and variance are calculated. Table below shows the preferences expressed by the sample of 44 people. Beginning from these, relative frequencies are calculated as a percentage. The calculation of the relative frequencies is based on the following formula:

$$F_r = \frac{Absolute\,Frequency}{Number\,of\,people\,interviewed}$$

In addition to the relative frequency, the average value is also calculated for each design variable to establish the order of importance. The formulas for the calculation of the mean and variance are respectively the following:

$$\mu[j] = \sum_{i=1}^{5} x_i F_{r_i} \quad \sigma^2[j] = \sum_{i=1}^{5} [x_i - \mu j]^2 \cdot F_{r_i}$$

where Fr = relative frequency, X_i = values of the Likert scale [1-5], j = design variable (Table 2).

Once data have been analyzed, it is now required to understand how this requirements affect eachother and how important they are. For this reason, the interrelation matrix is used. One possible use of the interrelation matrix is the evaluation of the relative importance between the variables. Also in this case we insert the needs in a matrix [NxN, N=variable] and assign the following values:

0 = If the row element has less importance than the column element.
1 = If the row element has the same importance as the column element.
2 = If the row element is of greater importance than the column element.

Table 2. Data collected from the survey

Questions	N of answers	1	2	3	4	5	μ	σ^2
Aesthetically good looking	44	0,0%	0,0%	15,9%	43,2%	40,9%	4,3	0,5
Cheap	44	0,0%	0,0%	18,2%	31,8%	50,0%	4,3	0,6
Space saving	44	2,3%	6,8%	27,3%	29,5%	34,1%	3,9	1,1
Easy to move	44	6,8%	29,5%	25,0%	25,0%	13,6%	3,1	1,4
Noiseless	44	0,0%	4,5%	9,1%	13,6%	72,7%	4,5	0,7
Easy to put away	44	18,2%	22,7%	38,6%	15,9%	4,5%	2,7	1,2
Number of velocities	44	2,3%	13,6%	56,8%	20,5%	6,8%	3,2	0,7
Multifunctional	44	0,0%	4,5%	15,9%	45,5%	34,1%	4,1	0,7
Energy saving	43	4,7%	2,3%	14,0%	44,2%	34,9%	4,0	1,0
Range of colours	44	9,1%	15,9%	40,9%	20,5%	13,6%	3,1	1,3
Flexible	44	2,3%	20,5%	31,8%	34,1%	11,4%	3,3	1,0
Powerful	44	2,3%	4,5%	34,1%	45,5%	13,6%	3,6	0,7
Innovative material	44	6,8%	9,1%	22,7%	40,9%	20,5%	3,6	1,2
Safe and reliable	44	9,1%	0,0%	11,4%	34,1%	45,5%	4,1	1,4
Lightweight	44	11,4%	13,6%	36,4%	22,7%	15,9%	3,2	1,4
Quality and protection	44	0,0%	0,0%	11,4%	38,6%	50,0%	4,4	0,5
Eco-Friendly	44	2,3%	6,8%	15,9%	31,8%	43,2%	4,1	1,1

It can be easily highlighted that the needs such as:

- Lack of noise
- Quality
- Cost
- Design
- Multi-functionality
- Energy efficiency

Table 3. Importance of customer needs

	Noiseless	Quality and protection	Cheap	Aesthetically good looking	Multifunctional	Safe and reliable	Eco-Friendly	Energy saving	Space saving	Powerful	Innovative material	Flexible	Range of colours	Number of velocities	Lightweight	Easy to move	Easy to put away	TOTAL	TOTAL [1-10]
Noiseless	1	2	2	2	2	2	2	2	2	2	2	2	2	2	2	2	2	33	10
Quality and protection	0	1	2	2	2	2	2	2	2	2	2	2	2	2	2	2	2	31	9
Cheap	0	0	1	2	2	2	2	2	2	2	2	2	2	2	2	2	2	29	9
Aesthetically good looking	0	0	0	1	2	2	2	2	2	2	2	2	2	2	2	2	2	27	8
Multifunctional	0	0	0	0	1	2	2	2	2	2	2	2	2	2	2	2	2	25	8
Safe and reliable	0	0	0	0	0	1	2	2	2	2	2	2	2	2	2	2	2	23	7
Eco-Friendly	0	0	0	0	0	0	1	2	2	2	2	2	2	2	2	2	2	21	6
Energy saving	0	0	0	0	0	0	0	1	2	2	2	2	2	2	2	2	2	19	6
Space saving	0	0	0	0	0	0	0	0	1	2	2	2	2	2	2	2	2	17	5
Powerful	0	0	0	0	0	0	0	0	0	1	2	2	2	2	2	2	2	15	5
Innovative material	0	0	0	0	0	0	0	0	0	0	1	2	2	2	2	2	2	13	4
Flexible	0	0	0	0	0	0	0	0	0	0	0	1	2	2	2	2	2	11	3
Range of colours	0	0	0	0	0	0	0	0	0	0	0	0	1	2	2	2	2	9	3
Number of velocities	0	0	0	0	0	0	0	0	0	0	0	0	0	1	2	2	2	7	2
Lightweight	0	0	0	0	0	0	0	0	0	0	0	0	0	0	1	2	2	5	2
Easy to move	0	0	0	0	0	0	0	0	0	0	0	0	0	0	0	1	2	3	1
Easy to put away	0	0	0	0	0	0	0	0	0	0	0	0	0	0	0	0	1	1	0

are the most important as they are the most appreciated by the sample interviewed. This means that in the design phase, these variables must be considered in larger parts. This last table is the official matrix that will be included in the QFD in the section dedicated to the needs (VOC) and the relative importations are normalized on a 1-10 scale.

3.4. Technical Response to Customer's Needs

At this stage, the project group meets, translates and deploys every customer need in one or more technical performance measures or functions performed by the product. For each need, one or more technical performances are identified. Needs are established starting from the tree diagram previously built. To be well defined, each performance measurement should be characterized in the following ways:

- The unit of measure (the metric) must be defined;
- The direction of goodness must be defined (More the better, Less the better, Target is best);
- Avoid excessive details;
- For each feature find target value.

The following table depicts a series of technical parameters that deploys customer needs of the case study (Table 4).

3.5. Relationship Matrix

It is necessary at this point to relate the technical answers to the needs of the client through the relationships matrix. Voice of Designer are placed on the columns while Voice Of Customer are identified on the rows. The procedure for its drafting is as follows:

- The matrix is prepared by listing the VODs on the columns and the VOCs on the rows;
- Numerical report evaluations are assigned to:
    - No relationship = empty box, meaning that every change of the VOD does not lead to an appreciable change in customer satisfaction;
    - Weak relationship = value 1, A broad change in the VOD leads to a slight change in satisfaction;
    - Medium Relationship = value 3, a large change in the VOD leads to an appreciable change in customer satisfaction;
    - Strong relationship = value 9, even a small change in the VOD leads to a wide change in customer satisfaction. These numeric values indicate how technical requirements help meet line requirements;
- The weighted sums by columns/rows of the assigned values are made, assuming the importance parameter as weight.

Precisely, in this passage lies one of the most peculiar characteristics of the QFD: it highlights the multiple impacts. Through a production that sums the individual products in each column (absolute weight needs x impact), we will obtain the ranking of importance of the individual characteristics. Next table shows the relationship matrix of the fan (Table 5).

Table 4. Example of technical requirements of the innovative fan

Requirements from the third level	*Technical requirement*	*Measure Unit*
Powerful	-Electric Engine type -Power supply -repeat per minute -Number of blades -Blade dimension -Engine power -Blade material -Air flow	- AC; DC - AC, DC - RPM -# -cm
Number of velocities	-Number of buttons -Engine type -Engine power	-#
Safe and Reliable	-Failure rate -MTTF -CE marking	-FIT [failures/10⁹h] -h -#
Quality and protection	-Quality marking -Compliance with design project -Technical assistance	
Lightweight	-Material -Dimensions -Weight -Engine type	 -cm -kg -DC/AC
Space saving	-Dimensions -Easy to assembly -Multifunctionality	-cm -numero preassemblati
Easy to put away	-Easy to assembly -Dimensions	-% -cm
Easy to move	-Weight -Dimensions -Easy to handle -Material	-Kg -cm
Eco-Friendly	-Energy saving -Engine power	-Kwh -W
Energy saving	-Energy saved -Engine type -Advanced materials -Engine power -Blade material	-Kwh

Table 5. The relationship matrix

	Importance	Engine type	Electric supply	Repeat per minute	Blade material	Number of blades	MTBF	Noise	Air flow	Blade dimension	CE marking	Design	Engine power	Dimensions	Materials	Degree of freedom	Angle of inclination	Weight	Led lamp	Thermal resistance	colours	Aroma diffuser	Assembly readiness	total
Noiseless	10	9	9	9	9	9		9	3	9		3	9		1					1				
Quality and protection	9	1			3		9				9	1			9				1			1		
Cheap	9	1			9	3		9				3	9	3	9			1	9	9	3	9		
Aesthetically good looking	8	3								9		9		9	9	1	3	1	9		9	9	1	
Multifunctional	8											1		3		3	3		9	9		9		
Safe and reliable	7			3							9	3			3									
Eco-Friendly	6	3			1				3				3		9									
Energy saving	6	9	3	3	9	9		3	9				9						3	9		3		
Space saving	5	3	1			9				9		3	9	9			3						3	
Powerful	5	9	9	3	9	3		3	9	3			9											
Innovative material	4											9			9			9						
Flexible	3													3		9	9	3					1	
Range of colours	3											3			3						9			
Number of velocities	2	1		3				3		3														
Lightweight	2				3									9	9			9						
Easy to move	1													9	5		1	9					9	
Easy to put away	1													9		3	3	3					9	
Technical importance	Abs	266	158	150	309	231	81	210	147	228	144	227	333	213	385	62	96	92	252	217	126	252	44	4223
	Rel. %	[illegible]	3,74	3,55	[illegible]	[illegible]	1,92	4,97	3,4809	[illegible]	3,41	[illegible]	[illegible]	5,04	[illegible]	1,47	2,27	2,18	[illegible]	[illegible]	2,98	[illegible]	1,04	100
Unit of measurement		AC/DC	AC/DC	RPM	#	#	h	Db	CFM	cm	%	[1-5]	KW	cm		#	°	Kg			#		%	

The absolute importance for each column is: ∑ (importance × relationship coefficient [0,3,9]).

The fan relationship matrix shows us the absolute and relative importance of the design variables. From the graph below, it is possible to identify:

- **Red:** Rather critical variables to which more attention must be paid;
- **Yellow:** Important but not particularly critical variables;
- **Green:** Non-critical variables that do not require particular improvements.

The following bar chart illustrates criticality of the technical variables (Table 6).

3.6. Interrelationship Matrix

In this matrix, which constitutes the roof of the house of quality, the technical characteristics are compared to each other. The roof highlights the relationships between each pair of technical answers. This helps to analyze the effects of changing a product feature, allowing the designer to determine the trade-offs between the characteristics and to recognise any problems. Very often, the VOD influence each other. The following matrix illustrates the relation between al technical parameters previously defined (Table 7).

3.7. Benchmarking

The benchmark is a continuous process of measuring products, services, processes through comparison with the best competitors or leading companies. It indicates the comparison between the company's performances and those of the "excellent" competitors. The Table below illustrates a technical benchmark between existing products in order to understand which one is better and to better improve actual parameters to satisfy the customer. Moreover, also a non-technical benchmark has been carried out to understand which of the existing product better satisfies the customer needs (Table 8).

Table 6. Technical critical variables

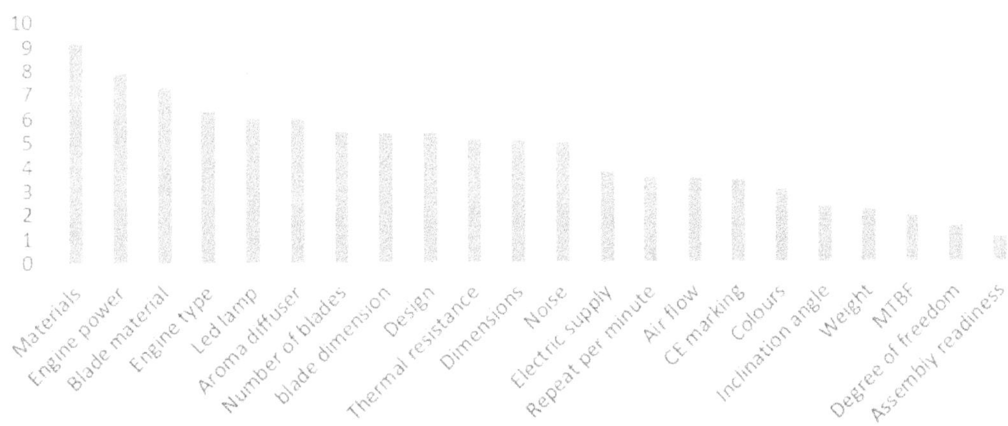

Table 7. Interrelationship matrix

	Assembly readiness	Aroma diffuser	Thermal resistance	Colours	LED lamp	Weight	Inclination angle	Degree of freedom	Materials	Dimensions	Engine power	Design	CE marking	Blade dimension	Air flow	Noise	MTBF	Number of blades	Blades material	Repeat per minute	Electric supply	Engine type
Engine type						O				XX	XX			X				XX		X	XX	
Electric supply		X			X					XX	XX			XX				XX				
Repeat per minute									X	X	XX			OO	X	XX						
Blades material										OO	OO				XX	OO		XX				
Number of blades										XX	XX			X	XX							
MTBF													X									
Noise										XX												
Air flow											X			XX								
Blade dimension						XX			X	XX	XX											
CE marking																						
Design	X			XX	XX				XX													
Engine power																						
Dimensions	O	XX			XX	XX	X															
Materials	X			XX		XX																
Degree of freedom	O																					
Inclination angle	O																					
Weight	O																					
LED lamp	O			X																		
Colours																						
Thermal resistances																						
Aroma diffuser	O																					
Assembly readiness																						

XX = Strongly positive
X = Positive
O = Negative
OO = Strongly negative

Table 8. Technical benchmarking

	Engine type	Electric supply	Repeat per minute	Blade material	Number of blades	MTBF	Noise	Air flow	Blade dimension	CE marking	Design	Engine power	Dimensions	Materials	Degree of freedom	Inclination angle	Weight	LED lamp	Thermal resistance	Colours	Aroma diffuser	Assembly readiness	
Measurement unit	AC/DC	AC/DC	RPM	#	#	h	Db	CFM	cm	%	[1-5]	KW	cm		#	°	Kg			#		%	
Ardes AR5EA40 EASY 40	AC	AC	1245	PLAS	3	490000	60	66,67	40	100	1	55	42 x 24 x 52	PLAS	2	35	2,4	no	no	1	no	45	
Rowenta VU2630 Turbo Silence Desk	AC	AC	1050	PLAS	5	490000	42	80	40	100	2,5	70	48 x 14,5 x 46	Allum	2	40	2,96	no	no	1	no	60	
Honeywell HCE100 Mini Thermoheater	AC	AC	650	PLAS	/	490000	50	68	4	100	4	250	11,5 x 14,7 x 21,6	PLAS	0	10	0,58	no	si	5	no	100	
Fancii fan USB	DC	DC	250	PLAS	/	490000	30	20	3	100	4,5	usb	12 x 10 x 10	PLAS	1	8	0,28	no	no	2	no	100	
Technical importance	266	158	150	309	231	81	210	147	228	144	227	333	213	385	62	96	92	252	217	126	252	44	Abs
	6,30	3,74	3,552	7,317	5,47	1,918068	4,973	3,481	5,399	3,41	5,375	7,885	5,04	9,12	1,47	2,3	2,18	5,97	5,14	3	5,97	1	Rel %

To conclude, we must establish those characteristics that have a strong relationship with the needs of the customers, or that have a low technical performance or that are strong sales factors. In particular, the green boxes show the technical variable that need to be improved for the innovative deskfan. For this reason, the innovative fan will:

- Use a DC motor to avoid noise;
- Meet CE marking prerequisites;
- Have 4 blades as an optimum trade-off between noise and air flow;
- Use innovative plastic material to be lightweight.

Overall, once all matrixes are built and filled, it is possible to build the house of quality which translates customer needs into general guidelines for the design process which will be developed in the next chapter. The final house of quality is shown here below (Frizziero et al., 2018; Donnici et al., 2018) (Table 9).

Table 9. Final QFD matrix

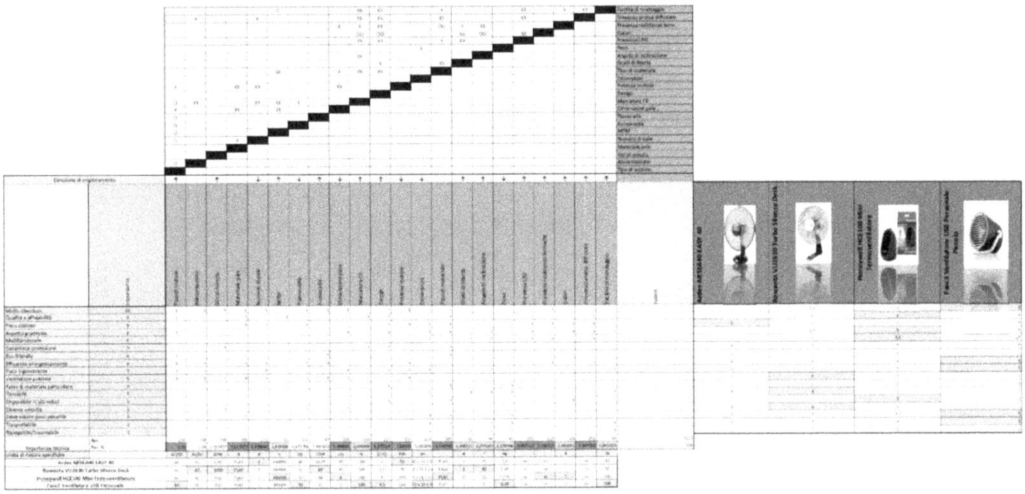

4. DESIGN: CAD & CAE

In the Design phase, it is necessary to develop the technical project and a virtual model of the fan. In this regard, techniques adopted by companies will be analyzed, such as the integrated systems Computer-aided-Design, Computer-aided-Engineering and Computer-aided-Manufacturing (CAD-CAE-CAM).

4.1. Concept Design and Initial Sketches

The initial idea is performed freehand, analyzed and improved to be transformed into a preliminary project in which the conceptual solution identified is developed in detail. Many times, the starting point is a sketch on white paper. Figure 6 shows initial ideas of the project.

Figure 6. Initial sketches

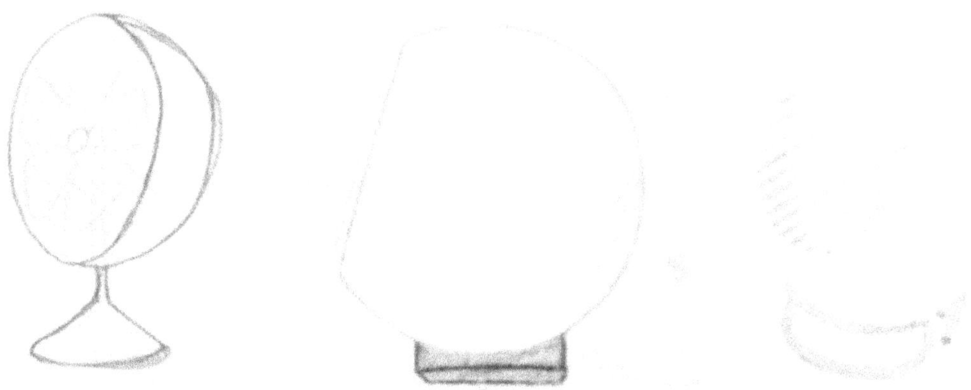

4.2. Executive Project

This phase involves the use of CAD systems and the first project drawings are drafted. Assembly drawings that contain very general information and only maximum dimensions are represented. Sub-assemblies are analyzed and CAE simulation systems are used to simulate the structural, kinematic, thermal and dynamic behavior of mechanical components. In particular, the finite element analysis (Finite element analysis FEA) determines the static and dynamic responses of the components at different load conditions. It is now necessary to introduce all components that have been designed:

Central body: This component must guarantee the necessary force to support the motor and withstand the thrust force of the impeller which rotates at 1500 RPM. An inner "spider" structure has been designed to guarantee support despite the spherical shape of the fan. On the upper part there is a threaded hole for the designed button and the lower part, however, has a 4mm hole to ensure the passage of the current cable for the power supply. Special holes and grooves have been created to facilitate air entry (Figure 7).

Finally, the body has inside the 10mm grooves to accommodate the LED strip and the thermal resistance. The body has a diameter of 150mm. It can be easily noticed that, in order to ensure an innovative design, the central body can't be manufactured easily by traditional techniques such as injection molding or milling. It is necessary to adopt additive manufacturing. Finally, 2mm holes were designed to allow powder out during manufacturing process.

Figure 7. Central body

Support: The central body rests on a spring support made in such a way as to guarantee the greatest number of degrees of freedom. The part subjected to greater mechanical stress has been reinforced (Figure 8).

Grid: To ensure good safety levels, a 5mm grill template has been designed. Holes also are made on the template to assemble it to the central body using self-tapping M2 screws (Figure 9).

Fan: The fan that occupies the central part has a maximum encumbrance of 110mm, therefore widely contained in the body. The propellers have an inclination of 30 ° to facilitate the flow of air but without generating excessive turbulence, ensuring greater silence. In the central part a threaded hole has been realized to guarantee the coupling with the 3.17mm diameter motor shaft (Figure 10).

Cap: In order to avoid mechanical vibrations and block the translation of the motor and of the fan, a "plug" has been designed with 3mm teeth to accommodate, by interference, the crankshaft support (Figure 11).

Engine-Fan-Body Assembly: The central part of the body, where the engine is located, is the part most subject to mechanical stress, due in part to the thrust force of the fan, on the other by the weight force of the engine and its generated torque. To minimize vibrations and secure the motor, a relief is designed at the end of the motor housing to generate interference, as a "vice" effect (Figure 12).

4.3. CAE Analysis

Once all components are modelled on a CAD software, it is possible to develop on a CAE software (as Autodesk Inventor) a Finite Element Analysis to evaluate the parts subject to greater mechanical stress. The finite element analysis is a mathematical technique designed to analyze and simulate the behavior of organs subjected to efforts, which was born in the 60s. The term efforts refers to mechanical, electrical

Figure 8. Support

Figure 9. Front grid

Figure 10. Fan

Figure 11. Cap

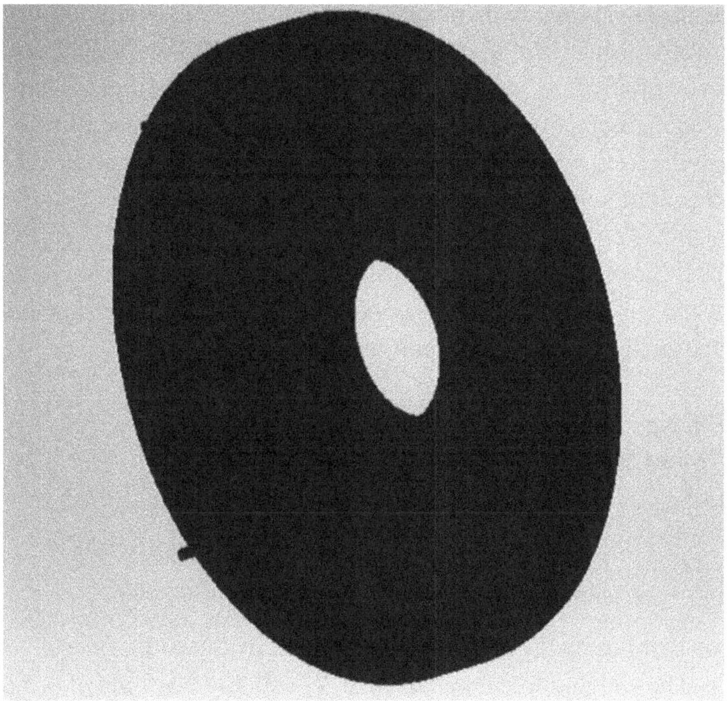

Figure 12. Engine-body assembly

or thermal stresses. This type of analysis breaks the structure under examination into smaller substructures, called finite elements. The finite elements and their interactions are converted into an equation and solved mathematically. Among the advantages in the adoption of numerical simulation systems it is possible to find: behavior prediction, raw material reduction, early validation, What-if scenario analysis.

Let's now analyze some components. The FEA analysis was developed in the Autodesk Inventor environment. The material used for all components is the Nylon PA 12, introduced by HP, whose Yield Strength Rp is 48 MPa and it has a density of 1,01g/cm3.

Support: The support is the most stressed component. In fact, it is required to consider the weight of all the components and apply a force on this support and analyze the deformations. Now we evaluate the weights of the pieces given by CAE software:

- **Body:** 0.196 Kg
- **Engine (From Catalog):** 0,303 Kg
- **Fan:** 0.014 Kg
- **Grill:** 0.041 Kg
- **Cap:** 0.010 Kg

The total weight on that lies on the support will be given by the sum of these.

Weight = 0.564kg. The weight force is obtained as W = 0.564 * 9.8 m / s^2 = 5.53N. Now we apply this force on the surface of interest. The results are obtained in Figure 13.

By applying the gravitational force and the load, and after fixing the constraints of the support, it is possible to notice that the area subjected to the greatest effort is the part highlighted in red. However, the maximum effort recorded is 3.778MPa, while the yield strength of PA12 is 48MPa. Therefore, we are still in the elastic field, far from the breaking load. In addition to the Von Mises factor, displacement must also be considered. Since both the structure and the material are very elastic, an important displacement can be noticed in the area furthest from the vertical support. A displacement of 8mm has been registered, allowing a spring effect (Figure 14).

Body: The body is the most fragile and difficult part to analyze. The "spider" supports of the motor housing are thin and risk breaking if subjected to high mechanical stress. In the FEA analysis of this component, both the thrust force of the fan and the motor weight force were calculated. Engine's torque is neglected:

- **Motor Weight:** 0.303 kg, which becomes W = 2.97N
- **Pushing Force:** the "actuator disk theory" of the fluid dynamics introduced by Rankine was exploited. This theory, considering null all aerodynamic and mechanical frictions, calculates the power P to generate a thrust T with the following:

$$P = \sqrt{\frac{T^3}{4\rho A}}$$

where:

Figure 13. Von Mises stress on the support

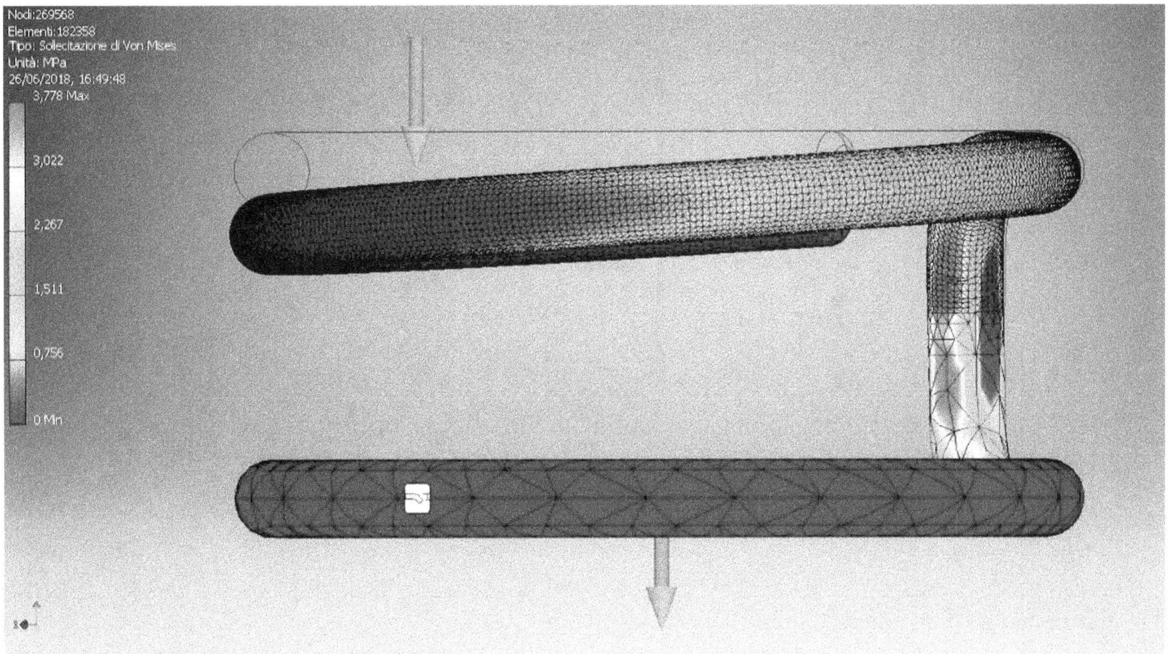

Figure 14. Displacement analysis of the support

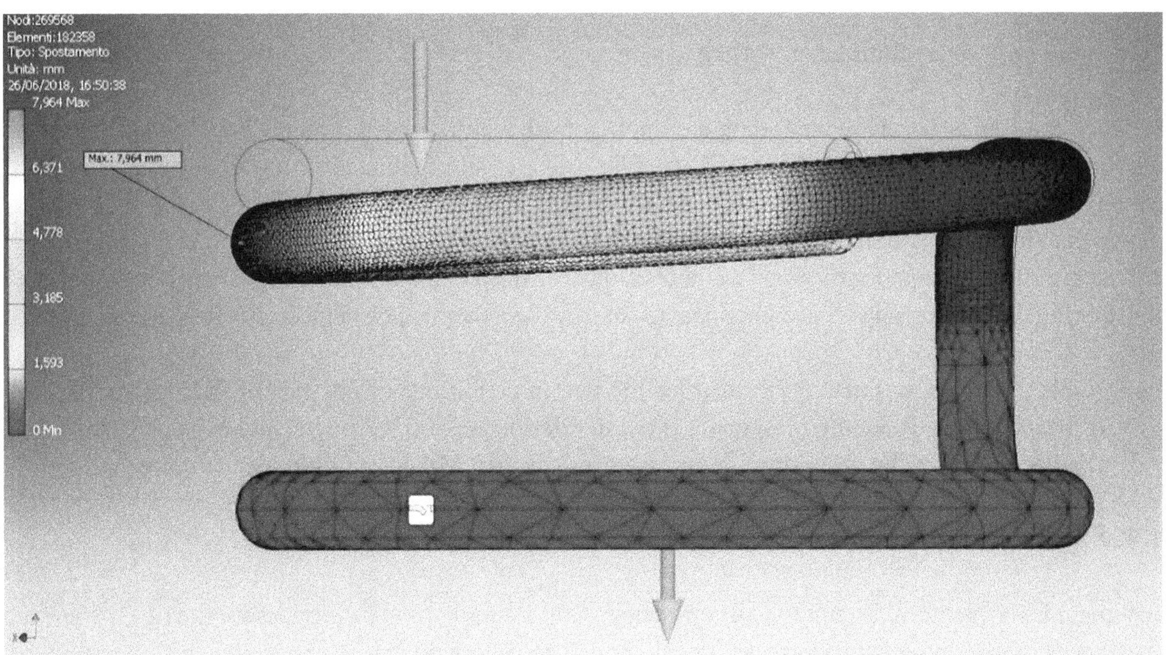

Figure 15. Forces acting on the central body and stress analysis

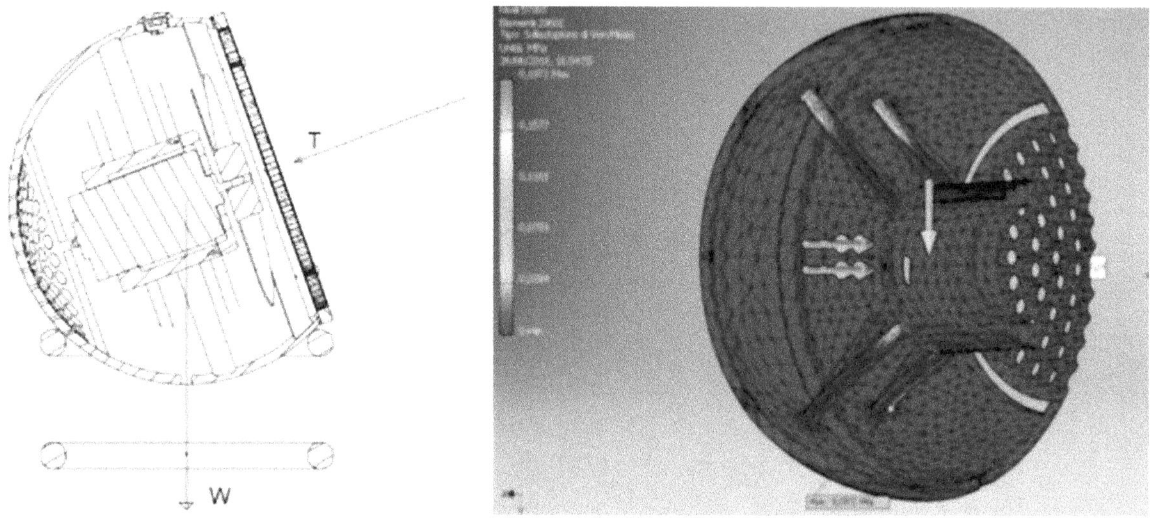

$$P = \frac{1}{2} \times \rho A v^3 \text{ and } T = \frac{1}{2} \times \rho \, (Cs^2\text{-}Cu^2)$$

ρ = air density (1,22kg / m3 at ISO conditions)

A = disk area

Cu = 5 m / s Cs = unperturbed air = 0.12 m / s

Once the forces are calculated, they are applied to the surface of the motor housing and pressure generated had been calculated: 0.00065MPa. This value is approximated in the FEA environment of Autodesk to a minimum cautionary value of 0.001MPa. Let's now analyze the results obtained. It can be shown that the maximum stress obtained is 0.1972 MPa with a material Yield Strength Rp of 48 MPa. In this case, unlike the support, the displacements along the horizontal axis are of the order of microns and considering the application of the component (domestic use), they can be neglected (Ghezzo et al., 2008).

Fan: A FEA analysis was also performed for the fan. In conditions of absence of friction and lack of resistance of the fluid, the component is put in rotation at 1500 RPM to evaluate its consequences. Displacements in the order of microns and stress of 0.19 MPa are highlighted (Figure 16).

4.4. Final Assembly

To conclude the technical feasibility, all components are assembled and catalogues for the LED, Thermal resistances, wires and manufacturer are considered. The following image shows the final assembly as a virtual model (Figure 17).

Figure 16. Stress and displacement analysis of the fan

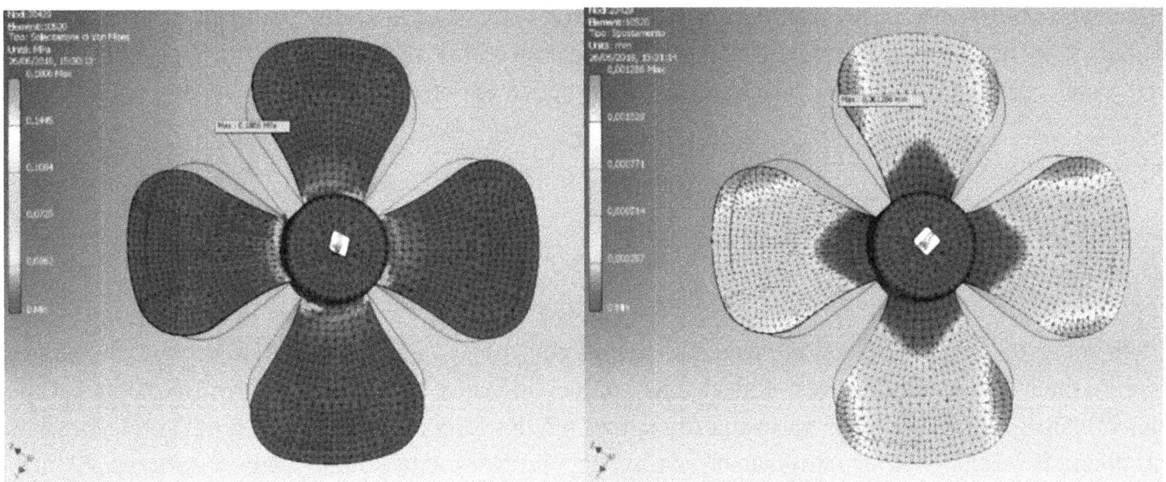

Figure 17. Final assembly

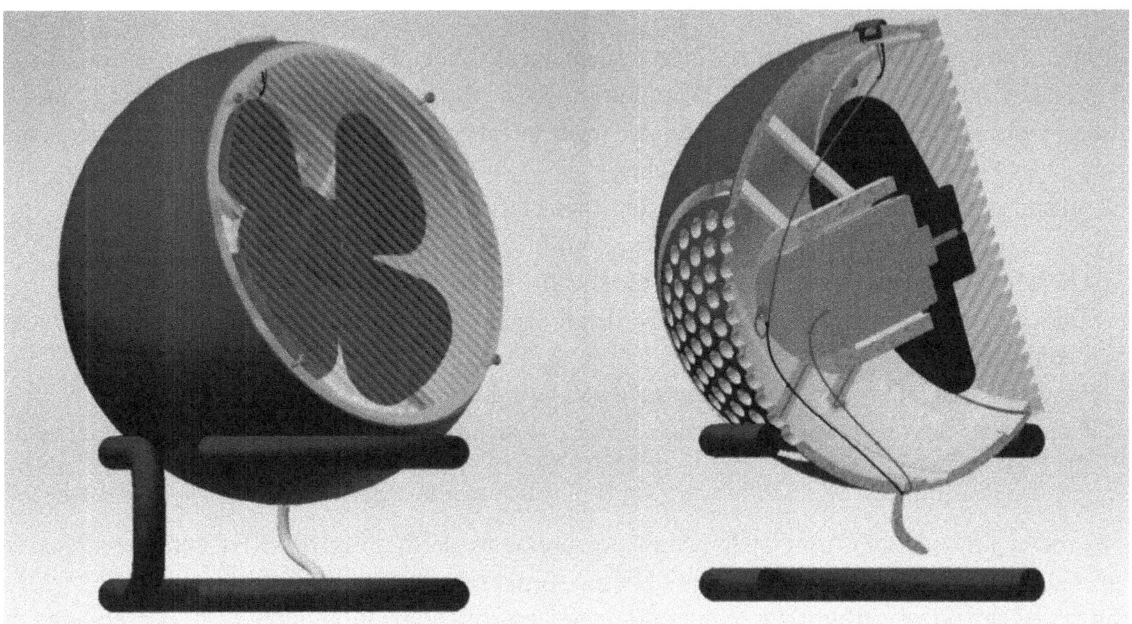

5. VALIDATE: DESIGN FOR ADDITIVE MANUFACTURING AND MULTI JET FUSION

In this last phase, it is stated with certainty that the finished product confirms the expected results. It is possible to produce prototypes to ensure that the product is in line with the required characteristics. Moreover, an FMEA analysis is developed to identify the weaknesses of the project very quickly. In this last chapter, validation and prototyping techniques will be treated by additive manufacturing technology (de Amicis et al., 2018).

Considered one of the enabling technologies of the industry 4.0, Additive manufacturing is the creation of three-dimensional objects, starting from a virtual 3D model. The digital model is produced with dedicated software and subsequently elaborated to be then realized layer by layer, through a 3D printer. It is a concept easy to automate. The main areas of application of AM technologies are:

- Rapid prototyping;
- Indirect production of molds and tools;
- Direct production (Rapid Manufacturing);
- Spare parts.

In order to reduce production costs and prototyping times, additive manufacturing technologies replace the traditional techniques of mechanical processing since the latter work by removing material step by step, generating a lot of waste and sometimes require complex code generation (G-code for CNC machine). RP techniques operate on an inverse conceptual basis, ie by adding material with the possibility of obtaining complex shapes.

5.1. Multi Jet Fusion by Hp: Voxel-Jet

Multi jet fusion is one of the latest additive manufacturing technologies introduced by HP in 2016. It offers a faster and cheaper process for 3D printed parts. The finished parts achieve better quality surface finishes, high resolution detail and more reliable mechanical properties than other 3D printing technologies, such as FDM (Fusion deposition modeling).

It all comes down to four main elements that characterize MJF:

- Greater speed thanks to the processing of an entire layer;
- Higher quality of the parts thanks to the detail agents; V Voxel-Level multi material and multicolor capability;
- Materials that are economically accessible and have lower environmental impact;
- Competitiveness with mass production. Unlike slow processes of some technologies (dot-to-dot like FDM).

By moving the recoater from top to bottom (according to the figure), HP MJF technology is started by placing a thin layer of powdered material in the work area. Afterwards, the cart containing the TIJ array passes the powder from right to left and then deposits agents through the area. If necessary, the processes of energy deposition and chemical agents are combined in a second continuous passage of the trolley from left to right. This process continues, layer by layer, until a complete part is formed. HP's proprietary synchronous architecture can print over 30 million drops per second across every inch of the work area (Figure 18).

In HP MJF, fusion and detail agents control the fusion of selected Voxels (3D elementary printed unit) and edges between the merged and unfused regions in each layer. Using HP Transforming Agents, HP is able to control different properties on its Voxels, allowing to obtain different mechanical properties, different colors and transparency in the same product (Figure 19).

To conclude the briefing on the breakthrough technology, it is necessary to introduce the cooling station (processing station). Once the 3D printing phase is complete, it is needed to remove the cart

Figure 18. Multi jet fusion process

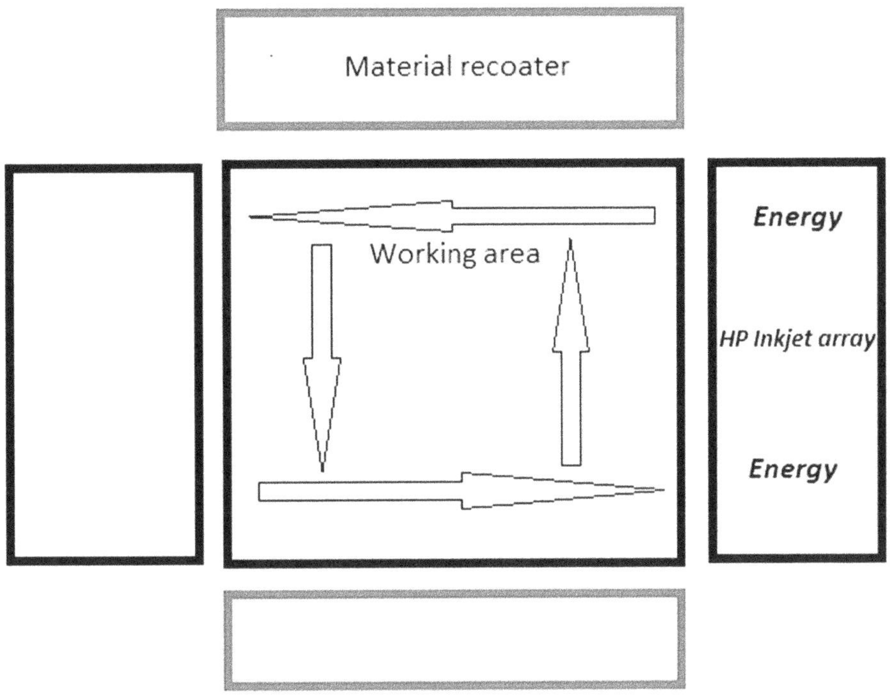

Figure 19. HP detail and fusion agents

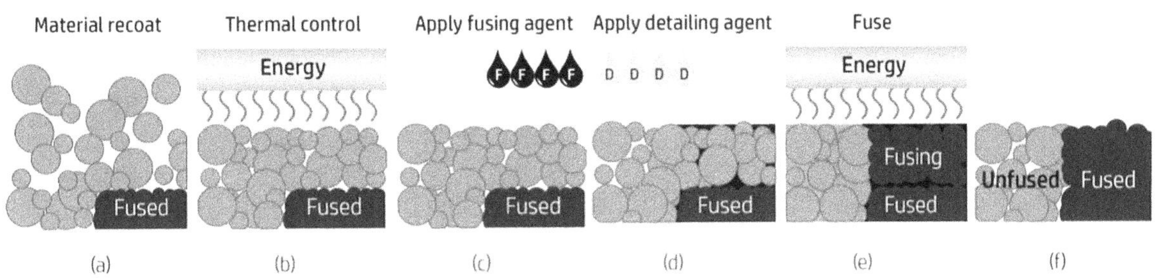

that contains the printed part and move it to the cleaning and cooling station. This phase is essential to avoid any deformation of the pieces. In fact, the cooling speed affects the flatness of the object. After the required cooling time (even faster than that required for Selective Laser Sintering), a vacuum tube system will automatically remove most of the dust. The processing station software allows a very high product recovery: 80% of the used powder is mixed with 20% of new material for the next processing. The machine can recover up to 96% of the unused material. Thereafter, it is possible to make post-processing operations such as coloring or improving the quality of the surface.

5.2. Nylon PA 12

Nylon PA 12 is a material introduced by HP to be used in Multi Jet Fusion. It is a powder polymer such as PA 11. They both have good mechanical properties and have less impact on the environment. Moreover, it has a good chemical resistance and good behavior at low temperatures. The table below resumes its characteristics. Nylon PA 12 has been selected for the innovative fan (Table10).

5.3. Design for Additive Manufacturing

Design for additive manufacturing (DFAM) is a type of Design for X applied to this new technology, full of opportunities in terms of design. It is a tool that optimizes functional performance and other product lifecycle considerations, such as manufacturability, reliability and cost. This process emerges thanks to the enormous design freedom provided by additive technologies. Typical DFAM methods or tools include topology optimization, multi material design, mass customization and lattice structures. As stated, the DFAM should "maximize product performance through the synthesis of forms, dimensions, hierarchical structures and material compositions, subject to the capabilities of AM technologies".

The DFAM can be seen as a version of the Design for Manufacturing and Assembly pushed to the limit. In the DFMA, which was introduced in the 90s to improve the production of components, the most common solutions are:

- Minimize the number of components;
- Remove the connections.

However, it is easy to understand that by decreasing the number of components, the cost of assembly decreases but the cost for geometric complexity increases exponentially. For this reason, the DFM presents this technological limit. The DFAM, instead, goes further thanks to the freedom of design offered by additive manufacturing. This philosophy focuses mainly on three issues (see Figure 20):

1. **Topology Optimization:** The Layer-by-Layer process allows to have complex sections up to the maximum allowed resolution. Moreover, it is possible to define the Shape Re-design: it means that it is possible to optimize the shape of the product solely based on the functions it meets. Topology

Table 10. Mechanical properties of PA 12

Powder melting point	187° C
Particle size	60 μm
Density	1,01 g/cm3
XY tensile strength	48 MPa
Z tensile strength	48 MPa
Elongation at break XY	20%
Elongation at break Z	15%
Shore Hardness D	80

Figure 20. Design for additive manufacturing

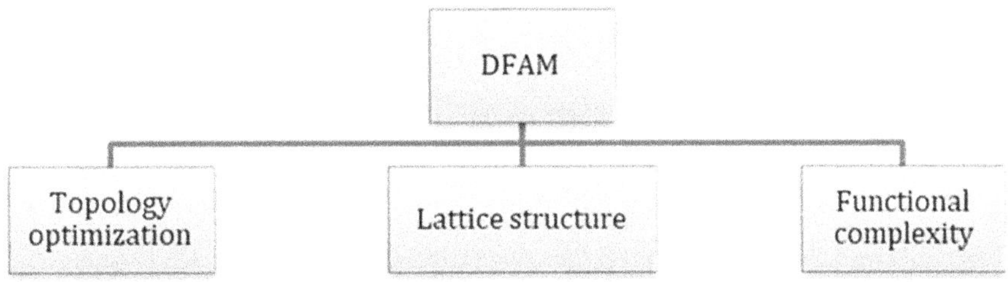

can be optimized in terms of stress and functionality and not in terms of technological limits. As for the fan, we have already seen that the central body can't be produced by conventional technologies apart from additive manufacturing;

2. **Lattice Structure:** It is possible to have a control of the different levels of structure. It generates much lighter and less expensive structures. Although the compact structures are stronger, the specific strength of the samples of latex structures is higher: they provide a higher force per kilogram of material. The reticular structures provide sufficient strength with reduced mass. During the project, both full and lattice structures were designed;
3. **Functional Complexity:** It is possible to print assembled parts in one operation, avoiding assembly costs.

As for machining, HP suggests some general rules to apply during the design process to avoid printing defects. The most important are:

- **Wall Thickness:** Printing resolution is strongly anisotropic. However, HP suggests wall thickness of 0,3 mm for XY planes and 0,5 mm for Z planes;
- **Lattice Structure:** Since HP uses powder selective technology, the not melted powder need some allowance to be removed. For closed components, it is necessary to leave holes of 2mm diameter, to let powder go out during cleaning;
- **Functional Complexity:** When printing assemblies, it is compulsory to allow at least 0,5 mm clearance between components;
- **Part Orientation:** As we already said, the process is strongly anisotropic and cooling strongly affects the shape of components, it is important the orientation of surfaces to reduce cooling defects.

5.4. Additive Manufacturing Process: Printing an Innovative Fan

Additive manufacturing is based on 8 steps that are independent from the technology used. Let's apply the following phases to the fan:

1. **CAD:** All processes, even non-additive ones, start with the definition of a mathematical model on a CAD software. This has the function of describing the object geometrically. We already have the 3D model of the fan, previously designed (Figure 21);
2. **STL Convert:** (Standard Triangulation Language) This file is a simple way to describe an object geometrically. The STL extension discretizes the solid model into a series of triangles as it happens

Figure 21. CAM software adopted during the case study

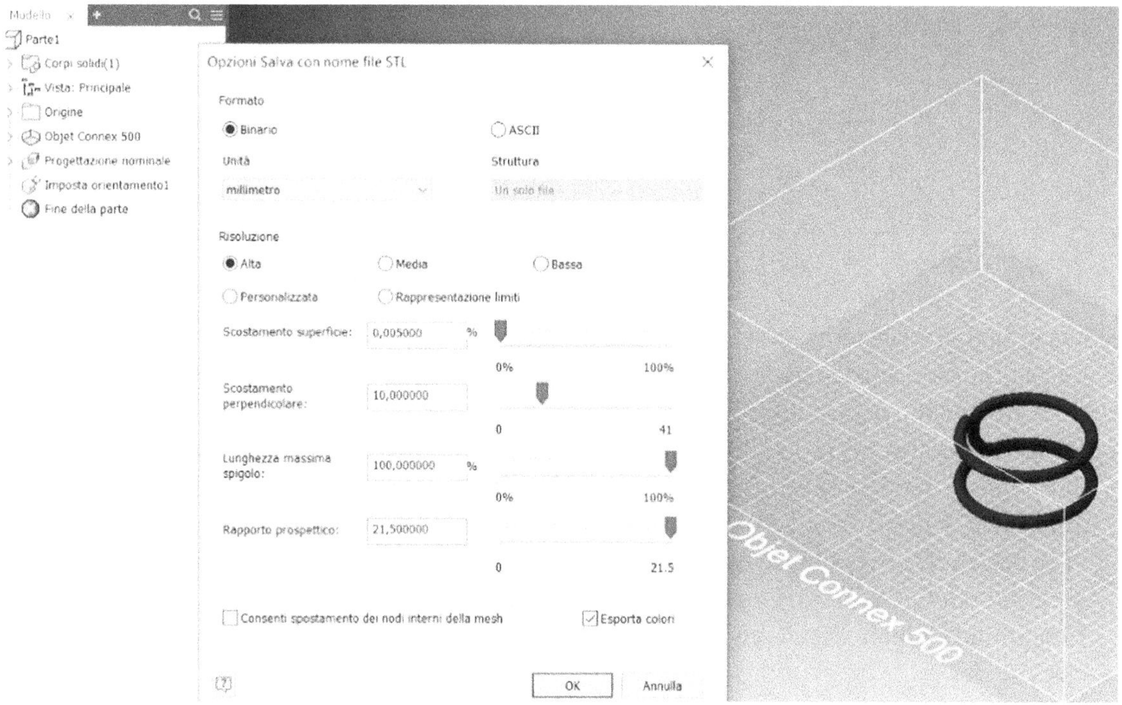

for the meshes (Elements in the FEA analysis. Moreover, sometimes it is required to repair STL file because triangulation is not accurate. Using Autodesk Netfabb, it is possible to manipulate STL files (Figure 22);

3. **File Transfer to Machine:** Once the STL file has been created, it can be directly sent to the AM. It is the machine itself that translates the STL file into G-Code. At this point, it is possible to analyse the fan by CAM software and to optimise the process. The following picture shows a Wall Thickness test that depicts which surface can be damaged during printing. The green surface means that the walls are thick enough. Finally, on the CAM software it is possible to simulate the whole process and understand which are the critical phases (Figure 23);
4. **Machine Set-Up:** Once all tests are passed, it is possible to set up the machine. In this case, the picture shows how many layers were generated and the total height of the print. Moreover, the machine display shows also the layout of components during machining. It is possible to print objects one inside the other without compromising the final result (space optimization). Before analysing the printing process, it is important to understand the difference between lattice structure and full structure of the innovative fan. During the case study, both structures were printed (Figure 24).

Figure 25 shows the difference in weight of the support of the fan, designed previously. Before applying the lattice structure, the total volume of material was 67,08 cm^3. Then, by introducing the lattice structure, the weight changed to 39,06 cm^3, saving almost 30 cm^3 of material used.

5. **Build:** The images on the right show a thin layer of powder placed on the working area and the image below shows the application of transforming agents and energy to merge powder particles

Figure 22. Netfabb STL file repair

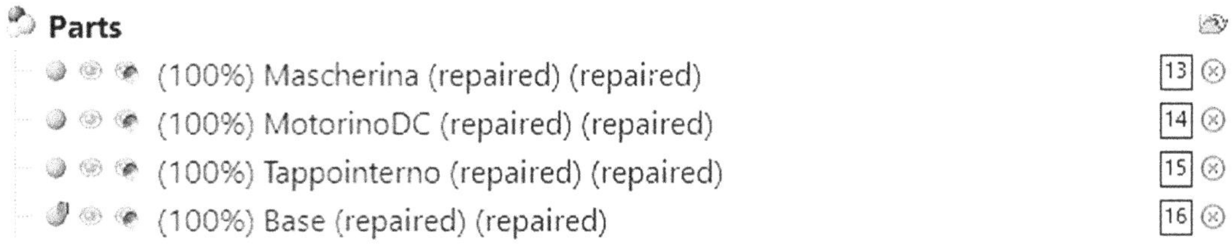

Figure 23. Wall thickness test

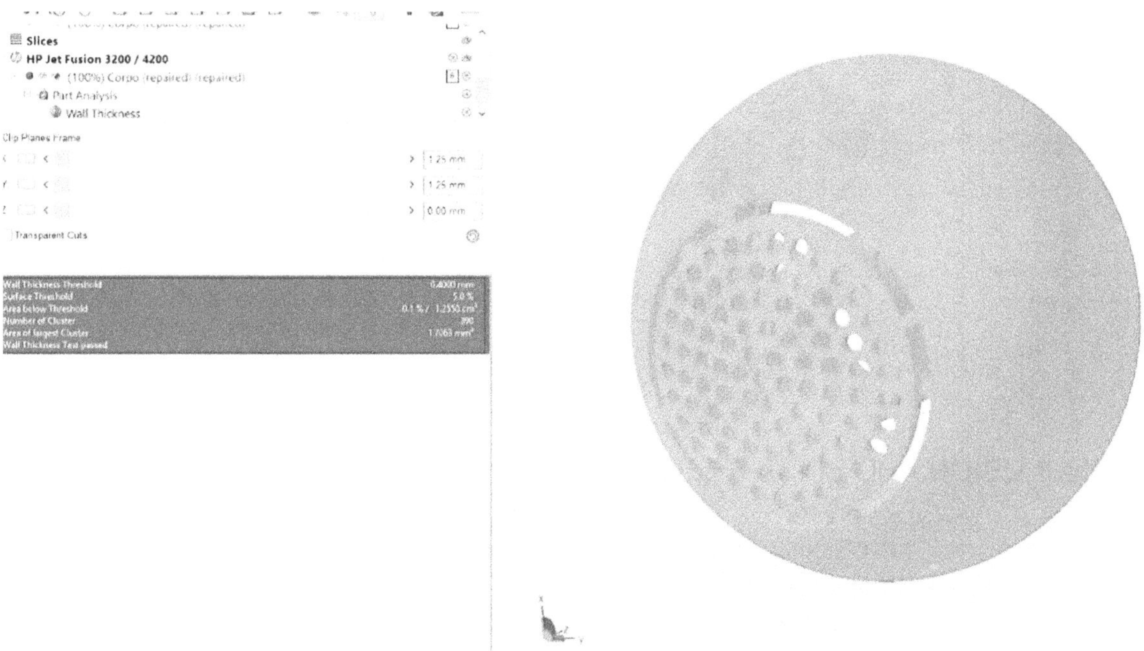

together. It is clear from the image the difference between the lattice structure (white arrow) and full structure (black arrow) (Figure 26);

6. **Remove:** After the printing process, which lasted 6 hours and 21 minutes, the build unit is transferred to the processing station for the cooling and cleaning of the components. The components are initially totally immersed in the powder that needs to be recovered. The recovered material ends up in a cardboard container, ready to be reused in the subsequent processing (Figure 27);
7. **Post-Process:** Finally, there were several post-processing available, but to limit costs, only sandblasting was performed. This machine allows inserting the pieces from the left side and handling the compressed air and sand nozzles for complete cleaning (Figure 28);
8. **Final Product:** Finally, the two variants were produced. Externally they look the same. However, the left one is printed using a full structure, whereas the right one is printed using a lattice structure (Figure 29).

Figure 24. Machine set-up

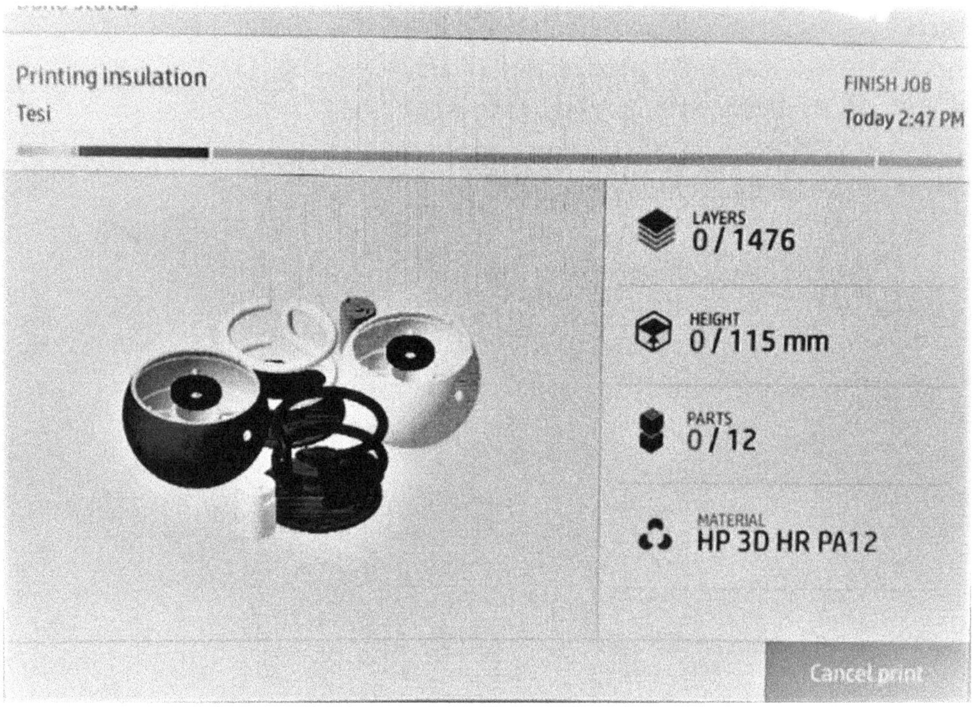

6. CONCLUSION AND FINAL RESULTS (COST ANALYSIS)

In this last paragraph an economic feasibility will be performed in order to consider the costs of the components and the workings. The processing costs are:

- 0.08 €/g for variable costs (material);
- 0.20 €/g for service cost or cost allocated by amortization.

The components have the following costs:

- DC motor from supplier: 8€/pcs, without considering the discounts for larger lots;
- Led: 2€/m from supplier catalogue. On the inner circumference of the sphere = € 0.45;
- Electrical components: a cost of 2€/fan is estimate.

Neglecting labour costs and electricity costs, it can be noted that, simply by adopting the lattice structure, it is possible to reduce production costs by several tens of euros. The most significant component is that linked to the company's service cost [0.20 €/g]. However, companies who have purchased this technology are able to allocate fixed costs to higher production volume, thus lowering costs. Nevertheless, production costs remain very high and the real challenge in the next few years is to reduce them considerably to allow for real mass production.

Figure 25. Material used comparison

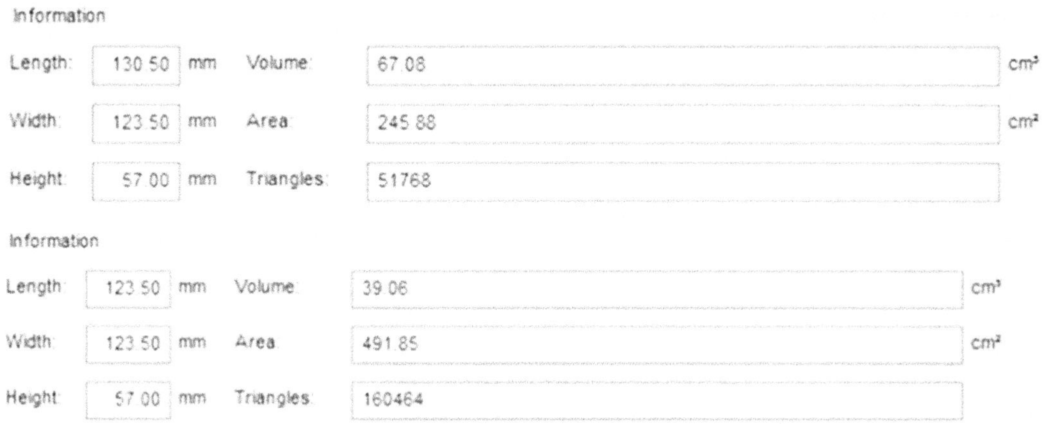

Figure 26. Comparison between lattice and full structure during printing

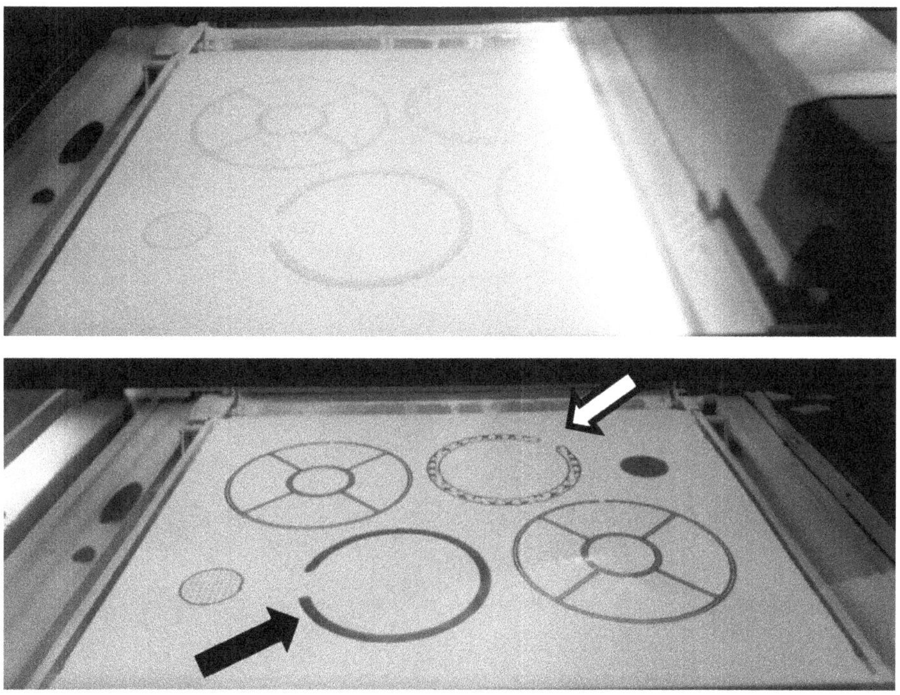

The main advantage of this technology is the ability to reduce costs simply by reducing the amount of printed material. In fact, starting from the basic 3D model, better solutions could be adopted (Topology optimization), such as reducing the thickness of the walls or creating pockets in the housings. This, in addition to the lattice structure, would have allowed to further reduce the amount of material required and, therefore, to reduce costs. However, in this article, no further structures or design variants have been developed. In the Validation phase, the emerging Multi Jet Fusion technology presented several advantages including:

Figure 27. Processing station and material recovery

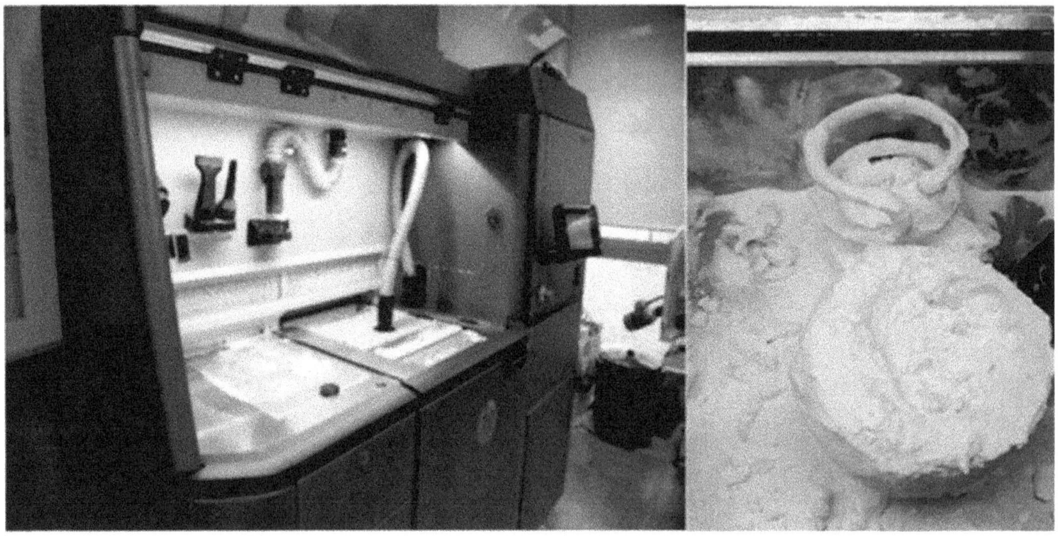

Figure 28. Sandblasting machine used to clean the fan

- Processing speed (6 hours);
- Accessible costs for the machines (300 000 €);
- Better quality due to the detail agents;
- Freedom of Design, in fact the fan body would have been particularly difficult to achieve with conventional techniques;
- Possibility to reduce costs simply by printing less material. The better the design, the lower the cost.

Figure 29. Final result achieved

Table 11. Cost analysis

	Lattice structure	Full structure
Total weight [g/pz]	253	320
DC engine [€/pz]	6	6
LED [€/pz]	0,45	0,45
Electrical components [€/pz]	2	2
Service cost [€/pz]	50,6	64
Powder cost [€/pz]	20,24	25,6
Total Cost [€/pz]	79,29	98,05

However, it is essential to discuss also the main drawbacks we came across in this case study: to begin with, costs are relatively high if we consider service cost. Moreover, some components were deformed by the cooling process: the fan hole (2 mm) does not fit the motor shaft. Therefore, tolerances are not precise. Furthermore, only few materials were allowed in Multi Jet Fusion, such as PA 12 or PA11.

To resume, only two solutions were developed: a full structure as per design and a lattice structure. By adopting the latter, we managed to reduce the costs of a few tens of euros. Unfortunately, in this article, it was not possible to design a better solution that would optimize costs (by topology optimization). For this reason, it would be interesting to push the design for Additive to the limit in a future case study, perhaps creating pockets, reducing the thicknesses to a minimum or generating different topologies thanks to the Generative Design software and algorithms. Moreover, as a continuation of this project, it would also be interesting to produce another series of parts with a complex design or to produce other design variants to minimize costs and to check their geometry and tolerances with the help of automatic coordinate machines (Reverse engineering). It would be appropriate to consider in the costs analysis also:

- Energy consumed;
- Operator costs;
- Dimensional tolerances and machining tolerances.

Compare them with existing estimates can be very useful to understand the real possibilities of these technologies to replace them with traditional processes already existing.

REFERENCES

Baril, C., Yacout, S., & Clément, B. (2011). Design for Six Sigma through collaborative multiobjective optimization. *Computers & Industrial Engineering*, *60*(1), 43–55.

Caligiana, G., Liverani, A., Francia, D., & Frizziero, L. (2017). Mechanical device with five-arms tested on service. *Far East Journal of Electronics and Communications*, *17*(2), 327–349. doi:10.17654/EC017020327

Caligiana, G., Liverani, A., Francia, D., Frizziero, L., Donnici, G., & Integrating, Q. F. D. (2017). Integrating QFD and TRIZ for innovative design. *Journal of Advanced Mechanical Design, Systems and Manufacturing*, *11*(2). doi:10.1299/jamdsm.2017jamdsm0015

Chirone, E., & Tornincasa, S. (2011). Disegno Tecnico industriale 1. Torino.

de Amicis, R., Ceruti, A., Francia, D., Frizziero, L. & Simoes, B. (2018). Augmented Reality for virtual user manual. *International Journal on Interactive Design and Manufacturing*, 689-697.

de Frutos, G. M. (2015). Product development process for additive manufacturing. Considerations on the design of final products for 3D printing [Thesis]. Universidad tecnica de Lisbona.

Donnici, G., Frizziero, L., Francia, D., Liverani, A., & Caligiana, G. (2018, May). Increasing innovation of a new transportation means using TRIZ methodology. *JP Journal of Heat and Mass Transfer*, *15*(2), 341–370. doi:10.17654/HM015020341

Donnici, G., Frizziero, L., Francia, D., Liverani, A., & Caligiana, G. (2018). Project of inventive ideas through a TRIZ study applied to the analysis of an innovative urban transport means. *International Journal of Manufacturing, Materials, and Mechanical Engineering*, *8*(4), 1–24.

Donnici, G., Frizziero, L., Francia, D., Liverani, A., & Caligiana, G. (2018, January 1). TRIZ method for innovation applied to an hoverboard. *Cogent Engineering*, *5*(1), 1–24. doi:10.1080/23311916.2018.1524537

Filippini, R., Ulrich Karl, T., & Eppinger Steven, D. (2007). *Progettazione e sviluppo prodotto*. McGraw-Hill Education.

Francia, D., Caligiana, G., Liverani, A., Frizziero, L., Donnici, G., & Printer, C. A. D. (2018, February 1). A QFD and TRIZ integrated design solution for large size open moulding manufacturing. *International Journal on Interactive Design and Manufacturing*, *12*(1), 81–94. doi:10.100712008-017-0375-2

Freddi, A. (2005). *Imparare a progettare, Principi e metodi del progetto concettuale per lo sviluppo della creatività industriale*. Pitagora.

Frizziero, L. (2018). Conceptual design of an innovative electric transportation means with QFD, Benchmarking, TOP-FLOP Analysis. Far East Journal of Electronics and Communications, 18(1), 189-205.

Frizziero, L., Francia, D., Donnici, G., Liverani, A., & Caligiana, G. (2018). Sustainable design of open molds with QFD and TRIZ combination. *Journal of Industrial and Production Engineering*, *35*(1), 21–31.

Ghezzo, F., Giannini, G., Cesari, F., & Caligiana, G. (2008). Numerical and experimental analysis of the interaction between two notches in carbon fibre laminates. *Composites Science And Technology*, *68*, 1057-1072, doi:10.1016/j.compscitech.2007.07.023

Gibson, I., & Rosen, D., W. & Brent, S. (2010). *Additive Manufacturing Technologies: Rapid prototyping to Direct Digital Manufacturing*. doi:10.1007/978-1-4419-1120-9

Grandi, A. "Gestione dei progetti d'innovazione", Bologna, McGraw-Hill Education, 2017 Hp website, 3D printing, http://www8.hp.com/it/it/printers/3d-printers.html

HP. (n.d.). Printing Manual HP4200. Retrieved from http://www8.hp.com/h20195/v2/GetPDF.aspx/4AA6-4894ITE.pdf

Ogawa, S., & Piller, F. (2006). Reducing the risks of new product development. *MIT Sloan Management Review*.

Schilling, M. A. (2010). *Strategic management of technological innovation*. Tata McGraw-Hill Education.

Tao, W., & Leu, M. C. (2016, August). Design of lattice structure for additive manufacturing. In *International Symposium on Flexible Automation (ISFA)* (pp. 325-332). IEEE.

Vaneker, T. H. J. (2017). The role of Design for Additive Manufacturing in the successful economical introduction of AM. *Procedia CIRP*, *60*, 181–186.

This research was previously published in the International Journal of Manufacturing, Materials, and Mechanical Engineering (IJMMME), 9(2); edited by J. Paulo Davim, pages 1-32, copyright year 2019 by IGI Publishing (an imprint of IGI Global).

Chapter 4
Clash of Cultures:
Fashion, Engineering, and 3D Printing

Jennifer Loy
Griffith University, Australia

Samuel Canning
Griffith University, Australia

ABSTRACT

In 2012, a Belgian company called Materialise hosted a fashion show featuring designs from a worldwide millinery competition. The featured pieces were paraded down a catwalk by professional models, and an overall winner chosen. What made this fashion show unusual was that the attendees were predominantly clinical and industrial engineers, and the host was a specialist engineering and software development company that emerged in 1990 from a research facility based at Leuven University. Engineers and product designers rather than fashion designers created the millinery and the works were all realized through additive manufacturing technology. This chapter provides an example of how fashion design has become a creative stimulus for the development of the technology. It illustrates how disruptive creativity has the potential to advance scientific research, with the two worlds of engineering and fashion coming together through a collaboration with industrial design. The chapter highlights the challenges and possible implications for preparing trans-disciplinary research teams.

INTRODUCTION

Additive manufacturing, now more commonly known as 3D Printing (3DP), began as a prototyping technology in the mid-1980s. After gaining traction in product design, architecture and engineering as a resin based visual modeller, this technology has matured with the development of new materials and equipment. These have opened up new areas for research and possibilities for production applications, as detailed by leading researchers in the field, Gibson et al (2014), Lipson (2013) and Anderson (2012). Although the technology is now starting to impact across almost all fields, from animation to construction, over the twenty years of its initial development the majority of research in the area of direct manufacturing for 3DP has been concentrated into two main themes. The first is medical, with specialists in areas such

DOI: 10.4018/978-1-5225-9624-0.ch004

a regenerative medicine, orthopaedic surgery and artificial heart development recognising a myriad of potential applications for the technology, from the bioplotting of cells to titanium implants. The second is industrial, with applications for 3DP explored in commercial applications such as for automotive and aerospace, with particular advances in light-weighting and material performance. Yet scientific research into additive manufacturing has been slow to capture attention both in terms of funding and in the eyes of the public as well as potential industry partners. In the last three years, fashion design has proved to be a vehicle for all three. 3D Printed fashion has caused disruptive change in the development of research in this field. The inspiration provided by fashion designers has provided a basis for innovation that has been realized by industrial designers and engineers in recent practice. This chapter discusses the value of trans-disciplinary research practice based on teams that include both scientific researchers in fields such as engineering and medicine, as well as creative researchers from design and art. It also outlines some of the challenges facing these teams that arise from a clash of cultures in values, approach, practice, aims and outcomes. It then goes on to more specifically provide an example of collaborative practice where fashion and engineering, and then fashion, industrial design and engineering, come together. The example highlights the role of the industrial designer in the collaboration, in providing a practical link for creative practitioners interested in working with 3D Printing technology, but without the understanding of design for 3DP or technical skills to achieve it on their own, to realize their ideas. However, this example highlights how the complexities of design for process in relation to 3DP mean the industrial designer has to evolve the design as part of the process, rather than providing a form of 3D documentation after the design has been completed, which challenges dominant practice in a field such as fashion design. The objective of this chapter is to illustrate that for collaborations around this cross-disciplinary technology to work, and for new trans-disciplinary research practice between creative and scientific researchers that transcend conventional practices to emerge and be effective in new ways, the different roles within the team have to be interrogated and ground rules established to avoid a clash of cultures on many different levels.

BACKGROUND

According to Milton and Rodgers (2013), "Unlike scientific research, design research is not concerned with what exists but what ought to be. Research in a design context breaks with the determinisms of the past; it continually challenges, provokes and disrupts the status quo" (p.11).

Research into additive manufacturing, now more commonly known as 3D Printing, has been embedded in scientific research communities since its inception. Dominated by materials science, mechanical engineering, robotics and medical research, it has developed incrementally, resulting in additional forms of the technology, such as material jetting and selective laser melting, and an increasing range of material properties, such as the bimetals developed by researchers at NTU. Progress in the field has been disseminated in research journals that are predominantly materials and technology focussed and that research has tended to be technical, as in the example by Qain (2015) on metal powders for additive manufacturing in the Journal of Minerals, Metals and Materials.

Although multidisciplinary, the roots of research in this area cannot be characterized as cross disciplinary. Yet studies on the organisation of research practice are suggesting that innovation in research is stimulated by multidisciplinary teams and by trans-disciplinary thinking, where trans-disciplinary research is characterized as working in new areas between disciplines. The argument is that a trans-disciplinary approach provides opportunities for synergy, where the combined abilities of the researchers involved bring new dimensions to the work (Cacchione, 2015). The need for this synergy in scientific research at this time has been highlighted by Wageman et al (2012), who suggest that archetypal teams need to be altered in the face of changes in the current discourse on the nature and form of collaboration. Just as more business leaders are recognising that innovation is stifled by over planning (Kaplan, 2012), the idea of a stable research team membership, with a collective consciousness built on a shared "transactive" memory with routines and processes traditionally seen as beneficial to efficiently creating quality output, is being challenged (Lewis, 2003, p.587).

In relation to the development of research in additive manufacturing, the disciplinary knowledge has been built through narrowly focussed research projects to address specific problems. However, this assumes that the problem to be solved is based on the right questions. With increasingly complex situations to address, problem framing is becoming a more essential skill in research, as well as the ability to view the specifics of a problem within a much broader context. This requires different views on the same subject, and a willingness and ability of the researchers involved to step outside the boundaries of their expertise to address the interdisciplinarity of complex problems. Moving ideas on, beyond an incremental change approach to encompass the possibilities of step change, requires structures that support creative and divergent thinking.

Research teams need to be built to support innovation. If not, the outcomes will remain incremental and opportunities for new thinking and new practice will be missed, along with new directions for further research. However, even in business, where innovation is overtly valued, predictability and control are part of conventional operations management. Researchers from the University of Virginia found the management of a team that is intent on stimulating innovative practice, required agile thinking that worked positively with ambiguity. They concluded that uncertainty created opportunities and that it was essential to push the team outside their comfort zones, and to help them learn to embrace risk and failure as a way of providing the context for disruptive innovation (Kaplan, 2012).

In the case of additive manufacturing, researchers based at Materialise, a commercial offshoot of Leuven University, found that the inherent mindset driving research and development in additive manufacturing was focussed on incremental, discipline specific technical research. This was highlighted as a concern, by Materialise CEO Leys (2015) during an Inside 3D Printing industry conference in New York in 2015, in relation to shifting current research focus from prototyping to direct manufacturing as a primary use for the technology. Leys called for the incremental development dominating research to be disrupted by new thinking to radically change practice and stimulate ideas that were not hampered by the body of knowledge and conventions of the industry up to that point. As Kaplan (2012) observes, more and more leaders now recognize that they must proactively disrupt conventions and practice in order to create innovation, and that detailed planning documents not only stifle disruptive innovation, but actively represent liabilities to creating innovative outcomes. This requires being open to adopting unconventional practices that currently fall outside the scope of traditional training for team leaders in business and research. If design research, as defined by Milton and Rodgers (2013), is disruptive to conventional practice, approach and ideas, then the leaders of research teams that include designers must be open to that possibility. The challenge for scientific researchers is to be able to work collaboratively with designers in a culture of risk and failure more dynamic than their research approach is conventionally likely to be. The challenge, therefore, is to provide the stimulus for creative thinking and innovative practice, within a field that is grounded in scientific theory research.

CREATIVE TECHNOLOGIES AND BARRIERS TO TRANSDISCIPLINARY RESEARCH

Design research also differs from scientific research, for the most part, in that it is concerned with the plausibility and appropriateness of proposals, whilst scientific research is concerned with universal truths. Design research tends to produce knowledge that can be defined as trans-disciplinary and heterogeneous in nature and that which seeks to improve the world (Milton & Rodgers. 2012 p.).

Within engineering and design education, there exists an uneasy truce between the critical and the creative. The Australian Curriculum advocates a combined approach, and Engineers Australia calls for more creative thinking in engineering education, and whilst there are exceptional Universities such as Olin, Stanford and MIT that openly embrace creative thinking in their curriculum, the classical model of engineering education begins with a solid grounding of theory, and is driven by the scientific method of enquiry. This allows little scope for creativity. As students progress through their engineering degree, they are increasingly engaged in practical learning, but this is built on predetermined theoretical foundations and so the students approach their applied work constrained – and supported - by their learning development.

In the same way, product design students are traditionally taught manufacturing technologies successively from conventional technologies that are well established in the industry, through to the most recent developments. This means that when they do engage with the most advanced manufacturing technologies, they approach them with a heritage of working with the constraints and opportunities of conventional manufacturing technologies. Working in teams within a discipline can be difficult enough, broadening a research team to include researchers from disciplines that have fundamentally different understandings of what research is, how it is conducted, what the outcomes will look like and how they will be evaluated, increases that challenge. Graeme Pullin (2011), whose research is in the field of design for disability, argues that this discipline has too long been dominated by the sciences, and this has resulted in very basic, unimaginative, solutions. He argues for the disruption of conventional scientific practice in the field by the involvement of designers at a very early stage and in conjunction with all technologies being developed. His reasoning is that scientific research has been restricted to the development of mechanical outcomes, and that this approach fails to meet the needs of the individuals for which the assistive technologies are designed for, and that a human centred approach would not only address non-compliance issues, where the user refuses to use the product, but also innovations in design for ability that would take disability research in more positive directions. He is critical of legislation, built on a legacy from the UK National Health Service in the 1930s that was not reformed until the 1970s (and even then only to a limited extent) that restricted the styling 'allowed' for assistive technologies such as eyewear. Eyewear designers Cutler and Gross have spent 30 years on the front line of the revolution in eyewear where their aim was to change it from a 'medical necessity to key fashion accessory' (Pullin, 2011 p.21). Pullin observes that although fashion and discretion are not opposites - fashion can be understated and discretion does not require invisibility - there is a tension between the two and for some in the medical field, fashion is the antithesis of good design. Nowadays, spectacles are seen as worn, rather than carried, which causes a shift in thinking from a medical model to a social model. Positioning them as an item of clothing, rather than a product, design writer Per Mollerup said of spectacles now "what others see is more important than what you see yourself" (Pullin, 2011 p.19) Customers are consumers rather than patients, which is a significant design distinction, and Cutler and Gross argue that to keep that way of thinking in societies' mind, they need to constantly push the boundaries of what is expected from eyewear, challenging styling conventions and questioning understandings of taste and acceptable practice.

Ross Lovegrove designed a hearing aide that was featured in Blueprint magazine that aimed to be seen as jewellery, rather than a medical product. It was produced from a carbon composite and drew on positive imagery from technology, such as Bluetooth. Rather than aiming for invisibility, this hearing aide was designed to be seen, and by not being restricted to the need for discretion, the technology fitted into the aid could provide enhanced performance and ease of maintenance. Designer Sam Hecht took this one step further and designed for ability, by creating hearing aides with an array of microphones that support multidirectional hearing, surpassing normal human hearing.

Pullin (2011) points to the current omission of designers in scientific research teams now for the development of assistive technologies, and suggests that designers can rethink the questions being asked about the fundamental purposes of products and should therefore be included. He argues that even products for medical purposes should be designed for the emotional as well as the practical needs of the user. This is supported through studies on the use of sleep apnoea masks, where the emotional response of the user to how they feel about wearing the mask and being seen wearing the mask is leading to a high degree of non-compliance. In addition, designers are concerned through their training with human machine interaction to a larger degree than engineers or clinicians, and are therefore more able to work on the performance of the mask during actual use, taking into account multiple factors, rather than its performance on a statistical and theoretical level. According to Pullin, in a recent major US research project, the team invitation was sent out to engineers, technologists, and clinicians, but not industrial designers or interaction designers, fashion designers or sculptors. The attributes of prosthetics affect the image its wearer is projecting, but more than that, a design approach to problem solving for a functional need takes into account situational design thinking, and includes experience mapping and systems thinking as advocated by Kumar (2012) and draws on the body of knowledge on the psychology of interaction design detailed in product design research into behaviour (Weinschenck, 2011) that will impact the fundamentals of what the design intent really should be, and what will impact the design development. A purely mechanical response to problem solving limits innovation, and because it draws on a narrow discipline with a shared 'transactive' memory built up over time, the likely outcome will be an incremental change rather than an innovative one. Including designers in research teams that are traditionally restricted to scientific enquiry is a way for a research team leader to embrace disruption that can lead to innovation.

Environmental researcher, Janine Benyus (2002), is a leading advocate of research teams that bring together scientists and designers within the field of biomimicry. The study of nature systems as biomimicry is an approach increasingly supported by theorists and practitioners, such as the author of Shark's Paintbrush (Harman 2014). However, Benyus works very specifically on creating diverse research teams that bring together scientific enquiry, design understanding and creative thinking. She argues that bringing designers and scientists together with a specific problem in mind, to explore nature together, stimulates innovation based on nature's systems that would not be identified by single discipline researchers alone with the same stimulus. One of the reasons given for including designers in research teams is their divergent thinking ability. Designers are trained to mind map and brainstorm, and to create a safe environment for risk taking in the expression of ideas. Risk taking in design is not excluded by the need for a functional, heavily structural outcome. Architectural risk taking is a strong example of studio practice informed by the structural, but not reduced by it. The work of innovators such as Zaha Hadid and Frank Gehry, who argue for unconventional thinking (Cohen et al, 2003) demonstrate that a conceptual rethink of conventions within a practice inspired by new material and technological innovations is not at odds with the need to provide engineering solutions and in the last two years there have been breakthrough designs where the outcomes illustrate the transformative potential of 3D Printing technology, such as demonstrated in the work of the engineering team at Airbus, but the predominant body of work in the field has been in substitute products and tooling, rather than disruptive innovations.

This low-key progression of applications in the engineering and industrial design field may be because the developmental pathway for the technology has come out of rapid prototyping, rather than as a new technology. Interest has traditionally been focussed on its use in supporting conventional manufacturing techniques and reducing time to market through providing an ability for designers and engineers to more quickly iterate designs during a traditional industry process approach rather than in new design opportunities. Exemplars for how the technology could impact not only design development and outcome for a company, but the business model of organisations, are still rare and the majority of outcomes in the field are still incrementally developed progressions.

In contrast, art and fashion are disciplines founded on expression, innovation and creativity. The often-transient nature of the output, its response to context, and the accessibility of its materials and technologies have produced a discipline that expresses risk in its exploration of form. Without the weight of longevity and more functional responsibility that can constrain industrial thinking, these disciplines have the potential to take more risks with 3D Printing and provide alternative directions for development that can then inform their industrial counterparts.

Digital media have provided revolutionary forms of creative expression through a myriad of methods, from the screen-based, with its abstractions and hyper-real macro and micro-scenarios, time-based, self generating, fast evolving imagery and media that have evoked consideration of new kinds of space and infinite zones, the potential future for which, in creative terms, is shifting, flexible, complex and unbounded (Braddock Clarke, Harris, 2012, p.229).

The work of Fung Kwok Pan provides an example of creative technologies research and disruptive practice. Pang is digital artist and designer who works with coding to create interactive online media to provoke a form of co-production between the designer and the consumer. An example of his work with additive manufacturing / 3D Printing, is his Supabold FluidVase series. This work was based on the idea of water splashing into a container, and capturing the movement of the water as a freeze frame that was then 3D Printed. The intent driving the work was very clear, and this provided the brief for the technical development work needed to meet that intent. Because the driver came from an expressive intent, the technical research to meet that intent was not based on previous work, but took the development of practice in new directions, driving innovation in the field. Again, because the driver was expressive, the supporting work was not constrained to a single discipline but worked across art, IT, design and engineering to meet the needs of the project, drawing on ideas and practice from across those disciplines.

Another example of additive manufacturing creative research stimulus comes from the work of Assa Ashuach. As well as his work on UCODO and Digital Forming, which had a user interaction element, as in Pan's work, Ashuach has also worked directly with additive manufacturing / 3D Printing to explore the potential of the technology to support innovation in design applications. This work has resulted in innovations that exploit the characteristics of the technology and draws technical development research along with it, such as was necessary for his AI light and AI stool. By starting with a creative design intent, rather than a technical development research intent, the work is arguably more likely to cross disciplinary boundaries and inspire innovations that lead to developments within technical research areas.

Fashion and Engineering

The young, Dutch, fashion designer, Iris van Herpen is the unlikely champion of disruptive innovation in the development of applications and practice in advanced manufacturing technologies. Unlikely, because her work until recently was known for its avant-garde nature, not its technical construction or materials science. Champion, because van Herpen's interest in additive manufacturing – 3D Printing – to create textiles and dress forms has raised the profile of this technology and, more than that, forced a re-evaluation of the technology that was contained and controlled by the dominant players that had established its roots in engineering science (Figure 1).

The first clash of cultures to occur in relation to a shared working practice of fashion designers and engineers using 3DP as a creative technology, has been in relation to the perception and appreciation of the relative roles within the design. Iris van Herpen's work broke the technology out of the constraints

Figure 1. Escapism dress by Iris van Herpen and close up of detail of Escapism dress
Photo: Loy, 2012.

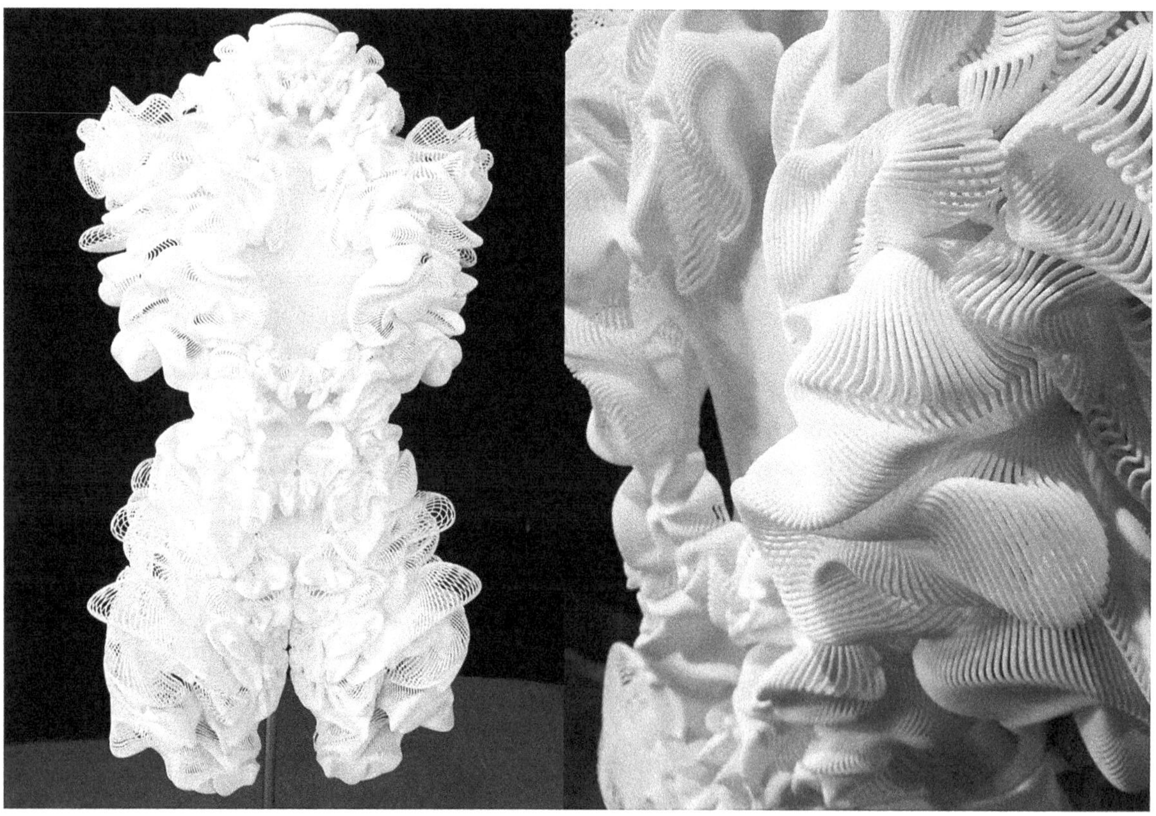

of conventional manufacturing and rapid prototyping it was constrained by, in using it as an expressive medium. She re-imagined what could be possible, and challenged industrial designers and engineers to keep up with her vision, which faced considerable technical barriers. However, in order to realize her ideas, van Herpen needed to work closely with 3D designer and architect, Daniel Widrig. In 2011, Daniel Widrig, who now heads up a research cluster at UCL and is a multi award winning designer and architect working across sculpture, furniture design and architecture (including experience with Zaha Hadid), collaborated with Iris van Herpen and engineers in Materialise on the creative and technical development of an innovative dress that pushed the boundaries of research in 3D Printing organic forms and flexible structures. The Escapism dress was named one of the fifty best inventions of the year by Time magazine. The dress provided a radical departure from convention in additive manufacturing research and proved a catalyst for designers to see its possibilities as a creative technology.

Yet, from a collaborative point of view, the accreditation of the dress to the designer has been focussed on the creative inspiration, with the 3D designers role initially presented at the level of a technician. This ignores the reality of working with 3D printing. In conventional fashion design, to be able to translate an initial sketch into the construction of form is integral to the design development, such as demonstrated in the work of Nakamichi (2011). Whilst a technical specialist might take the pattern cutting of Nakamichi and build the final form, the construction detailing is a vital part of the expertise of the fashion designer. It is not enough to be able to provide an initial sketch, the designer's role includes how a creative inspiration could be realized in the same way that an industrial designer would be expected to design for conventional manufacturing, such as injection moulding. Yet 3D printing has emerged out of the engineering disciplines, and fashion designers are not yet trained to build designs within solid modelling CAD software, or understand the constraints and opportunities of working with 3DP. The difficulty is that, just as CAD itself has shifted from a documentation tool, to be part of the design process, the design role within the development of digital fashion must extend to include the design for process portion of the activity. One of the challenges to collaboration in relation to bringing together disparate disciplines, such as fashion and engineering, is that the conventions of each have not developed in conjunction, and are not necessarily recognized or understood by both parties.

In 2012, Materialise hosted a competition for the design of 3D printed headwear inspired by the innovations demonstrated by Iris van Herpen and Daniel Widrig. Twenty designs from around the world were chosen. These designs demonstrated the differences between forms possible prior to the development of the technology and those now achievable. The designs were not, however, created by fashion designers, but by engineers and industrial designers. As a result, the aesthetic, and even the functional conventions, of fashion design were not addressed – for example most of the headwear had to be hand held as attachments or detailing for additional fixing had not been included. It is likely that fashion designers did not have the solid modelling computer skills, or the understanding of design for 3DP needed for the competition. Equally it is possible that the competition did not reach the social media channels frequented by fashion designers, as the competition was advertised through the Materialise web site, which may not have attracted fashion designers at that time. Alisa's Curls was one of the twenty chosen for exhibition in Belgium, and then in the US. It is an example of organic modelling using solid modelling software. The forms were challenging to construct using Solidworks (an industrial modeller rather than a creative visualisation tool), and the construction itself demonstrated the shift from component construction to a complex, single piece outcome not possible to construct using a mould or subtractive technology (Figure 2).

Figure 2. Progression of Alisa's Curls
By Alisa Newey, Jennifer Loy and Samuel Canning, 2012.

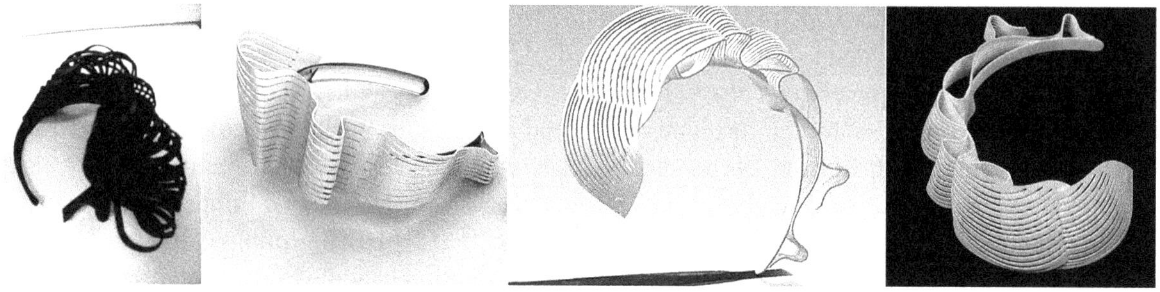

However, more than that, the design and technical advances made through this piece in terms of the mind shift in thinking from a conventional technology approach, to one that allowed for an organic, multilayered object to be created, that genuinely exploited the opportunities created by 3DP, was stimulated by the inclusion of a designer with a background in textiles into the team of industrial designers taking on the project. This was an experiment to mimic the working practice of van Herpen and Widrig. Newey was asked to create a headpiece using her normal practice. She chose to develop a prototype using cut leather. The industrial designers then interpreted that idea using Solidworks, and then evolved it within the program itself to exploit the characteristics of the software and design for 3DP. The final piece was printed as one part.

This example of the new working practice relied on the initial forms being created unconstrained by the technology, but equally it relied on the industrial designers knowledge of designing for 3DP, both of the appropriate solid modeling software and of the technology, to evolve the design, rather than be constrained by attempting to mimic the textile form. The textile provided the inspiration for a creative opportunity that created a step change in the way the industrial designers approached and stretched the capabilities of the software and printing technology, but the creative design involved in the development of the piece did not start and finish with the original inspiration, but rather became a design collaboration. The CAD work was not a documentation of the textile, but rather, used as a design tool in developing the design further. This new relationship between the creative and the technical provides the opportunity for significant, rather than incremental change, driven by meeting the needs of creative practice as a starting point. However, it relies on the recognition of the design role as being collaborative. If the textile had been provided as a blueprint, with no understanding of the technology or designing for that technology, or of working with solid modeling software, it would likely have resulted in a literal translation of the textile that failed to address significant design issues.

The 2012 competition and work with van Herpen and Widrig, led the engineers at Materialise to begin a collaboration with Malaysian fashion designer, Melinda Looi. The brief was to design a collection that demonstrated the artistic potential of the technology. The resulting collection was inspired by birds, and based on drawings that Looi provided to the engineers. In terms of impact, the collection was very successful. It was launched in a packed runway show in Malaysia that was attended by Malaysian Royalty, and in addition to the medical, engineering and industrial design audience that Materialise attracts to its conferences, there was also considerable interest from the media, including fashion media. The show then traveled, including to India, China, Australia and Europe. In terms of challenging what was possible to create with the technology and raise the profile, not only for engineers and industrial designers, but also for creatives. it was very successful.

However, the pieces themselves had problems that impacted both their display and their longevity. More significantly, the designs that worked on paper, has not translated into fashion pieces, but rather into pieces of sculpture that whilst having visual impact in photographs, could not be claimed to be wearable fashion. This had been an issue with some of the previous pieces of van Herpen's work too, which Materialise had been unable to easily take down the runway. Essentially, the engineers had worked to translate the drawings into 3D Printed forms, but that translation had been very literal, and this caused issues. As an example, there was a skirt form that was drawn to have movement and lightness, to shimmer as the model walked. It was modeled by an Engineer literally as it was drawn without the design being evolved to retain the intent, or redesigned to make the most of the opportunities provided by the technology, whilst avoiding its limitations. Nor was it redesigned to exploit the characteristics of sold modeling software. The result was solid and heavy printed when it was printed out. This meant it was very difficult for the model to 'wear' on the runway, it did not move as the fashion designer intended, and it was perceived as an artwork, rather than a genuine challenge to the fashion industry about what could be possible with the technology (Figure 3).

Another example was a pair of articulated wings. These were visually effective as they photographed well, but they had been constructed as they were drawn, whereas they needed to be designed to work with the movement of the model with a good understanding of how to design for the technology. The feathers were attached by narrow filaments, which were vulnerable and easily broken off. The material characteristics needed to be taken into account as well as the joints redesigned to create an improved connection. The reality is that this technology means that a genuinely 3-dimensional attachment could have been made, using light weighting techniques for which this technology is so appropriate. This could have created new detailing on the wing forms which would have had the potential not only to function better, but also demonstrate the possibilities of the technology. By not designing for the technology, it looks as if it is less successful than using conventional means of creating those joints, although this is not the case. The articulations of the skeleton also proved problematic and again, the ability of the tech-

Figure 3. Details of 'Birds' by fashion designer Melinda Looi
Photos: Loy, 2013.

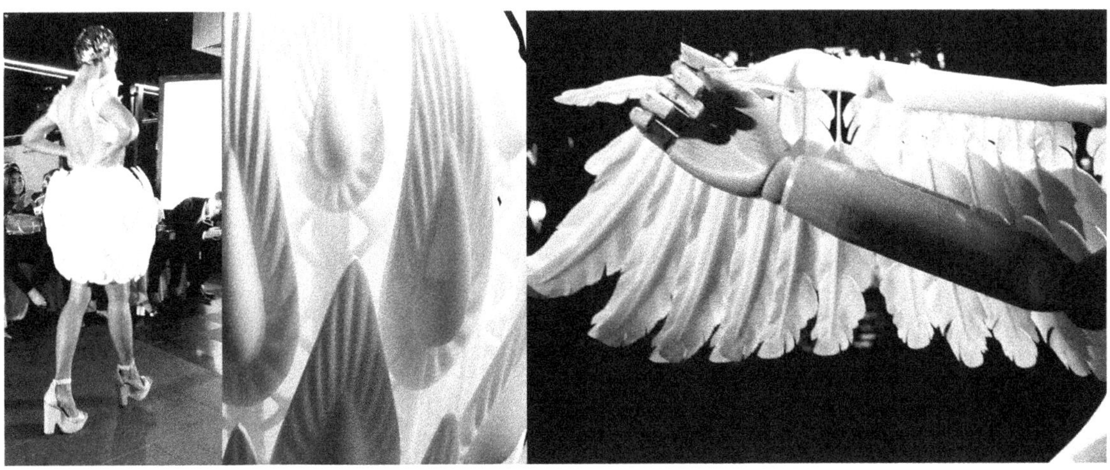

nology to create innovations in skeletal forms was not demonstrated here. Designing wings using 3DP should result in an entirely new set of forms, made possible by new ways of making. The shoes were another example. The sketch on paper was recreated exactly, with no redesign for the opportunities of the technology, or consideration of materials. 3DP is best known in engineering circles for optimized topology. The majority of the engineering examples of using the technology successfully are in this area. The shoes provided a perfect opportunity for using optimized topology, yet instead the drawings were literally translated and the resulting products unwearable. Models had to be taught how to walk down the runway wearing them so that they would not fall, and straps had to be added that undermined the designer's vision. As this is the Engineer's area of expertise is would have been expected that this would have been a design element in which the engineers and fashion designer would have collaborated on together. However, it is likely too that there was a clash of cultures. Fashion design is a discipline where designers are very protective of their personal vision and working practice is very different between a fashion house and an engineering office. Creating a dialogue on the design elements in a shared language with the best will in the world would have been a challenge. Overall, the resulting pieces of the initial collaboration were dramatic, and captured the spirit of the design intent, but the problems with the final pieces highlighted the need for an additional team member who understood specifically developing the design for process, and who could work with the human interaction element of the applications, and the visual elements of the outcomes, whilst also working with the technical aspects of the project. This led to the collaboration described in the practice-led enquiry example.

Practice-Led Enquiry

This example of practice-led enquiry describes a project designed to stimulate new mindsets and creativity in additive manufacturing research and the challenges involved in meeting that intent. The aim was to contribute to disrupting the current dominance of incremental, technical development in engineering-focussed research in additive manufacturing and raise the profile of 3D printing as a creative technology. Materialise VP, Wim Michiels, provided the stimulus. He brought together a team of creative and technical collaborators to work on a full-length 3D Printed evening dress. The research team consisted of engineers from Materialise, specialists in additive manufacturing, Griffith University lecturer and industrial designer, Sam Canning, specialising in design for 3D printing, and fashion designer Melinda Looi, who provided creative direction. The focus for this evaluation of practice is the challenges to this approach and in particular the role of the industrial designer as the bridge between fashion and engineering. Canning was invited to join the collaboration following the work he exhibited in Malaysia in 2013 based on his research into maximising design applications based on 3D printing. Canning's 'Peacemaker' included over 250 individual 'feathers' of graduating size, which are individually hinged. As part of meeting the requirements of designing for 3D printing, Canning created the hat in two parts, designed to be printed together within a limited build space of 200mm x 200mm x 200mm. The dashed line in the figure shows the line of the bounding box around the feathers with a 0.5mm clearance. This maximizes the build volume and accuracy of the print process. This design demonstrates a sound knowledge of the intricacies and opportunities of designing for process using SLS technology. As an illustration of the differences in approach between fashion and engineering, the fashion category that the Peacemaker was entered in was 'captive parts', demonstrating a clear engineering, rather than a fashion bias. 'Peacemaker' was named after the Native American Indian Legend of the Great Peacemaker, otherwise known as Deganawida (Heavenly Messenger) and his historical associations with peace and the white dove (Figure 4).

Figure 4. Details of Peacemaker
By Samuel Canning, 2013.

The addition of Canning to the research team brought the design for process knowledge to the team, but added an additional element in relation to collaborative practice, as the discipline practice, ideas and knowledge for industrial design are different again from fashion design and engineering. The significance of the role of the industrial designer is documented as providing the missing link between abstract creativity and engineering construction. The main barriers to collaborative practice between scientific and creative practitioners identified through this project are highlighted and recommendations for future practice are discussed.

Example of Practice

In this project example, the fashion designer provided a theme for the dress. This was based on coral, with the idea that this provided the opportunity to show detail and complexity in the dress, demonstrating the technical potential of the technology to create complex assemblies. The sketches shown were developed by the fashion designer independently and sent to the industrial designer to translate the intent into a product (Figure 5).

There was no reference to the technology in the sketches themselves, or are any specific details on the translation of the intent into a product. The sketches were intended to be expressive, rather than technical and guide the overall impression the design should make on the viewer.

The dress drawings and the photographs of coral were then given to the industrial designer. The designer undertook initial research in three parts by:

Figure 5. Drawings and inspiration by fashion designer Melinda Looi

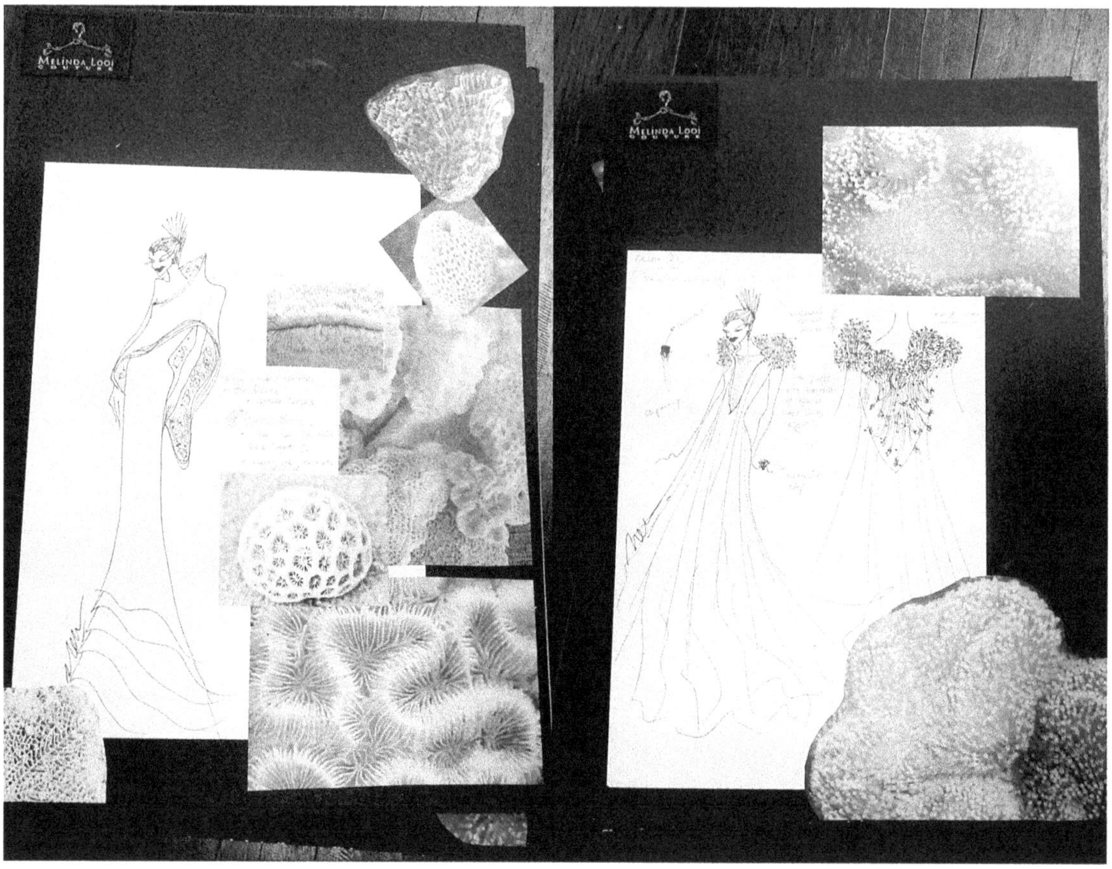

- Interrogating the marine life influence with the technology in mind.
- Exploring the capabilities of the technology to support forms that could relate to the theme,
- Researching material properties and characteristics in relation to the project theme,
- Bringing an understanding of how to design for 3D printing to the stimulus material, the designer redirected the initial stimulus material to retain the intent, but work with the technology.

Research involved testing of materials and process. Test pieces were designed to look at the properties and potential of the material, and structures for the project whilst exploring how to translate the creative intent into a design intent (Figure 6).

For example, a wave structure, consisting of narrow fibres of 1 mm diameter, is intended to mimic the way sea grass moves underwater, whilst testing if the material will be self-supporting. The material tested was Materialise TPU, which is a functional elastomeric material that is strong and durable, and is processed using selective laser sintering. Further test pieces on self-supporting structures designed to test the flexibility of the material, and the level of fluid movement of multiple entities included tapered filaments where the resolution at the base was within standard guidelines, but at their tip was finer than standard printing allowed. Additional models were intended to test the flexibility of the TPU material

Figure 6. Lattice detailing
Canning, 2015.

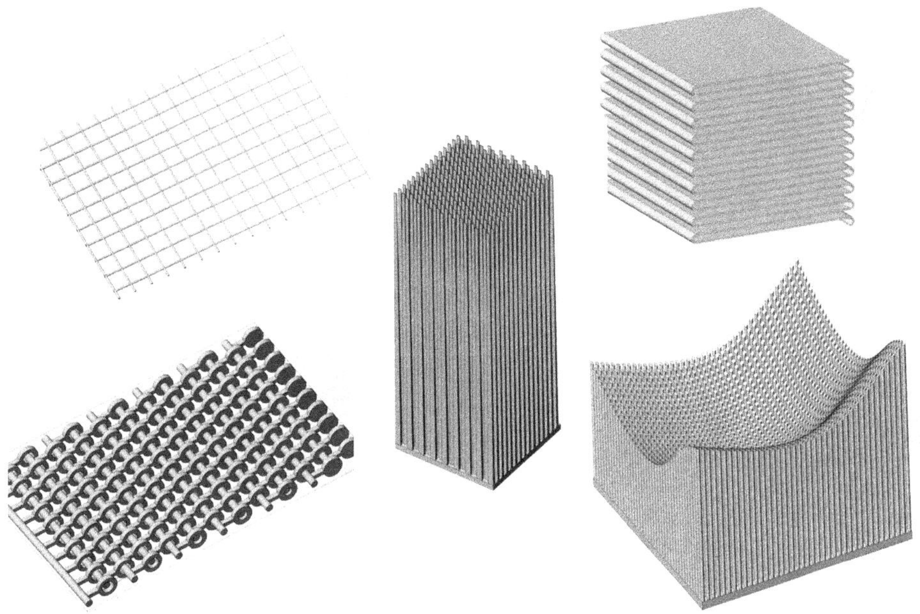

and its ability to make thin sheet like structures. At 0.8mm the leaf green structure shown was at the edge of the recommended thickness for this process and designed to test the consistency of the thin structure and its ability to create a concertina-like mechanism (Figure 7).

The performance of the material and process for interlocking parts was considered critical to the product, and therefore rows of hinged elements were tested, with the tolerances (clearance between elements) reduced on each row to ascertain the optimum tolerances for functional hinges. Other test pieces included a grid structure looking at printing resolution. The grid was graduated, with a thickness at one end well within the standard print resolution for the process, with the opposite end only 0.15mm diameter, which was beyond the standard print resolution for the process to test what would happen with the material and technology at its boundaries, where it started to fail. These test structures were converted to stereolithography (STL) format and sent to Materialise for printing. Alongside the testing and research into TPU more conventional materials and alternative processes were evaluated for the project including polyamide. Polyamide proved to be more robust and structurally stable than TPU in this context. This was particularly relevant because of the forms of the structures that were developing alongside the materials testing in response to the coral research. For this reason in addition to the research into TPU material concepts for rigid structures printed in Polyamide or a similar end use material were developed. Technologies considered were stereolithography, selective laser sintering and fused deposition modeling (both using a standard multihead extrusion printer and a material jetting system for UV cured resin with a wax support structure). In each case, the physical design would need to be different for each process. All 3D printing technologies have their idiosyncrasies and operate in a different way, and each requires a different approach. Many processes print in materials that are considered non-functional. This means that the materials used for printing are not durable or otherwise viable for an end use scenario. These

Figure 7. Grid patterning
Canning, 2013.

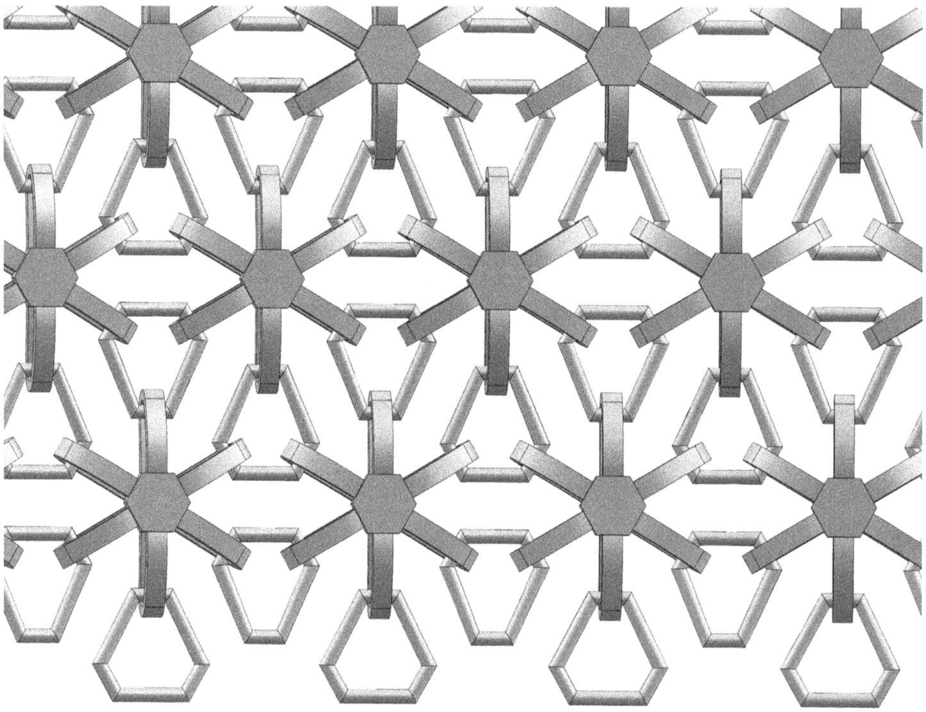

materials include those prone to rapid discoloration and degradation when exposed to UV light, which quickly become brittle. For this reason they were eventually rejected as unsuitable for this project. Selective laser sintering (SLS) proved capable of creating parts from a suitable end use material (polyamide/nylon) for the process. SLS also generates no support material as it uses the powder surrounding the part, not fused by the laser, as support (Figure 8).

The figure shows the sketch development of rigid interlocking structures based on stylistic references taken from marine structures. Numerous ideas were developed and three were selected to be taken into CAD for further development as they demonstrated the capabilities of the technology in different forms that all drew on the coral reference material. The hexagon style structure was based on the geometry of coral polyps. This structure has six sided structural elements and is linked by separate joiners, each linking three of the six-sided features (Figure 9).

An alternative structure was one in which a robust structure was created that did not use separated links between the main parts, but instead used features on the parts themselves. Developments of the structural design elements led to open lace like structures designed for maximum flexibility. This design contained ring elements each linking three individual elements. In this design each individual segment is linked to twelve separate parts. This made the structure strong despite its fragile appearance, which was in line with the original creative intent. In order to expedite the process of evaluation these structures were printed on a UV cured resin, material-jetting printer. This type of printer does not produce similar material properties to the Polyamide (the chosen print method and material) but that was not considered

Figure 8. Drawing detail
Canning, 2015.

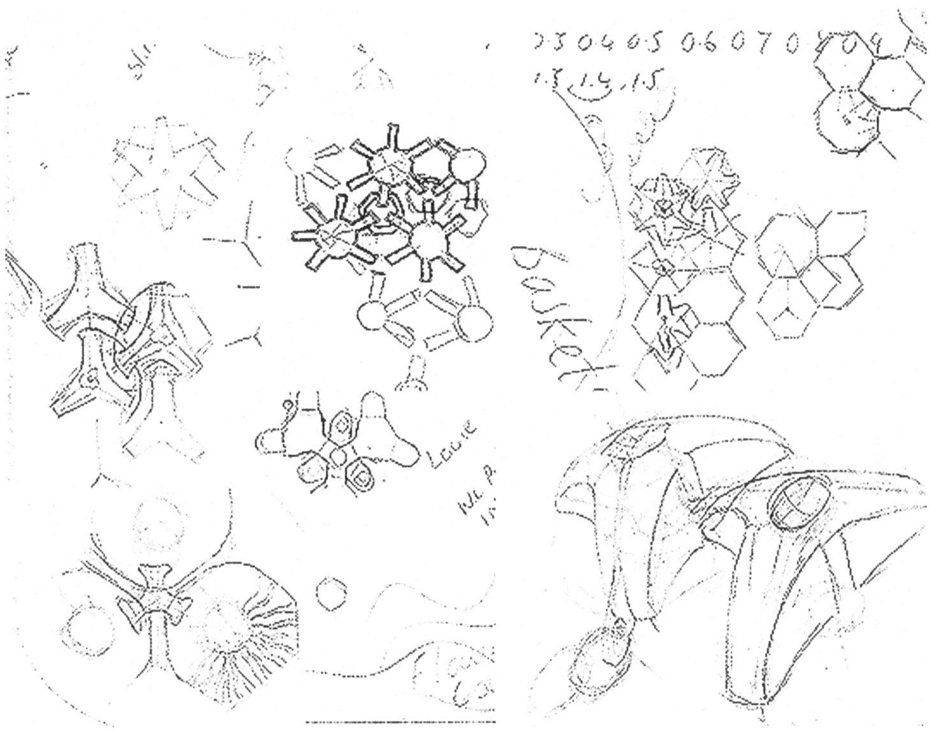

Figure 9. Structural detailing
Canning, 2015.

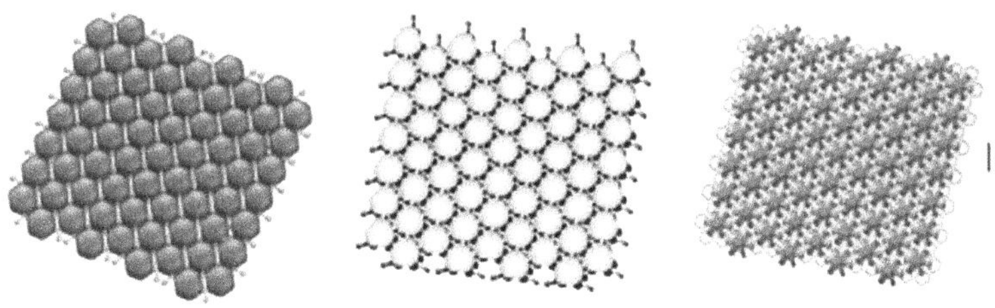

relevant for the purpose of these samples that were essentially to evaluate the flexibility of the actual structures themselves and their aesthetic relevance to the project aims (Figure 10).

The resultant printed structures were sent to the fashion designer for evaluation. These were discussed between the fashion designer and the industrial designer, with two selected for further development. In this context, the role of the industrial designer has some parallels with a textile designer who specializes in constructed textiles (such as woven, knit, mixed media) and particularly performance specific textiles, as discussed by McQuaid (2005), where the application impacts the construction to meet onsite requirements. However, as evidenced by the work outlined by Braddock Clarke and Harris (2012), the developments in digital textile design are challenging traditional discipline boundaries. In this case, the

Figure 10. Close-up constructed textiles
Canning, 2015.

most significant elements that develop, rather than replicate, practice in relevant constructed textile design fields (such as digital knitting), relate to parametric and bespoke modelling. The industrial designer brought an understanding of the constraints and opportunities of designing for 3DP that integrated the constructed textile elements into the bespoke 3D forming of the overall design with non-uniform elements that could be used for successive, staggered layers. A high level of design and artisanship was required to achieve this outcome.

A critical factor in deciding what to make, what material to make it in, and what process to use was the build volume available. All 3D printing technologies are restricted to the volumetric capacities of their respective machines. Materialise had an SLS machine with a capacity of 650 X 330 X 560mm dedicated to SLS Polyamide. Although in conventional manufacturing, this could be a small build size, compared to for example the capacity in rotational moulding, in 3DP terms, this is a large build volume. It allowed for a significant amount of 3D geometry to be built in a single build, allowing for the dress to be printed as a single part. Materialise had dedicated a much smaller machine to its TPU material that was around the size of a shoebox and unsuitable for the scale of the project. As one of the aims expressed by Materialise and the industrial designer was to print the assembly as a single part, the SLS using polyamide was chosen for the final print, which then dictated the material. This impacted the characteristics possible to work with and the industrial designer had to adjust the design construction to match these characteristics whilst retaining the fluidity of the garment (Figure 11).

The driver for the fashion designer was to create a distinctive fashion icon that utilized the technology to achieve her creative vision. In order to achieve this, Looi decided to finish the dress with hundreds of Swarovski crystals. In order to achieve the housing for these crystals, the industrial designer thickened the detailing on the components from 5mm to 8mm. Although the design was then functional, it became visually heavy and the team decided that it needed to be visually lightened to still capture the essence of the original vision. This required a redesign of the components themselves, so that they then could house the crystals but retain the visual lightness of structure. This design contained a pierced front section to allow the back of the crystal to protrude into the heart of the structure. This design development reduced the overall thickness of the structure to less than 5mm. At the same time patterned features present on the upper surface of the individual elements was increased in size. The original patterning is on the right in the photograph, which even at close inspection is difficult to distinguish. Fine features and patterning like this are extremely vulnerable to even fine machine variations. At this point in the process the industrial designer had completed the initial research phase of the project, including the initial test pieces.

Figure 11. Printed out test pieces
Canning, 2014.

Figure 12. Body scans for customization

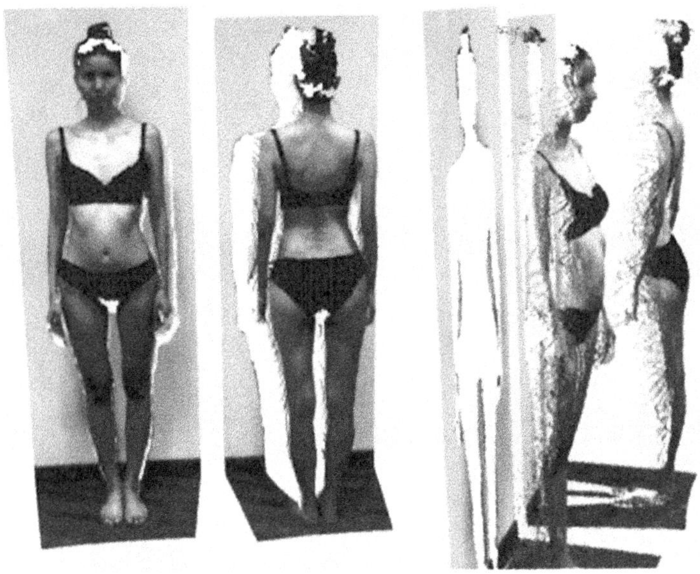

The next stage of the project was to source a model to tailor the design to fit. Looi organized for her chosen model, Eleen, to be digitally scanned by an Italian scanner, Daniel Siri, whilst on an assignment in Switzerland (Figure 12).

The point cloud data from the original scan was sent to Australia and processed at Griffith University by industrial designer and scanning specialist, Chris Little. Little translated the data into a 3D CAD solid model suitable to use as reference data in SolidWorks. The 650 x 330 x 560mm spatial constraints of the SLS Polyamide process could now be viewed in their correct context. The grey transparent bounding box represents the available build volume for the project. All 3D geometry for the dress had to be fitted into this area which meant that when the dress was modeled in SolidWorks, it had to be folded in on itself and rolled. This was a significant challenge (Figure 13).

The sketch in Figure 13 shows the final hand sketches of the dress design prior to the commencement of computer modelling. This sketch was returned to constantly as a reference point whilst modelling in Solidworks. From the early development of the rigid structures the industrial designer explored the capabilities of the 3D CAD modelling software in order to develop modelling strategies. Understanding how to work with the capabilities and opportunities of the software was an important element of the designer's role in responding to the geometric possibilities that 3DP provided. Computer modelling is no longer used purely as a documentation tool, but is integrated into design process and influences the development of the design details themselves through the operations available in the tools within the program. Creating layers within the dress was a particular challenge as a layered framework was needed to support the individual components and a variety of approaches were tested. These strategies involved creating shapes that could successfully maintain in a controlled position the many thousands of individual elements in relation to each other. Five different framework strategies were tested. The optimum design for the framework allowed the elements to be contained but segmented, which would make them easier to control whilst the build space was taken into account in the print. To start the overall build in SolidWorks, an initial surface model was created, that would provide a reference model for the framework and internal details. This also provided a representation of the correct sizing for the dress (Figure 14).

Figure 13. Drawing based on scans
Canning, 2014.

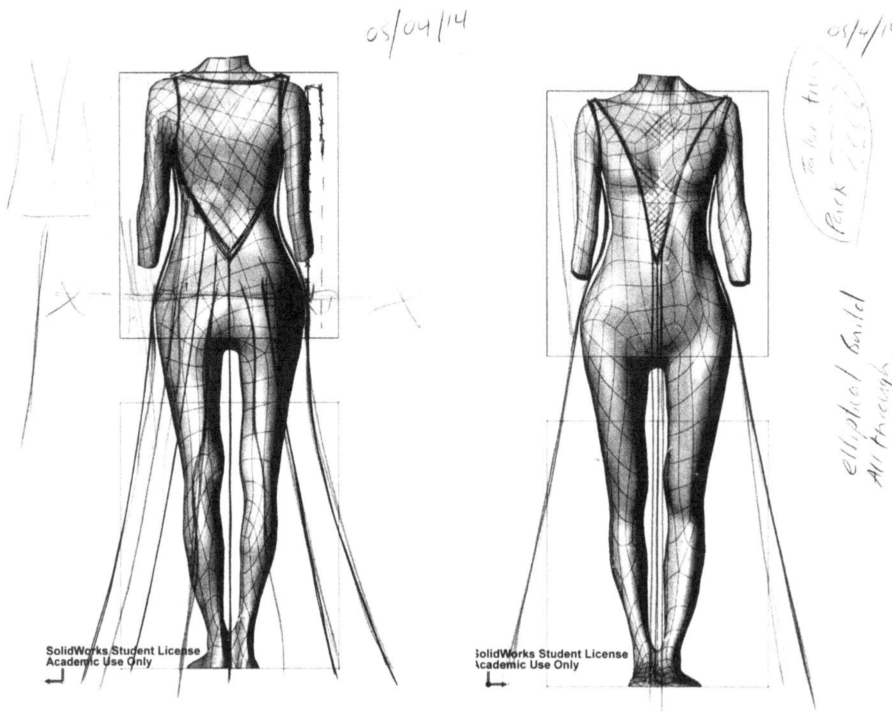

Figure 14. Initial development of dress form
Canning, 2014.

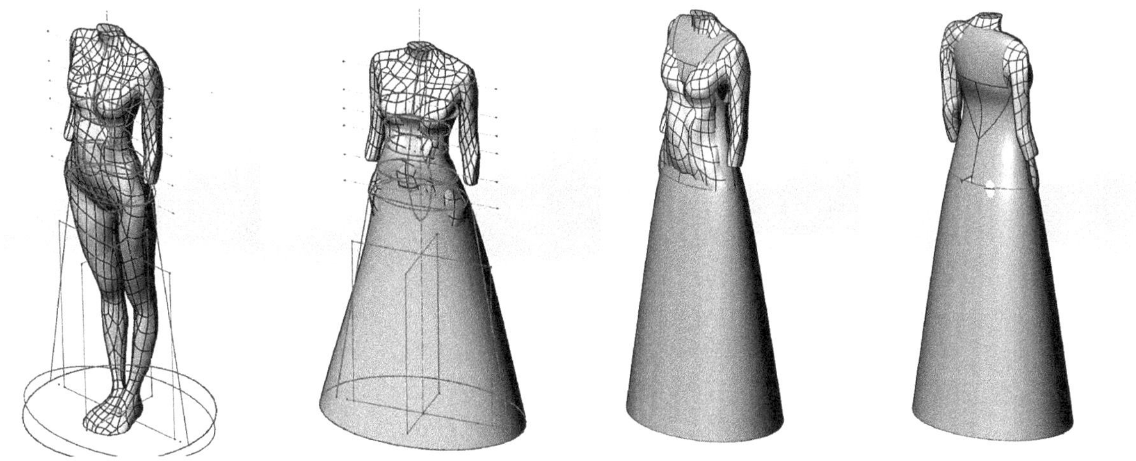

The location of the plunged back and front can be seen as split lines traced onto the surface model. Once the surface form was created, the framework and the detailing were gradually built into the model. As with the Peacemaker hat, the individual polyp elements are graduated in size to allow for the organic forms in the design. This was a complex challenge in relation to the dress as each layer has 62 polyp elements, and each horizontal layer is a different overall size. This created a smooth and almost imperceptible progression down the dress that curved the layers to the body form without needing the conventional fitting elements, such as darts. One of the complexities of this task was the decision to add crystals to the polyps, as this meant that the polyps could not be scaled, as the sockets had to remain exactly the same size to accommodate the crystal, irrespective of the outside dimensions of the polyp. This means each layer had to be recreated, as the geometry could not be re-used. Modelling of the dress components began at the waist section. The design then progressed down to the bottom of the build envelope and then returned to the waist and progressed up to the chest area (Figure 15).

One of the challenges of developing the design that the industrial designer had to take into account was the problem of file size and working a delicate balance between achieving the complexity in design required, and keeping the model so that it could be workable. When model file sizes become large or they contain large amounts of complex 3D geometry, the computer model will become slow to manipulate to the point where they can become impossible to edit. For this reason strategies need to be developed for generating large models. After development testing, it was found that a strategy of modelling five individual polyp layers at a time was the most workable solution, and then (as shown in the multicoloured layers in the figure), the layer groups were built into longer structures. Because of the problem with the limited build space, the layers had to be modeled in progression to the bottom of the build envelope and then turn back and work up to the top of the build space. Only once this stage was reached, could the design decisions relating to the neckline of the dress be addressed. A much looser, lace like version of a polyp lattice was created for the neckline. It needed to be very flexible to allow for upper body movements (Figure 16).

The fashion designer provided a sketch to indicate the form of the lower section of the dress. To capture this design intent, and create the fullness and flow of the lower half of the dress, the industrial

Figure 15. Designing the fold
Canning, 2014.

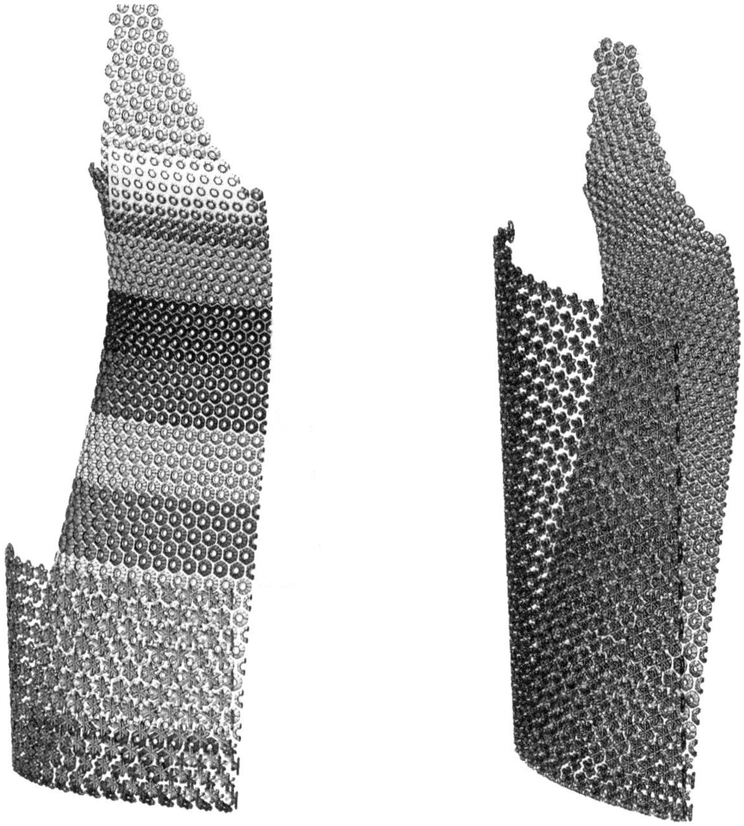

designer developed a visually lighter, more open star shaped polyp structure. This complex detailing was formed into a 3 dimensional mesh structure. The transition between the polyps in the top half of the dress, and those in the lower section was designed to take into account the curves of the body (Figure 17).

The shoulder straps represented another significant technical hurdle. Restrictions imposed by the build volume dictated that the straps could not be built in their upright state. A number of options were explored through sketching and decisions were made on the design by the industrial designer resulting in a design modeled in a folded state suitable for printing in the build space (Figure 18).

In order to complete the build and maintain a model that was useable in terms of its file size (not the build size) the dress had to be split into nineteen separate assemblies. In addition to the file size of the Solidworks files there is the added consideration of the file size of the STL files for printing which is different. This was a concern for the Materialise engineers who confirmed the maximum file size for an entire build on the machine was 8 Gigabytes. In general use, the machines are never used to build single models, rather the build envelope is packed with as many models as can fit inside the volume in order to make the build as cost effective as possible. In this regard the collation of the nineteen separate STL files was almost common practice. However, the industrial designer had to ensure that all of the files were exported with a common global origin. This point was critical for the alignment of the separate assemblies, manual alignment is impossible due to the number of connections that have to be made and

Figure 16. Development of neckline
Canning, 2014.

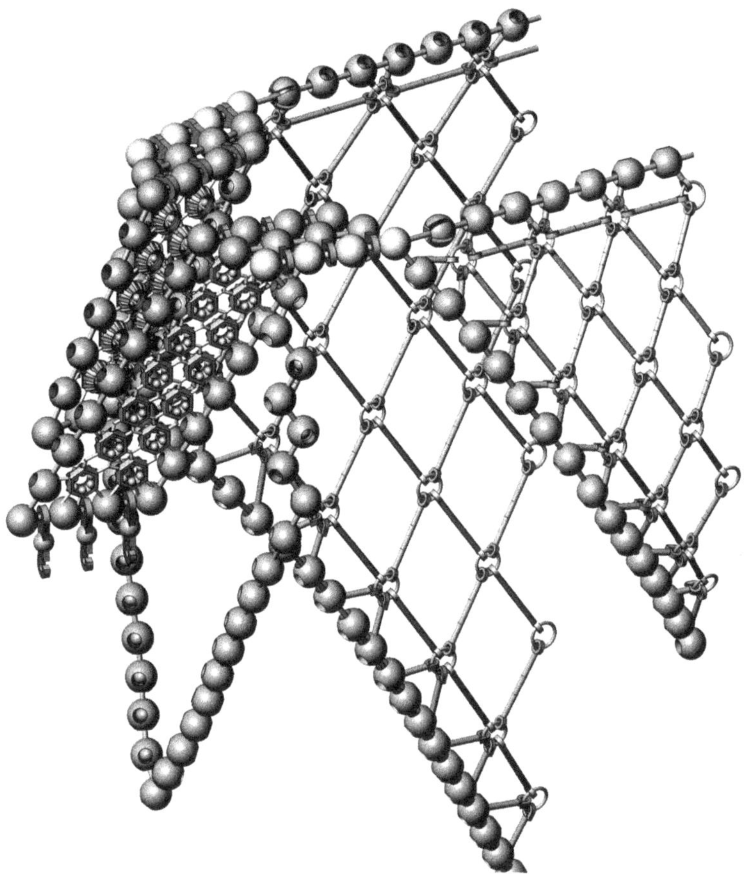

the tolerances essential to the 3D printing process that have to be maintained. The combined file size for the nineteen STL files for the dress was 7.14 Gigabytes. The final print model was sent to Materialise where the software engineers optimized the triangulation of the print in order to reduce the print time, which had to be balanced with any potential loss of fine detail, and to ensure a robust print. Vulnerabilities within the print, caused by, for example, the folding and rolling of the structure to fit the build space, were interrogated for possible re-enforcing. The print was then run, which took approximately four days. The Engineers then had to gently remove the print from the powder cake for post processing. The Engineers cleaned the excess powder off and the industrial designer then individually flexed the tiny components to release them. The dress was then sent to Looi, who dyed it, and attached the myriad of crystals to create the final finish. Each crystal had to be attached by hand, using tweezers, and glued in place.

Figure 17. Design of attachments
Canning, 2014.

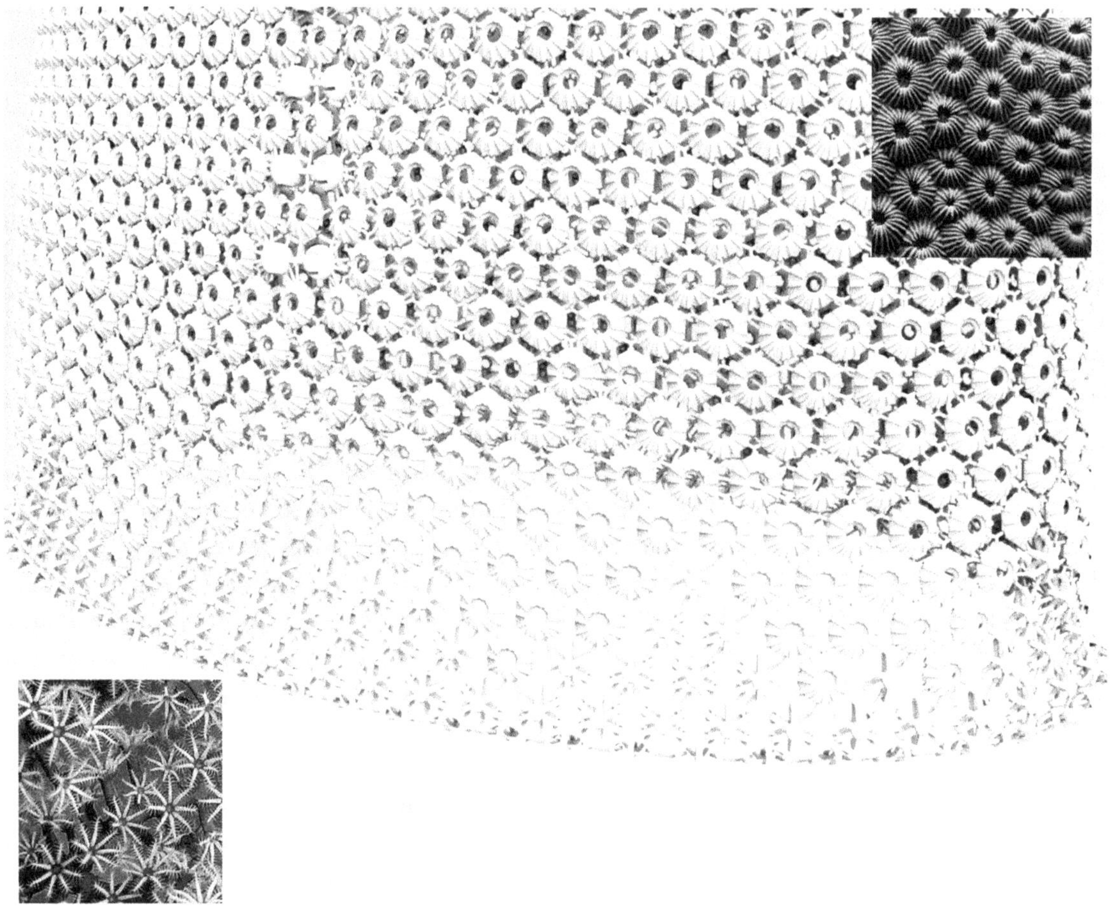

Review of Practice

The final outcome is an example of technical development driven by creative expression. By bringing the fashion designer into the team, the aesthetics drew the technical innovation behind it, as the industrial designer and engineers strove to meet the intent. This is therefore an example of disruptive practice for 3DP technical research, as the creative intent drove the technical development work. In this case, the project was executed in a linear progression through the disciplines, as a 'bouncing ball' project, that is one where the first team in a discipline works on a project, then their work is transferred into another project team from a different discipline, creating successive development, rather than as an example of the collaborative practice proposed by Benyus (2002). In this case, it was predominantly because the specialists were based in different countries, but it does provide a case study for the challenges involved in cross-disciplinary collaboration, and the aim of creating trans-disciplinary practice. Fundamentally, the project highlighted cultural differences in approach, values and ideas, which are summarized below with recommendations to mitigate the issues between the disciplines that could impede design development and the innovation that creative and technical collaboration is intended to stimulate.

Figure 18. Form development
Canning, 2014.

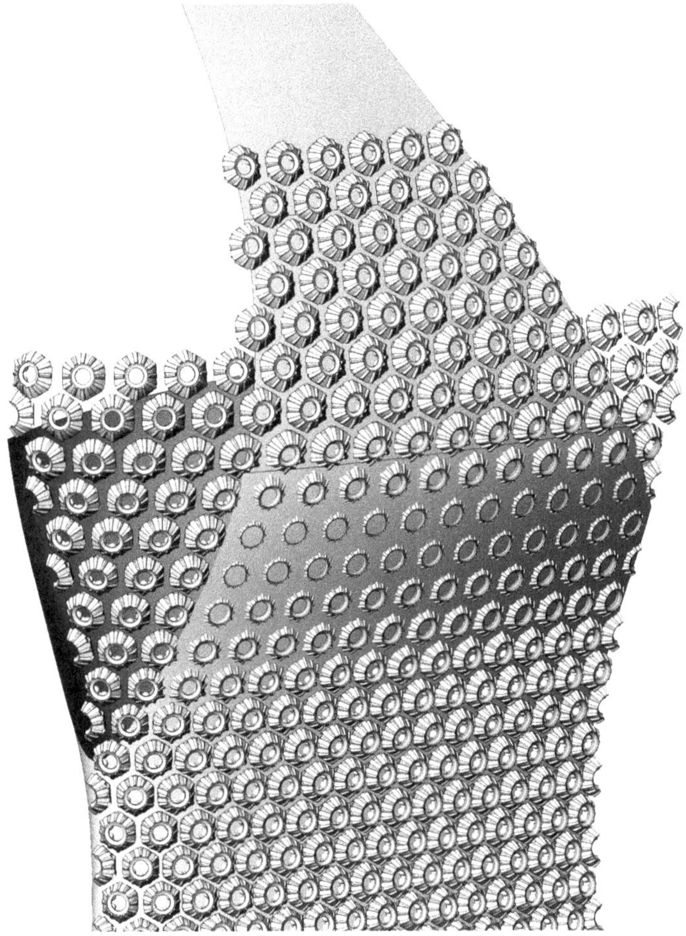

1. **Language:** Communication difficulties as the accepted definition of terms were different between disciplines, for example the word 'design' itself is interpreted very differently between disciplines.
    a. **Recommendation:** Prior to the start of the project, the team should be involved in a communication exercise to highlight differences in language understanding prior to the project. Key terms should be identified and discussed to achieve a common understanding between researchers. This may involve adding clarifiers to the terms that define the discipline bias they are being used with during particular discussions.
2. **Drivers:** The drivers for taking part in the project work will be different for the different disciplines. For example, the impact the dress would make on the peer groups for the different disciplines would be judged very differently by the different discipline members and critics of those disciplines which results in design decisions that are discipline specific. This is illustrated in this case of the dress, by the fashion designer choosing to add crystals as for the fashion Industry, the technical innovation of 3D printing the dress as a single part was less important than the overall visual impact. For the industrial designer, the driver was to maximize the technology, and therefore the crystals reduced

the value of the outcome, as they were not part of the print. Yet for the fashion designer, for the dress to gain the attention within a visual culture, where the technology is not widely understood, the crystals add value. For Materialise, the publicity value of featuring the dress on the Today show in the US, with 6 million viewers, was of less value to the industrial designer, than having it shown at the leading industry conference with less than 7,000 participants where its construction, rather than its visual impact would be better understood.

   a. **Recommendation:** Prior to the start of the project, clarify the drivers for the different team members and identify the vision for each discipline. The aims for each discipline in terms of publication of the outcomes should be identified and discussed.

3. **Values:** Every discipline has skills and understandings that it values. These are not necessarily known or understood by those outside that discipline. In addition, the amount of work, and the specific expertise needed for each aspect of a project such as this one, are not necessarily understood or valued by the other team members. A good example of can this is with scanning technology. The actual scan itself is a quick process and can be easily seen and understood by everyone, but the translation of cloud point data into usable geometry takes time and expertise, yet is invisible to those outside the process. Similarly the optimization of the geometry. With regards the industrial designer, the craftsmanship involved in developing the construction details for 3DP is currently not well understood or valued by those outside the discipline, who still see 3DP as a 'push button' technology. The dress, for example, required not only months of development work, but specifically over 400 hours of computer modelling time was logged during the development of the components and building the dress.
   a. **Recommendation:** Greater communication of the design and development process in real time, for example through project logs or a virtual studio would help each discipline value what is involved in each stage of the process.

For the multidisciplinary research team to be effective in creating innovation, mechanisms need to be put in place to support informed collaboration. For research teams that include creative as well as scientific researchers, this preparatory work is arguably more essential or opportunities may be missed. The reality is that different disciplines address research practice with very different interests, values and approaches. Sharing language, values and drivers will provide a starting point for a more genuine trans-disciplinary practice. For the potential of 3DP to provide opportunities for disruptive thinking in scientific research teams, by bringing in designers for traditionally scientific research based projects, to be realized, the lessons learned from the example of fashion and engineering need to be learned. The clash of cultures in cross-disciplinary practice between scientific and creative researchers needs to be understood and addressed.

Issues, Controversies, Problems

Working outside traditional discipline boundaries carries with it the risk of research work not being valued. Within a cross disciplinary project, there is an additional risk to the specific elements of the work not being understood across the disciplines, and the impact of the outcomes therefore diluted, as observed within engineering and architectural interactions by Zeiler (2014). If this is the case, the combined learning from project work is unlikely to be then developed further during follow up projects. Although the potential benefits of bringing together scientific research and creative research, as in the work of Benyus

(2002), there are significant challenges to practice. Between engineering and creative practices, such as in design and art, the clash of cultures is particularly evident and can stem from educational practices where values and organization of workflow are inculcated into students.

SOLUTIONS AND RECOMMENDATIONS

Just as graduate attributes include the ability to work in teams, the recommendation based on this research is for a commitment to education on working within unconventional cross-disciplinary teams, particularly for students intending to pursue a career in research. With the development of CDIO in engineering subjects, the gulf between design and engineering has narrowed, and the industrial design discipline can take on a role to proactively bring together more creative subject researchers, such as in art and fashion, with scientific based enquiry researchers, such as in Engineering and clinical practice. However, the importance of preparation in understanding the drivers, values, working practice and language of collaborators is highlighted through this study, and there needs to be an investment of time and effort in making those connections prior to, and during research projects for effective practice.

With regards research that relates to the technology, the mechanics and materials science aspects have attracted the interest of engineering researchers, as described by Gibson et al (2014), whilst social scientists, such as Anderson (2012) have explored the impact of the technology on society. However, Hahn et al (2014) identify a lack of literature on innovation for 3D printing applications. They emphasize that there should be more research in this area, as there are significant opportunities for the adoption of the technology in key sectors such as ‘aerospace, medical devices and implants, power generation, automotive manufacturing and the creative industries’ and predict that the technology will create alternative business models and practices and forecast a sector worth of $7.5 million US by 2020. As a disruptive technology, 3DP provides opportunities across disciplines, and this will equally impact research. Within Griffith University, new research collaborations are forming between disciplines as diverse as forensics, regenerative medicine and industrial design. Yet, there is likely to be a lack of understanding between team members both in terms of the discipline knowledge, drivers values etc, as well in terms of the potential of the 3DP and how to most effectively use it for specialist applications.

Leading disruptive innovation, however, is a process fundamentally laden with surprise, the core essence of uncertainty. Recognizing the potential power of surprise when we receive unexpected jolts to our strategies, plans, and assumptions, allows us to respond with purposeful agility - versus dismiss surprises as problems while concurrently disregarding the insights or messages they may contain. (Kaplin, 2012)

Working with 3DP in new fields of research within new collaborations (especially where they bring together creative practitioners and scientific researchers, such as in the teams for regenerative medicine and industrial design, or for fashion and engineering) will inevitably create some uneasy situations and unexpected outcomes. For the technology to achieve its disruptive potential, and researchers to discover where it can take their work, teams need to be open to new ways of thinking, and ideas and approaches very different to their own – which they need to be willing to engage with and learn to understand – and embrace the change. By deliberately working against the conventions of practice in established fields, without predefined objectives in exploratory work (Pink 2009, Ted talk), new research collaborations have the potential to provide new insights as provide starting points for trans-disciplinary practice.

FUTURE RESEARCH DIRECTIONS

Based on this study, further research is needed to develop a blueprint for working practice in research teams that bring together the creative and the scientific to stimulate innovation. 3D printing has created particularly interesting cross-disciplinary opportunities because it brings together such normally disparate fields. This provides a basis for the study of developments of innovations in collaborative practice where there are no pre-existing conceptions of standard relationships. The intention, for a follow-up collaboration on 3D printing across engineering and creative practice, is to invest in building cross cultural understandings between the disciplines through this work to enhance practice, and to develop collaborative tools to move from a linear, successive practice, towards a more genuine, real time collaboration informed by mutual understanding and shared – or at least understood - aims for the project outcomes for all involved. With regards to 3DP itself, with drivers from multidisciplinary teams, there should be a subsequent draw for the development of the technology that breaks the incremental progress of scientific research and allows for new applications research to take more of a lead. This will be supported by research into changing educational practice for new design thinking for advanced technologies, as discussed by Loy in Education Sciences (2014).

CONCLUSION

In 2014, Griffith University hosted two 3DP research forums. One was called '3D printing the Body, inside and out' and the other called 'Design and Manufacturing, 3D printing for the Creative Industries'. The multidisciplinary profile of the participants illustrated the potential for the technology to disrupt research teams, and bring together scientific and creative researchers to consider newly framed research questions and potentially stimulate innovation. The forums included research participants from engineering, industrial design, biomedical and art. One of the artists who provided an example of challenging scientific research practice using the technology was Professor Stelarc from Curtin University. His research work included a 3D Printed ambidextrous hand, which contributes to the discussion on Pullin's ideas for design meets disability, moving them further towards the aims expressed by Mullins on design for enhanced ability through assistive technology. In addition, with his anatomical work, (for example the scaffolding in an ear form he had surgically added to his left arm with an internet connection to transmit verbal communication), Stelarc (2013) provides a bridge between research into regenerative medicine and art through his interest in the use of bioplotting, a form of 3D printing that uses biomaterials (including cells).

The gulf between disciplines in terms of research drivers, values and even language is significant, and has the potential to derail attempts to bring together researchers for scientific and creative collaborations. What fashion has brought to scientific based research practice through 3D printing in engineering has been the stimulus for technical and applications innovation and highlighted the clash of cultures in research teams. Between fashion and engineering, the clash of cultures is clearly visible, and this has allowed for the identification of issues that may be more subtle within more closely related research teams but still need to be addressed. If the contrasting research approach between designers and scientists, that Milton and Rodgers observes, is deemed to be useful in research teams, then the example of fashion and engineering can provide a productive starting point for creating disruptive practice in scientific research to stimulate innovation. Working in teams is rarely straight forward, and there has been considerable research into how to operate within a team. If trans-disciplinary research is to be progressed, and long term relationships to develop, then the disciplinary research teams need to recognize their ethnocentric starting points, and invest in preparatory work to recognize and understand the values of their collaborators, build shared communication tools and articulate drivers and publication priorities prior to entering into collaborations.

More specifically in terms of project work that brings together fashion and engineering through 3DP, is the value of the role of the designer providing expertise in design for 3DP. This expertise is only starting to be developed. For the contrasting worlds of fields such as engineering and fashion to be able to come together to maximize the opportunities this new technology provides, the idea that it is a 'push button' technology needs to be dispelled, and designers (including fashion designers) and engineers, educated in working in the middle ground, where creative ideas are evolved and designed for the technology. This is when innovation will happen.

REFERENCES

Anderson, C. (2012). *Makers: The next industrial revolution*. New York, NY: Crown Business.

Benyus, J. (2002). *Biomimicry: Innovation inspired by nature*. New York, NY: Harper Perennial.

Braddock Clarke, S., & Harris, J. (2012). *Digital visions for fashion and textiles: Made in code*. London, UK: Thames & Hudson.

Braddock Clarke, S., & O'Mahony, M. (2008). *Techno textiles 2: Revolutionary fabrics for fashion and design* (2nd ed.). London, UK: Thames & Hudson.

Cacchione, P. (2015). Interdisciplinary research teams. *Clinical Nursing Research*, *24*(2), 119–120. doi:10.1177/1054773815574790 PMID:25748854

Cohen, J., Gehry, F., & Gehry, F. O. (2003). *Frank Gehry, architect*. New York, NY: Guggenheim Museum.

Gibson, I., Rosen, D., & Stucker, B. (2014). *Additive manufacturing technologies: 3D printing, rapid prototyping and direct digital manufacturing* (2nd ed.). New York, NY: Springer.

Hahn, F., Jenson, S., & Tanev, S. (2014). Disruptive innovation versus disruptive technology: The disruptive potential of the value proposition of 3D printing technology start-ups. *Technology Innovation Management Review*, *4*(12), 27–36.

Harman, J. (2014). *The shark's paintbrush: Biomomicry and how nature is inspiring innovation*. Ashland, OR: White Cloud Press.

Kaplan, S. (2012). Leading disruptive innovation. *Ivy Business Journal*. Retrieved July 22 2015 from http://iveybusinessjournal.com/publication/leading-disruptive-innovation/

Kumar, V. (2012). *101 design methods: A structured approach for driving innovation in your organization*. Wiley.

Lewis, K. (2003). Measuring transactive memory systems in the field: Scale development and validation. *The Journal of Applied Psychology*, *88*(4), 587–604. doi:10.1037/0021-9010.88.4.587 PMID:12940401

Leys, P. (2015). *The Killer Apps of 3D printing*. Paper presented at Inside 3D Printing, New York, NY.

Lipson, H., & Kurman, M. (2013). *Fabricated: The new world of 3D printing*. Wiley.

Loy, J. (2014). ELearning and eMaking: 3D printing blurring the digital and the physical. *Education Sciences*, *4*(1), 108–121. doi:10.3390/educsci4010108

Loy, J., & Canning, S. (2013). *Reconnecting through digital making. IDEN*. Sydney, Australia: Industrial Design Educators Network.

Loy, J., & Welch, D. (2013). A brave new creativity. *Art. Design and Communication in Higher Education*, *12*(1), 91–102. doi:10.1386/adch.12.1.91_1

McQuaid, M. (2005). *Extreme textiles*. Princeton, NJ: Princeton Architectural Press.

Milton, A., & Rodgers, P. (2013). *Research methods for product design*. London, UK: Laurence King.

Nakamichi, T. (2011). *Pattern magic 2*. London, UK: Laurence King.

Pink, D. (2009). *The puzzle of motivation*. Retrieved from http://www.ted.com/talks/dan_pink_on_motivation?language=en

Pullin, G. (2011). *Design meets disability*. Cambridge, MA: MIT Press.

Qian, M. (2015). Metal powder for additive manufacturing. *JOM*, *67*(3), 536–537. doi:10.100711837-015-1321-z

Stelarc, S. (2013). Aliveness & affect: Alternate art & anatomies. In P. Baler (Ed.), *The next thing: Art in the twenty-first century* (pp. 133–150). Fairleigh Dickinson University Press.

Wageman, R., Gardner, H., & Mortensen, M. (2012). The changing ecology of teams: New directions for teams' research. *Journal of Organizational Management*, *3*(33), 301–315.

Weinschenck, S. (2011). *100 things every designer needs to know about people*. San Francisco, CA: New Riders.

Zeiler, W. (2014). The difference in communication between architects and engineers and the effectiveness within integral design. In E. Bohemia, & ... (Eds.), *Proceedings of the EPDE 2014 16th International Conference on Engineering and Product Design Education* (pp. 238-243). Bristol, UK: Design Society.

KEY TERMS AND DEFINITIONS

Additive Manufacturing: Refers to a range of digital technologies that build objects in layers without the use of a mould.

This research was previously published in Creative Technologies for Multidisciplinary Applications edited by Andy M. Connor and Stefan Marks, pages 25-53, copyright year 2016 by Information Science Reference (an imprint of IGI Global).

Section 2
Modeling

Chapter 5
Cloud Based 3D Printing Business Modeling in the Digital Economy

Norman Gwangwava
Botswana International University of Science and Technology, Botswana

Albert U Ude
Botswana International University of Science and Technology, Botswana

Enock Ogunmuyiwa
Botswana International University of Science and Technology, Botswana

Richard Addo-Tenkorang
Aalborg University, Denmark

ABSTRACT

3D printing, also known as additive manufacturing, is becoming the industry standard for manufacturing and prototyping. Although the technology is very old, it gained a huge traction in the past two decades. 3D printing favors unique once-off orders (mass customization) in contrast to mass production. This calls for innovative business models in order to realize economic gains from the technology. Increased product innovations in the global economy also contribute to wide adoption of 3D printing unlike in the old days. A transition in the manufacturing field has brought e-manufacturing and now cloud-based manufacturing. Machines, including 3D printers, in the past were not Internet-enabled but modern designs have the capability of Internet connectivity. Cloud-based 3D printing is a new model of design that has a significant impact on today's entrepreneurs. This article focuses on a business case for a cloud-based approach in consumer product niches. A cloud-based 3D printing business model (3D-Cloud) is developed based on the business model canvas, which promises major breakthroughs in e-entrepreneurship innovation. The model uses a virtual community approach to bring together technocrats, enthusiasts, and shared 3D printer facilities of common interests, whilst promoting an enterprising spirit.

DOI: 10.4018/978-1-5225-9624-0.ch005

INTRODUCTION

3D Printing became a sudden Gold rush in the past two decades. This was inspired by rapid innovations in digitization and increased competitiveness among businesses. The pace of innovation across economic sectors also increased rapidly, and this brought a new challenge to enterprises, the need to hit the markets fast. The general consumer population is increasingly becoming more educated as more information becomes readily available through the cloud. Virtual communities and social networks have also led to the advent of new entrepreneurship approaches. Social entrepreneurship has become common in today's modern world. 3D printing (i.e., additive manufacturing) is an old technology that was progressing slowly but became a significant pillar of supporting innovation and quick time-to-market in many enterprises. Products can now be launched faster because the technology eliminates complex stages of product research and development. Where new tooling for new parts design would take several weeks or months to produce, 3D Printing has eliminated those stages. Big companies have gone further to using 3D printing to make ready-to-use parts, which can be used as spare parts replacements. This significantly reduces down time in many industries. However, by nature, 3D printing remains unsuitable for mass production. This calls for innovative business models to sustain the technology, whilst tapping from its unquestionable benefits. There are many proven business applications of 3D printing, which include new-product design/prototyping, spare parts replacement, 3D printed ornaments, children toys, education and many others such as the medical fields. However, business viability factors in many considerations which if not taken into account can result in unsustainable issues and subsequent failure of the initiatives. Adopting appropriate business models bring in viability and promote growth of the 3D printing technologies through proper appreciation and continued innovation.

3D printing is rapidly developing into an important but also disruptive technology with the potential for societal-wide change as witnessed with the introduction of the automobile, personal computer, the Internet and smartphones (Mills, 2015; Petrick & Simpson, 2013). The technology influences many processes in production, supply chain design, logistics, product life-cycle planning, and consumer behavior (Mellor et al., 2014; Berman, 2012; Bogers et al., 2016). Research has argued that this technology not only has profound effects on manufacturing businesses but also on society, which demands new corporate strategies and policies alike (Jiang et al., 2017). The article looks into details of various promising entrepreneurship applications of 3D printing, cloud computing technologies and their disruptions to the traditional business approaches. Innovative entrepreneurship models are then considered with the view of proposing new business models for adoption by the modern e-entrepreneurs in the digital economy. Practical scenarios are considered, and the case studies can be applied in any other 3D printing environment following the proposed template approach.

TRENDS IN 3D PRINTING

3D printing is a manufacturing method based on advanced technology that builds up parts, additively, in layers (Ventola, 2014). The 3D printing process starts with a 3D digital model, created using 3D software such as 3D CAD. After the expiration of the patent for the fused filament fabrication (FFF) technology of additive manufacturing (AM), where a single layer of polymer is deposited after another, there followed the release of the open-source self-REPlicating RAPid prototype 3D printer (RepRap) (Sells et al., 2007; Jones et al., 2011; Bowyer, 2014). The open-source hardware approach led to a rapid technical evolution, which resulted in aggressive cost declines and the emergence of desktop 3D printer market, dominated by various RepRap derivative machines (Gibb & Abadie, 2014; Rundle, 2014; Wohlers, 2016; Frauenfelder, 2013). 3D printing found many applications in rapid prototyping, manufacturing, arts and design, toy manufacturing, automotive, consumer electronics, education, architecture models and full-scale house printing, aerospace and defense, as well as medicine. The technology has been widely used by hobbyists to replicate 3D printers. Medical applications of 3D printing brought in a huge step from the virtual world to the physical world through bone reconstruction; 3D printed vital organs and dentistry. 3D printed models are useful for transferring information to the surgeon in a more informative way and allow for improved, more detailed surgical planning. They can help to illustrate intervention procedures to novice surgeons and patients, and can be useful for testing the procedure on patient-specific (PS) anatomy through the use of printing materials able to resemble the mechanical properties of bone (Auricchio, 2016). In the late 2000s, the cost of 3D printing began to be low enough (and quality high enough) to start directly manufacturing final products with 3D printers. Speed, quality, accuracy and material properties have developed to an extent that 3D printed parts can be made for final use (Gibson et al., 2010). This led to the third wave of adoption, generally referred to as Direct Digital Manufacturing (or DDM) or simply direct manufacturing, which implies an entirely digital production process, with end-products directly manufactured using digital (CAD) models and 3D printers, without molds, casts or machining. Advances in 3D printing materials, equipment, methods and techniques are already enabling on-demand, highly customized patient treatments (Mills, 2015). Table 1 shows different types of 3D printing technologies.

Characteristics (materials, speed, accuracy, safety, equipment costs, number of printing colors) of the different types of printing technologies influence the real-life applications. Typical applications are listed as follows (Scott, 2012):

- Small production runs

Table 1. The Seven 3D Printing/AM Process Categories (ASTM, 2014)

Process Type	Brief Description	Related Technologies	Materials
Powder Bed Fusion	Thermal energy selectively fuses regions of a powder bed	Electron beam melting (EBM), selective laser heat sintering (SLS), selective heat sintering (SHS), and direct metal laser sintering (DMLS)	Metals, polymers
Directed Energy Deposition	Focused thermal energy is used to fuse materials by melting as the material is being deposited	Laser metal deposition (LMD)	Metals
Material Extrusion	Material is selectively dispensed through a nozzle or orifice	Fused deposition modeling (FDM)	Polymers
Vat Photopolymerization	Liquid photopolymer in a vat is selectively cured by light-activated polymerization	Stereolithography (SLA), digital light processing (DLP)	Photopolymers
Binder Jetting	A liquid bonding agent is selectively deposited to join powder materials	Powder bed and inkjet head (PBIH), plaster-based 3D printing (PP)	Polymers, Foundry Sand, Metals
Material Jetting	Droplets of build material are selectively deposited	Multi-jet modeling (MJM)	Polymers, Waxes
Sheet Lamination	Sheets of material are bonded to form an object	Laminated object manufacturing (LOM), ultrasonic consolidation (UC)	Paper, Metals

- Small part size
- High-value products
- Products with high complexity
- Elimination of tooling

Due to more research directed towards 3D printing, the range of applications continue to grow. Specific markets currently saved by 3D printing are:

- Prototyping
- Tooling
- Direct part manufacturing
- Maintenance and Repair

Another dimension of application of 3D printing includes Humanitarian NGOs that are currently exploring the capabilities of 3-D printing to solve logistics issues. Desktop 3-D printers are unsuitable solutions to the challenges that they face. *Kijenzi* 3-D printers were designed and evaluated for rapid part manufacturing for two months at health facilities and other community locations in both rural and urban areas throughout Kisumu County, Kenya. They were proved to be successful for their ability to function independently of infrastructure, transportability, ease of use, ability to withstand harsh environments and costs (Benjamin et al., 2018).

Current and Future Applications of 3D Printing

3D Printing technology is used in manufacturing industries, the children toy industry (pre-school and primary education), secondary and higher education.

Children Toy Industry

The toy industry is among the biggest beneficiaries of the technology since toys tend to be small and made out of plastic, making it easy to 3D print them. Children can now customise their own unique toys by deciding on the features of the toys before they are 3D printed for them. Children can design their own custom dolls online and then order 3D printed models. Besides promoting creativity, it also speeds up and simplifies the manufacturing process of making toys, which in turn lowers the cost. According to Petersen et al (2017), the year 2020 toy and game market is projected to be US$135 billion.

Education Sector Applications

Teachers adopted 3D printing the technology looking for high-quality educational experiences for their students (Irwin et al., 2015; Grujovi´c et al., 2011; Schelly et al., 2015) as well as those looking for economic sustainable development with appropriate technologies (Birtchnell & Hoyle, 2014). The use of 3D printing in education was investigated by Slavkovsky (2012). Since physical models are important for hands-on active learning, 3D printing technology in education has been used since a while and considered for sustainable development, for secondary education in STEM projects as well as elementary mathematics education (Pearce et al, 2010; Berry et al, 2010; Lipson, 2007; Lacey, 2010). It is a well-known fact that the best way for students to learn and retain information taught is to apply it to real-life situations. 3D Printing technology is being used in the classroom to offer a highly engaging, hands-on way to teach STEM subjects as well as to expose students to technologies of the future from an early age.

Knill and Slavkovsky (2013) illustrate how 3D printing can help to visualize concepts and mathematical proofs. Models allow to bring mathematics closer to the public. The new 3D printing technology makes the realization of such tools more accessible than ever. Visualization is important for communicating mathematics and other scientific concepts.

The University of Winsconsin, department of mechanical engineering completely reworked all their lab activities so that they are based on a 3D printing platform (me-winter, 2016). The department purchased new FDM 3D printers for every lab workbench and then customized each desktop printer by installing automation hardware with touch panel interfaces and input/output electronics. All the industrial hardware is coupled to the 3D printer's functionality. These upgrades give students valuable experience using real-world automation equipment.

Schelly et al (2015) investigated the potential of open-source (OS) technologies in an educational setting, given the combination of economic constraints affecting all educational environments and the ability of OS design to profoundly decrease the cost of technological tools and technological innovation. The OS aspect proves to be particularly relevant in the educational setting, as it decreases cost of access as well as encouraging active participation and improvement through real experience with design and fabrication. 3-D printing technologies based on open-source design have transformative potential in an educational realm. Table 2 provides a list of practical applications of 3D printing by some major companies.

Cloud-Based 3D Printing and Mass Customization

Cloud Manufacturing (CM) is the use of manufacturing resources in a service-oriented way over the Internet. The use of Cloud Manufacturing can be viewed in two perspectives, where the first application is within an industrial environment for which CM provides a concept to embed, connect and utilise existing manufacturing resources, such as 3D printers, drilling, milling and other machines. The second application is for end-users that use AM/3D printing resources over the Internet instead of acquiring their own 3D printer (Baumann, 2016; Baumann and Roller, 2017). The second perspective is highly service-oriented, mainly targets end-users or consumers as the clients, and decentralises manufacturing. Cloud manufacturing enables service consumers to manufacture parts through Infrastructure-as-a-Service (IaaS), Platform-as-a-Service (PaaS), Hardware-as-a-Service (HaaS), and Software-as-a-Service (SaaS). Baumann (2017), proposes the concept, design and implementation of a cloud-based 3D printer service utilising existing 3D printer resources. The service is envisaged to help users collaborate and cooperate on tasks related to 3D printing. The hardware and software resources managed by of cloud-based service can achieve higher utilisation than independent and unshared resources. The concept of cloud-based services is alternatively known as online 3D printing. Professionals integrate their resources into online platforms for collaboration, sharing, ease of control and monitoring. Standalone 3D printing facilities have problems of high costs, limited capacity since individual centers may not have all the technology range, underutilizations, which result in opportunity costs. There is also a lack of collaboration and sharing of designs and knowledge about 3D printing technology. The concept of cloud-based 3D printing has been in existence through models such as commercial 3D Hubs and cloud platforms for 3D printer management. 3D Hubs is a network of manufacturing hubs, with services connected in over 140 countries. Benefits of

Table 2. Caption should be sentence case with no ending punctuation if only one sentence

Sector	Application
Aerospace	• Lockheed Martin built a 3D-printed part for NASA's Juno Jupiter orbiter, and has installed its first 3D-printed part in a military communications satellite, achieving cost reductions and cycle time reductions. Using 3D printing reduced design and production cycles, and Lockheed was able to make the prototype unit in 1.5 months instead of six months. Additionally, production time went from 12 hours to three hours. (Szondy, 2017). • Made in Space Inc., a company that aims to build big structures in space has set a record by earning a Guinness World Record for "World's Longest 3D Printed Nonassembled Piece. Using 3D printing could greatly expand humanity's space-exploration efforts by allow voyaging spacecraft to make what they need on the go (Wall, 2018). • Boeing expects to shave $2 to $3 million off each 787 Dreamliner's manufacturing costs by 2018 by using 3D-printed titanium. The company has teamed up with Norwegian company Norsk Titanium to create the first printed structural titanium components for a plane (Moon, 2017). •GE uses 3D-printed fuel nozzle on their bestselling Leap engine. The company produced the first fully 3D-printed working miniature jet engine capable of revving up to 33,000 RPM (GE Reports, 2017).
Automotive	• Ford creates complex molds in a fraction of the cost and time. • BMW makes ergonomic and lighter tools for its assembly line.
Medical and Healthcare	• Stratasys has partnered with medical institutes to print accurate replicas of patients' hearts to assist doctors in surgery preparation. • The nonprofit E-Nable facilitates access to low-cost prosthetics.
Logistics	• Maersk has explored installing a 3D printer aboard ships to manufacture spare engine parts on demand. • UPS is creating small factories in its warehouses to deliver parts as customers need them, reducing the need for large inventories.
Consumer Products	• L'Oréal uses the technology to produce living skin for product testing. • Nike uses 3D printing in their sports shoes. Nike's Flyprint sneaker uses a 3D printed intricate web that replaces typical textiles (Horaczek, 2018). • Vowsmith uses 3D printing in rapid mass customization of personalized wedding rings, driving fulfillment towards 72 hours after orders are received (3D Systems, 2015).
Construction	•The printing of full houses is one other leapfrog development. 3D printers are capable of extruding concrete, they make it possible to build different structures of varying complexity, from houses to bridges to skyscrapers. •Manufacturers of 3D printers for printing houses are Apis Cor (a Russian company), BatiPrint, Italian manufacturer WASP, KamerMaker (Ultimaker & DUS Architects), WinSun created the first 3D printed house in 2014, Silicon Valley-based company Cazza Construction and many others (Jamie, 2018).
Disaster Relief	• Students expected to reflect individually on course topics may be asked to maintain a blog. •The nonprofit Field Ready uses the technology to print humanitarian supplies in the field.
Wildlife Conservation	•Biotech start-up Pembient produces and sells artificial rhinoceros horns, seeking to curb poaching.

the network are fast turnaround times and competitive prices. Users upload their designs, specify their requirements and receive manufactured parts (3D Hubs, 2018). Other platforms, such as Materialise Onsite or 3D-Creation Lab, enable users to upload CAD files, which are used to manufacture 3D printed objects that are then shipped to the users. Major 3D printing companies (Sculpteo, Shapeways, Stratasys and 3D Systems) operate online marketplaces where designers can upload 3D models of objects that users can buy, have 3D printed and delivered. The companies have integrated direct manufacturing at the core of their business. A typical cloud platform for 3D printer management is 3DPrinterOS. Watkin. (2018) reports that 3DPrinterOS Partners with Microsoft to Make 3D Printing Easier. The partnership aims to use Microsoft's Azure cloud computing service to make the 3D printer management platform more powerful and scalable. Services provided by 3DPrinterOS include managing gcode, analytics, user and file management, and printer queues from a single interface (Stevenson, 2018). The company has been more focused on the education and institutional market, is now targeting enterprises. Cloud based platforms make it easy to maintain and deploy 3D. It enables printing on demand from any location worldwide. Cloud computing and the Internet of things (IoT) are transforming industries and creating new opportunities. Cloud and web service companies such as Microsoft Azure, Amazon Web Services (AWS), Ocean Drop and Google Cloud are providing opportunities to innovative businesses. A variety of building blocks such as database services, networking, and messaging are provided under one roof. Cloud platforms offered by such companies make it relatively easy to test new ideas, deploy, scale up and manage applications. The entry start-up costs for new businesses are low because of the innovative billing systems that are based on usage of selected services.

Cloud based services allow consumers to customize objects or designs uploaded on the platforms by other users using simplified web-based customization software, and order the resulting items as 3D printed unique objects. The process of manufacturing customized products at the efficiency of mass production while maintaining the low price allowed by mass production is called mass customization. In order to implement mass customization a product configurator, a well-planned configurable product platform and a flexible manufacturing technology are essential (Paulraj, 2014). It aims at offering customized products. The objects produced with mass customization are often called "made to order" or "built to order." Mass customization requires 3 essential elements (Recrosio, 2017):

1. Capturing your consumer's data efficiently (for this, you need to build the right interface);
2. Transforming this data to use it for your design;
3. Producing the right cost for your market.

Mass customization is essential for companies to stay competitive, and 3D printing is perfectly suited to such individualized products. Ultimately, this creates new business potentials and competitive advantages for companies, elevating their position on the value chain. Possible areas that can benefit from mass customization are the automation and robotics industry, patient-specific prosthetics, footwear and personalized jewelry. In automation and robotics, standardized components such as robotic arms meet tools such as gripping systems, which are tailored to each work-piece that has to be moved. Some innovative companies offering mass customization are Protos Eyewear (3D printed spectacles for end-users), ThatsMyFace (figurines that are customised with customers' face), Mymo, Zazzy, and Vowsmith (offering 3D printed jewelry).

Economic Implications of 3D Printing

3-D printing has notable economic and social implications. The technology offers the potential for job creation and enhanced entrepreneurship, as well as increased efficiencies in manufacturing. Other opportunities related to social impacts include improved access to products and services, increased opportunities for social inclusion, and reductions in workplace accidents (BSR, 2015). However, possible risks associated with 3D printing are job loss and/or the relocation of jobs due to automation and localized production. 3-D printing permits customization. This may allow for innovations that create more opportunities for social inclusion in consumer products. Table 3 shows the social implications and challenges of 3D printing.

The impact of rapid prototyping and rapid tooling has been found to be relatively limited compared to direct manufacturing and home fabrication, which have a highly disruptive potential. Although more value can be created, capturing value can become extremely challenging (Rayna and Striukova, 2016). Finding a suitable business model is critical. Two characteristics of 3D printing facilitate its disruptive

Table 3. Social implications and challenges of 3D printing

Opportunities	Challenges
• Cost and time savings in manufacturing	• Loss of jobs due to increased automation
• Increased consumer access to products and services	• Students transfer of jobs due to localized manufacturing
• Greater opportunities for social inclusion	• New health concerns such as toxic fumes from the technology
• Job creation and opportunities for entrepreneurship	• Access to 3-D-printed weapons
• Reduction in workplace accidents	• Need to update technical skills

Source: (BSR, 2015)

potential: Firstly, it enables direct production of physical objects from digital design data, which also provides new opportunities for freedom of design. Additionally, customized products can be manufactured without the high surpluses conventionally connected with one-of-a-kind manufacturing (Jiang et al, 2017). Current products or production processes can be substituted by additive manufacturing, which makes established companies to be anxious about new competitors (Gausemeier et al., 2011). Secondly, on the consumer side, additive manufacturing allows private and industrial users to design and produce their own goods, promoting the rise of the 'prosumer' idea (Birtchnell and Urry, 2013), which further enhances the competitive threat proposed by 3D printing to established firms.

CLOUD BASED 3D PRINTING BUSINESS MODELING

3D printing is undoubtedly commanding a lot of hype on social media, private and public organization, as well as the manufacturing industries. The technology has an impact on both social communities and the manufacturing sector. More opportunities for business and social good are embedded in the technology, but there is need for innovative business modeling in order to unlock the potential. Research in terms of product and process innovation through 3D printing has reached maturity and has been widely accepted in all the sectors of the economy. However, more sectors appreciate the technology just as an amazing trend but are battling to identify niche profitable markets for rolling out the technology. Business model innovation is the missing link in many sectors of the economy. A good business mode defines how a company creates and captures a value. Rayna and Striukova (2016) identify four phases of the adoption of 3D printing, notably, rapid prototyping, rapid tooling, digital manufacturing, and home fabrication.

Another notable hindrance to successful adoption of 3D printing is limited access to the 3D printers and knowledge of using them. 3D printers come in various models and sizes whose cost varies from low to extremely high. Since the technology is relatively not meant for mass production, it does not make economic sense to adopt every type of 3D printer. This is where cloud computing helps through cloud-based platforms for sharing resources, collaborating and managing 3D printers. The geographical distribution of 3D printing facilities is skewed in favor of big economies such as the European, American and Asian continents. Smaller economies are highly disadvantaged and there is need for a synergistic approach in order to tap the value of 3D printing technologies. The cloud manufacturing model enables service consumers to manufacture parts through Infrastructure-as-a-Service (IaaS), Platform-as-a-Service (PaaS), Hardware-as-a-Service (HaaS), and Software-as-a-Service (SaaS).

Business Modeling Process

According to Osterwalder et al (2005), a business model describes the value an organization offers its customers and illustrates the capabilities and resources required to create, market and deliver this value and to generate profitable, sustainable revenue streams. Figure 1 illustrates the flowchart for business modeling process, which is adopted in modelling cloud-based 3D printing business. The iterative individual steps will be illustrated in the actual modelling process.

The following questions are important when one begins to design a business model (MaRS, 2012):

1. How do you acquire customers? Briefly describe the steps involved, the amount of time required, the typical value of a deal and the stakeholders required (including the people on your side and the customers' side) to sign a new deal.
2. After you have landed a new customer, how do you plan to relate to that customer and manage the relationship (if at all)?
3. How do you charge your customers? What is your revenue model?
4. How much do you charge your customers? Can you calculate your revenues for the next month, quarter and year?
5. What assets are available to you or under your control?
6. Who are your key partners?
7. What key activities do you need to engage in to deliver your value proposition?
8. What are your fixed costs?
9. What are your variable costs? Can you calculate your total cost for the next month, quarter and year?
10. Does your revenue forecast demonstrate increased profitability toward the end of the forecast period?

The Business Model Canvas is the recommended and widely adopted tool for business modelling by entrepreneurs. According to Osterwalder & Pigneur (2010) the business model canvas is a visual design that simplifies the understanding and approach to business model analysis. The tool consists of nine interrelated building blocks as illustrated in Figure 2. Literature agrees on a broad consensus around four critical components of a business model: value proposition, value creation, value capture, and value delivery (Shafer et al, 2005; Teece, 2010; Osterwalder et al., 2005; Holm et al., 2013; Hvilsom, 2012). Figure 3 summaries the key components of a business model.

Cloud Based 3D Printing Model

The business model developed herein was inspired by multisided platforms (MSPs) and open-source business models. Multisided platforms (MSPs) are technologies, products or services that create value primarily by enabling direct interactions between two or more customer or participant groups (Hagiu and Spulber, 2013). Most platforms bring together service providers and service takers, but the platforms do not necessarily provide these services themselves. Other platforms connect sellers and buyers of products, but they do not sell the products themselves; they do not own the inventory, let alone manufacturing any of the products sold (Uenlue, 2017). Prominent examples of MSPs and the participants they connect are shown in Figure 4. Open-source refers to the principles and methodology of open source software

Figure 1. Business Model Process (Source: MaRS, 2012)

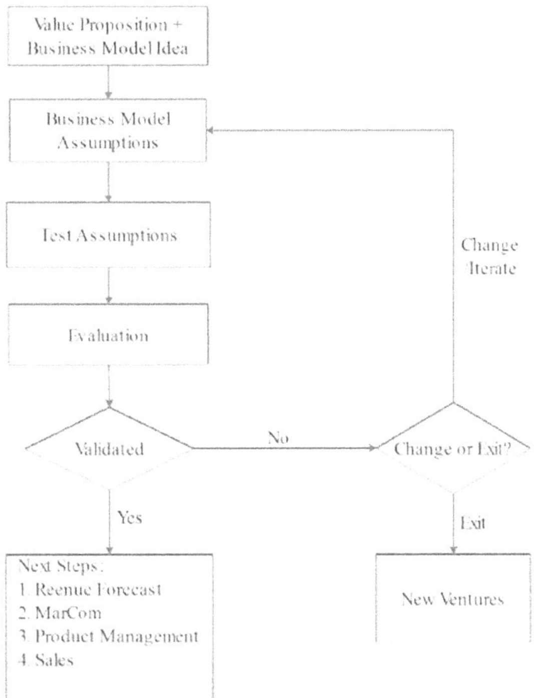

Figure 2. The business model canvas (Source: Osterwalder & Pigneur, 2010)

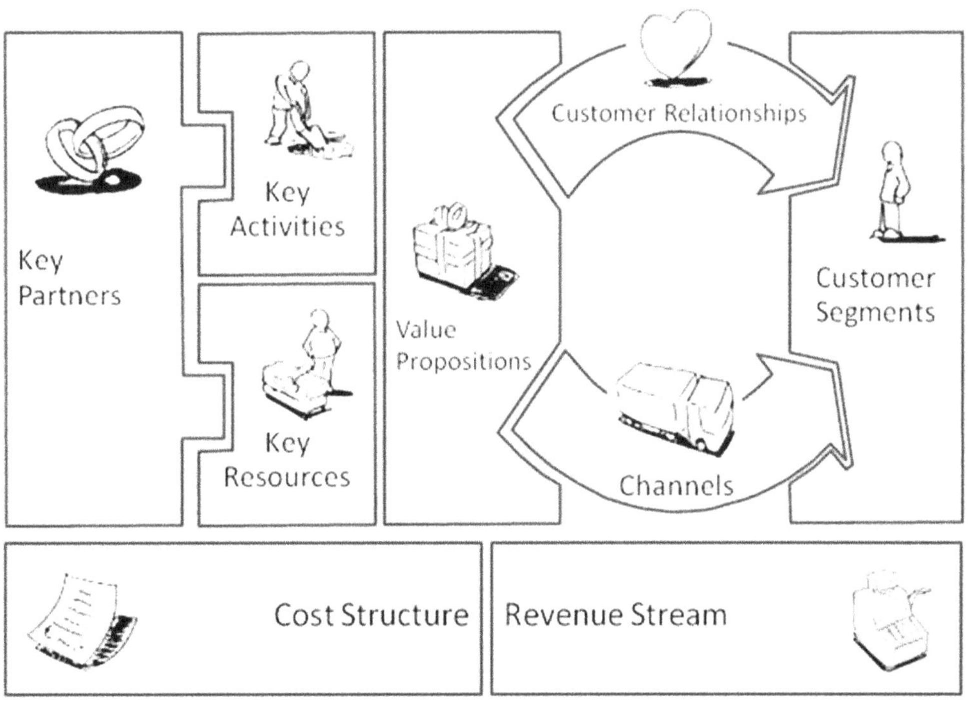

Figure 3. Key components of business models (Source: Rayna and Striukova, 2016)

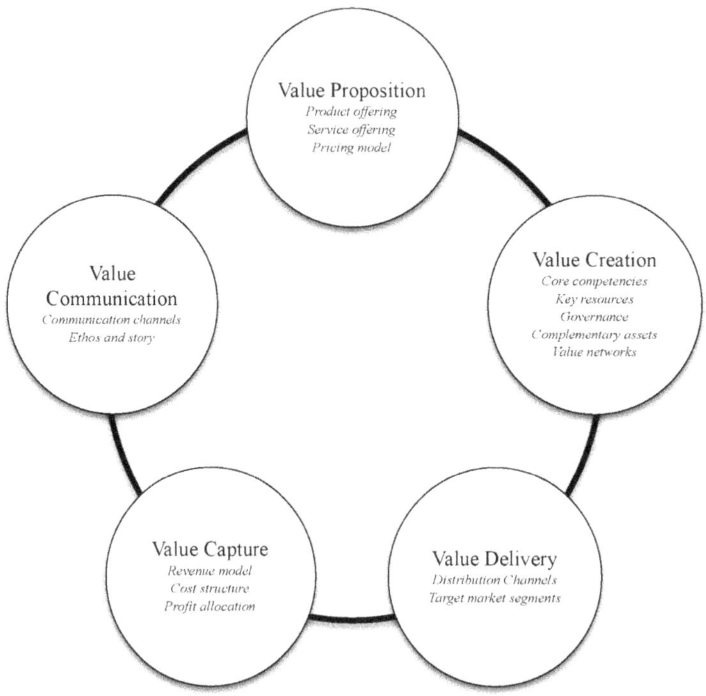

(OSS) and open source hardware (OSH) that are continuously improved, modified and redistributed. The first essential element of an open source business plan is to maximise the use value of the intellectual property in question (in this case, 3D printers and open-source cloud platform). Collaboration and resource sharing by a community of users/co-developers enhance the use value of 3D printer facilities. Cloud computing provides many business models that can be adopted in order to 3D print parts; Infrastructure-as-a-Service (IaaS), Platform-as-a-Service (PaaS), Hardware-as-a-Service (HaaS), and Software-as-a-Service (SaaS).

Key Partners

The cloud-based 3D printing business (3DCloud) has four categories of key partners; maker community (co-designers/co-developers), manufacturing industry, technology partners (providing web service platforms and online payment gateways), and those that help the business to be an established company (e.g. investors and lobbyists).

Key Activities

Platform businesses rely upon positive indirect network effects in order to achieve competitive advantage. Key activities for 3DCloud should revolve around enhancing positive indirect network effects whilst reducing negative ones. The activities are listed as follows:

1. Smoothening interactions and or/ collaborations;

Figure 4. Prominent examples of MSPs and the participants they connect (Source: Uenlue, 2017)

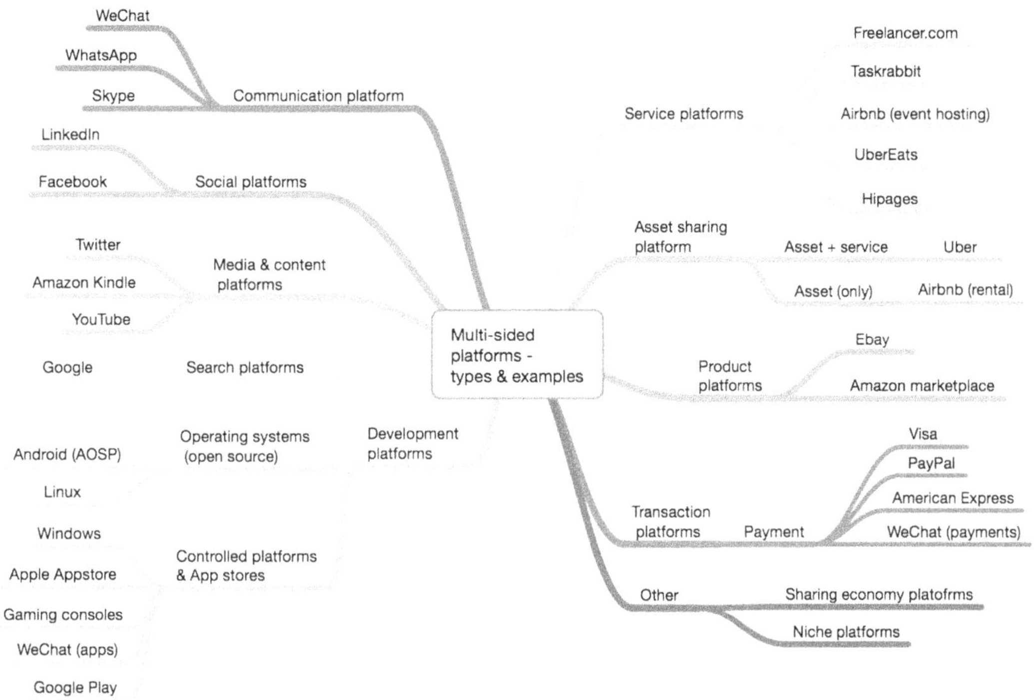

2. Scale beyond critical mass by encouraging more membership;
3. Engage the participants so that they hook on the 3D printer network and cloud platform. All types of participants should feel valued and welcome in the 3D printing community;
4. Refine value proposition through offering discounted rates to community members and regular platform users. There should be special rates for maker communities and industry type stakeholders;
5. Analyze data and improve quality of business offering by mitigating all types of inconsistences, poor quality of workmanship and service delivery;
6. Enhance the 3D printing network by looking out for complementary value propositions (e.g. makerspace 3D printer financing, new customer segments, etc.).

Key Resources

Key resources for 3DCloud are the 3D printers, network effects and skilled personnel to operate the machines and continuously improve the network platform. The quality and capacity of printers is very important in order to offer variety of printing technology and quality of printouts, whilst the network platform ensures smooth management and control of common printing facilities. CAD file generation, storage and retrieval, manipulation, the algorithms and the capability to analyse and gain insights are essential. The complexity grows with the size of the network. Other resources include various apps that enhance access to the service, architecture of the cloud platform, 3DCloud as a brand that needs to be promoted, and venture capital to keep the business growing.

Value Proposition

A product's value proposition is a statement of the functional, emotional and self-expressive benefits delivered by the brand that provide value to the target customer (Aaker, 1996). 3D-Cloud is a multi-sided platform (MSP), and as such it has to have a value proposition to both sides, the maker community as manufacturing industries. Value propositions for the maker community are that of Shared 3D printers and printing facilities, flexible 3D printing capacity, collaboration and openness. The maker community are able to generate income through specialized designs, printing for online stores, using the 3D-Cloud as a provision for digital economy jobs, ease of membership to the network, enjoy low startup costs, and resolution of any issues arising from the ecosystem. Manufacturing industries benefit from fast cycle times when they order 3D prints through a robust network of 3D printer facilities, high quality of workmanship and reliable printers that can handle the work at low costs, convenience, and easy transactions. The propositions fall into what economists call reduction of search costs and transaction costs, which is the predominant feature of multi-sided platforms.

Customer Segments

3DCloud benefits from segmenting both sides of the market, i.e. makerspaces and manufacturing industry. A combination of classic market segmentation combined with highly-targeted micro segmentation is used. Maker communities can be micro-segmented into demographic, socio-economic, micro-geographic, behavioural and situational analysis. Manufacturing industries can also be micro-segmented according to usage patterns, type of usage (prototyping, parts repair, consumer product), industry segment (jewelry, fashion, toys, manufacturing, aerospace, automotive, medical, construction), economic, and micro-geographic analysis.

Customer Relationships

Four elements make up 3DCloud's customer relations, i.e. maker community, industry, general public, and standards/regulatory authorities. Makers & industry share the following concerns:

1. Management of any liability and risk from design through to manufacture;
2. Management of undesirable practice (on both sides, maker community and industry) and improve rules of use and continuous engagement;
3. Expeditiously dealing with customer issues in an amicable manner;
4. Transparent costing and quotations;
5. Respect right to customer data protection, security and privacy;
6. Portraying good corporate image through social media and other forms of publicity;
7. Ability to generate income through 3DCloud as a means to provide employment in the digital economy era.

The general public and regulatory authorities have concerns for:

1. Communal/ social footprint
2. Adherence to standards
3. Safety and environmental concerns

Channels

Initial awareness and customer acquisition can be rolled out through free maker fares and exhibitions at the shared 3D printing facilities or separate venues for showcasing 3D printed wares. Demonstrations may also be made to industries as well as giving special rates to new industry type clients. Mass media campaigns may be used to encourage maker communities to subscribe to 3DCloud. Other channels of awareness and publicity include word of mouth, social media platforms, local campaigns, apps / app stores and any other communication channels.

Cost Structure

Customer acquisition costs (CAC) constitute the biggest cost element for 3DCloud business model, which is the case with many other online platforms. As many platform-based 3D printing businesses join the space, the major weapon for winning the battle becomes the customer acquisition "subsidies" (on both makers' and industry's sides). Another huge cost factor is the weighted average cost of capital (WACC). Other elements of the structure are the legal and settlement costs, lobbying, and technology development (R&D) costs.

Revenue Model

Revenue model for 3DCloud comes from 3D printing fees charged to the industry and 'prosumers' who are part of the maker community. The low costs of printing in the 3DCloud network comes from cost savings to the makers and industry on 3D printers' capital costs, no factory license and inspection and employee entitlements. Collaborative and shared resource approach eliminate the need to own 3D printers by individuals and industries. The complete 3DCloud Business Model Canvas is shown in Figure 5, whilst the actual model illustrating the flow of major activities is shown in Figure 6.

CONCLUSION

The article looked into major trends of 3D printing (i.e., additive manufacturing - AM). Literature has revealed wide practical applications of the technology across all economic sectors. 3D printing is transforming the ways of doing business in the current digital economy. Manufacturing has become democratised to the extent of making critical printouts such as prosthetic legs, space shuttle parts, food and houses. There is a general consensus about lack of business model innovation, which has the rigor to make 3D printing adoption by the corporate world a major economic boom. 3D printing has been widely accepted and is making significant impacts as a socially relevant technology, as well as a manufacturing

Figure 5. 3DCloud Business Model Canvas

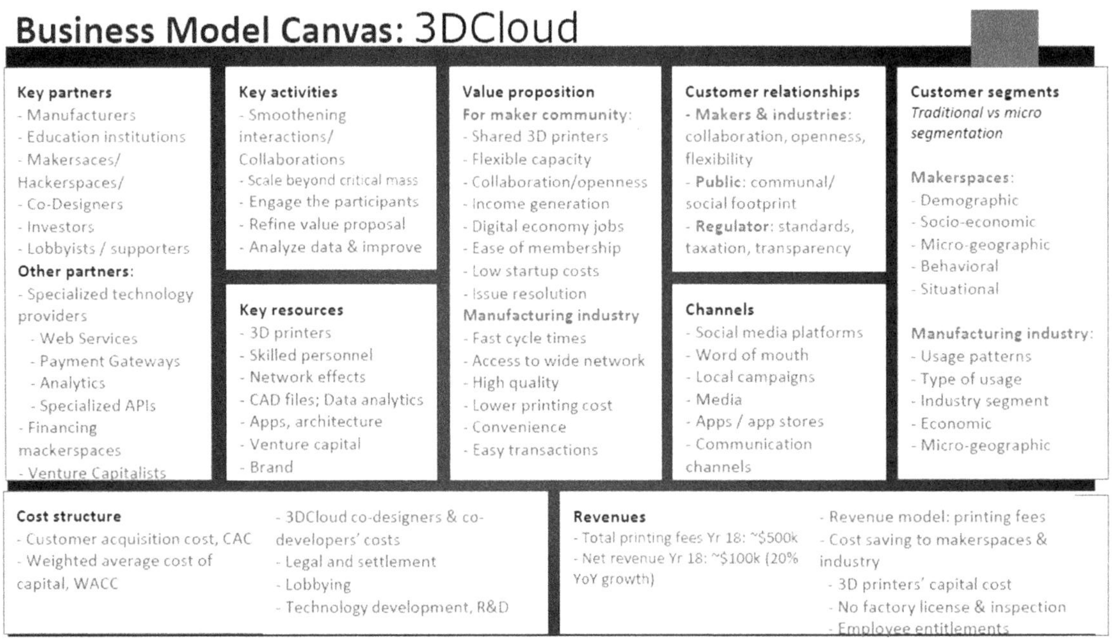

Figure 6. 3DCloud Business Model

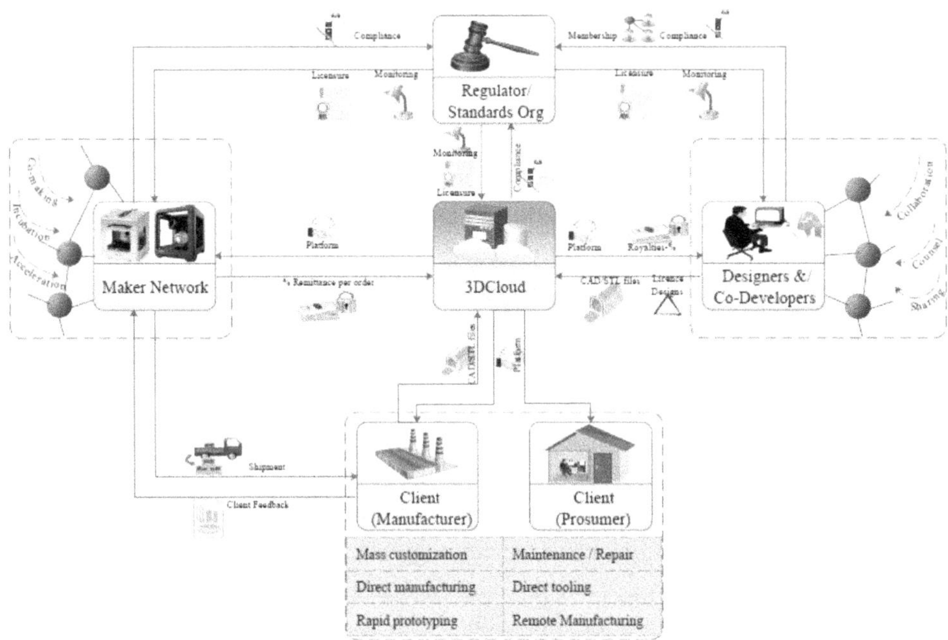

technology. Previously, the technology was mainly used for rapid prototyping purposes, but significant developments have happened, sufficient enough to make the technology production ready. Mass customization is another interesting niche area where the technology proves most suitable than dedicated mass production machinery. Viable ideas that featured in most parts of the article and literature include printing on demand, 3D printing centre and designing 3D printing files.

Inspired by Multi-Sided Platform (MPS) businesses, the article developed a cloud-based 3D printing business model (3D-Cloud). The Business Model Canvas was used as a toolkit for modelling 3D-Cloud. 3D printing business adopts an open business model with a consumer-centric value logic as opposed to the traditional manufacturer-centric business approach. Localized manufacturing is becoming more viable and a preferred option because of 3D printing. 3D-Cloud is a combination of a cloud platform for centralised control and management of collective 3D printer facilities, and a network of physical 3D printers. Key partners are the maker community (co-designers, co-developers, education institutions), manufacturing industry and promoters or lobbyists. The maker community and the manufacturing industry are the major market segments. The developed model caters for the needs of today's customers that are living in a digital world. Digitization promotes mass customization where customers can instantly download apps and customize their preferred product designs. The 'prosumer' concept has been noticed to be growing reality in today's digital economy; new generations delight in instant gratification when they design and make products for own use. Cloud based 3D printing platforms and networks of physical 3D printing facilities promote collaboration, openness, sharing 3D printers. Benefits to the user community are income generation, means of employment in the modern digital economy, ease of membership, flexible 3D printing capacity and low start-up costs.

The cloud manufacturing (CM) model, which is inspired by cloud computing (CC), promotes a customer-centric service type of business models. This creates more value and enhanced innovation. Merging cloud-based service-oriented business models with the social community approach (maker community) encourages rapid innovation of best practices, technology development and service delivery. There are various success cases of online platforms that are centred on a similar model. Open-source can be coupled with innovative monetization models to encourage more participation. Dispersing creativity brings significant benefits to society. This can be achieved collectively through participation from maker community, government, NGOs, corporations and venture capitalists (for financial muscle). Methodological approaches such as design thinking, camps, workshops, webinars and exhibition shows compliment the efforts to enhance collaboration, exchange of knowledge and the contagious, as well as the cumulative effect of innovation.

REFERENCES

Aaker, D. A. (1996). *Building Strong Brands*. New York: The Free Press.

Auricchio, F., & Marconi, S. (2016). 3D printing: Clinical applications in orthopaedics and traumatology. *EFORT Open Rev*, *1*(5), 121–127. doi:10.1302/2058-5241.1.000012 PMID:28461938

Baumann, F. (2017). 3D Printing-as-a-Service for Collaborative Engineering. Retrieved from https://elib.uni-stuttgart.de/bitstream/11682/9356/5/baumann-thesis-17.pdf

Baumann, F., & Roller, D. (2016). 3D Printing Process Pipeline on the Internet. In *8th ZEUS Workshop*, Vienna, Austria (Vol. 1562, pp. 29-36). Retrieved May 14, 2018, from https://pdfs.semanticscholar.org/d8c0/ff8bd94a6df4468f64f8bc860cde2d074fe5.pdf

Baumann, F. W., & Roller, D. (2017). Additive Manufacturing, Cloud-Based 3D Printing and Associated Services-Overview. *Journal of Manufacturing and Materials Processing*, *1*(15). doi:10.3390/jmmp1020015

Berman, B. (2012). 3-D printing. The new industrial revolution. *Business Horizons*, *55*(2), 155–162. doi:10.1016/j.bushor.2011.11.003

Birtchnell, T., & Hoyle, W. (2014). *3D Printing for Development in the Global South: The 3D4D Challenge*. Basingstoke, UK: Palgrave Macmillan. Retrieved May 18, 2018, from http://ro.uow.edu.au/cgi/viewcontent.cgi?article=2298&context=sspapers

Birtchnell, T., & Urry, J. (2013). 3D, SF and the future. *Futures*, *50*, 25–34. doi:10.1016/j.futures.2013.03.005

Bogers, M., Hadar, R., & Bilberg, A. (2016). Additive manufacturing for consumer-centric business models: Implications for supply chains in consumer goods manufacturing. *Technological Forecasting and Social Change*, *102*, 225–239. doi:10.1016/j.techfore.2015.07.024

Bowyer, A. (2014). 3D Printing and Humanity's First Imperfect Replicator. *3D Printing and Additive Manufacturing, 1*(1), 4–5.

Business for Social Responsibility (BSR). (2015). 3-D Printing: Sustainability Opportunities and Challenges. Retrieved April 25, 2018, from https://www.bsr.org/reports/BSR-Report-3D-Printing-Sustainability-Opportunities-Challenges-2015.pdf

Department of Mechanical Engineering, University of Wisonsin-Madison. (2016). Winter 2016 Newsletter. Retrieved from https://www.engr.wisc.edu/app/uploads/2016/04/me-winter-2016-newsletter.pdf

Ford, S., & Minshall, T. (2017). 3D printing in teaching and education: A review of where and how it is used. Retrieved May 5, 2018, from https://www.researchgate.net/publication/320617391_3D_printing_in_teaching_and_education_A_review_of_where_and_how_it_is_used

Frauenfelder, M. (2014). Make: Ultimate Guide to 3D Printing 2014. Sebastopol, CA: Maker Media.

Gibb, A., & Abadie, S. (2014). *Building Open Source Hardware: DIY Manufacturing for Hackers and Makers* (1st ed.). Boston, MAA: Addison-Wesley Professional.

Grujovi'c, N., Radovi'c, M., Kanjevac, V., Borota, J., Grujovi'c, G., & Divac, D. (2011). 3D Printing Technology in Education Environment. In *Proceedings of the 34th International Conference on Production Engineering*, Nis, Serbia (pp. 29–30).

Hagiu, A. (2006). Pricing and Commitment by Two-Sided Platforms. *The Rand Journal of Economics*, *37*(3), 720–737. doi:10.1111/j.1756-2171.2006.tb00039.x

Hagiu, A., & Spulber, D. (2013). First-party content and coordination in two-sided markets. *Management Science*, *59*(4), 933–949. doi:10.1287/mnsc.1120.1577

Holm, E. J. V. (2015). Makerspaces and Contributions to Entrepreneurship. *World Conference on Technology, Innovation and Entrepreneurship. Procedia: Social and Behavioral Sciences*, *195*, 24–31. Retrieved from https://www.sciencedirect.com/science/article/pii/S1877042815036460 doi:10.1016/j.sbspro.2015.06.167

Horaczek, S. (2018). Nike hacked a 3D printer to make its new shoe for elite marathon runners. *Popsci*. Retrieved from https://www.popsci.com/nike-3d-printed-sneakers

Hvilsom, L. (2012). *A study of understandings and interpretations of business models and a single case study of Liz Claiborne* [thesis]. Copenhagen Business School. Retrieved from http://studenttheses.cbs.dk/bitstream/handle/10417/3850/line_hvilsom.pdf?sequence=1

Irwin, J. L., Oppliger, D. E., Pearce, J. M., & Anzalone, G. (2015). Evaluation of RepRap 3D Printer Workshops in K-12 STEM. In *Proceedings of the 122nd American Society for Engineering Education-ASEE Annual Conference and Exposition*, Seattle, WA. 10.18260/p.24033

Jamie, D. (2018). The 11 Best 3D Printed House Companies – 3D Printing Entire Buildings. https://www.3dnatives.com/en/3d-printed-house-companies-120220184/

Jiang, R., Kleer, R., & Piller, T. F. (2017). Predicting the future of additive manufacturing: A Delphi study on economic and societal implications of 3D printing for 2030. *Technological Forecasting and Social Change*, *117*, 84–97. doi:10.1016/j.techfore.2017.01.006

Jones, R., Haufe, P., Sells, E., Iravani, P., Olliver, V., Palmer, C., & Bowyer, A. (2011, January). RepRap-the Replicating Rapid Prototyper. *Robotica*, *29*(01), 177–191. doi:10.1017/S026357471000069X

Knill, O., & Slavkovsky, E. (2013). Visualizing mathematics using 3D printers. In C. Fonda, E. Canessa & M. Zennaro (Eds.), Low-Cost 3D Printing for science, education and Sustainable Development. International Centre for Theoretical Physics.

Koren, Y. (2010). *The Global Manufacturing Revolution: Product–Process–Business Integration and Reconfigurable Systems*. Hoboken, NJ: Wiley. doi:10.1002/9780470618813

Lacey, G. (2010). 3D printing brings designs to life. *Tech Directions*, *70*(2), 17–19.

LaSelle, R. (2017). Meeting mass customization demands with 3D printing. *Jabil*. Retrieved from https://www.jabil.com/insights/blog-main/meeting-mass-customization-demands-with-3d-printing.html

Lipson, H. (2007). Printable 3D models for customized hands-on education. *Paper presented at Mass Customization and Personalization (MCPC)*, Cambridge, MA.

MaRS. (2012). Fundamentals of entrepreneurial management: Workbook 2 Business Model Design. Retrieved from https://www.marsdd.com/wp-content/uploads/2012/12/Business-Model-Design-WorkbookGuide.pdf

Mellor, S., Hao, L., & Zhang, D. (2014). Additive manufacturing: A framework for implementation. *International Journal of Production Economics*, *149*, 194–201. doi:10.1016/j.ijpe.2013.07.008

Moon, M. (2017). Boeing uses first FAA-approved 3D-printed parts for the 787. https://www.engadget.com/2017/04/11/boeing-faa-approved-3d-printed-metals-787/

Osterwalder, A., & Pigneur, Y. (2010). *Business Model Generation, A Handbook for Visionaries, Game Changers, and Challengers*. Hoboken, NJ: John Wiley & Sons, Inc.

Osterwalder, A., Pigneur, Y., & Tucci, C. (2005). Clarifying Business Models: Origins, Present and Future of the Concept. *Communications of the Association for Information Systems*, *16*(1), 1–25. Retrieved from http://users.jyu.fi/~mieijala/luama/materiaali/1/Clarifying%20Business%20Models-%20Origins%20Present%20and%20Future%20of%20the%20C.pdf

Paulraj, T. (2014). Implementing Mass Customization Using SAP Variant Configuration and a 3D Printer [master's thesis]. Retrieved from https://www.tritonia.fi/fi/e-opinnaytteet/tiivistelma/6126/Implementing+Mass+Customization+Using+SAP+Variant+Configuration+and+a+3D+Printer

Pearce, J. M., Blair, C. M., Kaciak, K. J., Andrews, R., Nosrat, A., & Zelenika-Zovko, I. (2010). 3D printing of open source appropriate technologies for self-directed sustainable development. *Journal of Sustainable Development*, *3*(4), 17–28. doi:10.5539/jsd.v3n4p17

Petersen, E. E., Kidd, R. W., & Pearce, J. M. (2017). Impact of DIY home manufacturing with 3D printing on the toy and game market. *Technologies*, *5*(3), 45. doi:10.3390/technologies5030045

Petrick, I. J., & Simpson, T. W. (2015). 3D printing disrupts manufacturing: how economies of one create new rules of competition. *Research Technology Management*, *56*(6), 12–16. doi:10.5437/08956308X5606193

Piller, F., Weller, C., & Kleer, R. (2015). Business models with additive manufacturing: opportunities and challenges from the perspective of economics and management. In C. Brecher (Ed.), *Advances in Production Technology* (pp. 39–48). Springer International Publishing. doi:10.1007/978-3-319-12304-2_4

Ponfoort, O., Ambrosius, W., Barten, L., Duivenvoorde, G., van den Hurk, L., Sabirovic, A., & Teunissen, E. (2014). *Successful Business Models for 3D Printing: Seizing Opportunities with a Game Changing Technology* (pp. 34–60). Utrecht: Berenschot.

Pryor, S. (2014). Implementing a 3D Printing Service in an Academic Library. *Journal of Library Administration*, *54*(1), 1–10. doi:10.1080/01930826.2014.893110

Rayna, T., & Striukova, L. (2016). From rapid prototyping to home fabrication: How 3D printing is changing business model innovation. *Technological Forecasting and Social Change*, *102*, 214–224. doi:10.1016/j.techfore.2015.07.023

Recrosio, E. (2017). Benefits of 3D Printing: Mass Customization. https://www.sculpteo.com/blog/2017/01/11/benefits-of-3d-printing-mass-customization/

Ren, L., Wang, S., Shen, Y., Hong, S., Chen, Y., & Zhang, L. (2016). 3D Printing in Cloud Manufacturing: Model and Platform Design. In *ASME International Manufacturing Science and Engineering Conference* (Vol. 2). 10.1115/MSEC2016-8669

Reports, G. E. (2017). 5 Ways GE Is Changing The World With 3D Printing. https://www.ge.com/reports/5-ways-ge-changing-world-3d-printing/

Savonen, B. L., Mahan, T. J., Curtis, M. W., Schreier, J. W., Gershenson, J. K., & Pearce, J. M. (2018). Development of a Resilient 3-D Printer for Humanitarian Crisis Response. *Technologies*, *6*(1), 30. doi:10.3390/technologies6010030

Schelly, C., Anzalone, G., Wijnen, B., & Pearce, J. M. (2015). Open-Source 3-D Printing Technologies for Education: Bringing Additive Manufacturing to the Classroom. *Journal of Visual Languages and Computing*, *28*, 226–237. doi:10.1016/j.jvlc.2015.01.004

Schelly, C., Anzalone, G., Wijnen, B., & Pearce, J. M. (2015, June). Open-source 3-D printing technologies for education: Bringing additive manufacturing to the classroom. *Journal of Visual Languages and Computing*, *28*, 226–237. doi:10.1016/j.jvlc.2015.01.004

Scott, J., Gupta, N., Weber, C., Newsome, S., Wohlers, T., & Caffrey, T. (2012). *Additive manufacturing: status and opportunities*. Washington, DC: Science and Technology Policy Institute. Retrieved from https://cgsr.llnl.gov/content/assets/docs/IDA_AdditiveM3D_33012_Final.pdf

Sells, E., Bailard, S., Smith, Z., Bowyer, A., & Olliver, V. (2007). RepRap: The Replicating Rapid Prototyper-Maximizing Customizability by Breeding the Means of Production. In *Proceedings of the World Conference on Mass Customization and Personalization*, Cambridge, MA, October 7-10.

Shafer, S. M., Smith, H. J., & Linder, J. C. (2005). The power of business models. *Business Horizons*, *48*(3), 199–207. doi:10.1016/j.bushor.2004.10.014

Slavkovsky, E. (2012). *Feasibility study for teaching geometry and other topics using three-dimensional printers* [thesis]. Harvard University.

3DSystems. (2015). 3D Printing helps jewelry start-up ride the mass customization wave. Retrieved from https://www.3dsystems.com/learning-center/case-studies/3d-printing-helps-jewelry-start-ride-mass-customization-wave

Szondy, D. (2017). Lockheed Martin using 3D-printed parts in military satellites. *Newatlas*. Retrieved from https://newatlas.com/lockheed-martin-3d-printed-military-satellite/48776/

Teece, D. J. (2010). Business models, business strategy and innovation. *Long Range Planning*, *43*(2), 172–194. doi:10.1016/j.lrp.2009.07.003

Uenlue, M. (2017). Platform business model [part 1]: the most revolutionary business model right now. *Innovationtactics*. Retrieved from http://www.innovationtactics.com/platform-business-model-1/

Ventola, C. L. (2014). Medical Applications for 3D Printing: Current and Projected Uses. *P&T*, *39*(10), 704–711. PMID:25336867

Wall, M. (2018). Made in space sets Guinness World Record for Longest 3D-Printed Piece. *Space*. Retrieved from https://www.space.com/39790-made-in-space-3d-printing-guinness-world-records.html

Wohlers, T. (2016). *Wohlers Report 2016*.

This research was previously published in the International Journal of E-Entrepreneurship and Innovation (IJEEI), 8(2); edited by Charice Hayes, pages 25-43, copyright year 2018 by IGI Publishing (an imprint of IGI Global).

Chapter 6
User Acceptance of Technologies in Their Infancy:
The Case of 3D Printing Business Models

Timo Hartmann
Maastricht University, The Netherlands

Evelyne Vanpoucke
Université Libre de Bruxelles, Belgium

ABSTRACT

This paper adjusts UTAUT2 to study user acceptance of an early stage technology by using the example of 3D printing for home users. The authors' findings show that UTAUT2 is capable of measuring user acceptance and that social influence and effort expectancy are good predictors for 'intention to use' for early stage technologies. Nevertheless, there seem to be significant differences in the antecedents depending on the involvement levels and the DIY mentality of the user. Moreover, this research confirms and extends the applicability of UTAUT2 to early stage technologies while improving the authors' understanding of prospective 3D printing home users.

INTRODUCTION

According to Davis (1993), most system failures are a result of lacking user acceptance rather than poor performance. This lack of user acceptance is also reflected in the status of being one of the most researched streams in information systems (e.g., Benbasat and Barki, 2007; Venkatesh, Davis, & Morris, 2007). In the last decades, literature has described dozens of approaches explaining technology acceptance and its use. In an effort to unify these approaches, Venkatesh, Morris, Davis, and Davis (2003) summarized all important technology acceptance frameworks and defined a 'unified theory of acceptance and use of technology' (UTAUT), refined later as UTAUT2 by Venkatesh, Thong, and Xu (2012). This comprehensive framework is able to explain high variances of technology acceptance and use in both organizational and voluntary user environments. While the UTAUT framework has been tested for different technologies and in different contexts (e.g., Zhou, Lu, & Wang, 2010; Cheng, Yu, Huang, Yu, &

DOI: 10.4018/978-1-5225-9624-0.ch006

Yu, 2011; Raman & Don, 2013), no research confirms its applicability in early stages of the product life cycle, i.e., in early stages of the technology diffusion process (Rogers, 1995) and when the product just started the growth stage of the product life cycle (Taylor & Taylor, 2012). While technologies in later stages of the development cycle equal already known and proven technologies, technologies in early stages can be called 'technologies in its infancy' and might be different in terms of antecedents of user acceptance. The main difference between established and recently introduced technologies is the rarity of applied usage of the latter. Hence, prospective users have little to no technology experience and knowledge. Consequently, the intention to use a technology will be dominated by users' perceptions about the technology rather than experience or deliberated opinions. Because of that, we believe that hedonic motivations and facilitating conditions might be far more important antecedents than others. In addition, we also believe that there might be differences according to involvement level. As suggested by the Elaboration Likelihood Model, technology use might influence perceptions and consequently intentions to use (Petty & Cacioppo, 1984). We will adjust the UTAUT2 framework towards technologies in the infancy stage and will test its applicability in the context of 3D printing (3DP).

3DP is a promising technology in the early stage of its user life cycle. With an annual growth of 25%, a market size of 649 million USD in service and 52 million USD and 100,000 machines sold in domestic use, respectively, 3DP in the home environment is on the rise (Wohlers, 2014). The total market projection in 2018 including the industrial application is estimated to increase by more than four times to 16.2 billion USD in comparison to 2014 (Canalys, 2014). However, there is no widespread use yet while it is seen as a revolutionary and radical technology (Campbell, Williams, Ivanova, & Garrett, 2011). 3DP technology enables the making of three-dimensional solid objects of almost any shape based upon a digital model file. Instead of traditional subtractive manufacturing methods as cutting or drilling, additive manufacturing adds raw material layer by layer without waste (Campbell et al., 2011). There are many signs that 3DP will be used in private households in the next years, creating opportunities for the development of new business models. However, current knowledge about prospective users and its propagation are still in an early stage. We hope to predict user acceptance for this new early stage technology by using the UTAUT2 framework.

Additionally, we will test whether user acceptance might be different according to the required involvement level, i.e., involvement in the design and/or production process. Moreover, this will enable us to test user acceptance for different business models of 3DP, based upon principles of (i) self-design, and/or (ii) self-production. Self-design is about users who design parts themselves. Instead of owning a printer, these people order parts via online platforms or via local 3DP (community) stores (Bradshaw, Bowyer, & Haufe, 2010; Birtchnell, Urry, Cook, & Curry, 2013; D'Aveni, 2013; Petrick & Simpson, 2013). Self-production relates to end customers that download pre-designed files to print them out on an own 3D printer without creative involvement but for utilitarian purposes (Reeves, 2009; Bradshaw et al., 2010; Campbell et al., 2011; Berman, 2012; Birtchnell et al., 2013; D'Aveni, 2013). In alignment with these different involvement stages, we test whether 'do it yourself' users (Bunshaft & Laubner, 2013; Hsu, 2013) as described by Wolf & McQuitty (2013) might be more inclined to use this technology and might be interested in a broader use of this technology.

In summary, this research will address the following gaps: First, it will test the applicability of UTAUT2 to technologies in their infancy by exploring its potential and differences to later stage technologies. Second, although 3DP is promising and revolutionary, technology perceptions of prospective home users are unclear yet. As such, insights into preferences in user involvement will be analyzed. The findings of our research will add to our understanding of a user acceptance model for early stage technologies, but might also provide information on how to set up valuable future business models for this 3DP technology.

The remainder of this paper is divided into five sections. First, the existing literature on 3DP and technology acceptance theories are summarized. In section 2, a framework and hypotheses are developed. In section 3, the research method and design including the constructs of the framework are introduced. The data analysis and findings are discussed in section 4. In section 5, the results are interpreted. Findings and insights are formulated subsequently. Managerial and theoretical contributions as well as limitations and future opportunities of the study conclude the paper.

LITERATURE REVIEW AND HYPOTHESES

Definition and History of 3D Printing

Additive manufacturing originated from the ink-jet industry and has been developed in the 1970s (Dimitrov, Schreve, & De Beer, 2006; Bradshaw et al., 2010). Cognates are rapid prototyping (Silva & Rezende, 2013), rapid manufacturing (Bak, 2003; Reeves, 2009; Berman, 2012), additive layer manufacturing (Reeves, 2009), solid freedom fabrication (Séquin, 2005), and 3DP. The term additive manufacturing is the official industry standard term, whereas 3DP is the de facto industry standard because it is illustrative enough to attract laymen (consumers) in the near future (Berman, 2012). Consequently, the term 3DP is used as a synonym for additive manufacturing.

Many experts recognize the impact of 3DP on the global business. According to Silva and Rezende (2013, p. 6), a 'big wave is coming'. Although already introduced in 1996, printers for private use were not successful because of limited possibilities, poor production results and high printer costs (Wohlers, Gornet, & Caffrey, 2013). Nowadays, these limitations have disappeared, making 3DP an attractive technology for companies and consumers. Starting at a price of $499 (Wohlers et al., 2013), 3DP is an affordable technology for private use (e.g., Malone & Lipson, 2007; Bradshaw et al., 2010; Campbell et al., 2011, Mercer, 2014). Due to these changes, industry experts have little doubt about the promising development of 3DP in the future. Still, some challenges exist. There are issues concerning intellectual property (Bradshaw et al., 2010), output quality, consistency, and complexity for multi-material products (Stemp-Morlock, 2010; Campbell et al., 2011; Berman, 2012; Birtchnell et al., 2013; Fung, 2013), and technology burdens for inexperienced users (Chen, Djajadiningrat, Feijs, Fraser, Hu, Kyffin, & Steffen, 2013).

Review and Hypotheses

Venkatesh (2006) suggests the need for further research in understanding individual technology acceptance. There are several suitable technology acceptance theories. This paper makes use of UTAUT2 by Venkatesh et al. (2012). This framework serves to explain intention to use and subsequent usage behavior of a technology in a consumer context. Its previous version, UTAUT (Venkatesh et al., 2003), is based on a wide-ranging synthesis of earlier technology acceptance research. In total, eight models are combined to create a full-fledged technology acceptance model. Amongst others, constructs out of theory of reasoned action (TRA, Ajzen & Fishbein, 1975), technology acceptance models (TAM, Davis, 1989; TAM2, Venkatesh & Davis, 2000; TAM3, King & He, 2006; Schepers & Wetzels, 2007; Venkatesh &

Bala, 2008), theory of planned behavior (TPB, Ajzen, 1991), and innovation diffusion theory (IDT, Rogers, 1995) are used. Thus, the consolidated UTAUT is seen as a well-meaning and thoughtful framework (Bagozzi, 2007). Venkatesh et al. (2012) extended UTAUT to UTAUT2 by adding more predictors and by relating it to a consumer environment, making UTAUT2 capable to measure 3DP user acceptance.

The reason why UTAUT2 is more suitable than long-serving technology acceptance frameworks is found in the details. The TPB, an extension of TRA, is too general and has to be adapted in many factors such as technology context and antecedents comparable to UTAUT2. As Benbasat and Barki (2007) argue, the TAM and its extensions are too wide-ranging in many aspects so that a theoretical chaos of dozens of TAM extensions exists. Since UTAUT2 fits consumer environments, it will be used to test for user's intention to use an early stage technology.

This research tests user acceptance of 3DP, i.e., for an early stage technology. The user technology diffusion process by Rogers (1995) explains five innovation decision stages: (i) Knowledge; (ii) persuasion; (iii) decision; (iv) implementation, and (v) confirmation. This paper focuses on prospective users in one of the first two stages of a technology diffusion process: i.e. these potential users have a first knowledge about 3DP and have or have not been persuaded. As such, they are faced with making a decision to use or not to use the technology. More specifically, our respondents are exposed to this innovation for the first time. Thus, users only know the basics of the technology without being surrounded by a widespread use in the social environment. Due to the fact that the framework of Rogers (1995) applies to an individual's diffusion process, the community diffusion process of applications cannot be explained. This can be done by the product life cycle, which has four main time stages: (i) Introduction; (ii) growth; (iii) maturity, and (iv) decline (e.g., Taylor & Taylor, 2012). Since 3DP for home users is in the first of these four stages, namely introduction, only 'early adopters' are familiar with this technology (e.g., Rogers, 1995). According to this stream of literature, home 3DP is classified as a recently introduced market with, as yet, low distribution. (Wohlers et al., 2013; Silva & Rezende, 2013). Taking both classifications together, 3DP can be described as a 'technology in its infancy'.

Few studies that used the TAM framework to test user acceptance in the technology diffusion process find 'attitude towards the technology' (Tzeng, 2011) as well as perceived usefulness, adaptability, enjoyment, sociability, companionship, and behavioral control (De Graaf & Ben Allouch, 2013) to be the strongest predictors of intention to use. Looney, Akbulut, and Poston (2008) found self-efficacy to be a strong antecedent for usage intention. Some of these predictors are also incorporated in UTAUT2 (e.g., enjoyment is similar to hedonic motivation). In summary, these studies point out that user perception rather than gained experience seem to better capture acceptance of an early stage technology. Moreover, since there is no diffusiveness of technology application, UTAUT2 should measure usage intentions instead of real use behavior.

Applicability of UTAUT2 to Early Stage Technologies

According to Wilde, Hess, and Hilbers (2008), there should be no validity problem when asking users, who did not use the technology before, to evaluate their intention to use. Hence, it is assumed that UTAUT2 can be used to test for prospective usage intention.

In the UTAUT framework performance expectancy (PE), effort expectancy (EE), and social influence (SI) predict intention to use. Intention to use is defined as the intention to use the technology in the future (Venkatesh et al., 2012). PE is the extent to which using a technology can lead to benefits in performing specific activities. It is a major contributor to the intention to use (Venkatesh et al., 2003; Venkatesh et al., 2012). EE is defined as ease of use in performing activities (Davis, 1989; Moore & Benbasat, 1991; Thompson, Higgins, & Howell, 1991; Venkatesh et al., 2003). Finally, SI is the extent to which a user is influenced by its social environment that motivates the user to adopt the technology (Moore & Benbasat, 1991; Thompson et al., 1991; Venkatesh et al., 2003). PE and SI are also confirmed by the Elaboration Likelihood Model (1984) on which Bhattacherjee and Sanford (2006) based their findings of peripheral influence being especially important if there is no user experience, which is the case for 3DP.

The extended model UTAUT2 (Venkatesh et al., 2012) adds four variables to the UTAUT model to increase the predictability of the intention to use: Facilitating conditions (FC), hedonic motivation (HM), price value (PV), and habit. First, FC describe one's belief about the accessibility or availability of resources required to use a technology (Taylor & Todd, 1995; Brown & Venkatesh, 2005). For instance, the knowledge to use 3DP, the compatibility to existing resources like computers, the technical support, and the availability of printers, print shops, and raw materials can influence the intention to use 3DP. However, previous research about the impact of FC on intention to use shows contradicting findings. Some authors suggest that this relationship is strong (e.g., San Martín & Herrero, 2012), whereas other researches ignore this variable or find it to be insignificant (e.g., Chen, Wu, & Yang, 2008; Cheng et al., 2011; Borrero, Yousafzai, Javed, & Page, 2014). In accordance with San Martín & Herrero and the TPB, this research includes FC. Second, HM, the degree of intrinsic motivation, enjoyment, and happiness from using a technology, facilitates intention to use (Brown & Venkatesh, 2005). Third, PV, adopted by Dodds, Monroe, and Grewal (1991), represents users' tradeoff between discerned benefits and costs of technology use. If users think that using a technology is not reasonably priced, they will refuse to use it. Fourth, habit is 'the extent to which people tend to perform behaviors automatically because of learning' (Venkatesh et al., 2012, p. 161), so that users' addiction and routine contribute to higher intention to use. PE and HM were the best predictors of user acceptance in UTAUT2. However, since 3DP is not used yet, habit and usage behavior will not predict intention to use of early stage technologies. Instead of profound opinion making, perceptions about the prospective technology potential should deliver insights for intention to use. Thus, predictors' items will be adjusted to represent perceptions.

Since previous research assumed that there are no profound differences in user acceptance of early and later stage technologies, UTAUT2 should be capable of explaining intention to use a technology in its infancy. Thus, the following hypothesis can be formulated:

H1a: All UTAUT2 predictors (i.e., PE, EE, SI, FC, HM, and PV) will positively affect the intention to use.

In UTAUT2, age, gender, and experience moderate the impact of the antecedents on the dependent variable. Regarding age, several studies claim that older users have more difficulties in processing new information (e.g., Plude & Hoyer, 1985; Morris & Venkatesh, 2000) because of reduced cognitive and memory skills at increasing age (Posner, 1995). As such, age influences all relations between predictors and intention to use. In addition, gender differences for intention to use are significant in several studies (Gefen & Straub, 1997; Venkatesh & Morris, 2000). For instance, PE is influenced by gender and age in that the effect is stronger for younger men. Venkatesh et al. (2012) confirms this for all antecedents in UTAUT2. These gender differences can be exemplified by EE: In contrast to men, female users are more affected by the technology's ease of use. Also experience is seen as a predictor of intention to use, but is not relevant in the context of early stage technologies. In summary, UTAUT2 shows interacting effects of gender and age on the relation between FC, HM as well as PV and intention to use. More specifically, these interactions are significant either via a three-way interaction (age and gender at once) or via a four-way interaction (age, gender, and experience). Two-way interactions are not detected. Together, age and gender moderate PE, FC, and PV, while EE, SI, and HM are only moderated in combination with experience. Since experience cannot be measured in a technology in such an early stage as 3DP, it can be expected that the formerly tested interactions show significance, however without taking into account experience:

H1b: Age and gender will positively moderate the effect of the proposed predictors (i.e., PE, EE, SI, FC, HM, and PV) on intention to use.

Extending the Model to Fit the 3DP Context

The next paragraphs are to explain why it is valuable to extend the model in two ways: (i) By splitting up intention to use according to involvement stages, and (ii) by introducing DIY mentality as predictor and moderator.

For the purpose of testing and measuring general intention to use, the hypotheses above are developed with overall intention to use as dependent variable. However, many 3DP investigators have different opinions about prospective involvement stages, i.e., (i) Self-design, and (ii) self-production. For business entrepreneurs, it is interesting to see which technology perceptions and user characteristics drive these two involvement stages. Furthermore, especially for early stage technologies, it is valuable to assess the attractiveness of possible business models and thus different involvement stages. Therefore, both involvement stages will be included as dependent variables in the framework. However, UTAUT2 should also be capable of testing intention to use technologies, regardless of involvement stages (Venkatesh et al., 2012). According to Van Raaij and Pruyn (1998), customers can be involved in production during three stages: (i) Specification or design; (ii) use of input, production and realization, and (iii) consumption of outcome. Applying two described 3DP scenarios to the production stages, self-design starts at the first stage, whereas self-production only treats stages two and three. Splitting up the dependent variable is not new. For instance, Sapio, Turk, Cornacchia, Papa, Nicolò, and Livi (2009) improve the explanation of digital television adoption by testing acceptance for both informative and interactive content. The impact of the two predictors PE and SI are found to be different for these two applications.

There might be differences in HM according to the involvement stage. HM is the fun or pleasure derived from using a technology (Brown & Venkatesh, 2005) and is based upon the concept of perceived enjoyment of the TAM3 model (Venkatesh & Bala, 2008). This is especially true for early stages of a technology, since hedonically motivated people use technologies just because of its novelty or innovativeness (Holbrook & Hirschman, 1982; Wells, Campbell, Valacich, & Featherman, 2010). Using a technology as an end customer outside an organizational context indicates that users are intrinsically motivated (Davis, Bagozzi, & Warshaw, 1992). Self-design can only be realized if the consumer is creative to the extent that achievements differ from off-the-shelf solutions. This can lead to novel, innovative products that are unique and suit a person's needs. Since innovativeness, customization, and uniqueness partly represent building blocks of HM (Brown & Venkatesh, 2005), and innovation is also driven by consumer creativity (as 'solving capability', Hirschman, 1980), HM should be directly related to self-design. Moreover, Dahl and Moreau (2007) identify that creativeness, which is related to self-design, leads to a higher task enjoyment, which is an indicator of HM (Davis et al., 1992). In addition, users of early stage technologies might also experience HM because of using the technology in a utilitarian manner, i.e., for production purposes. This HM can be explained by the pride of working with the newest technology, especially by people who like learning opportunities. In a 3DP context, Bradshaw et al. (2010) believe that users value their own creations more in terms of design and/or production. As such, a strong link between HM and self-design, as well as self-production, can be predicted.

H2: HM will positively affect the intention to use for the different involvement stages of the user, i.e., self-design and self-production stage.

In the market, there are companies that see advantages in attracting user with a DIY mentality (DIY; Hsu, 2013). According to Bunshaft and Laubner (2013) and Birtchnell et al. (2013), 3DP can lift DIY to a new level by letting consumers repair broken products immediately without the need to buy a complete new appliance. More specifically, DIY customers search for a sense of empowerment, a craftsman identity, a need for uniqueness, and the feeling of being part of a community (Wolf & McQuitty, 2011; 2013). These needs can be fulfilled by using 3DP technologies. As such, we can hypothesize that a DIY mentality increases the intention to use 3DP:

H3a: DIY has a positive effect on intention to use.

SI is very important to DIY-users (Wolf & McQuitty, 2011). According to Franke, Schreier, and Keinz (2008), peer opinions are important to develop and evaluate a self-made product (design) that meets all expectations and leads to a purchase intention. Their study shows that social influence and feedback is especially important for users with no or little experience who want to take initiatives and take action, since active starters largely benefit from information given by others. While this study is talking about users, the same reasoning might hold for new technologies, i.e., more social influence and feedback are evaluated positively by users who want to actively use the technology. The same reasoning is followed by Fung (2013), who assumes that leaders are required to increase the intention to use a technology like 3DP. Thus, inexperienced DIY-users, who create alternative products (Wolf & McQuitty, 2013), will be more inclined to use new technologies if they receive expertise of peers on how to use the 'new tool'. Accordingly, SI will drive the intention to the technology more for DIY-users than for regular users. More specifically:

H3b: DIY will positively moderate the effect of SI on intention to use.

People who believe that 3D printing technology is fun will engage more in these 3DP technologies. However, for people who want to be involved as a DIY in this technology, we will see an even larger impact of HM on intended use. As such, the following hypothesis can be described:

H3c: DIY will positively moderate the effect of HM on intention to use.

In summary, an adjusted and extended version of UTAUT2 has been proposed to measure user acceptance of early stage technologies. The conceptual research model with its hypotheses is visualized in Figure 1. Constructs with solid lines represent the original UTAUT2 model, whereas dotted constructs are added to the model based on our literature review.

Figure 1. Extended framework with direct hypotheses according to UTAUT2

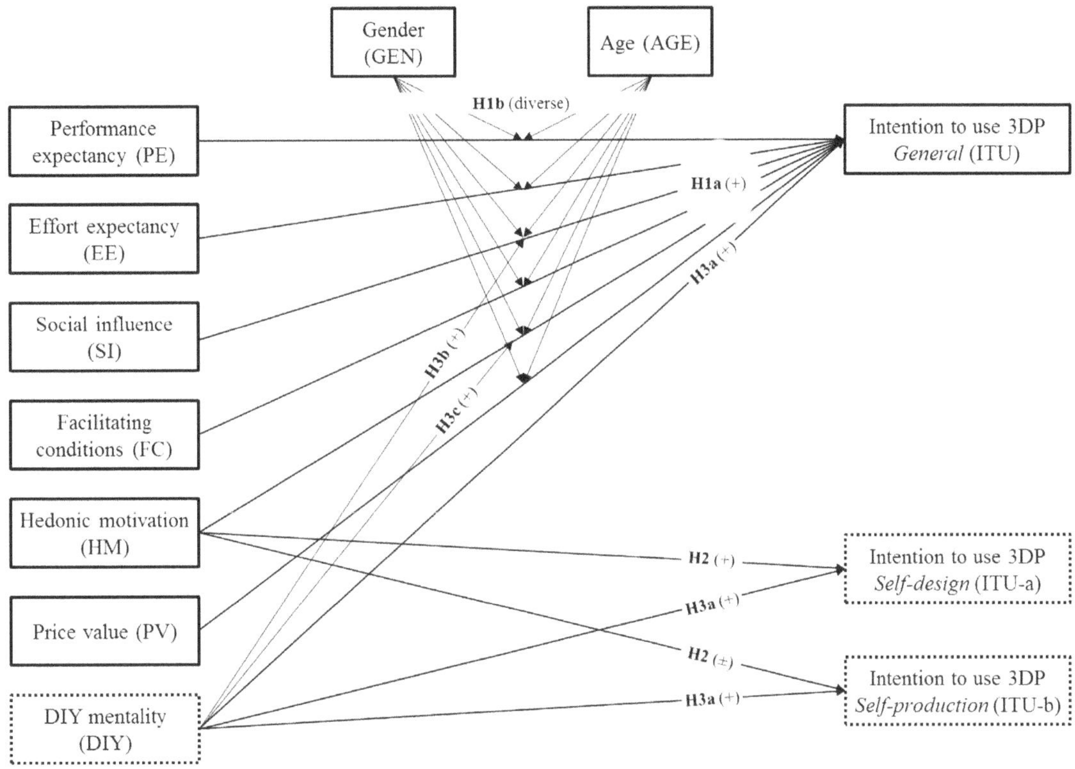

RESEARCH DESIGN

To reach a broad and diverse population, an online as well as offline questionnaire has been developed. At the outset of the online survey, a video and written explanation about 3DP and its future home potential were shown to ensure the same minimum knowledge base for all participants. Furthermore, the participants had to confirm their understanding of the 3DP context. The offline survey included a written, more detailed description about 3DP and a Q&A-part. To avoid preconceptions, participants were instructed to think of competitive products without further disadvantages regarding conventional manufacturing methods. The survey software 'Qualtrics Research Suite' was used to conduct the survey (Qualtrics, 2014). When the survey was completed, both on- and offline results were combined.

All of the scales were adopted from prior research studies and adapted to the context of 3DP. Especially the items of the UTAUT2 constructs had to be adjusted since the context of a never-before used and mostly unknown technology forced a change. Instead of experiences and opinions ('I find 3DP easy to use') the perception became central ('I think 3DP is easy to use'). Moreover, FC diverged in substance since required knowledge and resources cannot be judged by prospective users ('I have the resources necessary to use 3DP' became 'I would be willing to invest in the resources necessary to use 3DP'). DIY mentality consists of two sub-constructs. The main variable is reflective in first-order and formative in second-order. Since no content was changed and all items were already applied successfully in previous studies except for the FC construct, content validity was of no concern. All questions, except demographic factors, were rated on a five-point-Likert scale with the anchors 'strongly disagree' and 'strongly agree'. Age was measured in five levels, while gender was measured as a dummy variable. Table 7 in the Appendix shows the constructs' items and their sources.

The questionnaire was set up in English and has been reviewed for content validity by five academic researchers, specialized in information management. This has been peer-reviewed by graduate students to enhance the applicability. The target group is limited to Germans since Germany is one of the most developed countries and thus is expected to be one of the early adapter-countries of 3DP (Deloitte Global Services Limited, 2012). Consequently, the questionnaire has been translated to German. Translation equivalence is ensured by translating bidirectional (e.g., Brislin, 1986), which was done by a researcher and a professional translator. A few translations have been adapted so that content validity is assured for the translated survey. Additionally, the survey has been checked by a group of people who did not have prior knowledge about the subject.

To decrease non-response bias, participants could enter a lucky draw to win various 3D printed objects with a total value of about 50 €. For that purpose, participants were able to name their email address at the end of the survey.

Sample choice is crucial since it can be biased. The aim of this study is to ask a wide range of people to adequately represent potential users, i.e. early stage adopters of 3D printer technology, which are often younger and more educated persons (Rogers, 1995). The survey design and method were aligned to the overall goal of getting diversified responses. Gregori and Baltar (2013) test for biases in surveys spread via Facebook for business studies. There should be no significant disadvantages regarding non-response bias or access barriers. Moreover, the ease of use of filling out a survey during leisure time seems to be the key to getting a high response rate. Hence, this paper's online survey was shared twice on Facebook, LinkedIn, and Xing. To decrease the impact of sample bias due to access barriers of online platforms, 292 mails were sent to a diversified group. This group is to represent demographic groups that were not reached via social platforms. The offline survey aimed at reaching people that were, demographically seen, not in the focus of other acquisition means. To avoid repeated participation, Qualtrics placed a preventing cookie on the participant's browser. Double registrations (mail or IP address) were removed from the survey.

In total, 213 people started filling out the online questionnaire, with 138 participants finishing it. The offline survey reached 45 persons, of which 30 responded. In total, of 168 registered responses, 150 were complete, useful, no double entry, and answered by Germans. The average age was 33.6 years, 53.3% of the counted respondents were men and 46.7% were women. The median of the online questionnaire duration was 11:48 minutes. Furthermore, answers were tested for non-response bias (Armstrong & Overton, 1977) by comparing the last 25% of the participants and the remaining participants. The result of the two sample t-test indicated no significant differences for all indicators and demographics. Only the indicators DIYE4 and EE1 showed significant means at 5%-level, whereas there was clearly no significance of equal variance. Thus, nonresponse bias can be ruled out.

Although most participants were at the age between 18 and 35 years and had a high educational background, the German society is well represented in gender and marital status (GENESIS-Online, 2010). Descriptive information of the sample is shown in Table 1.

Table 1. Descriptive information of the sample

Construct	Option	Responses		Percentage	
		Total	Valid	Survey	Germany 2010
Age	17 or less	2	2	1.33%	16.32%
	18–25	47	42	28%	9.43%
	26–35	48	48	32%	11.91%
	36–45	14	11	7.33%	14.81%
	46–55	25	24	16%	15.95%
	56–65	17	14	9.33%	11.80%
	66 or more	10	9	6%	19.78%
Gender	Female	77	70	46.70%	50.90%
	Male	86	80	53.30%	49.10%
Education	Secondary school level I	30	26	17.30%	62.50%
	University entrance diploma	46	45	30%	24.50%
	College, Bachelor degree	38	38	25.30%	4.70%
	Master or Diploma degree	43	35	23.30%	7.10%
	Postgraduate studies	1	1	0.70%	1.10%
	Other	5	5	3.30%	30.90%
Marital status	Now married or civil union	53	46	30.70%	43.00%
	Widowed	6	6	4%	7.20%
	Divorced	9	9	6%	8.00%
	Separated	6	6	4%	-
	Never married, single	89	83	55.30%	41.80%

DATA ANALYSIS

Statistical Analysis

To verify the hypotheses, partial least squares regression (PLS) was used. The main benefit of this method is that it can handle small sample sizes, several interaction terms, and higher-order constructs. Moreover, PLS maximizes the explained variance of all dependent variables, thereby supporting prediction-oriented research (Chin, 1998). In addition, PLS enables researchers to test complex predictive models (Chin, Marcolin, & Newsted, 2003). SmartPLS 2.0 was used to conduct bootstrap analyses (Ringle, Wende, & Will, 2005).

According to Chin's (1998) 'ten times'-rule of thumb, the minimum sample size for analyzing our model should be 90, which was exceeded.

Descriptives of the constructs are shown in Table 2 and 3. Construct validity and reliability analysis of the outer measurement model were done prior to running the regression. First, reflective constructs were tested. Convergent validity and indicator reliability were demonstrated since all factor loadings are above 0.70 (Hulland, 1999, cf. Table 2). Additionally, convergent validity was supported by the fact that each construct's average variance extracted (AVE) exceeded the threshold of 0.50 (Bagozzi &

Table 2. Reliability and validity measures of reflective constructs

Construct	Item	Loading	Mean		SD[a]		CR[b]	α[c]	AVE[d]
			Item	Construct	Item	Construct			
PE	PE1	0.87	3.81	3.47	1.23	1.20	0.93	0.89	0.81
	PE2	0.93	3.47		1.19				
	PE3	0.91	3.15		1.19				
EE	EE1	0.95	3.55	3.46	1.23	1.19	0.96	0.95	0.86
	EE2	0.93	3.48		1.07				
	EE3	0.92	3.34		1.21				
	EE4	0.93	3.48		1.22				
SI	SI1	0.96	2.65	2.71	1.25	1.26	0.96	0.93	0.88
	SI2	0.94	2.63		1.27				
	SI3	0.91	2.85		1.25				
FC	FC1	0.89	2.89	3.13	1.34	1.28	0.89	0.84	0.68
	FC2	0.90	3.31		1.31				
	FC3	0.77	3.11		1.24				
	FC4	0.72	3.21		1.25				
HM	HM1	0.96	3.88	3.78	1.23	1.22	0.96	0.94	0.90
	HM2	0.93	3.62		1.21				
	HM3	0.95	3.85		1.20				
PV	PV1	0.96	3.23	3.16	1.31	1.28	0.97	0.95	0.90
	PV2	0.95	3.19		1.21				
	PV3	0.95	3.05		1.33				
ITU	ITU1	0.98	2.97	2.98	1.55	1.49	0.98	0.96	0.93
	ITU2	0.97	3.09		1.55				
	ITU3	0.95	2.88		1.37				
ITU-a	ITU-a1	0.98	2.85	2.86	1.48	1.45	0.98	0.97	0.95
	ITU-a2	0.98	2.90		1.51				
	ITU-a3	0.96	2.82		1.38				
ITU-b	ITU-b1	0.97	2.65	2.68	1.23	1.25	0.97	0.95	0.91
	ITU-b2	0.97	2.73		1.31				
	ITU-b3	0.93	2.65		1.19				
AGE	-	-		3.62		1.60	-	-	-
GEN	-	-		0.47		0.50	-	-	-
EDU[e]	-	-		3.51		1.14	-	-	-
MAR[f]	-	-		4.21		2.22	-	-	-

[a]SD = standard deviation;[b] CR = composite reliability; [c] α = Cronbach alpha; [d] AVE = average variance extracted; [e] EDU = education; [f] MAR = marital status.

Table 3. Reliability and validity of second order constructs

Construct	Item	Loading	ß[a]	Mean		SD[b]		CR[c]
				Item	Construct	Item	Construct	
DIY	DIYE1	0.84	0.58	3.60	3.33	1.31	1.32	0.94
	DIYE2	0.86		3.88		1.20		
	DIYE3	0.85		3.70		1.17		
	DIYE4	0.90		3.90		1.19		
	DIYE5	0.82		3.37		1.20		
	DIYE6	0.80		3.31		1.19		
	DIYL1	0.87	0.54	3.31		1.27		
	DIYL2	0.93		2.94		1.39		
	DIYL3	0.93		2.92		1.31		
	DIYL4	0.89		2.83		1.25		
	DIYL5	0.90		2.87		1.41		

[a]ß = path coefficient;[b] SD = standard deviation; [c] CR = composite reliability.

Yi, 1988). Discriminant validity was evaluated by the latent construct correlations and the square root of their specific AVE. The condition for this measure is fulfilled since all square roots of AVE were larger than any related correlation (Fornell & Larcker, 1981, cf. Table 4). Furthermore, since all values of composite reliability (CR) and Cronbach alpha (α) were above 0.70, internal consistency is assured (Cronbach, 1951; Bagozzi & Yi, 1988, cf. Table 2). The same tests were performed for the second order construct of DIY. The two-stage approach of Ringle, Sarstedt, and Straub (2012) was taken to model the construct in SmartPLS. The formative path coefficients (ß) of the two sub constructs were 0.58 and 0.54, which showed strong significance at a 1%-level ($p<0.01$; e.g., MacKenzie, Podsakoff, & Jarvis, 2005).

Table 4. Correlation and multicollinearity

Construct	VIF	1	2	3	4	5	6	7	8	9	10
PE	2.59	**0.90**[a]									
EE	2.49	0.44[b]	**0.93**								
SI	1.79	0.42	0.42	**0.94**							
FC	3.21	0.69	0.61	0.51	**0.83**						
HM	3.58	0.70	0.67	0.42	0.70	**0.95**					
PV	1.89	0.54	0.48	0.52	0.52	0.56	**0.95**				
ITU	-	0.66	0.38	0.54	0.71	0.63	0.54	**0.97**			
ITU-a	3.01	0.58	0.35	0.55	0.67	0.61	0.51	0.82	**0.97**		
ITU-b	1.77	0.51	0.22	0.44	0.57	0.40	0.34	0.71	0.56	**0.95**	
DIY	2.38	0.47	0.55	0.47	0.56	0.64	0.56	0.59	0.64	0.29	**0.78**

[a]Diagonal elements are square roots of AVE;[b] Off-diagonal elements are correlations between constructs.

In addition to the outer model, the inner structural equation model was tested for multicollinearity and common-method variance (CMV). First, multicollinearity does not cause any concern, since all variance inflation factors (VIFs) were between 1.8 and 3.6 (cf. Table 4), and thus below the critical threshold of ten (Henseler, Ringle, & Sinkovics, 2009). Second, CMV is the amount of spurious correlation between variables generated by using the same survey to measure all variables (e.g., Podsakoff, MacKenzie, Lee, & Podsakoff, 2003). Harman's single-factor test checks if the majority of the variance can be explained by a single factor (Podsakoff et al., 2003). The test indicated that all items together account for a total of 45.03% of the variance. This is lower than the critical threshold of 50% and also lower than the variance explained using separate constructs (68.47%, cf. Table 5, model 2). Thus, common-method bias should not be a concern.

Six separate models tested the support for baseline UTAUT2 (direct effects only (model 1)), baseline UTAUT2 (direct and moderated effects (2)), extended UTAUT2 with self-design (both direct (3) and direct + moderated effects (4)) and extended UTAUT2 with self-production (both direct (5) and direct + moderated effects (6)).

Table 5 indicates the results of predicting intention to use 3DP in general (ITU), for self-design (ITU-a), and for self-production (ITU-b). Cohen's f^2 was computed to check each construct's effect size, which is classified by Cohen (1988) as: (i) Small (0.02); (ii) medium (0.15), and (iii) large (0.25). All significant main effect variables and interaction terms showed values between 0.03 and 0.18 (cf. Table 6). Thus, a satisfying power was ensured.

Bootstrap analyses were performed by resampling 2,000 times. To understand the pattern and results of path and bootstrap analysis, split-sample analyses were conducted. As shown in Table 5, the basic UTAUT2 was partly confirmed. Regarding the direct impact on ITU (model 1), with the exception of PV and the partly confirmed HM, all predictors were significant at least at a 5%-level ($p<0.05$, $t>1.96$). EE showed a significant negative value ($p<0.01$, $t>2.58$). PV showed no significant impact on ITU ($p=0.64$, $t=0.47$). HM was significant in the basic model ($p<0.01$, $t>2.58$), but not in the model with interaction effects ($p=0.20$, $t=1.29$). However, it is significant with two moderators (AGE, DIY).

With standardized interaction terms and multi-group analyses (pooled standard error, model 2), several interactions showed significant paths, such as: EE x AGE, HM x AGE, SI x GEN, PV x GEN, PE x DIY, EE x DIY, SI x DIY, HM x DIY, PE x AGE x GEN, FC x AGE x GEN, and PV x AGE x GEN. However, all moderating variables also had direct predicting effects on ITU. Moreover, the results mainly supported the prediction of the basic applicability of UTAUT2 to predict consumers' ITU for a technology in its infancy. According to Chin's (1998) interval scales, a variance of at least 0.19 is weak and a variance of at least 0.33 is mediocre. If the R^2 is greater than 0.67, the coefficient of determination is substantial. As such, the variance in ITU explained by UTAUT2 with direct effects and both direct and moderated effects is mediocre and substantial at 63% and 68% respectively.

Models 3 and 4, dealing with ITU-a, showed similar patterns. However, PE and PV were not significant. There were significant direct effects of the moderators AGE and DIY on ITU-a. The coefficient of determinations of ITU-a were 59% and 67% for direct (mediocre) and both direct and moderating effects (substantial) respectively.

The coefficients of determination of ITU-b (model 5 and 6) were smaller than those of the first four models (41% and 48%). More specifically, HM and PV were not significant in predicting ITU-b. Regarding the moderators, only GEN showed direct contribution to the dependent variable.

Table 5. Structural model results: Baseline and extended UTAUT2

Independent variable or moderation effect	Baseline UTAUT2 (ITU)		Ext. UTAUT2 self-design (ITU-a)		Ext. UTAUT2 self-production (ITU-b)	
	1. B[a]	2. B+I[b]	3. B	4. B+I	5. B	6. B+I
Performance expectancy (PE)	0.17**	0.24**	0.02	0.09	0.2**	0.24***
Effort expectancy (EE)	-0.25***	-0.24***	-0.3***	-0.3***	-0.25***	-0.25***
Social influence (SI)	0.2***	0.16*	0.26***	0.2***	0.22***	0.23***
Facilitating conditions (FC)	0.4***	0.38***	0.4***	0.36***	0.46***	0.48***
Hedonic motivation (HM)	0.25***	0.16	0.35***	0.24***	0.02	-0.01
Price value (PV)	0.12	0.05	0.1	<0.01	-0.01	0.01
Age (AGE)		0.16**		0.16***		0.07
Gender (GEN)		0.13*		<0.01		0.26***
Do it yourself mentality (DIY)		0.2*		0.33***		-0.06
PE x AGE		0.04				
EE x AGE		0.13**				
SI x AGE		0.05				
FC x AGE		0.06				
HM x AGE		0.12**				
PV x AGE		0.04				
PE x GEN		0.07				
EE x GEN		0.13				
SI x GEN		0.15**				
FC x GEN		0.06				
HM x GEN		0.01				
PV x GEN		0.09**				
PE x DIY		0.14***				
EE x DIY		0.1**				
SI x DIY		0.15**				
FC x DIY		0.1				
HM x DIY		0.16**				
PV x DIY		<0.01				
PE x AGE x GEN		0.18*				
EE x AGE x GEN		<0.01				
SI x AGE x GEN		0.02				
FC x AGE x GEN		0.2**				
HM x AGE x GEN		0.02				
PV x AGE x GEN		0.19*				
R^2	0.63	0.68	0.59	0.67	0.41	0.48
Adj. R^2	0.62	0.66	0.57	0.65	0.38	0.44

Notes: * p<0.10, ** p<0.05, *** p<0.01.
[a]B = baseline model;[b] I = interaction model.

Table 6. Effect sizes of significant structural paths

Independent variable	Baseline UTAUT2 (ITU)		Ext. UTAUT2 Self-design (ITU-a)		Ext. UTAUT2 Self-production (ITU-b)	
	B	B+I	B	B+I	B	B+I
PE	0.03[a]	0.09	(0.00)	(0.01)	0.03	0.04
EE	0.08	0.10	0.10	0.10	0.05	0.05
SI	0.07	0.07	0.11	0.07	0.05	0.06
FC	0.16	0.18	0.14	0.14	0.13	0.16
HM	0.05	(0.04)	0.09	0.05	(0.00)	(0.00)
PV	(0.02)[b]	(0.02)	(0.01)	(0.00)	(0.00)	(0.00)
AGE		0.08		0.06		0.01
GEN		0.07		(0.00)		0.12
DIY		0.08		0.16		(0.00)

[a]Formula of f^2: $f^2 = (r^2$ complete model - r^2 without specific variable) / (1 - r^2 complete model); [b] Bracketed values have non-significant paths and are not relevant for measuring effect sizes.

Findings

The first part pertained to the applicability of UTAUT2 to early stage technologies in general (H1a) and the moderating effects of AGE and GEN without experience (H1b). With exception of the insignificant impact of PV and the opposite predicted effect of EE on ITU, all predictors were significant at a 5%-level. HM showed a strong significant path in the basic model but not with interaction effects included. According to Chin (1998), the two R^2 values of 63% and 68% were mediocre and substantial respectively. Thus, H1a was partially supported. The moderating effects of PE, FC, and PV in UTAUT2 were confirmed (AGE x GEN). EE, SI, and HM were expected to be significant via a three-way interaction of AGE and GEN. However, in fact all of them were only significant with either AGE (EE, HM) or GEN (SI). This was not identified in UTAUT2. Moreover, the effect of AGE on HM was significant in the opposite direction so that for older people the effect of HM on intention to use are stronger than for younger people. However, comparing the mean of HM for younger and older people shows that younger users show a significantly ($p<0.01$) higher HM in general, while there are no significant differences in ITU between these groups. Consequently, H1b was just partially corroborated.

The second part dealt with the extension of UTAUT2 to fit in the 3DP context. The first hypothesis treated the impact of HM on either ITU-a or on ITU-b. Since the first relationship was highly significant and the second was not, H2 was supported. The second group of hypotheses tested the impact of DIY as predictor and moderator. In general, DIY contributed to the explanation variance of ITU because the adjusted R^2 increased to 63% (DIY as predictor) and 66% (DIY as predictor and moderator) respectively. Supported by an effect size of 0.08, an enhanced explanation with DIY was reached. The predictive power of DIY on ITU, ITU-a, and on ITU-b was tested in H3a. The first two relationships were highly significant (ß=0.20, $p<0.10$; ß=0.33, $p<0.01$), whereas the third relationship was not significant (ß=-0.06, $p=0.35$). Thus, H3a was partly confirmed. H3b predicted the moderation of DIY on SI and ITU. The path coefficient was significant ($\beta=0.15$, $p<0.05$), supporting H3b. H3c predicted a stronger impact of HM on ITU for DIY-users. It showed a similar significance as H3b (ß=0.16, $p>0.05$), thereby confirming H3b. It is noteworthy that moderating effects of DIY on PE and EE were also found to be significant: The lower DIY, the higher the importance of PE and EE ($ß=0.14$, $p<0.01$; $ß=0.10$, $p<0.05$). These relationships were not predicted.

All in all, there are some paths which are not significant (PV→ITU, HM→ITU-b, DIY→ITU-b, several AGE/GEN-interactions). One path is significant in the opposite direction (EE→ITU) and two paths are significant although they were not predicted (DIY→PE/EE). One path (HM→ITU) is significant in the basic model, but not in the model with interaction effects.

DISCUSSION

Our main aim is to test the applicability of UTAUT2 to early stage technologies. As the variance explained of intention to use for our early stage technology approximates the overall findings of other UTAUT2 tests (e.g., Escobar-Rodríguez & Carvajal-Trujillo, 2013; Raman & Don, 2013; Mathur & Dhulla, 2014), it can be stated that UTAUT2 is also suitable to predict intention to use for technologies in early stages. These results complement other early stage technology acceptance tests by Tzeng (2011), De Graaf and Ben Allouch (2013), and Looney et al. (2008). Nevertheless, our model also shows some other interesting outcomes.

Although UTAUT2 predicts a positive relation of effort expectancy on intention to use (H1a), effort expectancy negatively impacts intention to use. This negative effect might be explained by the perceived difficulty to understand early stage technologies. Escobar-Rodríguez and Carvajal-Trujillo (2013) found no relevant relationship between effort expectancy for the online purchasing of airline tickets and intention to use, and explained this by the fact that users' experience no difficulties in using computer technologies. Our results go a step further and show that there is a negative effect, indicating that the intention to use technologies that require limited effort is low. This might be explained by the fact that more difficult technologies require a learning curve, which might form a barrier towards other users. This barrier might however signal that the technology creates a competitive advantage and as such increases intention to use. This outcome is interesting and requires further analysis.

In addition, price value did not explain a lot of the variance of technology acceptance. This can be explained by the difficulty to imagine prices, as participants have never been in contact with this technology. This contradicts earlier findings of Wilde et al. (2008), who do not expect validity problems for estimating the value of a technology when participants have never seen the technology before.

All moderating effects as hypothesized by UTAUT2 (i.e., H1b) were validated in our research context. Only the interaction between age and hedonic motivation led to unexpected results. According to UTAUT2, hedonic motivation should be a more important predictor for user intention for younger people, while our statistics showed the other way around. In UTAUT2, the tendency arose that younger people in combination with age and experience are triggered more by hedonic motivation. Chau and Lung Hui (1998) find that especially younger persons seek more novelties and innovativeness than older people. This prediction was confirmed after a descriptive analysis of younger and older users: On average, younger persons had a higher hedonic motivation level, while the average level of intention to use the technology was the same for both age groups. These findings might be explained by the fact that many younger people associated 3DP with fun and enjoyment, but might think that it is too early to use it, based on their better understanding of the technology of the 'digital natives' generation, whereas older people are blinded by the missing familiarity. An interesting secondary finding is that facilitating conditions seem to be the best predictor for prospective use. This means that users who had limited knowledge of a new technology would have to be willing to acquire knowledge and resources to run the technology. Since the content of this construct had to be changed to fit the early stage context, the result cannot be compared with the original UTAUT2 results for which performance expectancy, hedonic motivation, and habit were the best predictors. In the 3DP context however, effort expectancy and social influence seem to be the best predictors. This is supported by previous studies (Venkatesh & Davis, 2000; Venkatesh et al., 2003) that find effort expectancy and social influence to be especially strong predictors in the stage of low technology experience. In summary, the applicability of UTAUT2 to early stage technologies can be confirmed although there are differences among the strength of predictors.

The second aim of our research is to test whether an extension of UTAUT2 could increase the predictive value of the model for early stage technologies.

A first extension of the model is to split up general intention to use into use for self-design and self-production. Our analyses show a higher impact of hedonic motivation for self-design than for self-production. These findings are also supported by Sapio et al. (2009) who found differences in intention to use digital TV according to users' applications: i.e., informative or interactive use. Self-design might be seen as a less risky option due to the financial risk of owning the printer in case of self-production. In addition, performance expectancy did not contribute to the intention to use 3DP for self-design, but only for self-production. Self-designers might care less about the technology itself (i.e., how fast it is, and how it increases productivity) since they do not own the technology. They want to be creative and receive a good product, without caring about the production stage. A second extension of the model, i.e., the introduction of DIY mentality into the model, was confirmed to better explain the intention to use for early stage technologies. More specifically, by introducing DIY mentality, we were able to increase the predictive power of the model from 63 to 66%.

Our results also show three surprising, non-hypothesized results, as discussed in this paragraph. First, DIY influenced the two involvement stages, i.e., self-design and self-production, in a different way. More specifically, DIY mentality has a positive impact on self-design, but not on self-production. This might be explained by the higher creativity of people who like to design their own materials. People driven by DIY motivations – in this case unique products – like to create exclusive products and do not like to be restricted by predesigned objects, which are offered in case of self-production (Troye & Supphellen, 2012). In addition, at this stage DIY-users cannot be convinced solely by cost advantages of 3DP, since production prices are still too high, leading to little to no intention to use 3DP for self-production. Second, DIY mentality moderates performance expectancy and general intention to use. Especially for people with a low DIY mentality, performance expectancy and the perception of technology usefulness are of high importance. On the one hand, DIY-motivated persons are triggered by the economic advantage or by the access to products which cannot be produced with other technologies (Wolf & McQuitty, 2013). Hence, if the technology is able to fulfill the production requirements, it is irrelevant if the technology was evaluated as useful or not before. On the other hand, Venkatesh et al. (2003) predict that performance expectancy positively correlates with intention to use. Additionally, DIY mentality might strengthen this effect since this mentality requires people to be motivated to perform well. Finally, DIY mentality moderates the negative effect of effort expectancy on general intention to use. Moreover, for people with a low DIY mentality, effort expectancy has a stronger negative effect on intention to use than people with high DIY mentality, since people with a low DIY mentality have more burdens to overcome than people with a high DIY mentality. The latter group could have the above mentioned higher-ranking goal of saving money or having unique products, which makes effort expectancy quite unimportant. In sum, DIY mentality was able to enhance prediction of intention to use, especially in the case of self-design. This is the first study that empirically verifies this impact, which was predicted anecdotally by Hsu (2013). Moreover, the important role of DIY, age, and gender confirms the study of Yi, Wu, and Tung (2005), who find moderating as well as direct relationships between such personal characteristics, perceptions, and technology acceptance. All in all, different stages of intention to use and DIY mentality enhanced the understanding of user acceptance in the early stage technology 3DP.

Managerial and Academic Implications

Several managerial implications could be formulated based on our findings. These implications are important with respect to developing promising business models and appropriate strategies for early stage technologies, while enhancing our understanding of prospective users of early stage technologies, and more specifically potential 3DP users.

Our study shows how UTAUT2 could help marketers of early stage technologies to identify the right users and user requirements. In addition, by distinguishing between several involvement stages of 'intention to use', the usefulness of several business models could be evaluated. In the case of 3DP, prospective users were more interested in self-design than in self-production due to differences in hedonic motivation for these two stages. As such, businesses could benefit from offering easy-to-use software for self-design instead of just focusing on the production of 3DP items. It is also notable that especially women are interested in self-production so that marketers should try to offer cheap entry solutions and specifically target this segment for self-production initiatives.

In contrast to late stage technologies, social influence and effort expectancy are the most important predictors for intention to use. Less effort has to be put into convincing customers about the ease of use of the technology (negative impact of effort expectancy). Instead of ease of use, the social environment is more important. Thus, in the case of 3DP, neighbors, friends or local communities are crucial to get the business started and to convince prospective users to use the technology. This insight should be incorporated in prospective 3DP business models

Price value was not of much importance for prospective users of early stage technologies. Early stage adaptors are known for being less price conscious and willing to pay more for owning it earlier. As such, early stage technology marketers have to market the values, not as a cost-benefit decision, clearly. But this result also shows that other predictors are more important, indicating that marketers should attract users by offering for instance hedonic incentives. Hedonic motivation was clearly associated with self-design and not self-production. Marketers should try to inspire prospective users to be creative by attaching a brochure with realized creative and hedonic applications to the sales package. Moreover, especially for self-designers, print shops should adjust their price settings to some extent. Hedonic products such as art can be sold at a premium margin since people are motivated by hedonic reasons and attach a higher value to it. At the moment, shops calculate the product price on the basis of size, volume, and used material. In addition, our study shows that these self-design services should be offered through selling points with a large DIY customer base, such as construction stores or DIY stores.

Furthermore, we showed the importance of a person's mentality, i.e. DIY mentality, to adapt a certain technology. This insight should help marketers to define and optimize their marketing activities towards certain target groups.

Finally, personal characteristics (like age and gender) influence the acceptance of a technology, indicating that different cohorts of consumers attach different weights to various predicting factors. While in general, we could confirm UTAUT2 (Venkatesh et al., 2012), we also found some differences with respect to these personal characteristics. For instance, older women found price value more important than younger people or men. Thus, a market analysis about the estimated price value for different customer segments might be helpful to set up a strategy that is able to differentiate between customers. Besides, this paper identifies some unique results. It states that older people are motivated more by the hedonic benefits gained from using 3DP. To attract older people, hedonic products such as free design files, e.g. for parlor game tokens, can be offered in sales bundles. If these users get the feeling of being able to produce unique products, they could show it to their social environment, which is important to them. This can be done by letting them offer their customized designs to others on a web blog.

Finally, we tested the technology acceptance of a young technology. Within this young technology environment, we show that marketers have to address other conditions which would be applicable for matured technologies. For example, we found that DIY is a strong predictor for early stage technologies. In addition, the impact of peers seems to be much stronger for early stage technologies.

Limitations and Future Research

Our findings have a number of implications for theory development and future research.

First, the sample choice was limited to German users. Whether these findings can be applied to other countries is debatable. While limiting our sample towards Germany enabled us to increase the internal validity of our sample, it does not guarantee similar results in other geographical areas. Future research might take these geographical differences into account. Also the adaptation of the original UTAUT2 framework to an early stage environment might be seen as a limitation. For instance, facilitating conditions had to be changed in content to enable prospective users to respond. Habit, experience, and usage behavior were excluded. Future research could focus on these differences between early and later stage technologies. Here, the differences of antecedents should be revised and the subsequent use of the technology as moderator can be incorporated. It is especially interesting if digital natives will change their opinion with raising age. This test for differences could be done on the basis of 3DP. In addition, the framework was strongly related to the case of 3DP. This had the advantage of learning something about the characteristics and perceptions of prospective users, but the applicability to other technologies might be difficult. However, a test for DIY mentality in other technology acceptance contexts might be a first step to understand the generalizability. Lastly, we would recommend to analyze in-depth three- and four-way-interactions as done by the UTAUT2 model originally. This was not the focus of this study, but it would be interesting to see if the results are in line with past studies.

CONCLUSION

This study explores the predictors and influencing factors of technology acceptance of an early stage technology using the example of 3DP. This was done by using the UTAUT2 framework as described by Venkatesh et al. (2012). Since previous research focused on testing the UTAUT2 applicability to mature technologies, the question remains whether this model is also applicable for technologies in their infancy. By testing this model in the context of the emerging 3DP technology, we are able to explore this applicability of UTAUT2 for early stage technologies and to extend the existing model towards technologies in their infancy. Our research showed that UTAUT2, with some small adaptations, is a powerful framework for predicting user intentions for technologies in early stages. In addition, we extended the UTAUT2 model for young technologies by showing differences according to the user's involvement stage in the technology and the user's DIY mentality. For that, this paper not only applies UTAUT to another field and technology stage. Additionally, to the mentioned gaps it helps in two different aspects: First, we could improve the predictability of the current UTAUT2 model with those extensions. Second, the DIY mentality of users can help to predict different fields of application, which provides insightful information for marketers of 3DP technologies.

REFERENCES

Ajzen, I. (1991). The theory of planned behavior. *Organizational Behavior and Human Decision Processes*, *50*(2), 179–211. doi:10.1016/0749-5978(91)90020-T

Ajzen, I., & Fishbein, M. (1975). *Belief, attitude, intention and behavior: An introduction to theory and research*. Reading, MA, USA: Addison-Wesley.

Armstrong, J. S., & Overton, T. S. (1977). Estimating nonresponse bias in mail surveys. *JMR, Journal of Marketing Research*, *14*(3), 396–402. doi:10.2307/3150783

Bagozzi, R. P. (2007). The legacy of the technology acceptance model and a proposal for a paradigm shift. *Journal of the Association for Information Systems*, *8*(4), 244–254.

Bagozzi, R. P., & Yi, Y. (1988). On the evaluation of structural equation models. *Journal of the Academy of Marketing Science*, *16*(1), 74–94. doi:10.1007/BF02723327

Bak, D. (2003). Rapid prototyping or rapid production? 3D printing processes move industry towards the latter. *Assembly Automation*, *23*(4), 340–345. doi:10.1108/01445150310501190

Benbasat, I., & Barki, H. (2007). Quo vadis, TAM? *Journal of the Association for Information Systems*, *8*(4), 211–218.

Berman, B. (2012). 3-D printing: The new industrial revolution. *Business Horizons*, *55*(2), 155–162. doi:10.1016/j.bushor.2011.11.003

Bhattacherjee, A., & Sanford, C. (2006). Influence processes for information technology acceptance: An elaboration likelihood model. *Management Information Systems Quarterly*, *30*(4), 805–825.

Birtchnell, T., Urry, J., Cook, C., & Curry, A. (2013). *Freight miles: The impact of 3D printing on transport and society*. Lancaster, UK: Lancaster University.

Borrero, J. D., Yousafzai, S. Y., Javed, U., & Page, K. L. (2014). Expressive participation in internet social movements: Testing the moderating effect of technology readiness and sex on student SNS use. *Computers in Human Behavior*, *30*, 39–49. doi:10.1016/j.chb.2013.07.032

Bradshaw, S., Bowyer, A., & Haufe, P. (2010). The intellectual property implications of low-cost 3D printing. *ScriptEd*, *7*(1), 5–31.

Brislin, R. W. (1986). The wording and translation of research instruments. In W. J. Lonner & J. W. Berry (Eds.), *Field methods in cross-cultural research* (pp. 137–164). Beverly Hills, CA, USA: SAGE Publications, Inc.

Brown, S. A., & Venkatesh, V. (2005). Model of adoption of technology in households: A baseline model test and extension incorporating household life cycle. *Management Information Systems Quarterly, 29*(3), 399–426.

Bunshaft, A., & Laubner, D. (2013). 3D printing: Game-changing applications for consumers. *Design News*, 48-51.

Campbell, T., Williams, C., Ivanova, O., & Garrett, B. (2011). *Could 3D printing change the world? Technologies, potential, and implications of additive manufacturing. Strategic Foresight Report*. Washington, D.C., WA, USA: Atlantic Council.

Canalys. (2014). 3D printing market to grow to US$16.2 billion in 2018. Retrieved from http://www.canalys.com/static/press_release/2014/canalys-press-release-310314-3d-printing-market-grow-us162-billion-2018_0.pdf

Chau, P. Y. K., & Lung Hui, K. (1998). Identifying early adopters of new IT products: A case of Windows 95. *Information & Management, 33*(5), 225–230. doi:10.1016/S0378-7206(98)00031-7

Chen, C., Wu, J., & Yang, S. (2008). Accelerating the use of weblogs as an alternative method to deliver case-based learning. *International Journal on E-Learning, 7*(2), 331–349.

Chen, L.-L., Djajadiningrat, T., Feijs, L., Fraser, S., Hu, J., Kyffin, S., & Steffen, D. (2013). Design and semantics of form and movement. *Proceedings of DeSForM 2013*, Wuxi, CN.

Cheng, Y.-S., Yu, T.-F., Huang, C.-F., Yu, C., & Yu, C.-C. (2011). The comparison of three major occupations for user acceptance of information technology: Applying the UTAUT model. *I-Business*, 3(2).

Chin, W. W. (1998). The partial least squares approach for structural equation modeling. In G. A. Marcoulides (Ed.), *Modern methods for business research* (pp. 295–336). Mahwah, NJ, USA: Lawrence Erlbaum Associates.

Chin, W. W., Marcolin, B. L., & Newsted, P. R. (2003). A partial least squares latent variable modeling approach for measuring interaction effects: Results from a Monte Carlo simulation study and an electronic-mail emotion/adoption study. *Information Systems Research, 14*(2), 189–217. doi:10.1287/isre.14.2.189.16018

Cohen, J. (1988). *Statistical power analysis for the behavioral sciences* (2nd ed.). Hillsdale, NJ, USA: Lawrence Erlbaum Associates.

Cronbach, L. J. (1951). Coefficient alpha and the internal structure of tests. *Psychometrika, 16*(3), 297–334. doi:10.1007/BF02310555

D'Aveni, R. A. (2013). 3-D printing will change the world. [PubMed]. *Harvard Business Review, 91*(3), 34–34.

Dahl, D. W., & Moreau, C. P. (2007). Thinking inside the box: Why consumers enjoy constrained creative experiences. *JMR, Journal of Marketing Research*, *44*(3), 357–369. doi:10.1509/jmkr.44.3.357

Davis, F. D. (1989). Perceived usefulness, perceived ease of use, and user acceptance of information technology. *Management Information Systems Quarterly*, *13*(3), 319–340. doi:10.2307/249008

Davis, F. D. (1993). User acceptance of information technology: System characteristics, user perceptions and behavioral impacts. *International Journal of Man-Machine Studies*, *38*(3), 475–487. doi:10.1006/imms.1993.1022

Davis, F. D., Bagozzi, R. P., & Warshaw, P. R. (1992). Extrinsic and intrinsic motivation to use computers in the workplace. *Journal of Applied Social Psychology*, *22*(14), 1111–1132. doi:10.1111/j.1559-1816.1992.tb00945.x

De Graaf, M. M. A., & Ben Allouch, S. (2013). Exploring influencing variables for the acceptance of social robots. *Robotics and Autonomous Systems*, *61*(12), 1476–1486. doi:10.1016/j.robot.2013.07.007

Deloitte Global Services Limited. (2012). *Global Manufacturing Competitiveness Index 2013: Supplemental country analysis for top 10 nations* (C. Giff, Ed.).

Dimitrov, D., Schreve, K., & De Beer, N. (2006). Advances in three dimensional printing-state of the art and future perspectives. *Journal for New Generation Sciences*, *4*(1), 21–49.

Dodds, W. B., Monroe, K. B., & Grewal, D. (1991). Effects of price, brand, and store information on buyers' product evaluations. *JMR, Journal of Marketing Research*, *28*(3), 307–319. doi:10.2307/3172866

Escobar-Rodríguez, T., & Carvajal-Trujillo, E. (2013). Online drivers of consumer purchase of website airline tickets. *Journal of Air Transport Management*, *32*(0), 58–64. doi:10.1016/j.jairtraman.2013.06.018

Fornell, C., & Larcker, D. F. (1981). Evaluating structural equation models with unobservable variables and measurement error. *JMR, Journal of Marketing Research*, *18*(1), 39–50. doi:10.2307/3151312

Franke, N., Schreier, M., & Keinz, P. (2008). Complementing mass customization toolkits with user communities: How peer input improves customer self-design. *Journal of Product Innovation Management*, *25*(6), 546–559. doi:10.1111/j.1540-5885.2008.00321.x

=Fung, B. (2013, August 29). What happens when you mash up 3D printing and Amazon's same-day delivery? *The Washington Post*.

Gefen, D., & Straub, D. W. (1997). Gender differences in the perception and use of e-mail: An extension to the technology acceptance model. *Management Information Systems Quarterly*, *21*(4), 389–400. doi:10.2307/249720

GENESIS-Online. (2010). *German population statistics 2010: 12411*. Wiesbaden, DE: Statistisches Bundesamt.

Gregori, A., & Baltar, F. (2013). Ready to complete the survey on Facebook. *International Journal of Market Research*, *55*(1), 131–148. doi:10.2501/IJMR-2013-010

Henseler, J., Ringle, C. M., & Sinkovics, R. R. (2009). The use of partial least squares path modeling in international marketing. *Advances in International Marketing*, *20*(1), 277–319.

Hirschman, E. C. (1980). Innovativeness, novelty seeking, and consumer creativity. *The Journal of Consumer Research*, *7*(3), 283–295. doi:10.1086/208816

Holbrook, M. B., & Hirschman, E. C. (1982). The experiential aspects of consumption: Consumer fantasies, feelings, and fun. *The Journal of Consumer Research*, *9*(2), 132–140. doi:10.1086/208906

Hsu, J. (2013). How 3-D printing might revolutionize Amazon's same-day delivery. Retrieved from http://spectrum.ieee.org/tech-talk/robotics/diy/how-3d-printing-might-revolutionize-amazons-sameday-delivery

Hulland, J. (1999). Use of partial least squares (PLS) in strategic management research: A review of four recent studies. *Strategic Management Journal*, *20*(2), 195–204. doi:10.1002/(SICI)1097-0266(199902)20:2<195:AID-SMJ13>3.0.CO;2-7

King, W. R., & He, J. (2006). A meta-analysis of the technology acceptance model. *Information & Management*, *43*(6), 740–755. doi:10.1016/j.im.2006.05.003

Looney, C. A., Akbulut, A. Y., & Poston, R. S. (2008). Understanding the determinants of service channel preference in the early stages of adoption: A social cognitive perspective on online brokerage services. *Decision Sciences*, *39*(4), 821–857. doi:10.1111/j.1540-5915.2008.00215.x

MacKenzie, S. B., Podsakoff, P. M., & Jarvis, C. B. (2005). The problem of measurement model misspecification in behavioral and organizational research and some recommended solutions. [PubMed]. *The Journal of Applied Psychology*, *90*(4), 710–730. doi:10.1037/0021-9010.90.4.710

Malone, E., & Lipson, H. (2007). Fab@Home: The personal desktop fabricator kit. *Rapid Prototyping Journal*, *13*(4), 245–255. doi:10.1108/13552540710776197

Mathur, S. K., & Dhulla, T. V. (2014). Factors influencing professionals' decision for cloud computing adoption. *International Journal of Research in Advent Technology*, *2*(4), 397–401.

Mercer, D. (2014). *Home 3D printers: Market opportunities and obstacles*. Strategy Analytics.

Moore, G. C., & Benbasat, I. (1991). Development of an instrument to measure the perceptions of adopting an information technology innovation. *Information Systems Research*, *2*(3), 192–222. doi:10.1287/isre.2.3.192

Morris, M. G., & Venkatesh, V. (2000). Age differences in technology adoption decisions: Implications for a changing work force. *Personnel Psychology*, *53*(2), 375–403. doi:10.1111/j.1744-6570.2000.tb00206.x

Petrick, I. J., & Simpson, T. W. (2013). 3D printing disrupts manufacturing. *Research Technology Management*, *56*(6), 12–16. doi:10.5437/08956308X5606193

Petty, R. E., & Cacioppo, J. T. (1984). Source factors and the elaboration likelihood model of persuasion. *Advances in Consumer Research. Association for Consumer Research (U. S.)*, *11*, 668.

Plude, D. J., & Hoyer, W. J. (1985). Attention and performance: Identifying and localizing age deficits. In N. Charness (Ed.), *Aging and human performance* (pp. 47–99). London, UK: Wiley.

Podsakoff, P. M., MacKenzie, S. B., Lee, J.-Y., & Podsakoff, N. P. (2003). Common method biases in behavioral research: A critical review of the literature and recommended remedies. [PubMed]. *The Journal of Applied Psychology*, *88*(5), 879–903. doi:10.1037/0021-9010.88.5.879

Posner, R. A. (1995). *Aging and old age* (1st ed.). Chicago, IL, USA: University of Chicago Press.

Qualtrics. (2014). Qualtrics Research Suite. Provo, UT, USA: Qualtrics.

Raman, A., & Don, Y. (2013). Preservice teachers' acceptance of learning management software: An application of the UTAUT2 model. *International Education Studies*, *6*(7), 157. doi:10.5539/ies.v6n7p157

Reeves, P. (2009). *Additive manufacturing – a supply chain wide response to economic uncertainty and environmental sustainability*. Econolyst Limited.

Ringle, C. M., Sarstedt, M., & Straub, D. W. (2012). A critical look at the use of PLS-SEM in MIS Quarterly. *Management Information Systems Quarterly*, *36*(1), III–XIV.

Ringle, C. M., Wende, S., & Will, A. (2005). *SmartPLS*. Hamburg, DE: SmartPLS.

Rogers, E. M. (1995). *Diffusion of innovations* (4th ed.). New York, NY, USA: The Free Press.

San Martín, H., & Herrero, Á. (2012). Influence of the user's psychological factors on the online purchase intention in rural tourism: Integrating innovativeness to the UTAUT framework. *Tourism Management*, *33*(2), 341–350. doi:10.1016/j.tourman.2011.04.003

Sapio, B., Turk, T., Cornacchia, M., Papa, F., Nicolò, E., & Livi, S. (2009). Building scenarios of digital television adoption: A pilot study. *Technology Analysis and Strategic Management*, *22*(1), 43–63. doi:10.1080/09537320903438054

Schepers, J., & Wetzels, M. (2007). A meta-analysis of the technology acceptance model: Investigating subjective norm and moderation effects. *Information & Management*, *44*(1), 90–103. doi:10.1016/j.im.2006.10.007

Séquin, C. (2005). Rapid prototyping: A 3D visualization tool takes on sculpture and mathematical forms. *Communications of the ACM*, *48*(6), 66–73. doi:10.1145/1064830.1064860

Silva, J., & Rezende, R. (2013). Additive manufacturing and its future impact in logistics. *Management and Control of Production and Logistics*, *6*(1), 277–282.

Stemp-Morlock, G. (2010). Personal fabrication. *Communications of the ACM*, *53*, 14–15.

Taylor, M., & Taylor, A. (2012). The technology life cycle: Conceptualization and managerial implications. *International Journal of Production Economics*, *140*(1), 541–553. doi:10.1016/j.ijpe.2012.07.006

Taylor, S., & Todd, P. A. (1995). Understanding information technology usage: A test of competing models. *Information Systems Research*, *6*(2), 144–176. doi:10.1287/isre.6.2.144

Thompson, R. L., Higgins, C. A., & Howell, J. M. (1991). Personal computing: Toward a conceptual model of utilization. *Management Information Systems Quarterly*, *15*(1), 125–143. doi:10.2307/249443

Troye, S. V., & Supphellen, M. (2012). Consumer participation in coproduction: "I made it myself" effects on consumers' sensory perceptions and evaluations of outcome and input product. *Journal of Marketing*, *76*(2), 33–46. doi:10.1509/jm.10.0205

Tzeng, J.-Y. (2011). Perceived values and prospective users' acceptance of prospective technology: The case of a career e-portfolio system. *Computers & Education*, *56*(1), 157–165. doi:10.1016/j.compedu.2010.08.010

Van der Heijden, H. (2004). User acceptance of hedonic information systems. *Management Information Systems Quarterly*, *28*(4), 695–704.

Van Raaij, F. W., & Pruyn, A. T. H. (1998). Customer control and evaluation of service validity and reliability. *Psychology & Marketing (1986-1998)*, 15(8), 811.

Vargo, S. L., & O'Brien, M. (2007). Comparing through service: Insights from the service-dominant logic. *Journal of Retailing*, *83*(1), 5–18. doi:10.1016/j.jretai.2006.10.002

Venkatesh, V. (2006). Where to go from here? Thoughts on future directions for research on individual-level technology adoption with a focus on decision making. *Decision Sciences*, *37*(4), 497–518. doi:10.1111/j.1540-5414.2006.00136.x

Venkatesh, V., & Bala, H. (2008). Technology acceptance model 3 and a research agenda on interventions. *Decision Sciences*, *39*(2), 273–315. doi:10.1111/j.1540-5915.2008.00192.x

Venkatesh, V., & Davis, F. D. (2000). A theoretical extension of the technology acceptance model: Four longitudinal field studies. *Management Science*, *46*(2), 186–204. doi:10.1287/mnsc.46.2.186.11926

Venkatesh, V., Davis, F. D., & Morris, M. G. (2007). Dead or alive? The development, trajectory and future of technology adoption research. *Journal of the Association for Information Systems*, *8*(4), 267–286.

Venkatesh, V., & Morris, M. G. (2000). Why don't men ever stop to ask for directions? Gender, social influence, and their role in technology acceptance and usage behavior. *Management Information Systems Quarterly*, *24*(1), 115–139. doi:10.2307/3250981

Venkatesh, V., Morris, M. G., Davis, G. B., & Davis, F. D. (2003). User acceptance of information technology: Toward a unified view. *Management Information Systems Quarterly*, *27*(3), 425–478.

Venkatesh, V., Thong, J., & Xu, X. (2012). Consumer acceptance and use of information technology: Extending the unified theory of acceptance and use of technology. *Management Information Systems Quarterly*, *36*(1), 157–178.

Wells, J. D., Campbell, D. E., Valacich, J. S., & Featherman, M. (2010). The effect of perceived novelty on the adoption of information technology innovations: A risk/reward perspective. *Decision Sciences*, *41*(4), 813–843. doi:10.1111/j.1540-5915.2010.00292.x

Wicks, J. H., Reardon, J., & McCorkle, D. E. (2005). An examination of the antecedents of the consumer make-or-buy decision. *Journal of Marketing Theory and Practice*, *13*(1), 26–39. doi:10.1080/10696679.2005.11658536

Wilde, T., Hess, T., & Hilbers, K. (2008). *Akzeptanzforschung bei nicht marktreifen Technologien: Typische methodische Probleme und deren Auswirkungen. Multikonferenz Wirtschaftsinformatik* (pp. 1031–1043). Paderborn, DE: GITO-Verlag.

Wohlers, T. T. (2014). Printing and Additive Manufacturing State of the Industry. *Annual Worldwide Progress Report*. Fort Collins, CO, USA. *Wohlers Report*, *2014*, 3D.

Wohlers, T. T., Gornet, T., & Caffrey, T. (2013). History of additive manufacturing. In T. T. Wohlers (Ed.), *Annual Worldwide Progress Report*. Fort Collins, CO, USA.

Wolf, M., & McQuitty, S. (2011). Understanding the do-it-yourself consumer: DIY motivations and outcomes. *AMS Review*, *1*(3-4), 154–170. doi:10.100713162-011-0021-2

Wolf, M., & McQuitty, S. (2013). Circumventing traditional markets: An empirical study of the market-place motivations and outcomes of consumers' do-it-yourself behaviors. *Journal of Marketing Theory and Practice*, *21*(2), 195–210. doi:10.2753/MTP1069-6679210205

Yi, Y., Wu, Z., & Tung, L. L. (2005). How individual differences influence technology usage behavior? Toward an integrated framework. *Journal of Computer Information Systems*, *46*(2), 52–63.

Zhou, T., Lu, Y., & Wang, B. (2010). Integrating TTF and UTAUT to explain mobile banking user adoption. *Computers in Human Behavior*, *26*(4), 760–767. doi:10.1016/j.chb.2010.01.013

This research was previously published in the Journal of Organizational and End User Computing (JOEUC), 29(2); edited by Steven Walczak and Sang-Bing Tsai, pages 1-24, copyright year 2017 by IGI Publishing (an imprint of IGI Global).

APPENDIX

Table 7. Constructs, measures, and their origins

Construct	Corresponding items	Example studies
Performance expectancy (PE)	1. I think 3DP is useful. 2. Using 3DP could help me accomplish things more quickly. 3. Using 3DP can increase my productivity.	Venkatesh et al. (2003); Venkatesh et al. (2012)
Effort expectancy (EE)	1. Learning how to use 3DP will be easy for me. 2. My interaction with 3DP is clear and understandable. 3. I think 3DP is easy to use. 4. I think it is easy for me to become skillful at using 3DP.	
Social influence (SI)	1. I expect that people, who are important to me, will think that I should use 3DP. 2. I expect that people, who influence my behavior, will think that I should use 3DP. 3. I expect that people, whose opinions that I value, will prefer that I use 3DP.	
Facilitating conditions (FC)	1. I would be willing to invest in the resources necessary to use 3DP. 2. I would be willing to invest in the knowledge necessary to use 3DP. 3. 3DP is compatible with other technologies I use. 4. I can get help from others when I have difficulties using 3DP.	
Hedonic motivation (HM)	1. Using 3DP seems to be fun. 2. Using 3DP seems to be enjoyable. 3. Using 3DP seems to be very entertaining.	Venkatesh et al. (2012)
Price value (PV)	1. 3DP is reasonably priced. 2. 3DP is a good value for the money. 3. At the current price, 3DP provides a good value.	
Intention to use 3DP (ITU)	1. I intend to start using 3DP in the future. 2. The probability is high that I will try to use 3DP in the future. 3. The probability is high that I plan to start using 3DP frequently in the future.	Venkatesh et al. (2003); Venkatesh et al. (2012)
Intention to use 3DP for self-design (ITU-a)	1. I intend to start using 3DP for self-design. 2. The probability is high that I will try to use 3DP to self-design items. 3. The probability is high that I plan to start using 3DP frequently for self-design.	
Intention to use 3DP for self-production (ITU-b)	1. I intend to start using 3DP for self-production in the future. 2. The probability is high that I will try to use 3DP to self-produce items in the future. 3. The probability is high that I plan to start using 3DP frequently for self-production in the future.	
DIY mentality (DIY)	Economic benefit 1. I do my own home improvement projects to save money. 2. Performing my own maintenance saves me money. 3. By fixing things myself concerning my electronic devices I can spend money on other things. 4. I find that I can save much money by doing things myself. 5. When I make things myself, the money I save is important. 6. When I begin a project, I expect to save money by doing it myself. Lack of product availability 1. I make products myself so they match me or my home 2. I design things myself because stores often do not have what I want 3. To get the right size for things I need (e.g., in my home) I have to make it myself 4. To get products that are compatible with my home I make them myself 5. To get products that match the style of my home, I have to make things myself	Wolf and McQuitty (2013)

Chapter 7
Study on Different Graphic Representations in Architectural Heritage:
Digital and Physical Modelling

José T. Garfella-Rubio
Universitat Jaume I. Avinguda Sos Baynat s/n, Spain

Jesús Máñez-Pitarch
Universitat Jaume I. Avinguda Sos Baynat s/n, Spain

Joaquín A. Martínez-Moya
Universitat Jaume I. Avinguda Sos Baynat s/n, Spain

Jaume Gual-Ortí
Universitat Jaume I. Avinguda Sos Baynat s/n, Spain

ABSTRACT

In recent years, cutting-edge methods have emerged to gradually replace or be used with traditional methods to carry out graphic surveys of architectural heritage; modern topographic tools such as 3D scanners and specific software. In addition, the new technologies of additive printing and three-dimensional digital representations has made architectural heritage more accessible to the general public. The main objective of this study was to conduct an analysis of each of the methods, to determine their advantages and disadvantages, as well as to carry out a comparative study of the results obtained with each of them.

INTRODUCTION

The graphical representation of architecture has been a necessity for centuries, either as a sign of artistic expression or as a method of documentation and study. Sketches, drawings, maps, plans, engravings and paintings were the methods of graphic representation used for centuries (Garfella, Máñez, Cabeza,

DOI: 10.4018/978-1-5225-9624-0.ch007

& Soler, 2012) The advent of photography was a technical breakthrough as a method for documenting artistic and, more particularly, architectural heritage since it made it possible to capture images that were very similar to how scenes are perceived by the human eye, as well as allowing three-dimensional images and the photogrammetric rectification of buildings, which in turn enables us to take measurements directly from them (Lerma, 2002). These systems of representation have progressed exponentially with the development of both cutting edge optical and photographic optical equipment and the development of powerful computers and highly specific and advanced graphics software.

The purpose of this paper is to analyse the use of different graphic survey methods, both traditional and advanced, in the graphic surveying of architectural heritage applied to a case study, namely, the side Portal of "Nuestra Señora de la Asunción" in Vistabella del Maestrazgo in the province of Castellón (Spain). The volume and dimensions of this *retablo*-portal (Renaissance style from the XVII century) have made it an excellent testing ground for our study.

For the study, as a first step, we located several old photographs in historical archives and various publications.

Then, we carried out surveys using traditional methods, such as freehand sketches that were dimensioned manually with measuring tapes, distance meters, levels, plumb-lines and archaeologists' combs, and the later scaling of the façade from the field data. Anaglyph images were also created using both old scanned photographs and modern digital images.

Photogrammetric surveys were also carried out, which allowed us to rectify the item using specific software. Photogrammetric restitution of the façade was also performed using the Photomodeler program, which enabled us both to create three-dimensional images of this architectural element and to generate a cloud by spatial mesh points that define this element with precision.

Another technique used was to survey the façade by means of a 3D scanner. This was performed with the aid of topographic support equipment.

Finally, we employed SfM (Structure for Motion) systems or low-cost graphic surveying systems through the use of digital photographs and different commercial software applications that enable the user a simple way to generate, like the previous systems, and edit both three-dimensional images and meshes or point clouds.

In recent years, the physical and digital modeling of three-dimensional models has entered with some urge, all accompanied by a certain evolution of software and 3D printing equipment which have become significantly cheaper (Garfella, Máñez & Martínez, 2015).

Data management software has experienced a breakthrough in the sense of seeking the integration of information, and through a simple and accessible interface, proceed to modelling, making presentations and even making videos of the documented object, with scarce resources and from a graphic point of view.

The 3D printing has lately become a powerful tool, not only of display and spatial conception, but its potential goes beyond allowing the creation of prototypes in a quick and effective manner and even allowing the manufacture of items or unique elements at a reasonable cost. Its field of enlargement is quite wide from de artistic creation, the industrial designs, the engineering, the architecture, the archaeology and the biomedicine, only to mention some of the most relevant ones.

Another functioning of the models is to serve as a perceptive instrument so that people with visual impairment can have access to graphical and visual information. In the architectural heritage domain, there is a great deal of visual information, which is practically inaccessible for a blind person. The models used as a tactile perception tool allow these people with visual impairment to obtain information through the sense of touch.

Within the architectural heritage domain and the museology there are different approaches of study with reference to the development of devices directed to people with blindness or visual impairment. The tactile devices are used as didactic tools of support when visiting museums or spaces for cultural dissemination. We can establish the following types:

- Embossed sheets, draws and tactile diagrams;
- Scale models and embossed plans;
- Models.

In recent years, certain techniques have been developed for the rapid manufacturing of prototypes, in particular of additive manufacturing, commonly known as printing in 3D, which have made possible the realization of these models in a very effective manner (Gual, Serrano & Máñez, 2015).

In the field of architecture, several studies on the use of the architectural touch models produced through 3D printing have been carried out as the ones made by Voigt and Mr. Martens. From the vast amount of digitized buildings belonging to the architectural heritage, it is possible to develop tactile models oriented to everyone and especially to people with visual disabilities. Through them, it facilitates the understanding of constructed architectural space (Gual, Puyuelo & Lloveras, 2014; Gual, Puyuelo & Lloveras, 2015a; Gual, Puyuelo & Lloveras, 2015b; Gual, Puyuelo, Lloveras & Merino, 2012).

At the national level, there are also remarkable experiences as the one carried out from the Universitat Jaume I, in collaboration with the Universitat Politècnica de València and Universitat Politècnica de Catalunya, with diverse positive experiences in the field of tactile drawings produced by printing 3D (Voigt & Martens, 2006). As well as other experiences that aim at the development of models and tactile plans enabling the dissemination of local architectural heritage among people with visual problems and the general public (Máñez Pitarch et al., 2016).

The main objective of this chapter was to conduct an analysis of each of the methods on our testing ground, to determine their advantages and disadvantages, as well as to carry out a comparative study of the results obtained with each of them. We also sought to determine the degree of error or distortion each of them has by superimposing the models obtained for the same case study with the different methods, the suitability of using each of them depending on the item to be represented, and its volumetric conditions, location, setting, and so forth. The results thus obtained may be useful in subsequent graphic surveys both in the professional field of architecture, engineering and archaeology, and in the field of teaching.

BACKGROUND

In carrying out this study, the guidelines regarding the essential tasks to be observed when studying architectural heritage were followed, i.e. previous studies and/or the architectural survey. Annex B of "The Charter of Restoration" (which sets out instructions for the execution of architectural restorations), written in 1972 (Brandi & De Angelis, 1972), established that the drafting of projects involving the restoration of a piece of architecture must be preceded by a careful study of the monument. These studies, known as previous studies, should be performed from different approaches to take into account aspects such as its position within the territorial context, either in the urban or in the rural setting, its typological aspects, formal properties and appearances, and the characteristics of its construction and the systems used, etc. both of the original building and of any additions and modifications carried out in later times.

Moreover, previous studies should gather bibliographic, iconographic and archival research in order to collect as much historical data as possible.

Moreover, any restoration project should be based not only on a complete planimetric and photographic survey that is also interpreted under the metrological aspect, but also on the regulating lines and systems of proportions. Likewise, it must include a careful study to verify the stability conditions.

Any restorative or rehabilitative intervention on architectural and archaeological heritage requires, from our point of view, a survey which documents and provides rigorous evidence of its current state. The technologies used in these types of surveys oriented towards the geometric documentation of all kinds of heritage objects have advanced considerably in recent years. This has occurred despite the inexistence of paradigms on which to base the decisions concerning the protocols to be followed in the surveying process or the most appropriate techniques, depending on the type of object to be studied and the degree of definition required for each purpose.

In this sense the decision regarding the technology to be used should comply with the aims defined by the nature of the documentation that is required. The level of requirement may vary depending on the agents involved in each intervention (Barrera, 2006). Yet, as a reference to that level we could take into consideration the fact that the different degrees of documentation should be proportional to the historical interest of the architectural or archaeological element under examination, as well as being appropriate to the purpose of the study.

Therefore, the methodology followed in this study, as explained below, included the following activities:

1. Locate the building within the corresponding urban, territorial, social, historical and artistic context. For this, a thorough historical and geographical search was carried out;
2. Bibliographical, graphical and archival data about the building have been checked;
3. On-site collection of data about the buildings and their subsequent graphical representation, that is to say, a graphical survey. With the aim of capturing the dimensional, metric, geometric and graphic features of the construction fieldwork was conducted;
4. Finally, the results were compared and conclusions were drawn (Garfella, Máñez & Martínez, 2015).

With the use of active image (laser scanning) systems, and the use of passive systems (camera), it can be verified that the result of the data collection of the architectural elements has been positive. Regardless of the chosen method, we would be in a position to proceed in a simple way to the generation of digital models with texture very attractive to the eye and easy to handle. Then, we will be able to proceed to the physical printing of the model to scale with a certain degree of definition, even with the help of specific low-cost software.

In recent years, multiple data management and integration tools have been introduced, able to unite with precision, in a single model, the documentation obtained through structured point clouds (with all the points oriented according to their normals). Those clouds are obtained from scanners and models from automatic systems of photography by techniques from Structure from Motion (SFM) and later exported in a compatible format for printing.

With the evolution of 3D technology, a wide variety of 3D printers has proliferated in the market (mostly low-cost) capable of printing at a small scale and cost, models made in a simple and automatic way.

A COMPARATIVE STUDY OF DIFFERENT GRAPHIC SURVEYING METHODS APPLIED TO THE PORTAL OF "SAN MIGUEL" THE CHURCH OF "NUESTRA SEÑORA DE LA ASUNCIÓN" IN VISTABELLA

The Testing Ground

There is a church in the town of Vistabella del Maestrazgo, in Castellón (Spain), with two great Renaissance Portals. The one called lateral or secondary is the one taken as a testing ground, reference or model. This Portal, devoted to "San Miguel", has been greatly studied by this group of researchers from the Universitat Jaume I.

Geographical, Historical and Social Context

Vistabella del Maestrazgo is a small town located at the northwestern end of the province of Castellón, in the Valencian Community, Spain. The town is located about 1249 metres above sea level.

Some of the medieval walls and buildings, built after the Christian Reconquest from the Arabs by Jaime I of Aragon, are still standing in the town of Vistabella del Maestrazgo. The medieval village was developed following the city model that recalls the ancient Roman hippodamic, or grid plan cities, with rectangular roughly orthogonal blocks built upon a central cross, in the middle of which there is a square.

After the Christian reconquer, the village was owned by several lords. Until in 1303 Vistabella was sold by its lord, Guillem d'Anglesola, to the Order of the Temple. Upon the extinction of the Order of the Temple in 1312. In 1317 the Order of Montesa was created, at which point (and until the 19th century) Vistabella belonged to the Order of Montesa. In the fourteenth century a large territory in the north of Castellón took the name of Maestrazgo de Montesa, since it was under the government economic power and jurisdiction of the Order of Montesa.

In the early of the seventeenth century, a new plant temple was built in the village of Vistabella, following the Renaissance style that prevailed at that time: the Parish Church of "Nuestra Señora de la Asunción". The building is located in the northeast of the village outside the mediaeval town walls, in the "Plaça del Dau", which is the main square of the village. It was declared a Listed Building of Cultural Interest, classified as a monument, on 28th September 2007 (BOE, 2008).

From the outside, it can be seen to be a freestanding building with a rectangular layout, and having another small rectangle attached at the head. The main features that stand out in its southward-facing main façade are two portals and the bell tower. Particularly interesting is the Main Portal, framed by an impressive *retablo*-façade, sheltered by a pointed arch. Built of masonry and stonework, the main body is covered with a gable tile roof, the apse is covered with a five-water tile roof, and the header with a single-slope tile roof, which is lower than the previous constructions, as we can see on Figure 1.

Its interior consists of a main nave and chapels between five-section buttresses, an octagonal chancel and an ambulatory with a chapel and sacristy in the header. Both the nave and the chancel are covered with stellate ribbed vaults and the side chapels have simple ribbed vaults (Garfella, Máñez & Martínez, 2015).

Figure 1. General view of the main façade of the Church "Nuestra Señora de la Asunción" in Vistabella del Maestrazgo

Scope of the Study

The present study aims show the results obtained in a particular study case by comparing different methods of graphic representation that are used in architectural heritage to achieve a successful graphical representation of its elements and finally the physical modelling of the building.

When intervening on heritage buildings and archaeological remains, it is fundamental to understand the state they are in today and what it was like when they were built. Either in a preliminary study or in the project phase, a rigorous and accurate process of graphic documentation needs to be carried out before making any decision or undertaking any technical work on them.

The first step, prior to any fieldwork, is to find as much documentary and historical background material as possible, whether in the form of administrative documentation resulting from the possible drafting of projects or actions carried out on the element previously or through documents from historical files, bibliographic or church fabric books, among others. Then it is necessary to search for and collect all existing graphic documentation, through old photographs, engravings, or paintings and so on, with the aim of revealing the status of the element that we want to represent or document graphically.

From then on, and depending on the conditions of the setting and those of the actual object to be represented, the most appropriate method of working can be chosen for each particular case. In this study case, we have used the Portal of "San Miguel" as a testing ground to see the possible outcomes, and thus subsequently draw conclusions about the different systems of documentation used and how they can serve as a starting point or help when it comes to undertaking new works.

For this reason, a hybrid system of data collection was used, consisting in sharing traditional system processes and more up-to-date systems, using low-cost software or applications as far as possible for the management and post-processing of the collected data.

It is therefore becoming more and more common to demand that data collection in the documentation of survey elements be performed relatively quickly and with a high level of accuracy. The current trend is to use advanced graphic and geometric documentation tools by means of the mass capture methods (3D scanner) and terrestrial photogrammetry (Yastikli, 2007), all accompanied by the additional support of classical and more advanced topographical methods. This jointed or hybrid way achieves high rates of accuracy and cost-effective processes, thereby making this process even more sustainable.

The graphic documentation that is generally obtained and required for subsequent phases (project, damage assessment, enhancement, etc.) is usually focused on the conventional orthodox planimetric, scaled and two-dimensional dihedral representation of the object that has been measured (floor layouts, sections, elevations) on paper and, in some cases, with a complete absence of colour (i.e. in black and white). At the most, in certain cases, we can observe the presence of shades of grey or a hierarchy of line thicknesses or the use of colours in reference marks or shaded areas that are completed with the inclusion of colour or black and white orthophotos. However, technological and visual advancement make it increasingly more necessary to obtain, provide and record 3D data and then use them to create a three-dimensional or virtual model that graphically represents both the geometry of the building and the appearance of its different faces, shapes and colours, and even photorealistic computer graphics of transformed reality can be created. In this sense, the application of photogrammetry has progressed over the last few decades. Recent developments include capturing point clouds using a 3D Laser Scanner or Terrestrial Laser Scanning combined with photography (Garfella et al., 2012, Garfella, Máñez & Martínez, 2015).

Once the results from the point clouds have been obtained and the meshes created, a physical printed model at scale can be generated. These models can be made in volume conforming the three dimensions or as a bas-relief on a surface, creating a tangible image, especially suitable for the representation of surface architectural elements such as façades and hollows.

Methodology

Survey methods can be divided into three main groups: the first and most classical involves traditional techniques and the use of the simplest devices to measure the object directly. The second group consists of topographic methods and techniques. Finally, the third group includes methods based on photography (Martinez-Espejo, 2014):

- **Direct measurement methods:** Of all these methods, the measuring tape has been the most frequently used instrument, and it offers adequate quality if chained measurements are not performed. Metal and invar tapes generally offer better results because they are less subject to deformation due to stretching. Currently, manual laser meters are replacing measuring tapes, as they are being used increasingly more often, but the methodology is analogous. Besides the tape measures, we found other support tools that were also commonly employed, such as the plummet for defining the vertical, a water level to define the horizontal, and an archaeologist's comb for measuring mouldings;
- **Topographic methods:** These can be categorised in three groups according to the type of coordinates that they represent: planimetric (ordinate and abscissa referenced to a Cartesian system, triangulation and/or trilateration, or polygonation), altimetrical (using topographic levels or total stations) or the three-dimensional position directly (planimetric survey plus levelling taken by theodolites, total stations and 3D laser scanners);

- **Photogrammetric methods:** Depending on the data provided, these may be divided into planimetric surveys (monoscopic photogrammetric methods, stereophotogrammetric methods, simple rectification, mosaic and orthophotography) or three-dimensional surveys (monoscopic photogrammetric methods, stereophotogrammetric methods and chained rectification).

Of all these methods, we will focus on three-dimensional surveying with advanced tools, and especially on two types of surveys: "Image-based modelling" and "Range-based modelling". These two methods were chosen as they are currently the most widely used to obtain polygonal models with a high level of detail (Lerma, Cabrelles, Navarro, & Seguí, 2011).

Those instruments that allow the generation of a 3D image of the framed scene have to be established as three-dimensional sensors. These 3D images may be either the representation of the external surface of an object or the representation of everything that is inside the volume that is enclosed by this surface. In terms of their characteristics, the three-dimensional sensors can be classified in different ways. It is worth noting a particular type of sensor that uses non-ionised light radiation to explore the surface of an object, without any physical contact. Depending on the nature of the light used to take the measurement, we can speak of passive and active three-dimensional sensors. The passive ones are those that use the natural daytime light resulting from sunlight in open spaces or in closed environments using generic illuminators located in a random fashion with respect to the object. Active ones are, however, those that need encoded light to play a role in the measurement process.

In surveying with passive sensors, data is acquired through inexpensive and easily transportable systems. The images need a mathematical model to obtain information about the spatial position of the points of the object from two-dimensional images. To obtain accurate studies, a certain amount of experience is required in data collection and in processing the recorded information.

When surveying with active sensors, distances are surveyed directly and produce three-dimensional measurements of the surveyed object. Typical examples are the laser scanner (Time of Flight or triangulation) or beam projection systems (stripe projection system). The level of detail obtained by the distance sensors varies from tenths of a micron to centimetres. Generally, they are easy to use. Some of its drawbacks are that they are costly, bulky, specific to a distance interval and volume, and present problems with some materials. However, distance sensors are a frequently used and convenient working tool, especially in the case of not very experienced users, since they allow millions of points to be surveyed in a few minutes.

For this study case, we chose to use a hybrid system of data collection, consisting of both traditional and cutting-edge processes and systems. For the management and post-processing of the data collected, we used low-cost software and applications whenever possible. The traditional data collection methods allow us to obtain fast and accurate direct contact with the actual building or piece to be surveyed, thereby allowing their materials, shapes, imperfections, texture and colour (Garfella, Máñez, Martínez & Cabeza, 2013) to be felt – something we can hardly appreciate if this is not carried out in a direct sensory way (see Figure 2). However, the limits of these methods are those of human perception itself, and often have to resort to high travel costs and auxiliary means in order to endow a certain study with some degree of scientific rigour. Therefore, in this case we have used these traditional methods to make sketches and direct measurement with measuring tapes, profile combs, measuring tapes and manual laser distance meters to represent and measure the elements that are in the field of the human scale, without the need to employ unusual and expensive auxiliary media.

Figure 2. Examples of the traditional data collection method

Because of the technological means available to us nowadays, we did not want to miss the opportunity to improve the value of data collection by traditional and conventional methods using the topographic support based on the equipment we had. We have been able to provide topographic support to the manual fieldwork by collecting topographic data from various significant points on the model using a robotic laser-measurement calibrated total station with image and reflectorless measurement. This station includes among its main tools two Topcon iS-203 internal coaxial digital panoramic cameras with an apparent resolution equivalent to 4.8 megapixels, with an angular accuracy of ± 3" (1.0 mgon). This equipment ensured metric rigour in the measurements taken in the field. A single positioning of the equipment and a single reading by point radiation were performed in an attempt to eliminate as far as was feasible any possible errors due to the equipment or changes in position (see Figures 3 & 4). This topographic support was improved by giving the values of the local coordinates the overall position obtained by means of a Trimble model 5700, RTK GPS global positioning device, which is accurate to within centimetres (Garfella, Máñez & Martínez, 2015).

We also documented the Portal by means of mass non-intrusive data collection using a 3D terrestrial laser scanner based on flight time or pulses with a long-range invisible class one Topcon GLS 1500 laser. The reading rate of the Topcon is 30,000 points per second and has an accuracy of 4 mm at a scan range of up to 150 m and an angular precision of 6" (2.0 mgon), associated to a 2.0-megapixel integrated coaxially aligned digital camera. A single point was established from which to perform a frontal scan of the Portal. This point was located about 15 m away and a scan resolution of 1x1 mm was set at this distance. A normal field of view of 70° (± 35°) was set in order to cover the whole surface of the Portal. The scanner referred to above gives us a densely structured cloud of points. Because we only used one station position with a local coordinate system, no registration was needed. We can then transfer the data to a georeferenced system, supported by GPS data if necessary. The scanner gives us the local coordinate system (x,y,z), along with the intensity of the refraction of the laser beam that bounces off the surface of the object to be measured. In the lab and using specific software, it becomes possible to deduce, among other things, the colour index of each point with its RGB (Red, Green and Blue) or RYGCB (Red, Yellow, Green, Cyan and Blue) index. This is thanks to the photos taken coaxially by the equipment during data collection, in which the management software applies a colour pixel from the internal camera to each point in the cloud, since the point cloud and the photographs are linked because they have been produced at the same time during the process of measuring the piece. Therefore, we can obtain almost realistic models where the colour takes on its highest expression and, depending on the quality of the point cloud mesh and the pixel quality of the camera, we can accomplish effects with very realistic colours. This in turn will enable us to create a realistic virtual three-dimensional model that is very close to the virtual model we are looking for. The time needed to scan a surface of 6 x 10 m is 15', which produces a dense point cloud of 677,426 points (see Figure 5). The point cloud thus obtained

Figure 3. Topography data collection and coordinate data chart position on the floor layout as provided by Topcon Link v. 8.2. software

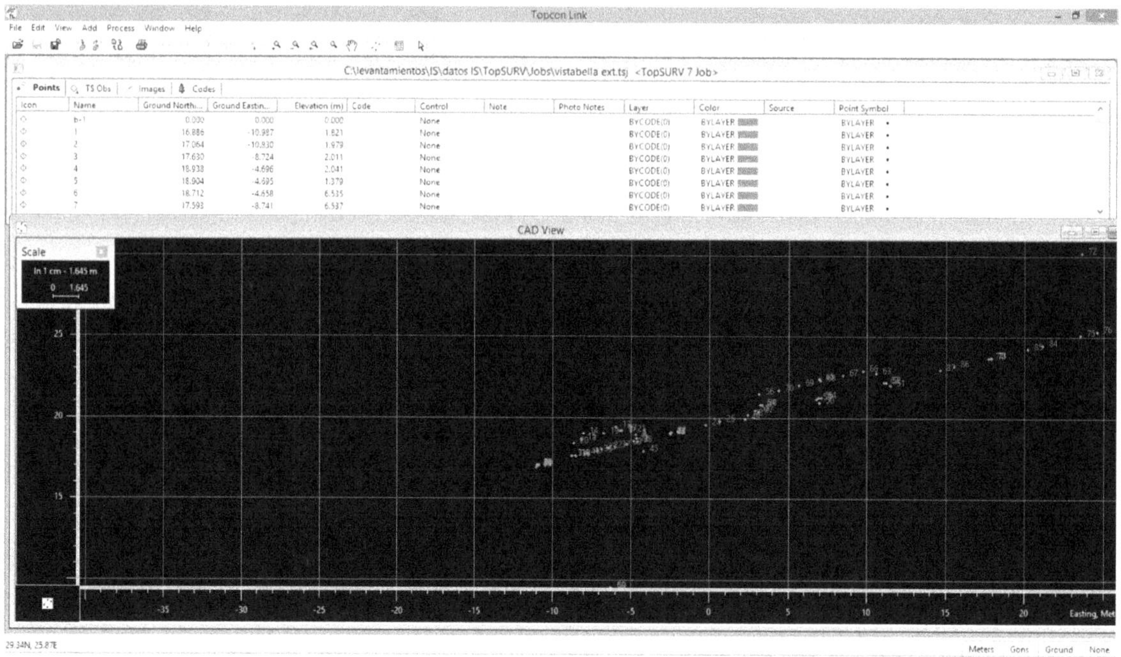

does not need any registration because we have only employed one station position and only one point cloud has been obtained.

In this case, point collection with both the total station and the scanner was performed without placing artificial targets with a known reflection index. Instead natural targets or support points on the building itself were chosen for use, determined by the operator's own experience.

The standard measurements or data collection were performed with the automatically activated two-axes angular compensator (horizontal and vertical), both for the scanner and for the total station.

The software provides us with meshes and an application with simple textures based on the construction of irregular triangular meshes, which in the case we are studying would cover about 1,272,266 units and last about three hours (see Figure 6).

Finally, the different data collections were completed with a 10.2-megapixel Nikon D-80 digital camera and a Nikon 5200 24-megapixel camera, fitted with a conventional lens between 18-135 mm, with a focal aperture of f/3.5-5.6 and Sigma wide-angle 8-16 mm lens, and focal aperture of f/4.5 to 5.6 mm.

With the photos captured with the 10.2-megapixel camera, plus the help of topography and all the notes made while performing the manual data collection, we were able to undertake the photogrammetric rectification of the façade with a certain degree of rigour. Prior to moving on to this stage, the camera and the focal opening were calibrated on graph paper, using a low-cost program. Thus, we were able to vectorise the Portal with some precision, although for the rectification work to be easily vectorisable it had to be carried out plane by plane both parallel and perpendicular to the portal (see Figure7).

Camera calibration and image orientation are critical processes. In the calibration phase it is important to distinguish between internal and external parameters of the cameras. The internal orientation parameters consist of: the constant of the camera (or focal length, f), position of the main point (x0, y0)

Figure 4. Image showing the collection of photographic data with two different zoom lenses, one more panoramic and the other more detailed, as shown by Topcon Link v. 8.2. software

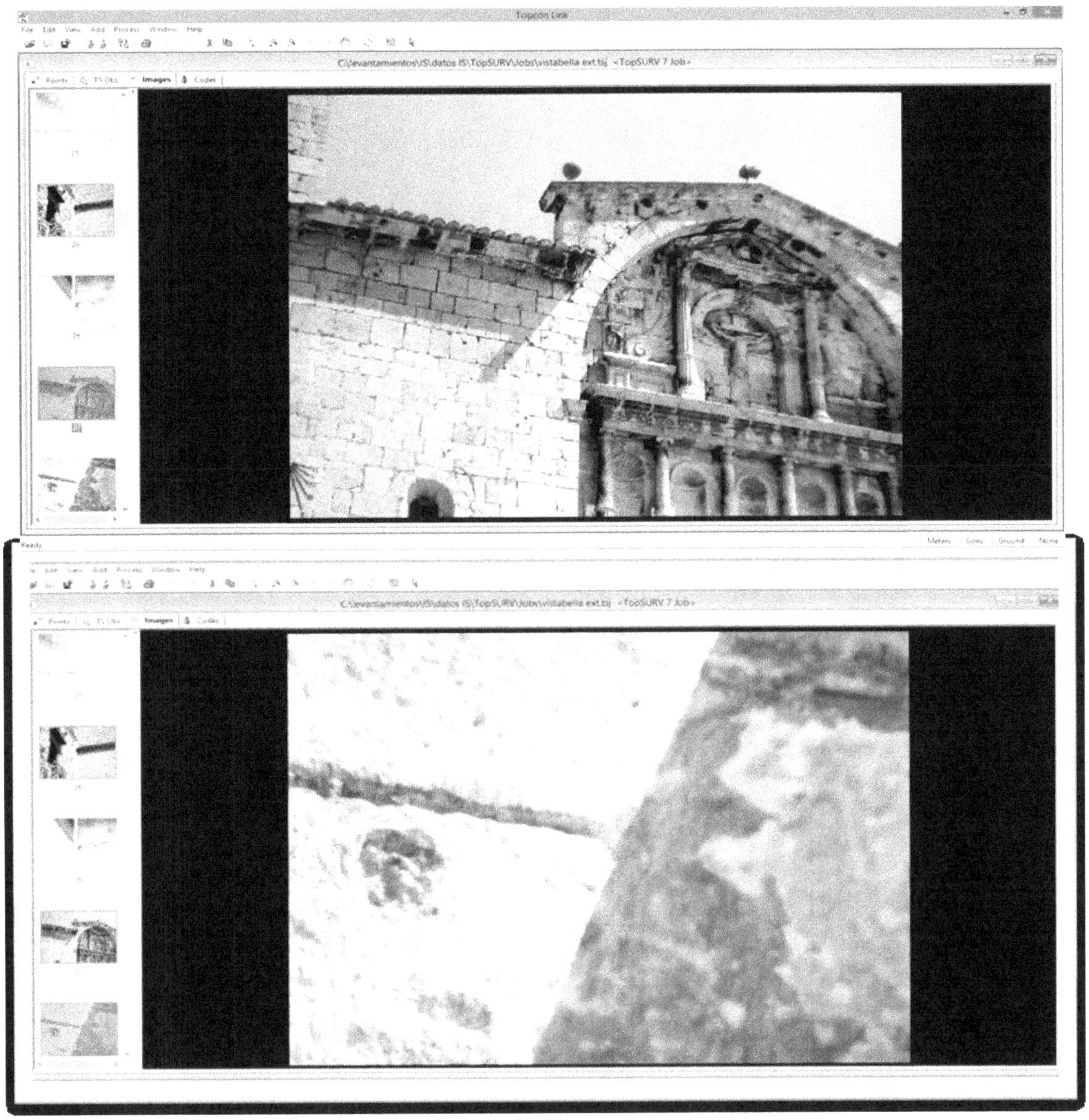

and several additional parameters used to model possible system errors due to, for example, lens distortion. The external orientation parameters consist of: the three spatial coordinates (position of the camera in space, generally indicated by X0, Y0, Z0) and three rotations of the prospectic centre of the camera in space (Pérez et al.,2007). Obtaining both internal and external camera parameters can generally be achieved using two images (relative orientation) or by a set of images (Bundle solution).

With at least two of the photos taken of the façade, one to the right and the other to the left, we can easily create a stereoscopic 3D view of the portal based on anaglyph images of the object aided by the free software application Anamarker (see Figure 8).

Figure 5. Image of the point cloud obtained using the scanner, grid step 1 x 1 cm and processed with Topcon Scanmaster software

Figure 6. Image of the mesh and texture of the point cloud with an irregular triangular mesh (TIM)

Figure 7. Ortho-image of the process of photogrammetric rectification using Axris software from the current status, its vectorisation, and rectified image

The next step was to document the portal by using cutting edge photographic capture systems, based on SfM (Structure for Motion), which consists in setting up point clouds or mesh models from groupings of the projections from different photographs by matching homologous points in the different exposures. In this process, no specific planning is needed, providing that the overlapping surface between the different photographs is large enough. This system is also known as Automated Digital Photogrammetry.

Figure 8. Anaglyph image of the façade created using Anamarker software

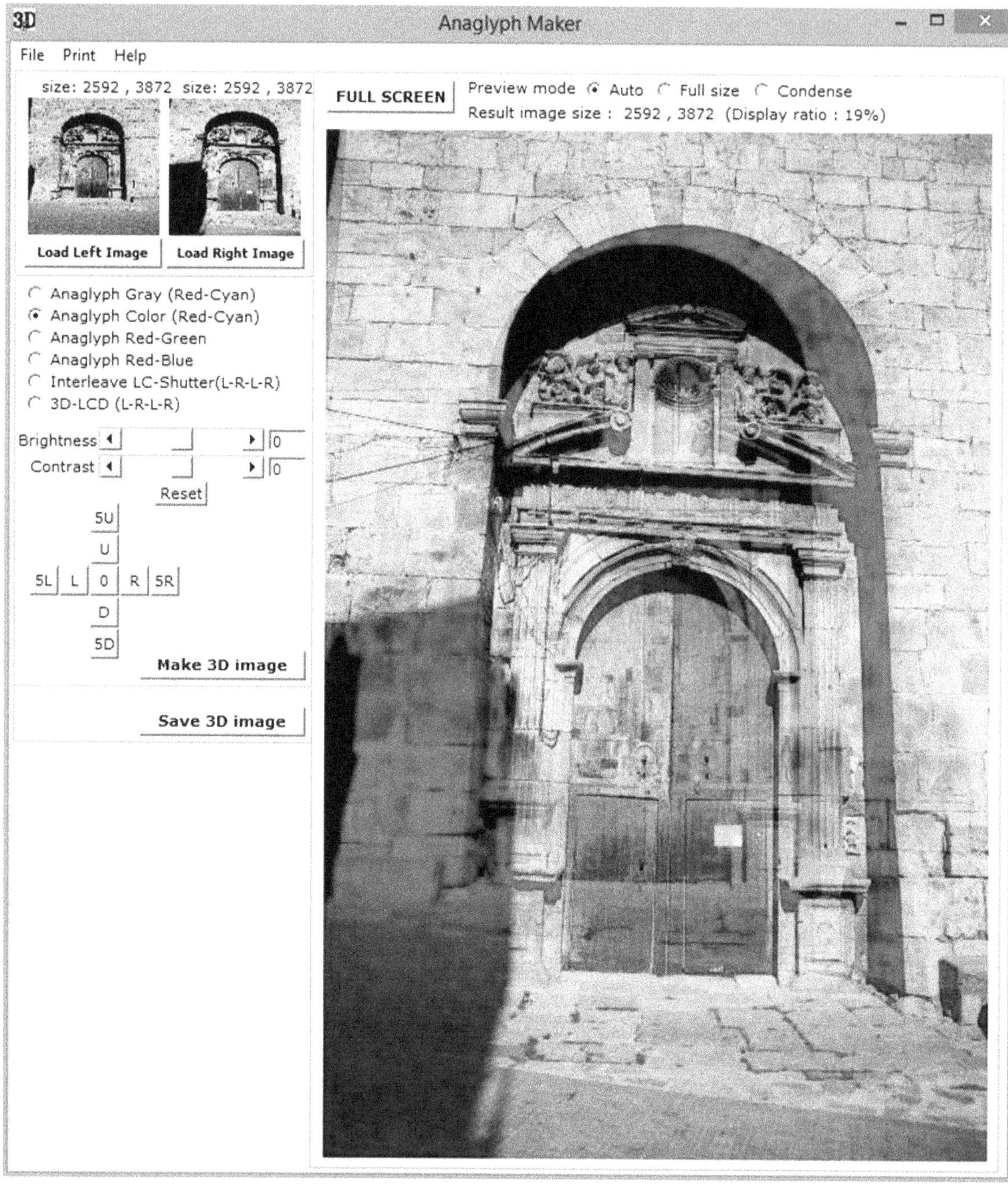

To achieve this, we used a commercial low-cost software application called PhotoScan by Agisoft and another free one with similar features, 123 Catch by Autodesk, as a testing ground.

With this purpose in mind, we took 22 digital photos of the area around the portal, from left to right, with the Nikon D-5200 with a resolution of 24 megapixels and a separation between photographic exposures of about one metre. First, we used PhotoScan to automatically orientate the 22 photographs,

Figure 9. View of the dense cloud of points obtained with the software Pothoscan from 22 photo captures

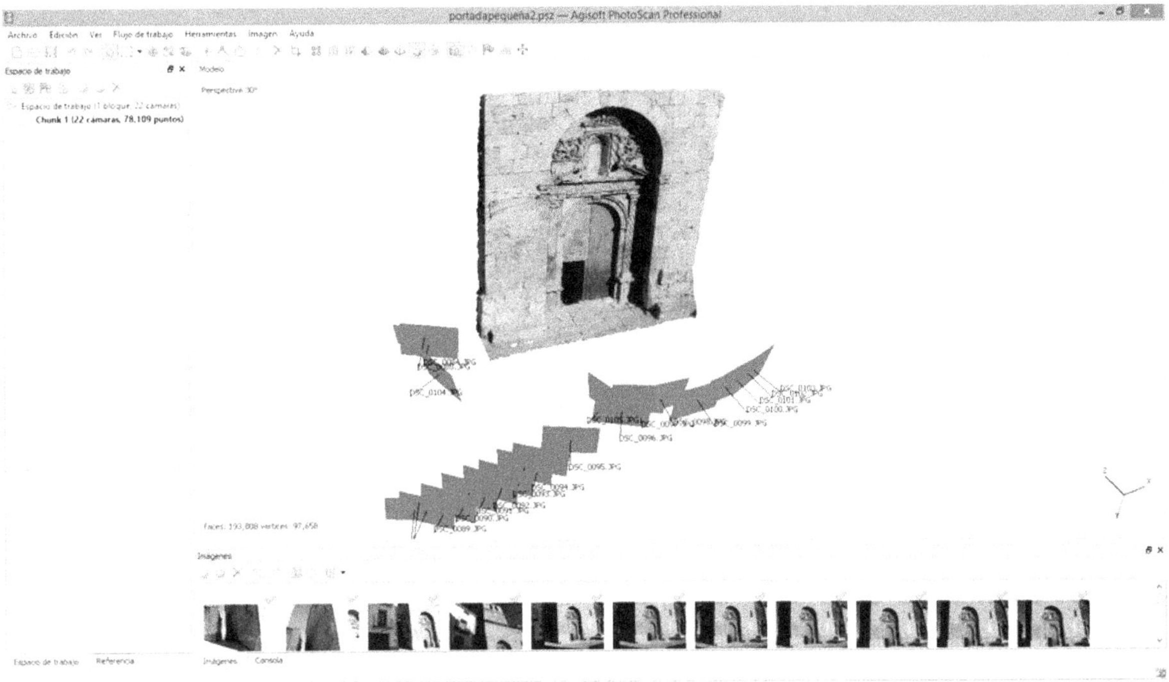

and then we went on to create a dense cloud of points that automatically provided us with 3,743,716 points (see Figure 9).

Once the dense cloud of points has been obtained, we can go on to carry out the structured meshing and texturing, which are also performed automatically, the result being 193,808 faces and 97,658 vertices. The software application offers the possibility of filling any holes in the mesh, although we did not make use of this feature so that subsequent comparisons among the different models would be similar and homogeneous (see Figure 10).

After meshing and texturing, we can automatically obtain different images of the portal by applying colour and lights (see Figure 11).

Finally, we export the point cloud obtained in (ASCII compatible) TXT format, where data include relative position (x, y, z), colour (R, G, B) and normal vectors in the three directions.

The next step is to process the same 22 photographs in the same way using 123 Catch software, which automatically aligns the photographs and creates points, meshes and textures. We have the possibility to select a working area and crop and clean it (see Figures 12 and 13).

Once the surface has been discriminated, we export the data thus obtained in OBJ format, which is compatible with many commercial mesh or solid applications such as MeshLab; on opening this free software application we checked the results and found that in this case we had obtained 73,030 points (see Figure14).

MeshLab is a very interesting free software application that has a powerful engine for moving and managing data on the screen and is capable of moving millions of points with ease almost instantly. Likewise, this application allows us to improve, soften and light the meshes and texture obtained, thus enabling us to construct a solid model that can easily be printed on 3D printers. Another important feature

Figure 10. Image of the texture obtained

Figure 11. Orthographic image of monochrome shading

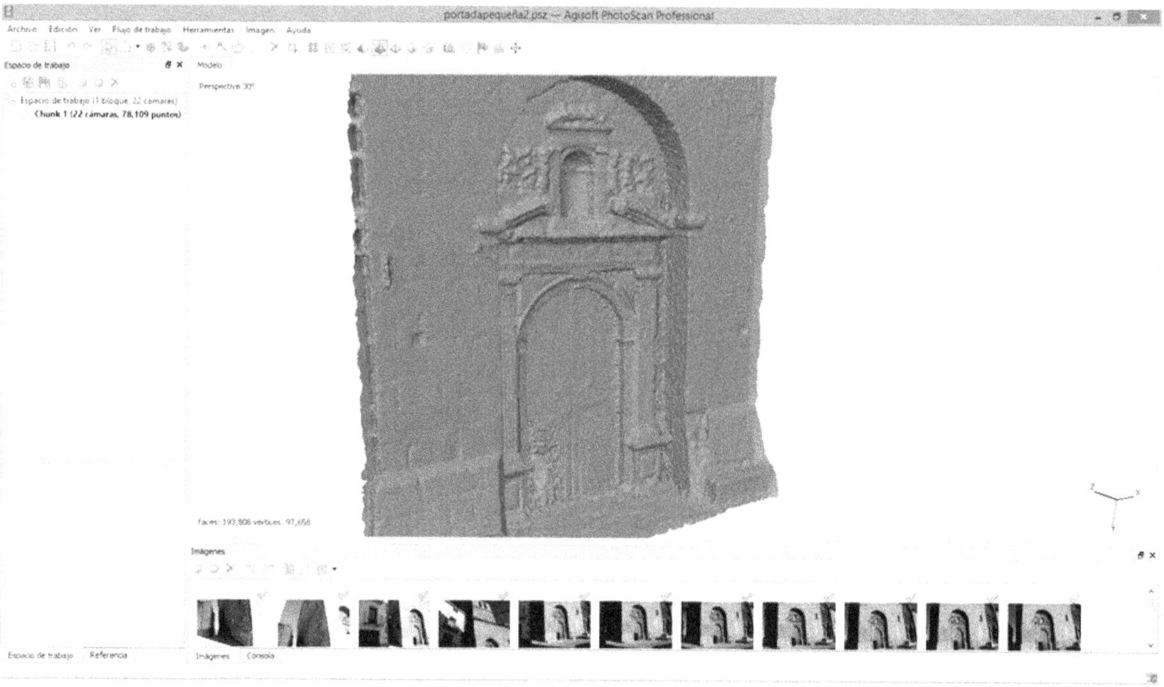

Figure 12. Image alignment and the texture obtained

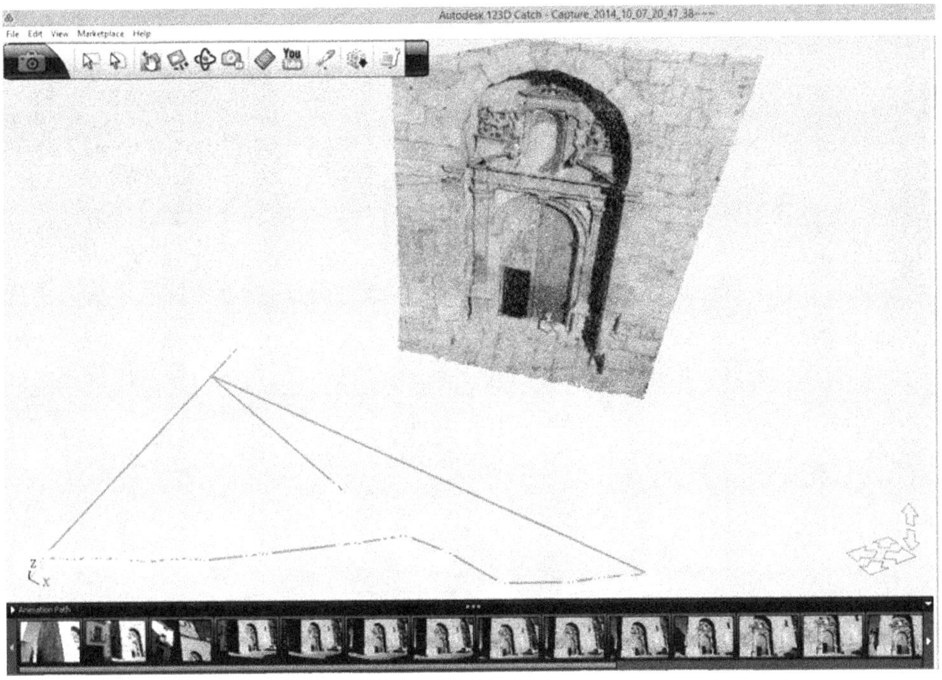

Figure 13. Image point cloud and mesh

Figure 14. Point cloud and mesh obtained with 123 Catch using MeshLab software

of this application is the number of formats that are supported for import as well as export. We therefore employ it as an interface for improving application data that are exclusive of measuring equipment, allowing us export to more common files (Garfella, Máñez & Martínez, 2015).

Method of Graphic Restitution

Modelling or 3D printing can be classified into two major groups, according to the form of creation of the model. One of them will be by addition, or Additive Manufacturing, (AM), consisting mainly in adding particles or elements to the object, to create the total volume. The other method is by subtraction; consisting in removing or shaping material to get the desired form.

The Creation of the physical models by addition has been carried out by different techniques of rapid prototyping of prototypes by the method of Additive Manufacturing, (AM), known as "3D printing". During the process of manufacture or construction of its parts, the material is deposited slowly so that it progressively creates the geometry of the piece; they are also known, as in layers manufacturing techniques (Máñez Pitarch et al., 2016 & Martínez-Moya, Gual-Ortí & Máñez-Pitarch, 2018). Three printing techniques have been used: printing by deposition of plastic thread plastic (FDM), stereolithography (SLA), and printing powder (3DP).

The FDM (Fused Deposition Modeling) printer is fed by rolls plastic filaments PET (polyethylene terephthalate) based, PLA (polylactic acid) or ABS (acrylonitrile butadiene styrene), which melts through an extruder head, which deposits it in its precise corresponding position, layer by layer, on the printing bed. After the machine is warmed, a strand of material 1.75 mm or 2.85 mm diameter is extruded on the platform through a nozzle that moves on 3 axes (x, y, z). The platform descends a level with each new layer applied, until the object is printed. During printing, depending on the object to print, the use of supports is required to improve the quality of the model. Its function is to support the protruding parts of the 3D model, given that, if not done, certain models can't be printed. These supports, in our case, have been built with the same material as the printed object, which once completed must be cut or removed. The used equipment has been the WANHAO Duplicator Printer i3Plus with a print speed of 50-250 mm and a thickness of 0.01 mm with strand of 1.75 mm in diameter has been used to print with wire. The maximum print size is 200 × 200 × 200 mm. The material used has been Green PLA plastic (see Figure 15).

The second experienced and used additive printer has been the one known as resin printer. Through the process of Stereolithography (SLA), it uses as printing material a liquid resin solidified under the effect of ultraviolet (light curing). This printer works downward, from top to bottom, in such a way that the photosensitive liquid is introduced on a base tray, which must be perfectly levelled, on which, in our case, a light beam is projected layer by layer, that by striking the photosensitive liquid solidifies and sets up the shape of the object to print. A proper arrangement of the part to be printed must be planned in order to make it freestanding during the printing process. For resin printing it has been used a 3D printer XYZprinting Nobel 1.0a with Stereolithography (SLA) technology. The printable area maximum size is 128 × 128 × 200 mm and the material used has been transparent photosensitive acrylic resin. A thickness of 0.05 mm has been used. (see Figure 16).

The powder printer used, as previously indicated, works creating the model through the addition of material, prior injection on a foundation of a binder material, which, pass after pass, gets attached and secured on a specific clogged dust tray. In successive passes, it forms and grows upright. These models are embedded in a block of loose powder, so at the end a removal of the material and removal of the interior of the model by manual means must be done, suction or blast of air as the case may be. These models are often designed with interior holes for the evacuation of the dust inside the model, leaving

Figure 15. Image deposition printer

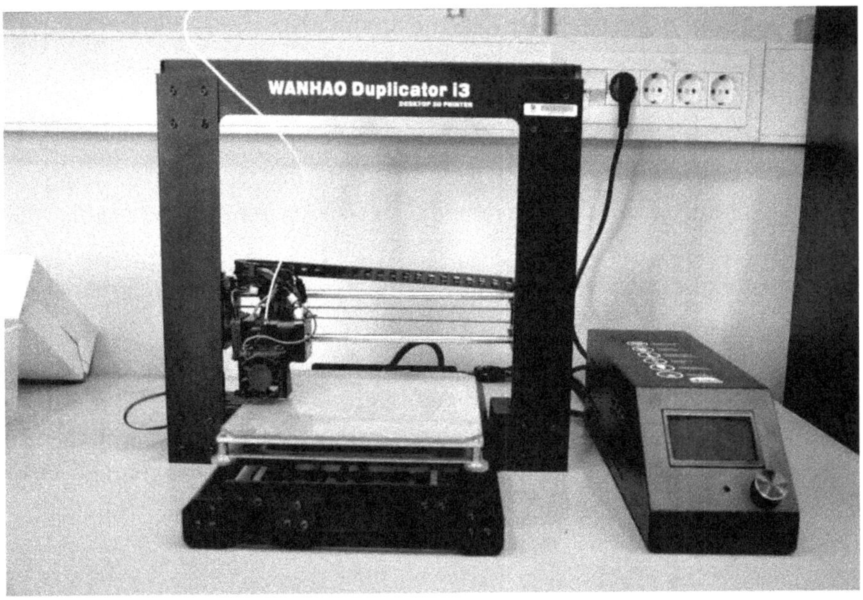

Figure 16. Resin printer

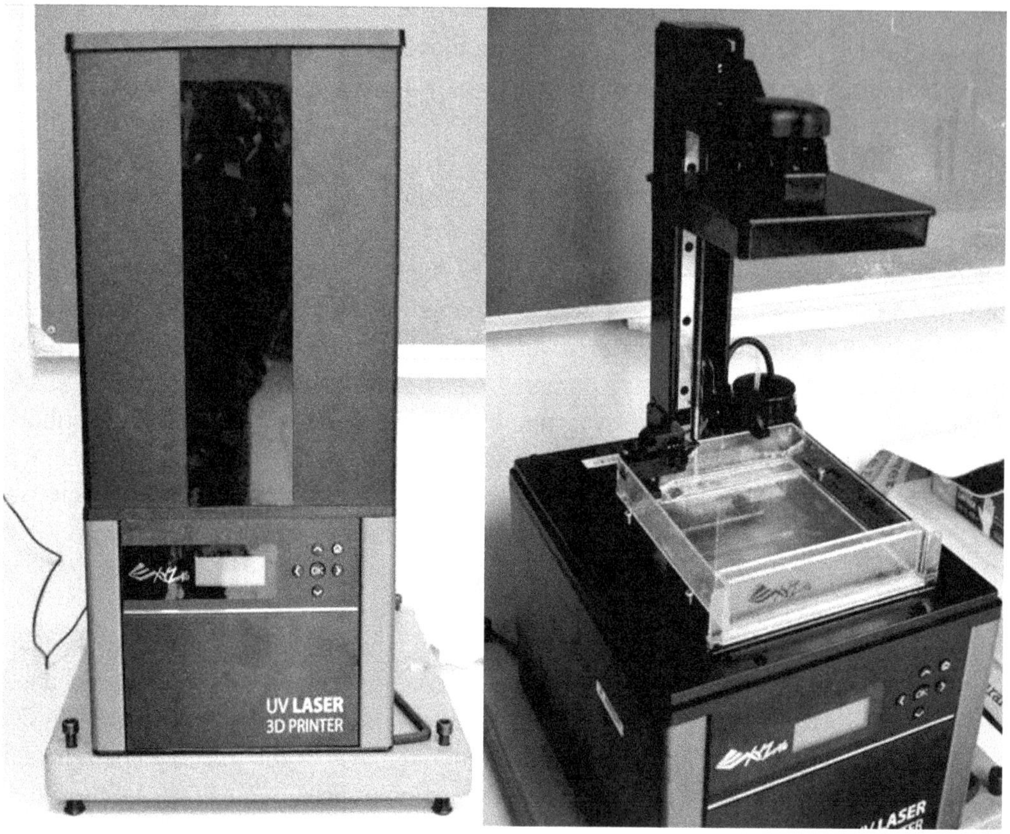

initially printed the exterior surface or the perimeter with a certain thickness, which allows optimizing the material. The used equipment has been the 3D Mod. Z 310 Printer. In this case the material is special powder printing zp® 150-caking zb® 60. The maximum print size is 280 × 220 × 200 mm and the thickness is 0.1 mm (see Figure 17).

Zprint team's own software, which transfers the digital format in STL or OBJ format to the printer, allows us to have a visualization of the part to be printed, it also allows us to move it, shift it, scale it and adjust it when necessary to the printing platform's size. It can even simulate the reproduction of the part (see Figure 18) and has a small rendering engine. We are currently investigating the use of atomized ceramic material.

Finally, a subtraction system has been used, consisting fundamentally in a numerically controlled table top milling machine.

With all the documentation procedures, it has been possible to generate a surface in three dimensions through the point clouds and the integration software with more or less depth. For that, before printing, a solid digital volumetric model has to be generated, able to be recognized by the various printers, usually in STL format. These models should be perimeter closed and have no holes. Depending on the printer used, those models may be completely solid on the interior or lightened only with a surface layer of finite thickness.

Figure 17. Image printer powder

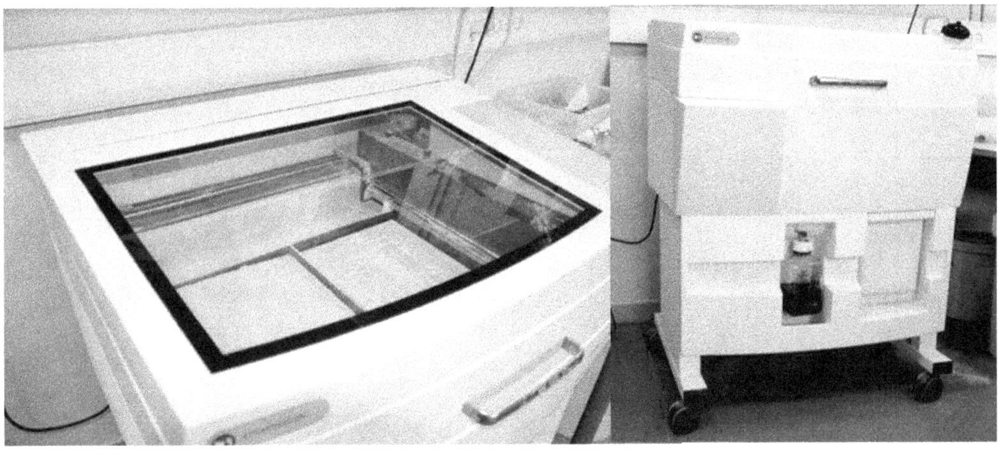

Figure 18. Image of the cover through the Netfabb program

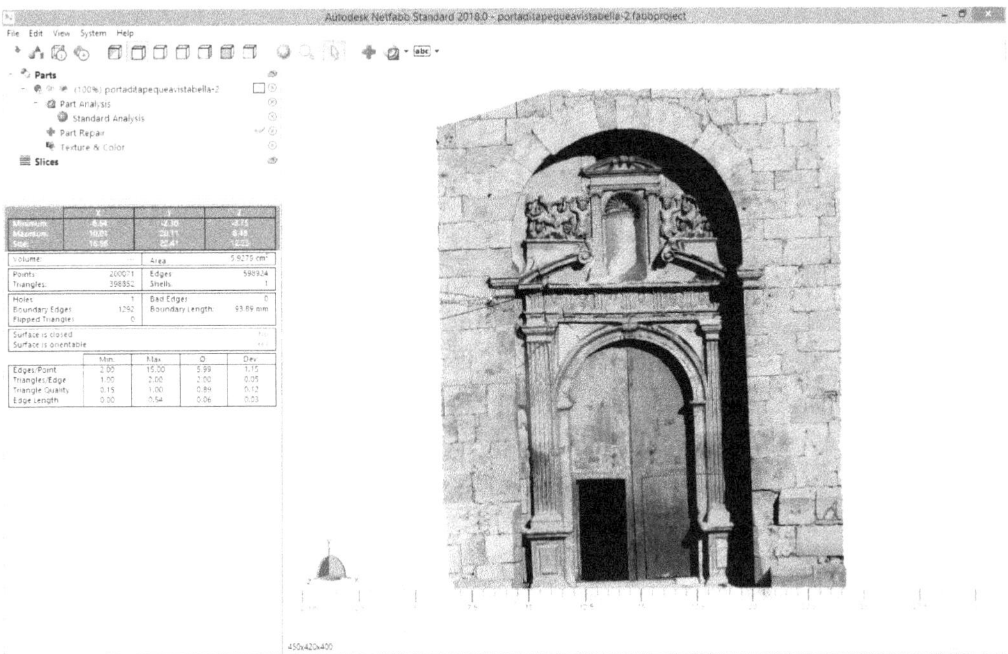

Prior to the printing process, it is suitable to proceed with a verification of the digital model, in order to check that there are no gaps in the surface and that the generated meshes are completely closed, as well as check that the faces of the mesh are oriented according to its normal. Otherwise, the printing process will fail. This verification process is carried out by means of specific software, in this specific case Autodesk Netfabb tool has been used. (see Figure 19).

The management software used to transfer the model to the printer has been the Cura (see Figure 20).

The management software used to transfer the model to the printer has been the Builder 3D (see Figure 21).

Figure 19. Image program management for the printer from dust

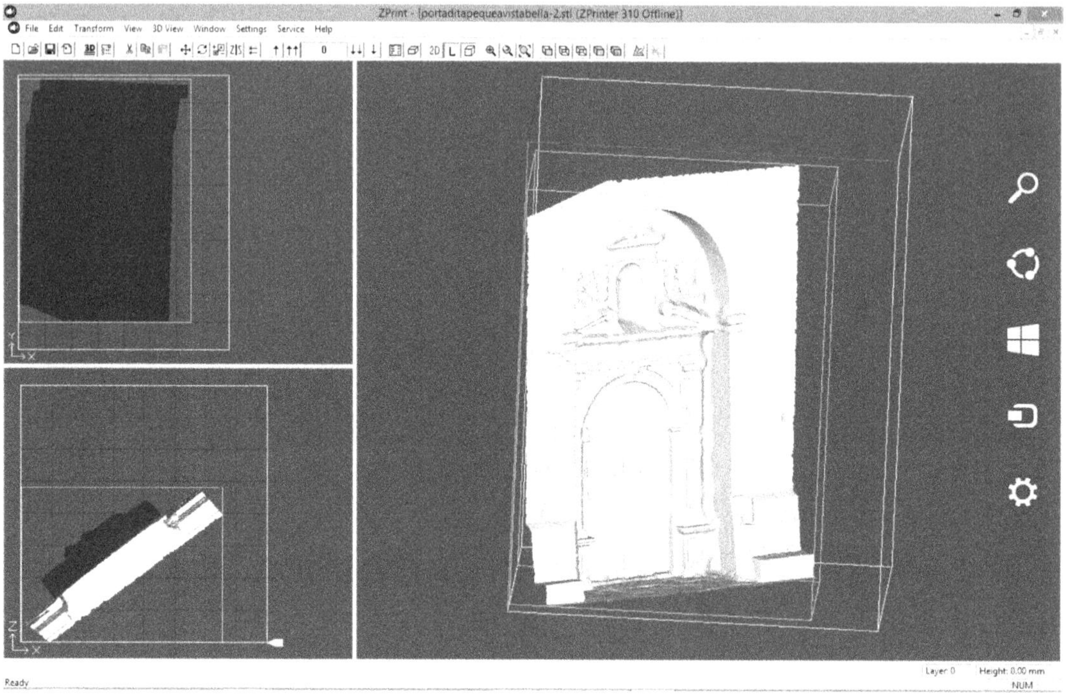

Figure 20. Screen shot of Cura software

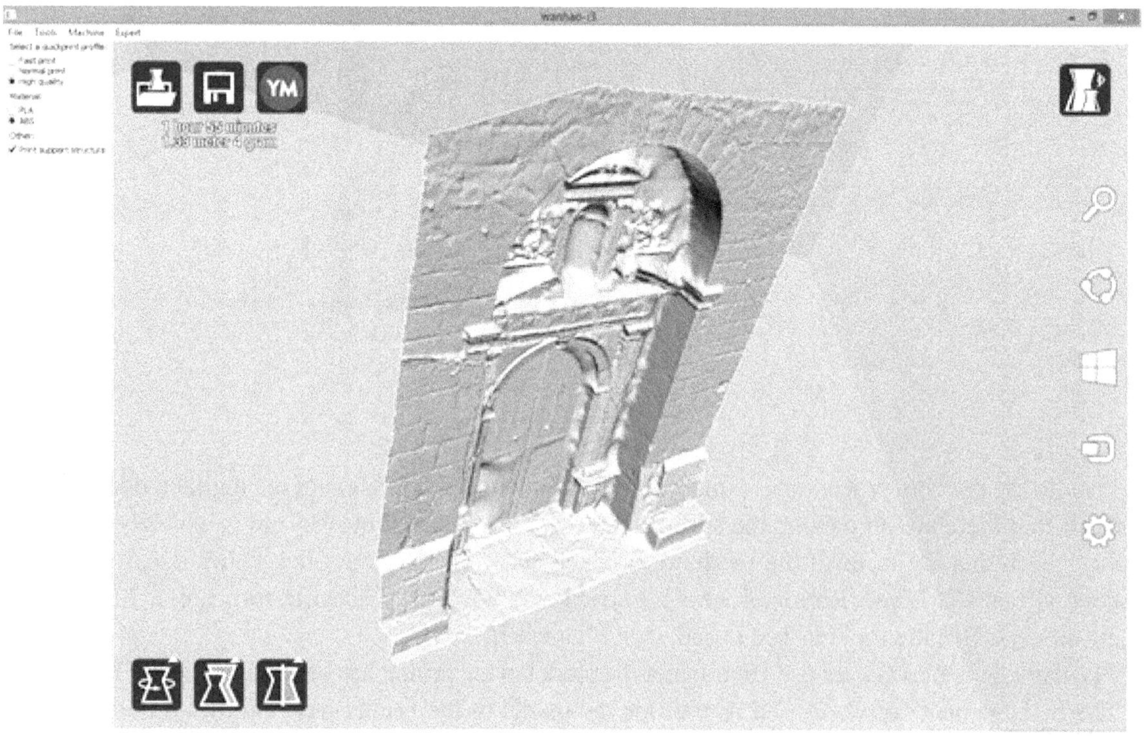

Figure 21. Management of the point cloud through the 3D Builder

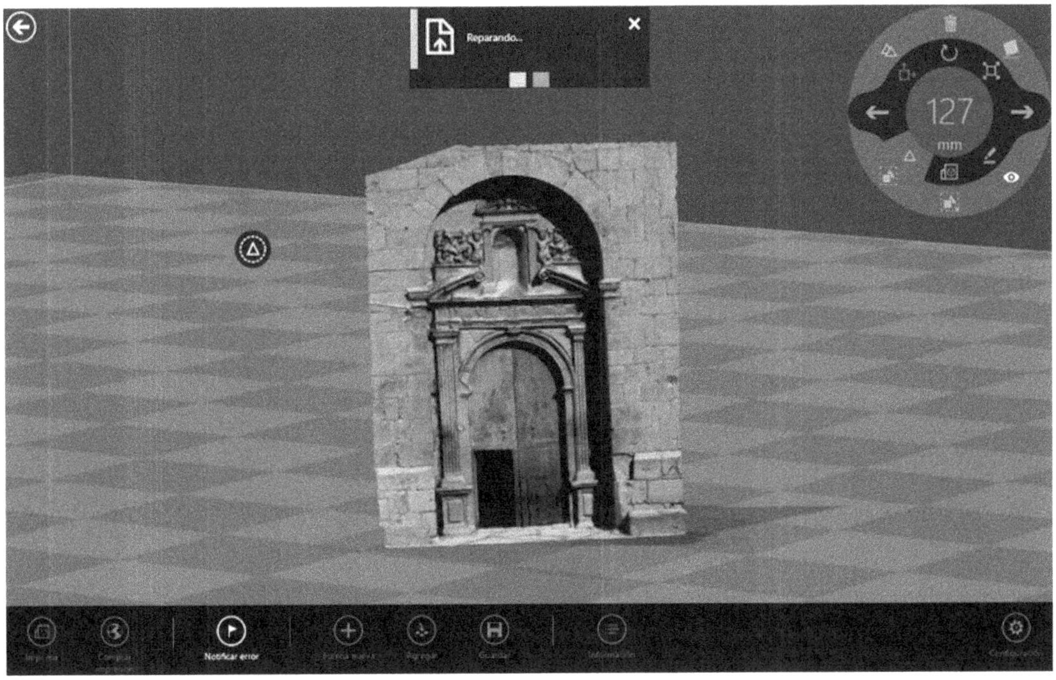

The subtractive printing, consisting of a milling cutter, directed under numerical control, that produces displacement in both horizontal directions and in depth through the drill or tool (milling cutter). From a piece of soft material, attached to the table, a series of slits are made in its interior by cutting or emptying the material in both horizontal axis and at certain depths, creating bass-reliefs with millimeter accuracy depending on the tool, it is not possible the conclusion of right angles, being all of them lightly blunt. The used equipment has been a three-axis milling machine Galileo (see Figure 22).

The software used for printing with this equipment has been FotoCam Software engineering PC, by means of a plug-in of Photoshop from a bitmap image.

With all of them, different printings have been made to check the quality, accuracy and results (see Figure 23 and Table 1).

NEW TRENDS

In relation to the application of new trends and once the accuracy of the procedures has been analyzed, a new system to verify the suitability of the making of the models has been employed, all of this using the same photographic resources previously used with the 22 photographs taken of the small door of the Church. This test has materialized with a commercial type platform such as is the tool Recap of Autodesk, successor and catalyst for the previous resources of the mentioned mercantile, as have been 123 Catch, Memento, and finally Remarke. This integrator software has a module to be able to collect information both through a scanner or cameras, using the module Recap Photo, also from Autodesk. The program is very complete and the point cloud management module from the laser scanner is able

Figure 22. Printer using numerical control milling

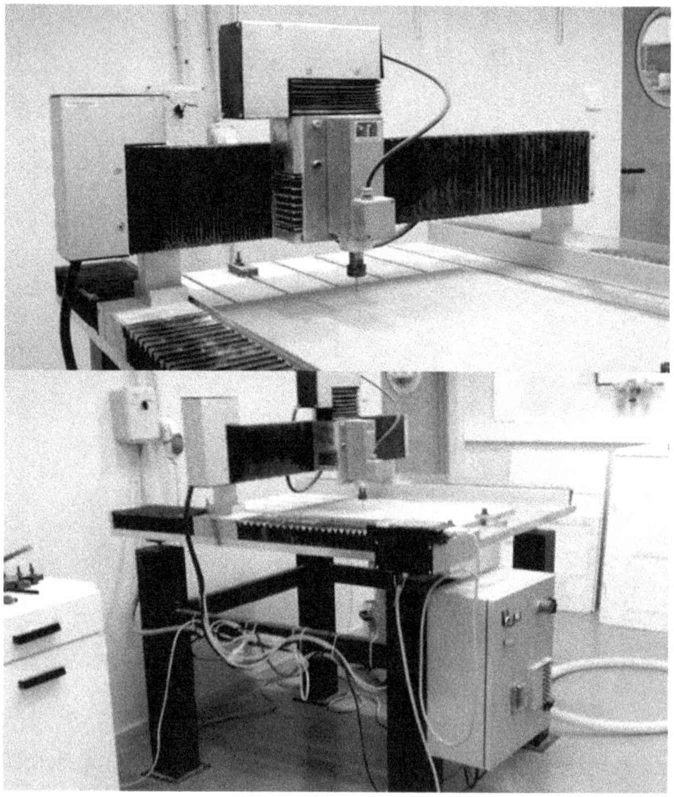

to automatically proceed to the automatic registration if the cloud is a structured cloud with all points oriented in their normal (see Figure 24).

The photo integration module, proceeds similarly to the Photoscan of Agisoft, given that automatically proceeds to the calibration of the camera, the orientation of the different photos and finally to the creation of a mesh of uniform triangles, to which the texture of the photo can be applied to. This way, very visually engaging virtual models are obtained (see Figure 25 and 26).

In this line of new trends, the graphic tool Sketchup Trimble Inc. has begun being used as a visualizer. For the realization of virtual reconstructions and efficient presentations (see Figure 27).

RESULTS

Finally, in this section, we will compare the differences among the various point clouds obtained through the two SfM systems and through the 3D laser scanner. We favoured the 3D laser scanner as the more reliable data as, to date, it is the most proven system. We chose CloudCompare as our reference software, which is also free.

Hence, first we will load the point cloud obtained using the 3D scanner (see Figures 28 and 29).

We then load the dense point cloud obtained using PhotoScan in the same project as can see in Figure 30.

Figure 23. Result of the different impressions of resin, yarn extrusion, milled and dust

Table 1. Comparative data of 3D printers

Results	Accuracy	Quality	Durability	Size	Time Printing	Cost	Overall Result	Color
Powder Printer	Very good	Good	Scarce	Large format	Medium-high	Raised by Material	Positive	White Monochrome
Thread Printer	Good	Regular to good	Very good	Medium size	Reduced	Economic	Positive	Monochrome Color
Resin Printer	Very good	Regular	Good	Small	Medium	Medium	Regular	Transparent Monochrome
Numerical control	Good	Good	Very good	Very large size	Media	Economic	Very good	Depending on Base material

Figure 24. The cloud photo points in RGB colour of the scanner with the Recap

Figure 25. Model obtained with Recap Photo

Figure 26. Model of meshing retrieved with Recap Photo

Figure 27. Model of meshing retrieved with Sketchup

Figure 28. Point cloud image obtained with the scanner by means of CloudCompare (Garfella, Máñez & Martínez, 2015)

Figure 29. Image of point cloud and texture obtained with the scanner by means of CloudCompare (Garfella, Máñez & Martínez, 2015)

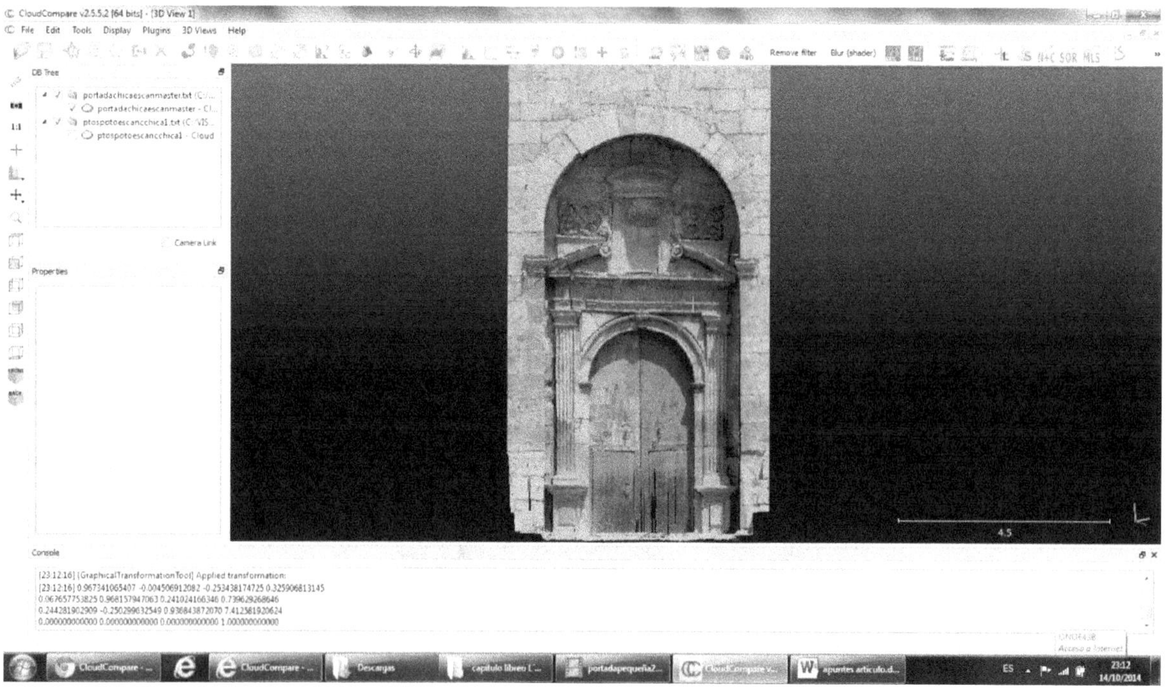

Figure 30. Point cloud image obtained with Pothoscan by means of CloudCompare (Garfella, Máñez & Martínez, 2015)

Once the two clouds of points have been loaded we compare them through the semi-automatic engine contained in the program itself in order to orient and, if necessary, scale the clouds by an automatic orientation and scaling system, an error of 1000^{e} - 20 being set as the defining parameter for all the models studied as we can see in Figures 31, 32 and 33.

When the two point clouds have been aligned by the software, the program itself offers us an array of data and statistics about the point clouds being compared. On examining the results, we can see that in the central part of the point clouds to be compared there is practically no difference among them, since this is the approximately 6x10m area captured with the terrestrial scanner, while in the peripheral areas of the point cloud the dispersion range is greater because no reference points can be found with which to compare their difference (see Figures 34 and 35).

Similarly, we now load the point cloud obtained with the 3D laser scanner and the mesh obtained by the 123 Catch software application (see Figure 36) and then align or pair the two models (see Figures 37 and 38).

From our viewpoint, the combined use of surveying with scanner and multi-photo systems notably reduces the time and the number of operators required. As much as it becomes possible to overlap different works, taking advantage of the time intervals in which the laser is scanning to take the photographs and draw the sketches. In this particular case, we have only set up one topographical station; if different stations are employed, the time increases significantly.

Finally, and once the digital model has been developed, we can continue with one of the models or meshes obtained to produce a digital solid model, which we can reproduce on a smaller scale as an inverse architecture (Celani & Milan, 2007), by preparing a tangible physical model either by modelling

Figure 31. Frontal image of the two textured point clouds obtained with the scanner and Pothoscan using CloudCompare (Garfella, Máñez & Martínez, 2015)

Figure 32. Frontal image of the two point clouds distinguished by yellow and red tones, before alignment and comparison using CloudCompare (Garfella, Máñez & Martínez, 2015)

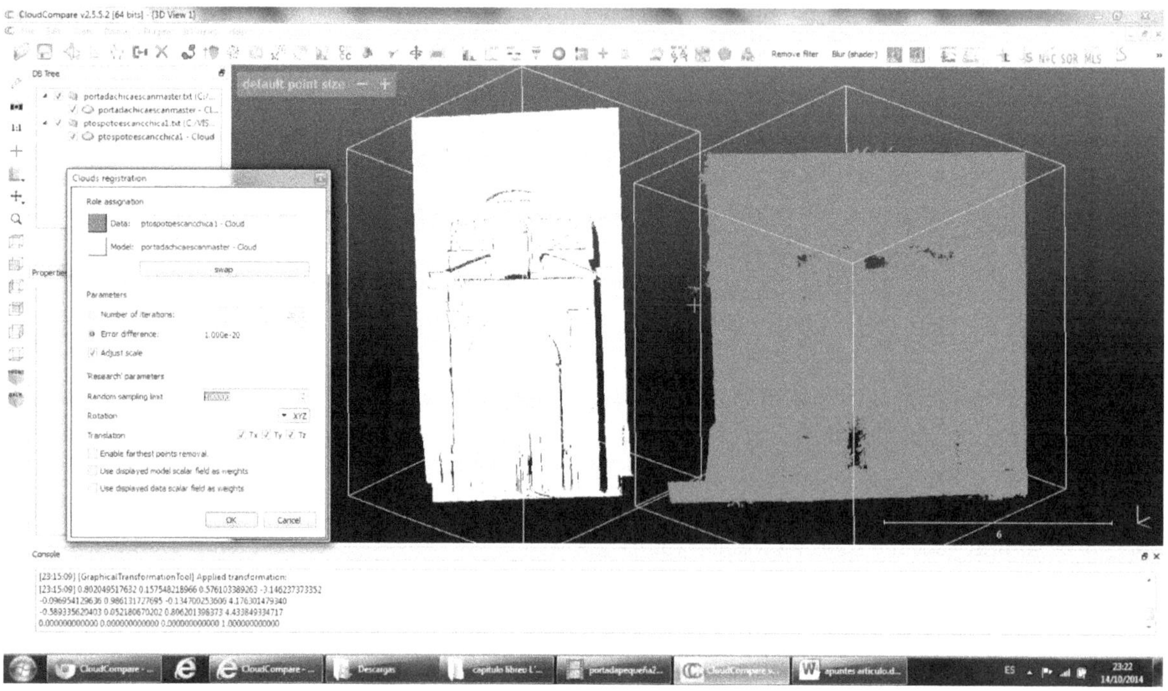

Figure 33. Frontal image of the two point clouds once they have been aligned for comparison using CloudCompare (Garfella, Máñez & Martínez, 2015)

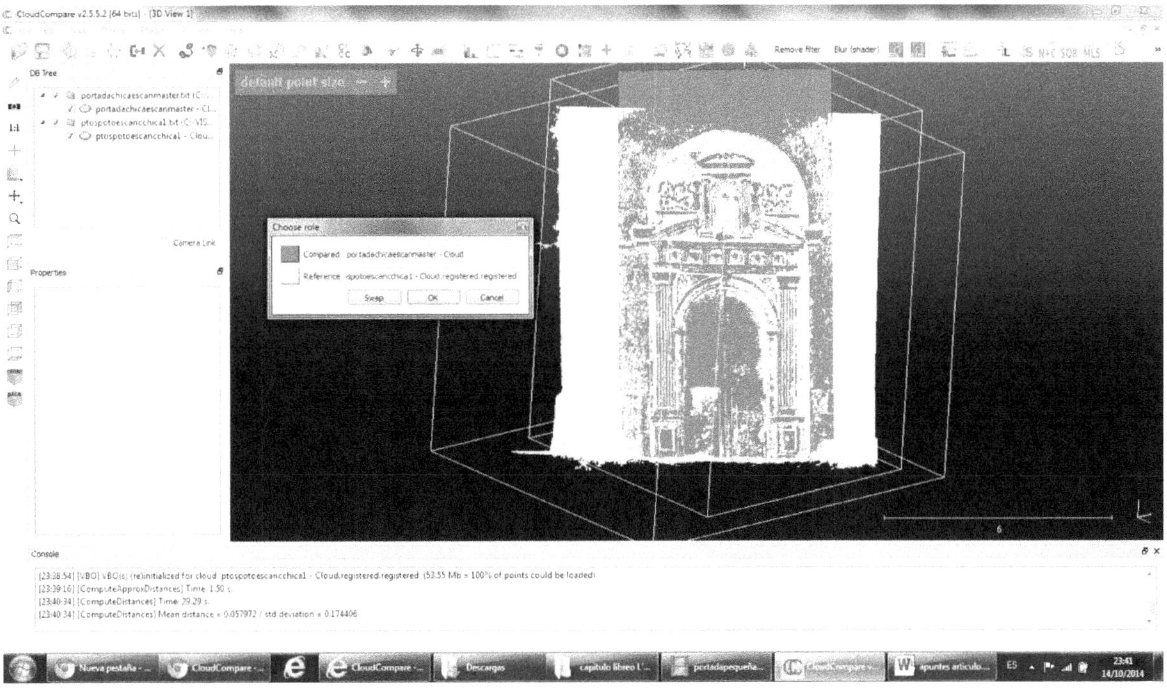

Figure 34. Frontal image of the two point clouds once they have been aligned, with the tone indicating the dispersion in terms of the distance between the two point clouds through CloudCompare (Garfella, Máñez & Martínez, 2015)

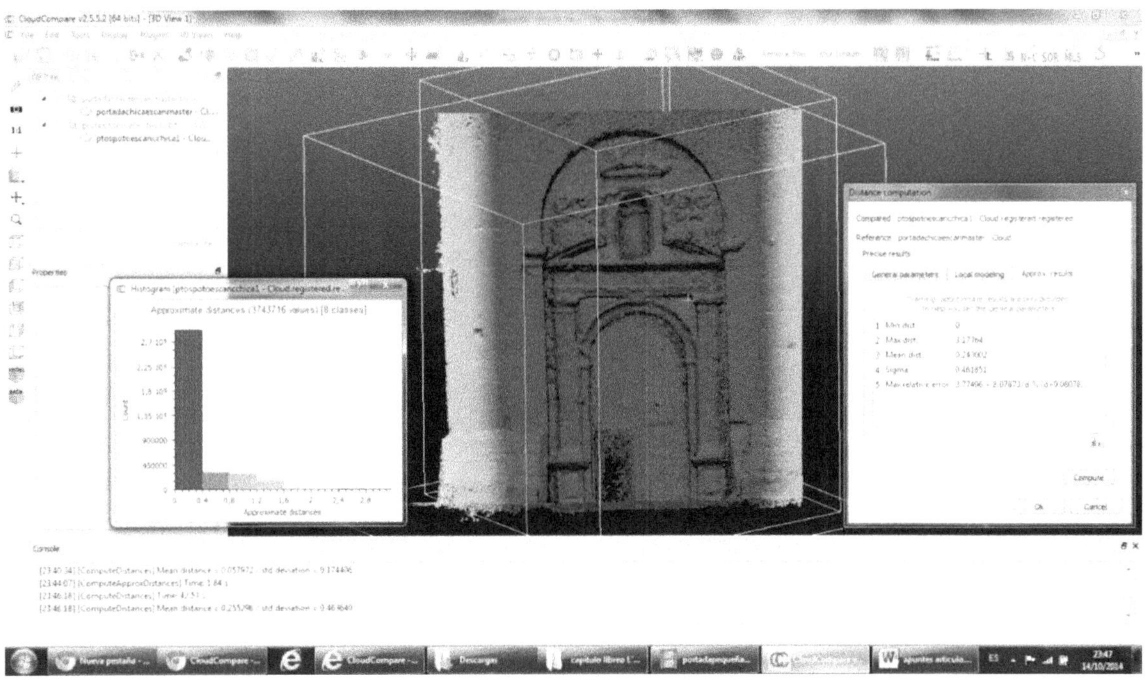

Figure 35. Image of the two clouds of points and different comparison parameters through CloudCompare (Garfella, Máñez & Martínez, 2015)

Figure 36. Frontal image of the two textured point clouds obtained with the scanner and 123 Catch using CloudCompare (Garfella, Máñez & Martínez, 2015)

Figure 37. Frontal image of the two point clouds once they have been aligned for comparison using CloudCompare (Garfella, Máñez & Martínez, 2015)

Figure 38. Image of the two clouds of points with different comparison parameters using CloudCompare (Garfella, Máñez & Martínez, 2015)

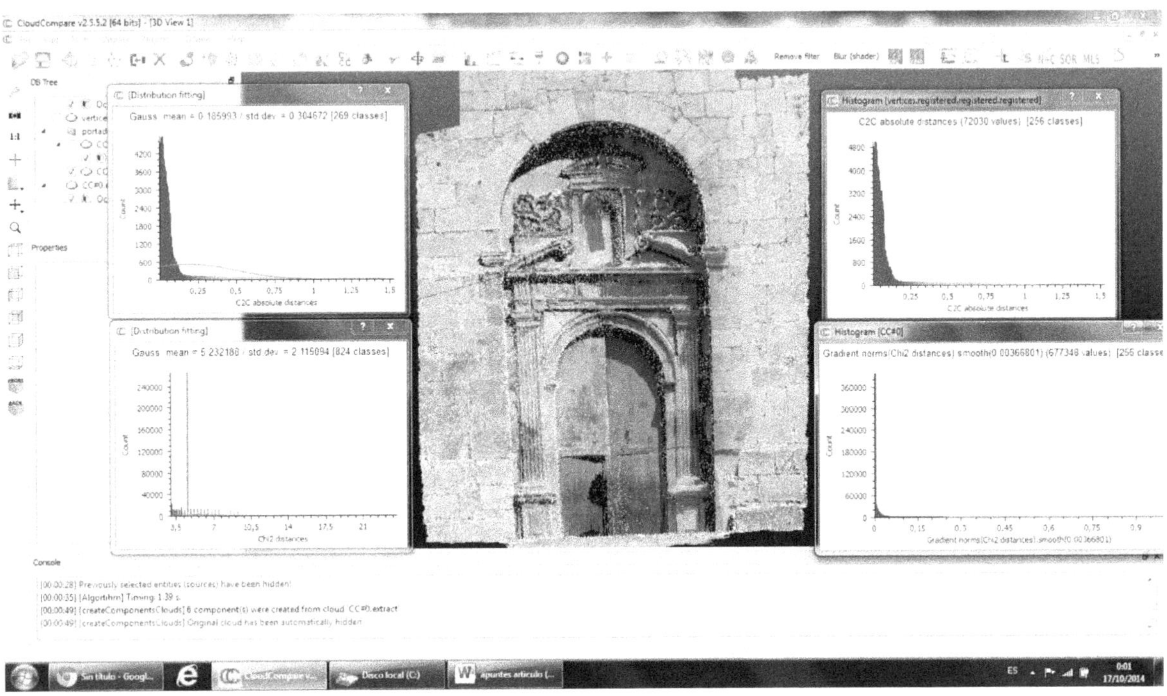

on a 3D printer or with numerical control equipment or milling (Gual, Puyuelo, & Lloveras, 2004) as we can see in Figure 39. In recent times these techniques have moved towards the physical impression of an almost realistic 3D model that can even include colour (Garfella, Máñez & Martínez, 2015).

In the case of the 3D models made with thread and photosensitive acrylic resin the results are acceptable, although the texture of the materials is not optimal for this type of architectural pieces, there also exist a great limitation on the print size. The advantage of these materials is their toughness and resistance to bumps and drops.

Another printing tests were conducted through the dust printer. In this case it is obtained a more architectural finish, with a texture closer to the original models (stone). The disadvantage of these models is its fragility. They can be treated with a consolidating, even though they lose their texture and original colour.

FUTURE RESEARCH DIRECTIONS

One possible future research line would be to study graphic methods that allow us to cover easily in the models obtained by 3D scanner, shadow areas not perceptible from the normal height, or in inaccessible places for them, by using photographic images produced by cameras installed in drones or small cranes, that allow us to raise the point of view and can view these elements (cornices, covers, bell towers) or access any area of difficult access (extrados vaults, narrow stairs, etc.).

CONCLUSION

According to everything outlined above, from the early steps of the architectural survey, we can conclude that all the systems are compatible with one another and that one should not necessarily exclude another. Instead, they can support or enhance each other by providing the improvements needed by another while

Figure 39. Image of a reproduction by extruder with red plastic thread and two physical reproductions of the Portal made by numerical control milling. On the left: in plastic, center: in plaster, and on the right: in wood.

having its own shortcomings offset, and this is why we cannot do away with the traditional collection of data even if it only involves the use of small sketches or diagrams. For a dimensional study, depending on the resources we have available to us, we will have to resort to traditional methods of direct or indirect measurement based on traditional or electronic topography, which will endow our work with a certain degree of geometric and dimensional accuracy.

As for the mass data collection systems used in this case, we must highlight the accuracy and amount of information provided by the terrestrial 3D scanner, although stationing time, travel requirements and moving equipment is rather laborious. Yet, in this case it has been cost-effective and necessary to achieve the aims of this study, because only a single station is needed. As a weak point of the system, we can highlight, for example, the high cost of both the measurement equipment and the management software.

In contrast, through the two tried and tested programs (PhotoScan and 123 Catch), multi-photo capture systems are very convenient as far as data production and subsequent processing are concerned. However, as we understand it, the weak point of this system lies in the total automation of the process without the operator being able to intervene in the process of control and data management. In contrast, their main advantage is their economy and ease of use.

On evaluating the quality of the metric results obtained, we can state that in the test conducted if we compare the different point clouds accomplished by the scanner (see Table 2), and after testing the other two point clouds from the multi-photo systems, the deviation or error is practically negligible. In consequence, the three clouds can be considered to provide adequate mass information about the object to be represented with a certain degree of accuracy and rigour.

A weakness of this kind of work is the invisibility or lack of perception of the parts of the objects to be documented. In this case, because of its size (8.5 m high), the Portal does not allow us to photograph the cornices or reliefs with standard means and equipment. Likewise, perhaps the scanner cannot scan certain areas that produce their own shadows. This is even more the case when a single station and a static system of measurement are used and there is a lack of reach in the case of the photographic cam-

Table 2. Table summarising relevant data

	Point Cloud Generated With Terrestrial 3D Laser Scanner	**Cloud Points Obtained With Photoscan**	**Cloud Points Obtained With 123 Catch**
Collection data speed	Good	Very good	Very good
Number of shots	1 station	22 photographs	22 photographs
Economic cost of the system	Expensive	Economic	Very economic
Time of process	Good	Very good	Very good
Precision	Very good	Good	Good
Point Number	677.426	3.743.716	73.030
Point Faces	1.272.266	193.808	
Results outside comparison range	Very good	Good $2,7x10^6$ points with less than 0,4	Good 4800 points with less than 0,25
Output Format	txt	txt	obj
Color	Good	Very good	Very good
Control parameters	Manual	automatic	automatic

eras. This issue has been solved in recent years with the use of drones that are able to reach or access any point that cannot be appreciated from the human standpoint.

Planning carried out prior to undertaking the fieldwork is, from our point of view, of vital importance because it is not only necessary to ensure the correct and complete collection of information, as happens with each of the systems separately, but also to avoid unnecessary redundancy thereby preventing obstacles, noise and shadows. If they are combined, an appropriate programme of work can be established for each situation.

Improvements are also seen when it comes to generating graphical documentation because being able to obtain textured photogrammetric surfaces means it is no longer necessary to perform this work with the data from the laser scanner, whose point clouds are used directly for the development of geometric studies, which in turn ensures the highest accuracy of the regulating lines of the construction elements.

Furthermore, the textured surfaces generated by photogrammetry are the best instrument for studying pathologies, as the data are recorded precisely because of the values for the brightness of the surface being photographed.

We can state although the new SfM systems cannot replace traditional systems such as 3d scanner by its rigor and precision as we can see on Table1. The data collection from these systems, through proper and controlled use of them, can provide us a sufficient degree of reliability in certain graphic surveys. Depending on the object to represent can be very useful to fill empty areas that it is not possible to document using the 3D scanner for its lack of accessibility. According to Table 1 data, PhotoScan gives us much information of the surveying object, 3.743.716 points obtain with PhotoScan versus 73.030 points of 1,2,3 Catch, although the distortion rate is lower in 1,2,3 Catch, the use of one or another tool depends on if must predominate work accuracy level or quality of the image to represent.

Powder printers often create heavy, unstable and soft models; the material used is excessively expensive.

The deposition printing, as fundamental problems, lacks proper resolution in relation to the other two printing systems by addition and needs support systems so that the printed models are freestanding. On the other hand, the models are rigid and stable.

The resin one, to be able to print correctly must be perfectly leveled, materials must be perfectly stored and protected so as not to alter the light sensitive conditions, these being very delicate and perishable. On the other hand, the models are of high quality and resolution, but large sizes cannot be achieved.

Numerical control printer (CNC) allows large format works, in material and more stable shape, but in bass-relief, very elaborate textures may be obtained, with the limitation right angles, which should be replaced by blunt ones.

The creation of physical models or mock-ups made by additive manufacturing techniques allow us to obtain physical models in a fast and effective way which enable us to be able to work in the field of inclusive design and heritage architecture, since there is numerous CAD digital information, which employed in a proper manner and, at the same time, simple and low-cost. They could serve as instrument for people with disabilities to access cultural content that otherwise would be inaccessible for them.

To complete the process, it is necessary to count with the collaboration of affected users to check at first hand the effectiveness of the models and to be able to correct or improve certain aspects such as the scale, texture, and explanatory texts among others.

In this case study case the physical mock-ups acquire special importance as a tool for the dissemination of the architectural element for the general public, as these architectural elements do not exist physically any longer and they are difficult to interpret for the general public from simple orthogonal views due to the geometric complexity of the piece.

The new technologies of digitization in construction through additive manufacturing (3D printing), allow us to currently represent a physical and tangible model to offer it as a resource for people with visual impairments, as well as for all those interested in the architectural heritage.

ACKNOWLEDGMENT

We wish to thank the Universitat Jaume I and the Department of Industrial Systems Engineering and Design for their collaboration and the staffing needed to carry out this work, to the researching project Cód. P1·1B2015-69, *El Renacimiento: impronta arquitectónica en la provincia de Castellón. huellas, rastros, trazas y vestigios* and to Cristina Fabuel for her great help and support.

REFERENCES

Barrera Vera, J. A. (2006). *Aplicación de tecnologías innovadoras en la documentación geométrica del patrimonio arquitectónico y arqueológico.* Unpublished doctoral dissertation, Universidad de Sevilla, Sevilla, Spain.

Brandi, C., & De Angelis, G. (1972). *Carta di Restauro, 1972 (Circolare n°117 del 6 aprile 1972).* Roma, Italy: Ministero della Pubblica Istruzione.

Celani, G. C., & Milan, L. F. M. (2007). Tactile scale models: three-dimensional info graphics for space orientation of the blind and visually impaired. In *Virtual and rapid manufacturing: Advanced research in virtual and rapid prototyping* (pp. 801–805). London, UK: Taylor & Francis Group.

Conselleria de Cultura i Esport. (2007). *Decreto 169/2007, de 28 de septiembre, del Consell.* España: Boletín Oficial del Estado.

Garfella Rubio, J. T., Máñez Pitarch, M. J., Cabeza González, M., & Soler Estrela, A. (2012, April). La documentación gráfica a través del empleo de metodologías avanzadas de fotogrametría y escáner 3D. In *Paper presented at the 4º Congreso de Patología y Rehabilitación de Edificios*, Colexio Oficial de Arquitectos de Galicia, Santiago de Compostela, Spain.

Garfella Rubio, J. T., Máñez Pitarch, M. J., Cabeza González, M., & Soler Estrela, A. (2014). Los sistemas de fotogrametría y láser escáner, en combinación con los métodos tradicionales en la documentación del patrimonio, para la obtención de una arquitectura inversa y su preservación. *Paper presented at the Congreso Latinoamericano REHABEND*, Santander, Spain. Universidad de Cantabria.

Garfella Rubio, J. T., Máñez Pitarch, M. J., & Martinez Moya, J. A. (2015). Comparative Study of Graphic Representation Methods on Architectural Heritage. In S. Brusaporci (Ed.), Handbook of Research on Emerging Digital Tools for Architectural Surveying, Modeling, and Representation, Hershey, PA: IGI Global. doi:10.4018/978-1-4666-8379-2.ch005

Garfella Rubio, J. T., Máñez Pitarch, M. J., Martínez Moya, J. A., & Cabeza González, M. (2013, June). Aplicación del color a las nubes de puntos, las texturas y su posterior modelización mediante toma de datos avanzada a través del escáner y la fotogrametría arquitectónica. *Paper presented at the X Congreso Nacional del Color*, Valencia, Spain. Universitat Politècnica de Valencia.

Gual, J., Puyuelo, M., & Lloveras, J. (2004, November). Three-dimensional tactile symbols relief maps for the visually impaired. In *4th World Conference on Design Research*. Delft, The Netherlands: International Association of Societies of Design Research and Delft University of Technology.

Gual, J., Puyuelo, M., & Lloveras, J. (2014). Three-dimensional tactile symbols produced by 3D printing: Improving the process of memorizing a tactile map key. *British Journal of Visual Impairment*, *32*(3), 263–278. doi:10.1177/0264619614540291

Gual, J., Puyuelo, M., & Lloveras, J. (2015a). Improving tactile map usability through 3D printing techniques: An experiment with new tactile symbols. *The Cartographic Journal*, *52*(1), 51–57. doi:10.1179/1743277413Y.0000000046

Gual, J., Puyuelo, M., & Lloveras, J. (2015b). The effect of volumetric (3D) tactile symbols within inclusive tactile maps. *Applied Ergonomics*, *48*, 1–10. doi:10.1016/j.apergo.2014.10.018 PMID:25683526

Gual, J., Puyuelo, M., Lloveras, J., & Merino, L. (2012). Visual impairment and urban orientation: Pilot study with tactile maps produced through 3D printing. *Psychology: Ambiental-Bilingual Journal of Environmental Psychology*, *3*(2), 239–250.

Gual, J., Serrano, J. & Máñez, M. J. (2015). Aplicación de la fabricación aditiva en la obtención de moldes para termoconformar gráficos tangibles orientados a personas con discapacidad visual. Tangible graphics using rapid prototyping techniques: volume as a design constituent. *The Spanish Digital Journal on Blindness and Visual Impairment, 66.*

Lerma García, J. L. (2002). *Fotogrametría moderna: analítica y digital*. Valencia, Spain: Universidad Politécnica de Valencia.

Lerma García, J. L., Cabrelles, M., Navarro, S., & Seguí, A. E. (2011). La documentación patrimonial mediante sensores de imagen o de barrido láser. In Documentación gráfica del patrimonio (pp. 108-117). España: Ministerio de Cultura.

Máñez, M. J., Gual, J., Garfella, J. T., & Martínez, J. Á. (2016). Renaissance-Style Architecture en El Maestrazgo: From Virtual to Tactile Models. In G. Amoruso (Ed.), Handbook of Research on Visual Computing and Emerging Geometrical Design Tools. Hershey, PA: IGI Global.

Martínez-Espejo Zaragoza, I. (2014). *Precisiones sobre el levantamiento 3d integrado con herramientas Avanzadas, aplicado al conocimiento y la conservación del patrimonio Arquitectónico*. Unpublished doctoral dissertation, Universidad Politécnica de Valencia, Valencia, Spain.

Martínez-Moya, J. A., Gual-Ortí, J., & Máñez-Pitarch, M. J. (2018). Physical scale models as diffusion tools of disappeared heritage. In G/ Amoruso (Ed.), *Putting Tradition into Practice: Heritage, Place and Design*. Milano, Italy: Springer. 10.1007/978-3-319-57937-5_69

Pérez, A., Lerma, J. L., Martos, A., Jordá, F., Ramos, M., & Navarro, S. (2007, February). Generación automática de orto fotografías verdaderas en Arquitectura. *Paper presented at the Séptima Semana Geomática de Barcelona*, Barcelona, Spain.

Voigt, A., & Martens, B. (2006). Development of 3D tactile models for the partially sighted to facilitate spatial orientation. In *Communicating Space(s): 24th eCAADe Conference Proceedings* (pp. 366-370). Volos, Greece: University of Thessaly.

Yastikli, N. (2007), Documentation of cultural heritage using digital photogrammetry and laser scanning. Journal of Cultural Heritage, 8, 423-427. doi:10.1016/j.culher.2007.06.003

This research was previously published in the International Journal of Computational Methods in Heritage Science (IJCMHS), 2(2); edited by George P. Pavlidis, pages 36-74, copyright year 2018 by IGI Publishing (an imprint of IGI Global).

APPENDIX A: ADDITIONAL READING

Herráez Boquera, J., Navarro Esteve, P. & Denia Ríos, J. L. (2010) Fundamentos y aplicaciones de la tecnología de Escáner Láser para documentación y restauración de patrimonio. In A. Alcántara Onofre & A.M. Lara Gutiérrez (Ed.), *Implementación de nuevos métodos de documentación y registro fotogramétrico digital para la protección y puesta en valor del patrimonio monumental mexicano* (pp. 17-41). México: Universidad Politécnica de Valencia.

Máñez Pitarch, M. J. (2014) La arquitectura religiosa renacentista en tierras del maestre: la iglesia de Nuestra Señora de la Asunción de Vistabella del Maestrazgo. Unpublished doctoral dissertation, Universidad Politécnica de Valencia, Valencia, Spain.

Mataix Sanjuán, J., León Robles, C. & Montes Tubío, F. de P. (2013) Fortalezas y debilidades de la técnica de levantamiento gráfico con escáner láser para la catalogación del patrimonio cultural Aplicación a la iglesia de San Francisco (Priego de Córdoba). *Revista EGA, 21,* 219-225.

APPENDIX B: KEY TERMS AND DEFINITIONS

3D Model: It is the three-dimensional representation of an object made on a computer with specialized software. It is made up of a collection of points in space connected by geometric entities such as triangles, lines, surfaces, etc. It can be obtained through the scanning of a real object or through algorithms that provide the necessary data for its creation.

3D Rendering: Computer graphics process to get 2D photorealistic effect images from 3D virtual models.

ABS: Acronym of Acrylonitrile Butadiene Styrene, it is a plastic term, used as fused filament for 3D printers.

AMF: It is a file format designed to support additive manufacturing processes, such as 3D printing. Its acronym comes from Additive Manufacturing File. Besides the model geometry, AMF files provide a lot of information about the model being printed, which does not need post processing to define data such as the model position in relation to the 3D printer, orientation, colours, material, etc.

Additive manufacturing: It consists in the formation of 3D objects by adding successive layers of material in order to form a piece with volume. It is a type of manufacturing that can be done with many different types of materials and is faster than conventional manufacturing techniques. When combined with digital production tools, high precision can be obtained. It includes, among others, the following processes: stereo lithography, fused deposition modelling (FDM), selective laser sintering, inkjet binder, light-curing resin, laminated objects manufacturing.

Blender: It is a 3D modelling software.

CAD: Acronym of Computer Aided Design. Normally drawing or virtual modelling software for 2D and 3D.

CAE: Acronym of Computer Aided Engineering. In 3D printing, it refers to the engineering aspects linked to 3D modelling from CAD. While in CAD it is a question of generating a model, in CAE the purpose is the analysis of its characteristics using engineering methods. CAE processes, thus, allows us to extract information to optimize the development and the cost of manufacturing by analysing the model characteristics, properties, material and geometry. Likewise, they allow to determine the viability and profitability, the quantity of materials, the times and costs of manufacturing.

CAM: It is the use of computer software to assist in the manufacture of a 3D design. It is an acronym for Computer Aided Manufacturing. In 3D printing we refer to CAD –computer aided design- regarding design and to CAM regarding manufacture. The design and manufacture with the help of the computer form a technology that encompasses many disciplines, such as graphic design, database management for design and manufacturing, robotics and computer vision, among others. Through the CAM, precise instructions based on G-Code are created to drive the machine and its tools.

CNC: It is the acronym of Computer Numerical Control, which is a process based on a series of coded commands, normally, on a G-Code file.

Cultural Heritage: The legacy of tangible and intangible attributes inherited by a society from past generations. In this case: buildings, monuments and landscapes.

Cura: It is a free slicer 3D software that can be used with Windows, Mac and Linux. It can open files with the extensions STL, 3MF and OBJ.

DLP: It is the acronym of Digital Light Processing. It is a type 3D printer that solidifies photosensitive resin by means of a light source that can be a projector, a halogen or LED bulb, or an LCD or UV LED plate. The DLP printing process is similar with that of the SLA but faster. As with SLA, the object can be created from the top down or from the inside of the tray upwards. In both cases, the surface of the resulting piece is smoother than with processes such as FDM or FFF, since the layers are less visible.

DXF: It is a CAD universal file format. It was developed by Autodesk and it is one of the most used in design, engineer and product development. DXF files can be imported in CAM software and, afterwards, be converted into G-Code.

Extruder: It is a part of the 3D printer that, fed by cold filament of the thermoplastic reel, extracts it in liquid state by heat, pushing it through a small nozzle.

FDM: It is the acronym of Fused Deposition Modelling. It is a 3D printing technique which consists in depositing flat layers of molten material superimposed on each other to get an object with volume. By extension, 3D printers that use this technique are called FDM printers. That is, they print from the fusion of plastic filament.

FFF: It is the acronym of Fused Filament Fabrication. It is another way of naming the FDM technique.

Focal length: The focal length of a lens refers to the distance between the lens itself and its focus point. The focal length of a lens determines the field of view of the scene. It is called wide angle lens because it captures a large part of the scene and has a very broad field of vision. A target of great focal distance is called a telephoto lens as it acts approaching a particular area of its scene.

G-Code: It is the language that the 3D printer understands. It is a programming language that is used to control automated machines and tools. In 3D printing, the code determines a trajectory from which the parts of the printer responsible for creating the piece move.

ISO: It is the index or level of sensitivity of a photographic film to light. The lower the ISO, the lower the sensitivity of the film, which requires a longer exposure, while a film with a high ISO only needs a brief exposure to light.

Laser Scanning: Data Collection of real-world objects by 3D scanner devices to construct a digital three-dimensional model.

Meshlab: It is a free mesh editing program to view, join, transform and repair STL, PLY, STL, OFF, OBJ, 3DS files. It also supports other file extensions and point clouds.

Montesa Order: Old christian military order, territorially limited to the old Kingdom of Aragon.

OBJ: It is a file format that contains 3D geometries and includes data such as the position of each vertex, and the normals and faces that make up each polygon. It is one of the main formats for the exchange of files between software and is often used as an alternative to the STL extension because it allows to incorporate information about colour, materials, etc. Its full name is Wavefront 3D Object.

Open source: It is the free distribution software that has a public code that can be downloaded, modified and redistributed by multiple developers. There is several open source software to create and modify 3D models for printing.

Orthophoto: Photograph geometrically corrected getting an orthogonal view that can be used to measure true distances.

Photogrammetry: The science of taking measurements from photographs by computational models.

Photopolymer: It is a photosensitive resin used in 3D printing by stereolithography or by DPL. That is, it is a type of liquid plastic that reacts to light solidifying. The reaction occurs under UV rays and at a certain wavelength.

Photosolidification: Also known as stereolithography, optical manufacturing, SLA or SL. It is an additive manufacturing process that is based on the solidification of resin by ultraviolet light. Three-dimensional objects are formed by the addition of superimposed layers on top of each other. Each layer is generated by directing an ultraviolet light laser to the sectors of the object where the addition of material is needed, which causes the resin to solidify creating a thin layer of solid material glued to the previous layer. As with DSL, the object can be created from the top down or from the inside of the tray up. In both cases, the surface of the resulting piece is smoother than with processes such as FDM or FFF, since the layers are less visible.

PLA: It is the acronym of Polylactic Acid. A biodegradable thermoplastic polymer used in filament form as a 3D printer material.

Point Cloud: A three-dimensional coordinate system that represent the external or internal surface of an object created by 3D scanners or photogrammetry.

Polyamide: Better known as nylon. It is a synthetic fibre used in 3D printing by laser sintering from a very fine white granular powder. The printed pieces are strong and slightly flexible, so they can withstand small impacts and withstand some pressure while bending.

Resin: It is a photopolymer compound that solidifies with light at a certain wavelength and is used as a printing material mainly used in 3D printers with SLA and DLP technology.

RGB: It means "red, green and blue". Red, green and blue are the 3 colours that monitors use to display images.

SketchUp: It is a 3D modelling software developed by the architectural company Trimble Buildings. It is available as a commercial version called SketchUp Pro and as a free version called SketchUp Make.

SLA: Also known as SL, stereolithography, optical manufacturing or photosolidification. It is an additive manufacturing process that is based on the solidification of resin by ultraviolet light. Three-dimensional objects are formed by the addition of superimposed layers on top of each other. Each layer is generated by directing an ultraviolet light laser to the sectors of the object where it is necessary to add material, which causes the resin to solidify and thus a thin layer of solid material is glued to the previous layer. As with DSL, the object can be created from the top down or from the inside of the tray up. In both cases, the surface of the resulting piece is smoother than with processes such as FDM or FFF, since the layers are less visible.

Spool: It is the support of the plastic filament roll used in 3D printers.

STL: It is the acronym of Standard Tessellation Language. It is a computer file format that defines the surface of 3D modelled objects and is used in 3D printers and numerical control machines (CNC). Most 3D modelling programs can export in this format, which does not include information about colour, textures or physical properties of the model which the OBJ format does.

Structure for Motion: Imaging technique that gets 3D structures from 2D image sequences coupled with local motion signals by specific software.

Texture: It is the term used in 3D design when referring to the colour or the image simulating a concrete texture applied on a 3D model surface.

Thermoplastic: It is a type of plastic that makes up the majority of 3D printing filaments by FDM, such as PLA, ABS, PVC, PE and nylon. When melted it becomes a fluid able to be extruded and that hardens quickly when losing temperature.

Vistabella del Maestrazgo: Municipality in the north-western end of the province of Castellon at Valencian Region, Spain.

Section 3
Processes and Techniques

Chapter 8
A Particle Swarm Optimization Approach for Minimizing GD&T Error in Additive Manufactured Parts:
PSO Based GD&T Minimization

Vimal Kumar Pathak
Malaviya National Institute of Technology, India

Amit Kumar Singh
Malaviya National Institute of Technology, India

ABSTRACT

This paper presents a particle swarm optimization (PSO) approach to improve the geometrical accuracy of additive manufacturing (AM) parts by minimizing geometrical dimensioning and tolerancing (GD&T) error. Four AM process parameters viz. Bed temperature, nozzle temperature, Infill, layer thickness are taken as input while circularity and flatness error in ABS part are taken as response. A mathematical model is developed for circularity and flatness error individually using regression technique in terms of process parameters as design variables. For the optimum search of the AM process parameter values, minimization of circularity and flatness are formulated as multi-objective, multi-variable optimization problem which is optimized using particle swarm optimization (PSO) algorithm and hence improving the geometrical accuracy of the ABS part.

INTRODUCTION

In recent times, due to emergence of advance manufacturing technologies, the market situation has become extremely competitive and volatile. To sustain in such competitive market, it is imperative to not only produce low cost product but also with high quality and shorter development time. To achieve the

DOI: 10.4018/978-1-5225-9624-0.ch008

aforesaid objectives in flexible manufacturing industries, they are forced to use advanced manufacturing tools and machines such as Rapid prototyping (RP).

Almost all the RP systems work on the principle of deposition of material in the form of thin cross-sectional layers. Each layer is built upon the previous layers until the complete 3D model is developed (Suresh & Narayana, 2016). Additive manufacturing (AM) is a rapid growing tool that belongs to RP technologies. Due to its ability to develop complex engineering parts and fast product development, it is gaining wide acceptance among different applications like biomedical (Pathak & Singh, 2016), automobile (Wiedemann & Jantzen, 1999), construction (Buswell, Soar, Gibb, & Thorpe, 2007) and many more. Although RP has shown competence in fast product development but its full- fledged use is hindered by limited materials available in market (Levy, Schindel, & Kruth, 2003; Pilipović, Raos, & Šercer, 2009). To improve its sustainability in market, its performance needs to be improved.

One way of improving the performance of any process is by suitably adjusting and optimizing its process parameters so that improved quality products may be fabricated (Kibria, Doloi, & Bhattacharyya, 2014; Molla & Manna, 2013; Silva Filho, Abrão, Paiva, & Ferreira, 2012; Balraj & Krishna, 2014). Many studies have been conducted to determine the optimum process parameters for RP using different conventional optimization techniques. Ahn et al. (2002) and Khan et al. (2005) applied design of experiment (DOE) approach to determine the influence of process parameters like raster angle, air gap and layer thickness on AM response. Thirumurthulu et al. (2004) used GA to determine the optimum part orientation in AM to minimize surface roughness and build time. Chockalingam et al. (2006) optimized stereolithography (SL) process parameters to achieve maximum part strength. Raghunath and Pandey (2007) applied Taguchi method to improve the accuracy of SLS parts through shrinkage. Sood et al. (2009) applied gray relational analysis to minimize the dimensional accuracy of AM parts. Sood et al. (2010) and (2012) studied the effect of process parameters on mechanical properties using RSM and sliding wear of the AM models. Phatak and Pande (2012) used GA to determine the optimum part orientation for improved part quality by minimizing the build time and material used. Guralla and Regalla (2014) used NSGA algorithm for optimization of part strength and volumetric shrinkage in the AM parts.

All the aforesaid studies suggest that researchers have correlates AM process parameters with the dimensional accuracy, surface roughness, strength, shrinkage improvement, mechanical properties and wear analysis in a part produced by AM. However, the correlation between process parameters and Geometric dimensioning and Tolerancing (GD&T) errors have not found yet for AM based parts. The current work thus bridges this gap and by establishing an empirical relationship between process parameters and GD&T error in parts manufactured by AM using regression analysis. Further, the analytical model is unified as multi-objective and optimized using PSO. So the current paper motivation is that industries want AM parts to be aesthetically appealing and at the same time to be functional for its sustainability. The minimized circularity and flatness error will help the AM parts to be functionally reliable and would make it more sustainable for engineering applications.

MATERIALS AND METHODS

The quality and accuracy of any part produced using AM process typically depends on its process parameters. This study considered four important process parameters bed temperature (A), nozzle temperature (B), Infill (C), layer thickness (D) to study their effects on circularity and flatness. The factors and their levels under which tests were carried out are given in Table 1. They are defined as follows:

- **Bed Temperature (T_b):** It is the temperature of printing bed. Its value changes for different material and adhesion of part depend on its correct setting.
- **Nozzle Temperature (T_n):** It is the temperature of the extruder.
- **Infill Density (I):** The quantity of material inside the part. Optimum material inside the part can improve part quality with less waste of material.
- **Layer Thickness (L):** It is the thickness of each deposited slice. It is known that lower the layer thickness, better the part quality.

However, the number of experiments could be large when considering more parameters and levels. Taguchi orthogonal array is a good method to reduce the number of experiments According to Taguchi's design of experiments, for four parameters and three levels L_9 Taguchi orthogonal array (Jiang, Li, Yan, Sun, & Zhang, 2010) was selected (Table 2). A 3D solid model of test part (Figure 1) is modelled in Creo parametric software and exported as stereolithography (STL) file. Further, STL file is imported in open source Slic3r AM software for G-code generation. The control parameters were set as per experimental plan (Table 2) and three parts per experiment were performed using Aha! R3D2 AM machine. The parts were manufactured using digital ABS material. Three readings for all the responses were taken per sample and average is taken and represented in Table 2. GD&T errors were measured using a computer controlled coordinate measuring machine (CMM).

MATHEMATICAL MODEL AND ANOVA ANALYSIS

Development of Mathematical Model

The AM parameter correlation with the circularity and flatness response were obtained using statistical regression analysis with the help of Minitab 16 software. During regression it is assumed that process parameters and response were linearly related to each other. The input factors of Bed temperature (°C), Nozzle temp. (°C), Infill (%) and layer thickness (mm) were used as the independent factors in the regression analysis. The final predictive mathematical model in terms of actual factors is as follows:

$$Circularity, C_1 = -0.0805 + 0.00291 * T_b - 0.000850 * T_n + 0.00210 * I - 0.0962 * L \quad (1)$$

$$Flatness, F_1 = -0.018 + 0.00131 * T_b - 0.000443 * T_n + 0.00767 * I - 0.332 * L \quad (2)$$

$$Flatness, F_2 = -0.291 + 0.00133 * T_b + 0.000558 * T_n + 0.00293 * I + 0.0757 * L \quad (3)$$

Figure 1. Test sample for GD&T error analysis. (All dimensions in mm)

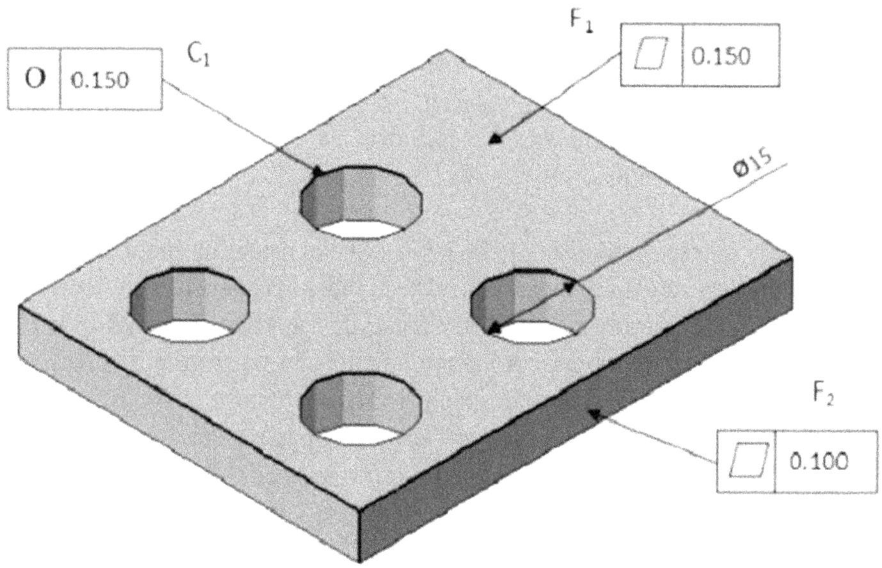

Table 1. Factors and their levels

Factor	Symbol	Level			Unit
		1	2	3	
Bed temp. (T_b)	A	110	125	140	°C
Nozzle temp. (T_n)	B	200	215	230	°C
Infill (I)	C	20	25	30	%
Layer thickness (L)	D	0.2	0.3	0.4	mm

Table 2. Experimental run using L_9 orthogonal array

Exp. No.	Factors				Responses		
	A	B	C	D	C_1 (mm)	F_1 (mm)	F_2 (mm)
1	1	1	1	1	0.0925	0.1251	0.0439
2	1	2	2	2	0.0810	0.1290	0.0735
3	1	3	3	3	0.0688	0.1221	0.1051
4	2	1	2	3	0.1350	0.1142	0.0845
5	2	2	3	1	0.1380	0.2176	0.0867
6	2	3	1	2	0.0964	0.0765	0.0823
7	3	1	3	2	0.1878	0.1950	0.1232
8	3	2	1	3	0.1426	0.1030	0.1049
9	3	3	2	1	0.1736	0.1959	0.1144

Checking the Model Accuracy

For testing the accuracy of developed model ANOVA analysis technique was used. According to ANOVA, if the calculated p-value is less than 0.05, the developed model is significant. Otherwise, p-value greater than 0.1 means the model is insignificant. However, for this study, the p-value is less than 0.05 for all three models. Hence, the developed models are significant (see Table 3, Table 4, and Table 5).

Furthermore, to test the adequacy of fitted regression model, determination coefficient (R^2) was determined. For present study, the value of R^2 and adj. R^2 for developed models is more than 80% and 70% respectively. This shows that these models are highly significant (Ramasamy, Gould, & Workman, 2002). The results of ANOVA analysis for three models were shown in Figure 3, 4 and 5. Additionally, the normal probability graphs of the response factors has been plotted as shown in Figures 2-4. The plot clearly shows that the regression model is fairly well fitted to the observed values as the residuals were located on a straight line.

Table 3. ANOVA for the Circularity (C_1) model

Source	df	Sum of Squares	Mean Square	F-value	p-value Prob > F	
Model	4	0.01360	0.00340	57.84	0.001	significant
Residual	4	0.000235	0.000059			
Total	8	0.013835				
R-Squared = 98.3%			R-Squared (adj) = 96.6%			

Table 4. ANOVA for the Flatness (F_1) model

Source	df	Sum of Squares	Mean Square	F-value	p-value Prob > F	
Model	4	0.018012	0.004503	21.27	0.006	significant
Residual	4	0.000847	0.0002117			
Total	8	0.018858				
R-Squared = 95.5%			R-Squared (adj) = 91.6%			

Table 5. ANOVA for the Flatness (F_2) model

Source	df	Sum of Squares	Mean Square	F-value	p-value Prob > F	
Model	4	0.044542	0.0111355	27.26	0.004	significant
Residual	4	0.0001634	0.0000408			
Total	8	0.0447054				
R-Squared = 96.5%			R-Squared (adj) = 92.9%			

Figure 2. Normal plot residuals for Circularity C_1

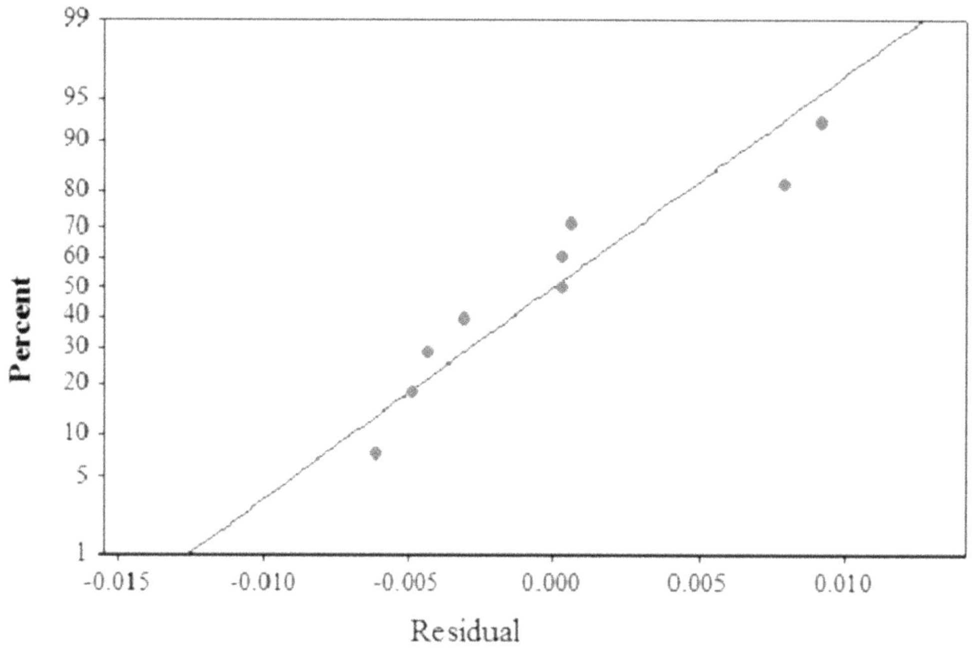

Figure 3. Normal plot residuals for Flatness F_1

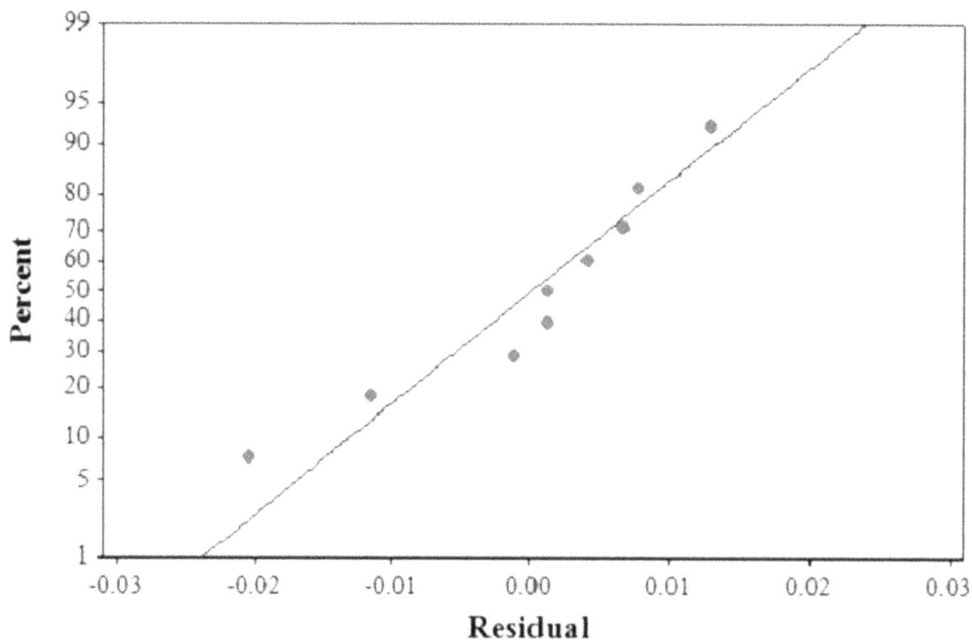

Figure 4. Normal plot residuals for Flatness F_2

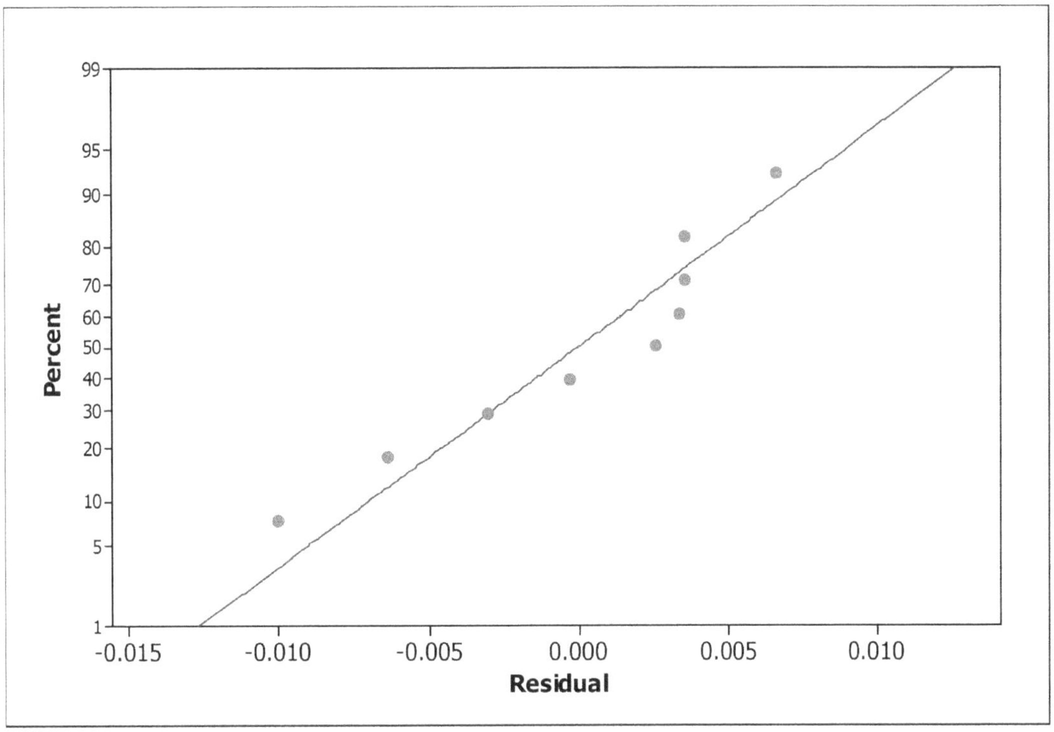

MULTI-OBJECTIVE OPTIMIZATION

The objective of the present study is to minimize circularity (C_1) and flatness (F_1 and F_2) errors, which is formulated using regression analysis technique. The objectives are function of decision variables viz. bed temperature (T_b), nozzle temperature (T_n), infill (I) and Layer Thickness (L). The empirical relation obtained between input and output parameters obtained from regression analysis were used as functional equations. The objective functions are given below:

Objective 1 = min. C_1
Objective 2 = min. F_1
Objective 3 = min. F_2

A weighted method was used for combining all objectives into a single objective. To overcome any significant difference in the different objectives the function corresponding to every AM performance output is normalized by dividing each function with their individual optimal values obtained in single objective optimization. Now, the weighted objective function now becomes:

Minimize $Z = \left(w_1 * f_1 + w_2 * f_2 + w_3 * f_3\right)$ (4)

where f_1 = Normalized function for C_1

f_2 = Normalized function for F_1

and f_3 = Normalized function for F_2 with the parameter feasible ranges:

$110 \le T_b \le 140$ (°C)
$200 \le T_n \le 230$ (°C)
$20 \le I \le 30$ (%)
$0.2 \le L \le 0.4$ (mm)

Here, w_1, w_2 and w_3 are the three weighting factors applied to normalized C_1, F_1 and F_2 respectively. The weighting factors were selected in such a manner that the sum of three is equal to one. Matlab R2014a was used to write program of particle swarm optimization for determining optimum process parameters with minimum circularity and flatness errors by multi-objective optimization.

The basic particle swarm optimization is a population based method suggested by Kennedy and Eberhart in 1995. PSO is modeled after the simulation of social behavior of birds in a flock (Kennedy, 2011; Eberhart & Kennedy, 1995). PSO is initialized by distributing each particle randomly in D-dimensional search space. The performance of each particle is measured using a fitness function which depends on the optimization problem. Each particle i represented by following information

x_i, the current position of the particle i
v_i, current velocity of the particle i
p_i, personal best position of the particle i

The personal best position is the best position that particle i has been so far. The fitness function is highest for that position of ith particle. Here, velocity acts like a vector which helps in guiding the particle from one position to another with updated velocity at every iteration. The personal best position of ith particle with dependence on time step t as:

$$P_i(t) = p_i(t) \qquad if\, f(x_i(t)) \ge f(p_i(t))$$

$$x_i(t) \qquad if\, f(x_i(t)) < f(p_i(t)) \tag{5}$$

New position and velocity for i^{th} particle is updated at every iteration and expressed as:

$$v_i(t+1) = v_i(t) + c_1 * r_1 (p_i(t) - x_i(t)) + c_2 * r_2 (g_i(t) - x_i(t)) \tag{6}$$

$$x_i(t+1) = x_i(t) + v_i(t+1) \tag{7}$$

r_1 and r_2 are two independent uniformly distributed random numbers within given interval [0,1]. c_1 and c_2 are two accelerating coefficients whose value are generally 2 each for almost all applications

(Shi, 2001), $p(t)$ is the best position parameter of an individual particle and $g(t)$ is global best position parameter of entire swarms. Shi and Eberhart (1999) introduced an inertia weight w into the velocity updating of the PSO that helps in controlling the scope of the search. Often, w decreases linearly from 0.9 to 0.4 over the whole iteration. High value of inertia weight helps in exploration whereas low value favors exploitation (Shi and Eberhart, 1998). The velocity updation with inertia weight is shown in Equation (8).

$$v_i(t+1) = w(t) * v_i(t) + c_1 * r_1(p_i(t) - x_i(t)) + c_2 * r_2(g_i(t) - x_i(t)) \tag{8}$$

For optimization the PSO parameters were given in Table 6.

After performing PSO, optimum process parameters were shown in Table 7, with different weighting factors. For the present study, case-1 provides optimum results with minimization of all GD&T objective functions. The minimized value of the objective function is 0.0761 with $w_1 = 0.35$, $w_2 = 0.35$ and w_3

Table 6. PSO parameters

S. No.	Parameters
1	Population size = 50
2	Max. iteration = 100
3	Simulation runs = 20
4	C_1, C_2 = 2.1, 2.1
5	w_{start}, w_{end} = 0.9, 0.4

Table 7. Optimal process parameters with different weighting factors

Control factors	Optimal AM condition			
	Case 1	**Case 2**	**Case 3**	**Case 4**
	w_1=0.35	**w_1=0.50**	**w_1=0.25**	**w_1=0.25**
	w_2=0.35	**w_2=0.25**	**w_2=0.50**	**w_2=0.25**
	w_3=0.30	**w_3=0.25**	**w_3=0.25**	**w_3=0.50**
T_b (°C)	133.37	114.39	134.93	111.17
T_n (°C)	201.32	214.01	226.25	226.40
I (%)	20.13	23.06	22.12	20.44
L (mm)	0.12	0.26	0.25	0.26
C_1 (mm)	0.0611	0.0794	0.0982	0.0828
F_1 (mm)	0.0692	0.0742	0.1167	0.0938
F_2 (mm)	0.0712	0.0851	0.1376	0.1201

= 0.3. Hence, case 1 is recommended as it give minimal values of circularity and flatness errors with process parameters as:

T_b=133.37°C, T_n =201.32°C, I = 20.13%, L = 0.12 mm.

CASE STUDY

A 30 x 30 x 30 mm cube with 10 mm hole was taken as a case study (see Figure 5 (a)), for experimental verification of the predicted optimized process parameters by PSO. Six cubes were printed by entry level AM machine with the same process parameters as shown in Figure 5 (b). The case study was measured

Figure 5. (a) Case study (b) Cubes printed with AM machine

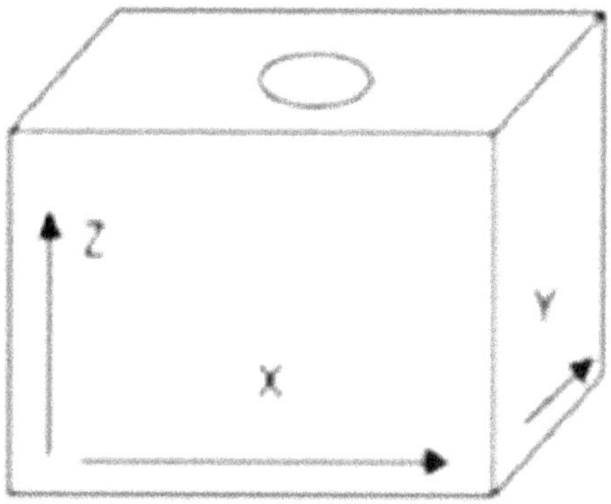

(a)

(b)

Table 8. Experimental validation of the developed model

	Predicted	Experimental				
T_b (°C)	133.37	133				
T_n (°C)	201.32	201				
I (%)	20.13	20				
L (mm)	0.12	0.12				
F (mm)	0.069	0.081	0.077	0.082	0.079	0.072
C (mm)	0.061	0.072	0.066	0.065	0.061	0.062

for flatness and circularity using computer controlled Coordinate Measuring Machine (CMM). One part was measured three times and average of the three trials was noted. The result for the case study part is shown in Table 8. As shown in Table 8, the predicted and experimental values of circularity and flatness are in proximity, which confirms that the developed model is correct.

CONCLUSION

The present study deals with the optimal AM part production with minimum GD&T error so that part build by AM can be functional and used in industrial applications. The main motivation of the current work is that industries doesn't invest in the technology that only produces aesthetic parts without being functional. With this point in mind, the geometrical accuracy of the parts are improved in the present work. For that, optimum process parameters were found out for minimum GD&T error. Linear analytical models were developed for circularity and flatness using regression technique in terms of process parameters. Further, the mathematical models were optimized by multi-objective optimization using PSO. At last, a simple case study is used for experimental confirmation of the developed model. The adopted methodology will help the AM users in producing parts with high geometrical accuracy and bring more sustainability in AM parts for engineering applications as a functional components.

REFERENCES

Ahn, S. H., Montero, M., Odell, D., Roundy, S., & Wright, P. K. (2002). Anisotropic material properties of Additive manufacturing ABS. *Rapid Prototyping Journal*, *8*(4), 248–257. doi:10.1108/13552540210441166

Balraj, U. S., & Krishna, A. G. (2014). Multi-Objective Optimization of EDM Process Parameters using Taguchi Method, Principal Component Analysis and Grey Relational Analysis. *International Journal of Manufacturing, Materials, and Mechanical Engineering*, *4*(2), 29–46. doi:10.4018/ijmmme.2014040103

Buswell, R. A., Soar, R. C., Gibb, A. G., & Thorpe, A. (2007). Freeform construction: Mega-scale rapid manufacturing for construction. *Automation in Construction*, *16*(2), 224–231. doi:10.1016/j.autcon.2006.05.002

Chockalingam, K., Jawahar, N., Ramanathan, K. N., & Banerjee, P. S. (2006). Optimization of stereolithography process parameters for part strength using design of experiments. *International Journal of Advanced Manufacturing Technology*, *29*(1-2), 79–88. doi:10.100700170-004-2307-0

Gurrala, P. K., & Regalla, S. P. (2014). Multi-objective optimisation of strength and volumetric shrinkage of FDM parts: A multi-objective optimization scheme is used to optimize the strength and volumetric shrinkage of FDM parts considering different process parameters. *Virtual and Physical Prototyping*, *9*(2), 127–138. doi:10.1080/17452759.2014.898851

Jiang, F., Li, J., Yan, L., Sun, J., & Zhang, S. (2010). Optimizing end-milling parameters for surface roughness under different cooling/lubrication conditions. *International Journal of Advanced Manufacturing Technology*, *51*(9-12), 841–851. doi:10.100700170-010-2680-9

Kennedy, J. (2011). Particle swarm optimization. In *Encyclopedia of machine learning* (pp. 760-766). Springer US. Eberhart, R., & Kennedy, J. (1995, October). A new optimizer using particle swarm theory. *Proceedings of the Sixth International Symposium on Micro Machine and Human Science MHS'95* (pp. 39-43). IEEE.

Kibria, G., Doloi, B., & Bhattacharyya, B. (2014). Parametric study and optimization of Nd: YAG laser micro-turning process of different grade of alumina ceramics based on Taguchi methodology. *International Journal of Surface Engineering and Interdisciplinary Materials Science*, *2*(1), 15–46. doi:10.4018/ijseims.2014010102

Lee, B. H., Abdullah, J., & Khan, Z. A. (2005). Optimization of rapid prototyping parameters for production of flexible ABS object. *Journal of Materials Processing Technology*, *169*(1), 54–61. doi:10.1016/j.jmatprotec.2005.02.259

Levy, G. N., Schindel, R., & Kruth, J. P. (2003). Rapid manufacturing and rapid tooling with layer manufacturing (LM) technologies, state of the art and future perspectives. *CIRP Annals-Manufacturing Technology*, *52*(2), 589–609. doi:10.1016/S0007-8506(07)60206-6

Molla, K. Z., & Manna, A. (2013). Optimization of electrochemical grinding parameters for effective finishing of hybrid al/(al2o3+ zro2) mmc. *International Journal of Surface Engineering and Interdisciplinary Materials Science*, *1*(2), 35–45. doi:10.4018/ijseims.2013070104

Pathak, V. K., & Singh, A. K. (2016). Assessing the challenges and issues in entry level additive manufacturing. *International Journal of Rapid Manufacturing*, *6*(1), 53–74. doi:10.1504/IJRAPIDM.2016.078746

Phatak, A. M., & Pande, S. S. (2012). Optimum part orientation in rapid prototyping using genetic algorithm. *Journal of Manufacturing Systems*, *31*(4), 395–402. doi:10.1016/j.jmsy.2012.07.001

Pilipović, A., Raos, P., & Šercer, M. (2009). Experimental analysis of properties of materials for rapid prototyping. *International Journal of Advanced Manufacturing Technology*, *40*(1-2), 105–115. doi:10.100700170-007-1310-7

Raghunath, N., & Pandey, P. M. (2007). Improving accuracy through shrinkage modelling by using Taguchi method in selective laser sintering. *International Journal of Machine Tools & Manufacture*, *47*(6), 985–995. doi:10.1016/j.ijmachtools.2006.07.001

Ramasamy, S. R., Gould, J., & Workman, D. (2002). Design-of-experiments study to examine the effect of polarity on stud welding. *Welding Journal, 81*(2), 19-S.

Shi, Y. (2001). Particle swarm optimization: developments, applications and resources. *Proceedings of the 2001 Congress on evolutionary computation* (*Vol. 1*, pp. 81-86). IEEE. 10.1109/CEC.2001.934374

Shi, Y., & Eberhart, R. (1998, May). A modified particle swarm optimizer. *Proceedings of the 1998 IEEE International Conference on Evolutionary Computation* (pp. 69-73). IEEE. 10.1109/ICEC.1998.699146

Shi, Y., & Eberhart, R. C. (1999). Empirical study of particle swarm optimization. *Proceedings of the 1999 Congress on Evolutionary Computation CEC 99.* (*Vol. 3*, pp. 1945-1950). IEEE.

Silva Filho, A. R., Abrão, A. M., Paiva, A. P., & Ferreira, J. R. (2012). Multivariate Optimization of the Cutting Parameters when Turning Slender Components. *International Journal of Manufacturing, Materials, and Mechanical Engineering*, *2*(4), 12–31. doi:10.4018/ijmmme.2012100102

Sood, A. K., Equbal, A., Toppo, V., Ohdar, R. K., & Mahapatra, S. S. (2012). An investigation on sliding wear of FDM built parts. *CIRP Journal of Manufacturing Science and Technology*, *5*(1), 48–54. doi:10.1016/j.cirpj.2011.08.003

Sood, A. K., Ohdar, R. K., & Mahapatra, S. S. (2009). Improving dimensional accuracy of fused deposition modelling processed part using grey Taguchi method. *Materials & Design*, *30*(10), 4243–4252. doi:10.1016/j.matdes.2009.04.030

Sood, A. K., Ohdar, R. K., & Mahapatra, S. S. (2010). Parametric appraisal of mechanical property of fused deposition modelling processed parts. *Materials & Design*, *31*(1), 287–295. doi:10.1016/j.matdes.2009.06.016

Suresh, G., & Narayana, K. L. (2016). A Review on Fabricating Procedures in Rapid Prototyping. *International Journal of Manufacturing, Materials, and Mechanical Engineering*, *6*(2), 23–40. doi:10.4018/IJMMME.2016040102

Thrimurthulu, K., Pandey, P. M., & Reddy, N. V. (2004). Optimum part deposition orientation in Additive manufacturing. *International Journal of Machine Tools & Manufacture*, *44*(6), 585–594. doi:10.1016/j.ijmachtools.2003.12.004

Wiedemann, B., & Jantzen, H. A. (1999). Strategies and applications for rapid product and process development in Daimler-Benz AG. *Computers in Industry*, *39*(1), 11–25. doi:10.1016/S0166-3615(98)00126-2

This research was previously published in the International Journal of Manufacturing, Materials, and Mechanical Engineering (IJMMME), 7(3); edited by J. Paulo Davim, pages 69-80, copyright year 2017 by IGI Publishing (an imprint of IGI Global).

Chapter 9
Improving Surface Integrity Using Laser Metal Deposition Process

Rasheedat M. Mahamood
University of Johannesburg, South Africa & University of Ilorin, Nigeria

Esther T. Akinlabi
University of Johannesburg, South Africa

Mukul Shukla
University of Johannesburg, South Africa & MNNIT Allahabad, India

Sisa Pityana
National Laser Centre, South Africa

ABSTRACT

Laser Metal Deposition (LMD), an additive manufacturing process (also known as 3-D printing) and a non-traditional fabrication process used for improving the surface integrity of components is presented in this chapter. In LMD, parts are manufactured directly from the 3-D Computer-Aided Design (CAD) model data. Complex parts can be produced in a single step, which is impossible with the traditional manufacturing methods such as casting, cutting, and turning operations. The major steps required in the production of parts using the laser metal deposition process are highlighted. The flexibility offered by the LMD technique makes it an important surface engineering technique. Composite parts or parts whose surfaces are made of composite materials can also be produced in a single step because two or more dissimilar materials can be handled simultaneously in the LMD process to produce parts. This is because the building of parts in LMD is achieved by the LMD machine following the detail described by the CAD model of the part being made. The processing parameters affecting the properties of laser metal deposited parts are described in detail. This chapter establishes the ability of the LMD in the production of complex and one of its kind parts, its ability to improve surface properties, repair high-valued parts, and reduce the buy-to-fly ratio in the production of aerospace parts. It also highlights the use of non-traditional finishing techniques on laser deposited parts to further improve the surface integrity of components. The chapter is concluded by presenting a laser metal deposited Ti6Al4V/TiC composite. The laser metal deposited Ti6Al4V/TiC composite was characterized through the microstructure, microhardness, and wear resistance, and it was found that the resulting deposits were fully dense and of improved surface properties when compared to the parent materials.

DOI: 10.4018/978-1-5225-9624-0.ch009

INTRODUCTION

Laser surface engineering deals with the modification of surface properties of a solid material using laser. The requirement of surface properties in some applications differs from that of the bulk material because of the working environment to which the surface will be subjected. Depending on the type of application, often the properties desired from the surface material are conflicting with what is required from the bulk material and the cost issue may restrict the use of these materials. When a cheaper material is used, there is a need to modify the surface of the material to meet the service requirements such as wear and corrosion resistance. This is because it is this surface of the part that interacts with the working environment and not the bulk material. The interaction between the surface material and the working environment (e.g. working fluid or contacting surfaces) can cause the surface to lose its integrity over time as a result of wear, corrosion and fatigue. Laser surface engineering involves altering the surface properties to reduce the effect of environmental degradation on the surface material over time. This is achieved by modifying the surface such that the surface can withstand the working environment without losing its integrity. Some of the areas where laser surface engineering finds its applications include: automobile, aerospace, biomedical, chemical, petroleum and power. These applications require surfaces that are wear and corrosion resistant amongst other properties. This chapter describes the use of laser metal deposition, an additive manufacturing technique for improvement of surface properties.

BACKGROUND OF LASER SURFACE ENGINEERING

Laser surface engineering, one of the many techniques employed for improving the surface of materials, is achieved by applying laser energy on the surface of the material or melting similar or dissimilar materials (such as metals, ceramic or composites) on the surface of the material in order to improve the properties of the surface of the bulk material. Laser surface engineering processes includes laser alloying (Mabhali et al., 2010; Zhang et al., 2011), laser cladding (Cai et al., 2007; Richter et al., 2004), laser heat treatments (Meng et al., 2012), and laser deposition (Liu & DuPont, 2003; Shukla et al., 2012). These are briefly explained in the following sub sections.

Laser Surface Alloying

Laser surface alloying involves the addition of alloying materials in form of powder or wire, which is melted using laser energy. The molten material mixes with the surface of the bulk material to create a homogeneous surface material with a different chemical composition and phases. An example of laser surface alloying is that of the AISI 316L stainless steel with ruthenium and nickel (Lekala et al., 2012). Another typical example is the laser alloying of aluminum mixed with nickel, titanium and silicon carbide to improve the surface hardness of aluminum (Mabhali et al., 2010).

Laser Heat Treatment

Laser can be used to selectively heat the material surface without melting with energy density sufficiently large enough to cause solid state transformation. Heat flows from the surface of the material being heated with laser into the bulk material leading to rapid surface cooling that is similar to quenching. For example, in steel, this high cooling rate causes solid state transformation of austenite to martensite resulting in a very hard surface. This process is known as solid state phase transformation hardening (Micheal et al., 2012). In some materials, phase transformation may not occur but the rapid cooling may cause the grains to be refined as a result of the bulk material acting like a heat sink and causing the surface hardness of the material to be improved.

Grain refinement can also occur when the laser heat is applied to the surface of the bulk material causing the surface of the material to melt and rapidly solidify while the bulk material acts as the heat sink. This rapid solidification causes a new microstructure to be formed because of the high cooling rate caused by the high energy transfer rate to the surface and the action of the highly collimated and focused laser beam.

The melting and subsequent rapid cooling lead to grain refinement in the topmost layer of the surface material. The cooling rate can be controlled by preheating the substrate before laser ablation thereby producing the refined microstructure or glazed (that is, amorphous) structure. The resulting surface is homogeneous in chemical composition thereby improving the corrosion and wear resistance properties of the surface being treated (Liu et al., 2013; Wu et al., 2003; Zhang et al., 2011). In biomedical applications, roughened surface may be needed in order to improve the bio-compatibility with the host tissue to aid the growth of the surrounding tissue on the implant. Lasers can also be used to create this surface roughness. Details of how lasers are used to improve the biocompatibility of implants can be found in Kurella and Dahotre (2005).

Laser Cladding

Laser cladding involves creating a surface layer that is completely different from the bulk material. Laser cladding is achieved by preplacing the cladding material(s) in powder form on the surface of the material to be cladded and the laser beam is scanned on the preplaced powder. The powder is melted and upon solidification, the cladded material forms a protective coating on the bulk material with improved surface properties. The chemical composition and microstructure of the cladded layer is completely different from that of the bulk material thereby providing the surface with the required protection which includes: oxidation protection, corrosion and wear resistance.

Laser Deposition

Laser deposition is similar to laser cladding, but instead of preplacing the cladding materials on the surface to be cladded, the materials are deposited directly and simultaneously as the laser is being scanned over the surface. The laser deposition is also useful in the production of composites and functionally graded materials that can be deposited directly on the surface of bulk material to improve the surface properties. More details of the laser deposition process are presented in a later section in this chapter. Laser surface engineering has the following advantages according to Mabhali et al. (2010) and Steen and Mazumder (2010).

Advantages of Laser Surface Engineering

- In laser surface engineering, the energy can be delivered in a controlled manner and to a localized area.
- The laser system, being an optical system can be manoeuvered to process a complex structure and hard-to-reach intricate parts.
- Power density can be controlled by controlling the laser beam size allowing rapid processing to be achieved without distorting the bulk material.
- Rapid processing rate in laser surface engineering results in rapid solidification which produces refined microstructure in the surface material.
- Coherent nature of the laser beam makes it possible to produce near net shape with controlled surface properties.
- Laser surface engineering is a tool-less and non-contact process.
- It requires no post machining operations.

An overview of the laser surface engineering technique has been presented in this section to recognize the role lasers can play in improving the surface integrity of engineering materials. Some of the laser surface engineering techniques were discussed, while the areas of application and advantages of laser surface engineering were also presented. The next sections will focus on the laser metal deposition process and its usefulness for improving the surface integrity of materials.

Laser Material Deposition Process

Laser Material Deposition (LMD) also known as Direct Metal Laser Deposition (DMLD) or Direct Laser Deposition (DLD) or Direct Metal Deposition (DMD) or Laser Powder Deposition (LPD) is a solid freeform fabrication or additive manufacturing process. It is based on the Laser Cladding (LC) technique that belongs to a class of the Additive Manufacturing (AM) technology known as Directed Energy Deposition (Scott et al., 2012). This technique is unique in the sense that it can be applied to repair high valued component parts that in the past were prohibitive or difficult to repair (Bergan, 2011). The ability to use different materials simultaneously makes it possible for the production of functionally graded materials. A schematic of the LMD process is shown in Figure 1.

Laser metal deposition produces better coating with minimum dilution of the deposited material and the substrate materials of the Parts produced using the LMD technique exhibit minimum physical distortion, better surface quality, near net shape and are fully dense. They have refined grain structure

Figure 1. Schematic of the laser metal deposition process

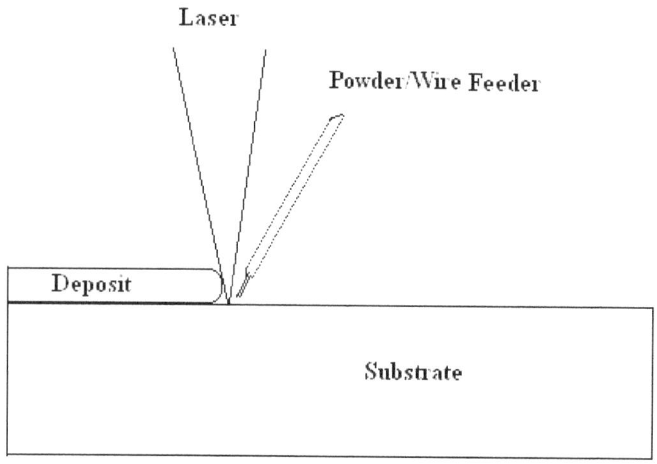

that determines the mechanical properties of the components produced. The parts that were discarded in the past as a result of not being wieldable can now be repaired using the LMD technology (Gasser et al., 2010). It enhances thermal control through its well-controlled heat affected zone. For repair of parts, LMD unlike the traditional repair methods (e.g. welding) which are destructive in nature is a safe repair technology especially on critical parts. This section focuses on the LMD technology; steps required for building components, processing parameters influencing the properties of the deposited surface and material characterization of laser metal deposited layers.

Laser Engineered Net Shaping is another example of the AM technology based on the laser metal deposition LMD technique as presented below.

Laser Engineered Net Shaping (LENS)

Laser engineered net shaping was developed at the Sandia National Laboratory in the mid-1990s (Kumar & Stucker, 2005; Paschotta, 2008). It uses a high power laser to melt the metal powder using laser beam on small spot at a time with the help of one or more lenses (Hedges & Keicher, 2002). Powdered or wired material is fed into the laser beam by powder delivery nozzle(s) or wire feeder respectively, the molten material solidifies very quickly which results in fully dense components with better properties thereby eliminating the need for heat-treatment after processing (Grylls, 2003). LENS is a very important technology that is used to produce a part with better control of property through the use of inert gas that shields the melt pool, and protecting it from atmospheric oxygen and nitrogen contaminations especially when dealing with reactive metal like titanium. Hence, better surface wetting is provided and better layer adhesion (Yan & Gu, 1996). In LENS process, it is also possible to dynamically change material composition leading to the production of composite and functionally graded parts (Liu & DuPont, 2003). The advantage of flexibility offered by additive manufacturing and laser technology has been employed widely for the production of functionally graded materials and for improving surface properties of materials (Meng, 2012; Obielodan, 2010).

The Laser metal deposition process can be achieved in five basic steps and these are common to all AM technologies and are briefly explained in the next section.

Steps in Laser Metal Deposition Process

Laser metal deposition processes involves five basic steps and they are briefly presented in the following.

- **Step One:** The first step involves receiving of CAD model Data of the part from AutoCAD, Solid Edge, Unigraphics, Pro Engineer, Solid Works etc.
- **Step Two:** The CAD model data received is converted into a Standard Additive Manufacturing File (AMF) format which is based on an open standard Extensible Markup Language (XML) that supports colour, texture, curved triangles, lattice structure and functionally graded materials (Scott, 2012). The old file format was Standard Triangulation Language (STL) file format which has recently been changed to AMF format by the ASTM International F42 committees on AM standards. This old file format does not support the functionalities supported by the AMF file format mentioned above. The AMF file represents a three dimensional surface of an assembly of planar and curved triangles containing the coordinate of their vertices. Conversion to AMF file format has been adopted as a standard for consistency.
- **Step Three:** The AMF file is prepared for the building process. The file is sliced into two-Dimensional (2-D) cross-section profile based on the geometry of the CAD model data and the required build orientation to be employed. Each AM machine has its own preparatory software for slicing the AMF file and it may also generate auxiliary support structure for the object to be built if required (Boboulos, 2010).
- **Step Four:** This step involves the actual building of the part by adding material using the 2-D slice section layer by layer until the part is completed. The first three steps are similar for all AM techniques but the building process (step four) differs from one laser additive manufacturing process to the other.
- **Step Five:** The final step involves the removal of the part and finishing operations. The object is removed from the machine and cleaned. The auxiliary support structures are removed if any; the part is polished, painted and heat treated, depending on the service requirements of the part. The finishing operations could be traditional or non-traditional. Some traditional finishing operations include: grinding, fine abrading, honing, abrasive belt polishing, elastic wheel polishing, steam spraying, and pill spraying. Some of these traditional finishing operations may not be suitable for laser additive manufactured parts because of the complexity of parts with intricate hard to reach or hidden channels thereby necessitating the use of non-traditional finishing operations. Some of these non-traditional finishing operations that are performed on LMD produced parts are presented in a later section.

Laser metal deposition process that belongs to Directed Energy Deposition class of AM technology has been presented in this section. Basic steps involved in laser metal deposition process, the importance of using LMD in the manufacturing of parts directly from 3-D CAD model data, in the repair of high value components and in surface modification (with composites and functionally graded materials) has also been highlighted. The properties of laser metal deposited parts are influenced by a number of processing parameters and these are the focus of the next section.

BASIC LASER METAL DEPOSITION PROCESSING PARAMETERS

The basic processing parameters that influence the properties such as hardness, strength, density, and surface quality of parts produced in LMD processes are grouped into three namely: Laser related (laser power, laser spot size, laser scanning velocity etc.), Process related (powder flow rate, shielding gas flow rate) and Material related (powder shape and size, material properties). These processing parameters are explained below:

Laser Related Processing Parameters

The laser related processing parameters include: the laser spot size, laser scanning speed, laser power, and the laser energy density. These parameters are explained as follows.

Laser Spot Size

the laser spot size is the diameter of the laser beam (for circular beam) or the dimension of the beam for (non-circular beam). The spot size has effect on the laser energy density, the surface finish, and the accuracy of the final part. The laser spot size is related to the Laser Energy Density (LED) through equation 1 according to Sentilkumara et al. (2009).

$$\text{LED (J/mm}^2\text{)} = \frac{\text{Laser Power (P) (kW)}}{\text{Beam Diameter (D) (mm) X Scanning Speed (V) (m/sec)}} \quad (1)$$

The smaller the laser beam diameter, the higher the laser energy density available to melt the powder or wire material. The smaller spot size is able to concentrate all the available laser energy on a smaller area as against a larger spot size that spreads the available energy over a larger area making it less concentrated and less intensive. Also, the smaller the spot size, the better the surface finish of the part produced. Since the melt pool size is limited by the laser spot size, hence the smaller the spot size, the smaller the track thickness. Larger spot size will produce lager track thickness and hence give rise to a more pronounced stair stepping effect which additive manufacturing is characterized.

Laser Scanning Speed

The scanning speed can be achieved in laser metal deposition process in two ways: either by moving the laser relative to the processed material or by moving the processed material while the laser source remained stationary. The scanning speed determines the available time the laser is able to interact with the materials being processed. The scanning speed is inversely proportional to the energy density as shown in equation 1. The lower the scanning speed, the higher the laser energy density. The lager the laser material interaction time, the larger the total energy made available to interact with the materials and hence the higher the energy density. At low scanning speed, more material is melted as a result of high energy density that is made available because of more time provided for laser material interaction. If the scanning speed is increased, less time is available for the laser to interact with the materials and hence, there may be unmelted powder particle when the scanning speed is too high. The scanning speed has a great effect on the material efficiency utilization (Akinlabi et al., 2012); it also affects the microstructure and the properties of the processed material (Mahamood et al., 2012).

Laser Power

Laser power is an important processing parameter in laser additive manufacturing because it determines the limiting capability of any particular laser based machine. The laser power is the only factor that is directly proportional to the laser energy density as presented in equation 1 above. The higher the laser power/energy, the higher the laser energy density. Laser power has also been found to influence the properties of the laser additive manufactured materials by Shukla et al. (2012).

Laser Energy Density (LED)

The laser energy density is an important factor in laser additive manufacturing as it determines the amount of energy that is made available to the processed material. The laser energy density is affected by three main factors according to equation 1, they are, the laser power, the scanning speed and the beam spot size. The LED can be made high or low by making the laser power high or low (direct relationship). LED can also be made high by reducing the scanning speed or the beam diameter or both. LED can also be reduced by increasing either the beam spot size or the scanning speed or even both. The overall LED that is made available to the processed material will determine whether the powder will be fully melted, partially melted or not melted. If the available total LED is too low, it may not even melt the processed material at all but just heat up the material, even if the laser power is very high. In some applications like laser heat treatment, melting is not required in such process. At a very high laser power and small spot size, if the scanning speed is too high, the processed material may not be melted at all because of a very short laser material interaction time as a result of high scanning speed. The total laser material interaction time (LMIT) is given by:

$$\text{LMIT} = \frac{\text{Total Track length(m)}}{\text{Scanning Speed(m/s)}} \tag{2}$$

Process Parameters

The process related processing parameters include the powder flow rate and the shield gas flow rate. Shield gases are usually employed during the LMD processes to protect the powder from environmental attack and also to protect the deposited part. The gas flow rate has an effect on the property of the part produced using the laser metal deposition processes. If the gas flow rate is too high, it causes high disturbance in the delivered powder making it difficult to get the powder delivered directly on the melt pool area. The balance should be found when selecting the gas flow rate to minimize this disturbance as well as providing enough shielding protection for the powder being carried. The powder flow rate also has a great influence on the properties of the laser metal deposited part in a similar way as does the laser power or the scanning speed. If the powder flow rate is too high, it may result in a situation where there are more materials being delivered than what can be fully melted by the available energy density thereby causing the part to have poor mechanical properties, poor surface integrity and poor surface finish. On the other hand, too low powder flow rate setting may result in a situation whereby little material is being delivered into the available energy density on the substrate in the melt pool area, causing the substrate to be more melted than required resulting in higher dissolution which is undesirable. Against these backgrounds, these factors need to be carefully set prior to deposition to achieve the desired results and to reduce scrap.

Material Properties

The material properties that affect the LMD processes include the powder shape and sizes, the substrate surface condition and the thermal properties of the materials (melting temperature, specific heat capacity, thermal conductivity etc). The shape of the powder particles also affects the rate at which the laser energy is absorbed or reflected. A spherical shaped powder particle is better than other shapes in laser metal deposition process because it aids energy absorption through the reflection of the laser beam to the neighboring powder particles. The smaller the powder particle, the better the density of the final produced part and the better the surface integrity, this is because larger powder particle size create more inter particle voids which could be difficult to remove even when the powder is completely melted. The substrate surface condition also has a great influence on the property of the resulting part or surface finish. Too smooth surface increases the reflectivity of the materials causing serious energy loses, therefore smooth surfaces should be prepared prior to laser additive manufacturing process. Also, too shining surfaces increases reflectivity and the surface needs to be coated where practicable, to reduce the reflection of laser beam and to aid laser absorption. The surface should also be cleaned and degreased prior to deposition to aid absorption of laser energy and also aid adhesion of the deposited materials.

The basic processing parameters affecting laser metal deposited parts has been discussed in this section. These parameters affect the final properties of the part being produced and there is need to carefully control these parameters in order to achieve the desired properties. After the laser metal deposition process, some parts may require secondary or finishing operations depending on the service requirements of the part. Traditional finishing operations may not be suitable on most laser additive manufactured part due to the complex nature of the part or hard-to-reach areas. Non-traditional finishing manufacturing processes are more attractive for these parts because of various advantages that they have over the traditional manufacturing processes. Some of these non-traditional manufacturing finishing operations are presented next.

NON-TRADITIONAL FINISHING OPERATIONS APPLIED TO LASER DEPOSITED PARTS

Finishing operation is usually the last manufacturing procedure that helps to further improve the surface finish and also to further improve the surface integrity of the processed parts. Non-traditional finishing operations include Electrical Discharge Machining (EDM), Chemical Machining (CM), Electrochemical Machining (ECM), Ultrasonic Machining (USM), Laser Beam Machining (LBM) and Water Jet Cutting (See Figure 2 for the schematics of these non-traditional finishing operations). Non-traditional finishing operation becomes attractive when traditional finishing is not feasible, not satisfactory, or not economical. Laser metal deposition technology is used to produce very complex parts that are difficult to produce using the traditional methods, performing finishing operations on those parts using traditional

Figure 2. Schematic of the non-traditional manufacturing methods (http://www.turnxon.com/articles.html) (a) EDM process, (b) chemical machining process, (c) ECM process, (d) USM process, (e) LBM process, (f) abrasive water jet machining

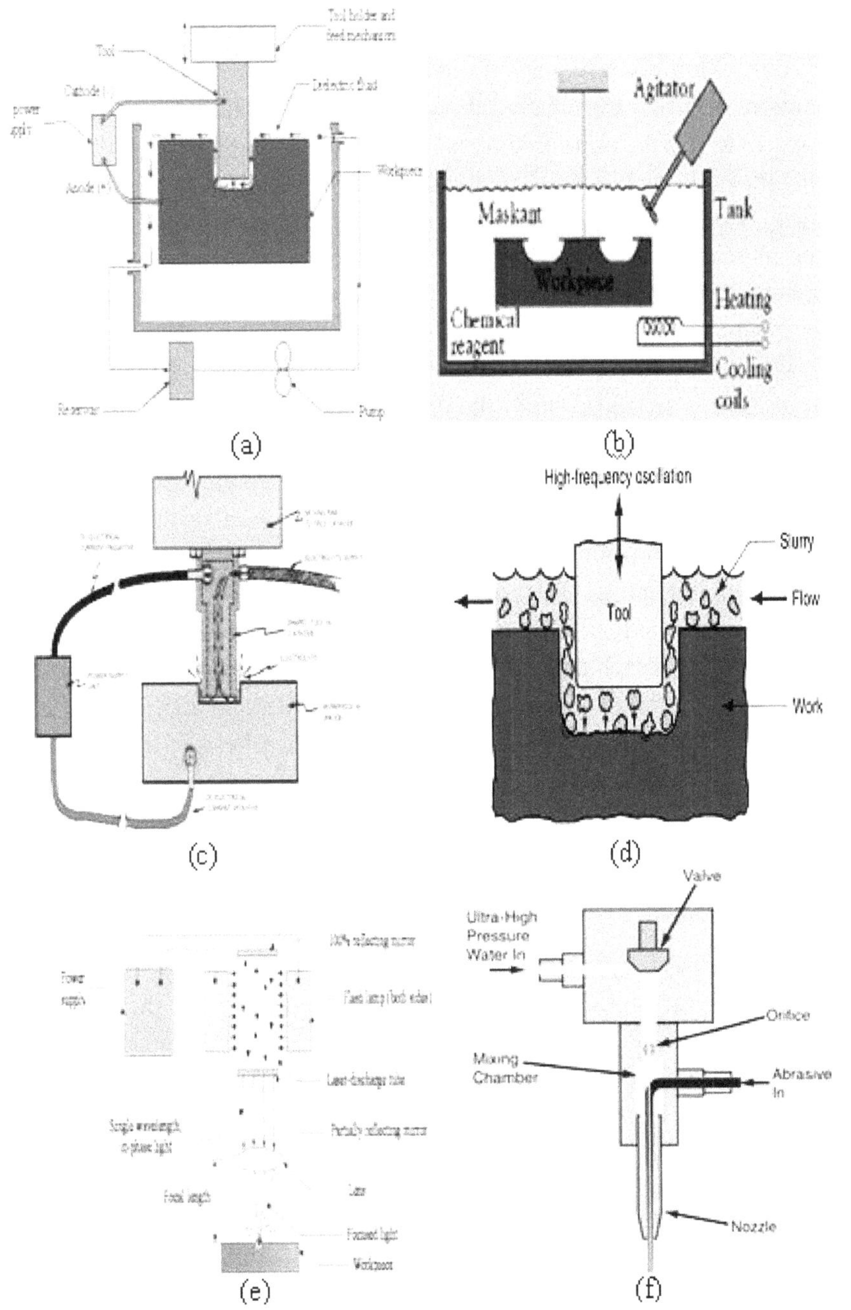

techniques is also difficult and uneconomical, therefore non-traditional finishing operations are the only viable and economical option especially when difficult to clamp or delicate or complex shaped parts are involved. It has been established in the literature that non-traditional finishing operation is capable of reducing operating noise of gear and bearing that were finished using traditional finishing operations and the life of the component can be increased ten times (Bang-Zong & Jion-jin, 2005) when these non-traditional finishing operations are used to process parts. Some non-traditional finishing operations that are performed on laser deposited parts are described below (Yao et. al., 2005).

Electrical Discharge Machining (EDM)

Electronic discharged machining is a non-traditional manufacturing technique that can be used to perform finishing operations on LMD manufactured components. It is a thermoelectric process that uses a series of electrical sparks to remove undesired parts like cling powder particle on the part, or the support structure used during the building process. The hardness of the material being processed is not a factor in EDM process when compared to the traditional metal removing operation which required that the cutting tool material must be harder than the material being cut. It becomes extremely difficult when trying to cut or machine a very hard material using traditional methods. Material removal is achieved in EDM process by applying high voltage across the gap between the electrode and the workpiece, which is very narrow, the schematic of the process is shown in Figure 2a. The high voltage induces an electric field in the insulating dielectric fluid (the dielectric fluid could be hydrocarbon oils, kerosene or deionised water) in the narrow gap between the electrode and the workpiece. This causes the metal particles that are suspended in the dielectric to concentrate at the points of strongest electrical field. With the potential difference between the electrode and the workpiece sufficiently high, the dielectric breaks down generating a transient spark discharges through the dielectric fluid.

The spark that is generated causes the removal of small amount of material from the workpiece surface and the process is repeated until the required amount of material is totally removed. EDM is very attractive with its ability to machine hard, difficult-to-machine materials, parts with complex, and irregular shapes and intricate internal shapes for aerospace and medical applications. No burrs are left in the machined surface because the dielectric fluid constantly flushes the removed chips away from the surface of the material being machined. Thin fragile and brittle components can also easily be machined without distortion. The major limitation in this process is that unwanted erosion and over cutting of material can occur.

Chemical Machining (CM)

Chemical machining (CM) is the controlled dissolution of surface material of the workpiece by means of a strong chemical reagent. The surface of the material to be removed is immersed in a chemical reagents or etchants such as acids and alkaline solutions. The material removal action is similar to corrosion action. The chemical weakens the exposed surface of the material thereby peeling the weakened layer which is subsequently dissolved into the solution and removed. The schematic of the chemical machining process is shown in Figure 2b.

Electrochemical Machining (ECM)

The Electrochemical machining (ECM) process is based on the principle of reverse electroplating. The workpiece is made the anode while the tool is made the cathode. The metal particles travel from the anodic material toward the cathodic material. The current in the electrolyte fluid carries away the depleted material before it reaches the cathode (machining tool). The shape produced on the workpeice material is a depression of the machining tool shape. The workpeice hardness is not a limiting factor in ECM, making it suitable for machining difficult-to-machine materials. Complex shapes can also be machined by this process regardless of the workpeice material hardness. The ECM tool is placed very close to the workpiece, a low voltage and high amperage DC current is passed between the workpiece and the electrode. The passage of the current causes the workpiece surface material to be depleted and removed by the electrolyte, the schematic is shown in Figure 2c. The ECM process is very attractive as a finishing operation in laser additive manufactured part because it can deburr difficult to access areas of parts. Also, very good surface finish can be achieved by the ECM process. The major limitation of the ECM process is that it cannot be used to machine sharp square corners or flat bottoms because of the tendency for the electrolyte to erode away sharp profiles. High equipment costs of ECM machine makes it application to be limited to highly specialised applications. Complex geometrical shapes can easily be machined with the ECM process.

Ultrasonic Machining (USM)

The USM process is a mechanical material removal process that is used to erode holes or cavities on hard or brittle workpiece by using shaped tools, mechanical motion, and an abrasive slurry. USM is very attractive for machining brittle materials and complex parts with intricate shape profiles that are difficult to machine using the traditional manufacturing processes. The hard particles in slurry that oscillate on the surface of the workpiece material repeatedly remove material thereby creating a cavity of a cross section identical to that of the tool. See Figure 2d for the schematic of the USM process. All materials, whether they are electrically conductive or insulators, they can be machined by this process. It does not require any chemical and it does not create any changes in the microstructures of the workpiece material and it produces virtually stress free machined surfaces. Parts that are machined using the USM have better surface finish. The limitation of the USM is that it consumes very high power and the material-removal rates are lower than the traditional fabrication processes. Also, very complex parts cannot be machined effectively using the USM process.

Laser Beam Machining (LBM)

Laser beam machining is a thermal material removal process that uses a high energy, collimated light beam to melt and vaporize particles material on the surface of the workpiece (see Figure 2e for the schematic of the process). High precision hole can be made using the LBM. Laser beam can also be used to weld metals and non-metals. They are very useful in electrical, automotive and aerospace industries to mention but few. High maintenance cost limits its use to high valued parts and it also requires cover gas during operation. It is not economical for high volume work and for drilling operation, there is limit to which the depth can go because of taper, and high capital is needed for the equipment.

Water Jet Cutting

Water jet cutting is a fast machining process. The support structures used during part build up in some laser metal deposited process can easily be cut off using the water jet cutting. The cutting edges are usually cleaned with minimal burr because no heat is applied during the cutting operation. Water jet cutting process involves the use of water under extremely high pressure flowing through a very small orifice or jewel to cut soft materials such as wood, plastics and aluminium. To be able to use this technology to cut harder materials, abrasive materials are usually added to Water. The main attraction of using this technology in finishing laser additive manufactured part is that it does not generate any heat during the cutting operation thereby retaining the properties of the workpiece material and this is especially useful for cutting materials that excessive heat may change the properties of the material. Water jet cutting does not induce distortion in the workpeice, it uses low cutting forces on workpiece, limited tooling is also required, and it produces little or no cutting burr, and produces better surface finish. The main limitation of water jet cutting is that a limited number of materials can be cut economically making their use limited to high valued parts for aerospace, automotive and electronics applications. The abrasive water jet cutting requires high capital and it is noisy in operation, the schematic of the abrasive water jet is shown in Figure 2f.

Additive manufacturing technology is a non-traditional and an interesting manufacturing system, the use of non-traditional finishing technique is most appropriate because of the complex and fragile nature of some parts (Sommer, 2000). The foregoing section explains some of these non-traditional finishing techniques. Laser Additive Manufacturing (LAM) is the most popular of all the AM technology and of the LAM technologies, only laser metal deposition process can be used to repair high value component parts (Bergan, 2009) with proven results. Parts that are not repairable in the past can now be repaired using the LMD. The next section will focus on material characterization of laser metal deposited Ti6Al4V, an aerospace alloy.

CHARACTERIZATION OF LASER METAL DEPOSITION OF TITANIUM ALLOY

Introduction

Titanium and its alloys exhibit a special combination of properties that made them to be desirable for critical and demanding industrial applications. The industries that use titanium most include: aerospace, chemical and energy industry (Peters, 2003). Titanium alloys have high strength to weight ratio, which is structurally efficient for critical, high-performance applications like jet engine parts and airframe components (Richter, 2004). Ti6Al4V is the most widely used titanium alloy for aerospace application because of the exciting properties of this alloy (Peters, 2003). Despite these properties of titanium, they are classified as generally difficult to machine material because of their high interaction with the cutting tool materials resulting in high temperature thereby galling the cutting tool and shortening the tool life (Wang & Ezugwu, 1997). Laser metal deposition provides an alternative manufacturing method as well as for repair of parts since it is a tool-less process. The following sections present, a study on the effect of laser power on the resulting properties of laser deposited Ti6Al4V.

Materials and Experimental Method

Material

A 72 X 72 X 5 mm 99.6% pure Ti6Al4V plate was used as a substrate in this experiment. The Ti6Al4V powder used in this study is of particle size range between 150 and 200 μm and of 99.6% purity. The substrate was prepared prior to the deposition process, this was necessary to allow the laser beam to be absorbed into the material. If the surface of the substrate is not prepared and if the surface is shining as a result of its smoothness, or if there is grease or dirt, the laser beam will be reflected as it is reaching the surface of the substrate and as such, most of the energy carried by the laser beam will be lost. To prevent this loses from occurring; the surface of the substrate was prepared prior to the deposition. Some of the surface preparation include: acid etching, sand blasting, and grinding. After the roughing of the surface, it was then cleaned with solvent mostly methanol, ethanol or acetone to remove the chips, dirt, and grease. In this experiment, the substrate was sandblasted and cleaned with acetone to remove the grease and dirts.

Experimental Method

The Laser metal deposition was achieved using a Kuka robot, carrying a 4.0 kW fiber Nd-YAG Rofin laser shown in Figure 3a. Also, attached to the robot is the coaxial nozzle for powder delivery from two hoppers shown in Figure 3b. The experimental set-up is shown in Figure 3c and the schematic of the process is shown in Figure 1a. The powder was delivered through argon gas which is used to protect the powder from environmental contamination, since Ti6Al4V is very reactive with oxygen. The deposit was also protected from the environment through the argon gas that was delivered into the glove box improvised through another set of nozzles. The robot was programmed to deliver the laser power on the substrate thereby creating a melt pool and then deliver the powder into the melt pool that creates the deposit track upon solidification.

The objective of the experiment was to study the effect of laser power on the properties of the laser deposited Ti6Al4V powder. The laser spot size was maintained at 2 mm at a focal length of 195 mm above the substrate. The scanning speed, powder flow rate and the shield gas flow rate were fixed at 0.005 m/s, 1.44 g/min and 4 l/min respectively. The laser power whose effect needs to be studied was varied between 0.8 kW and 3.0 kW. After the deposition process, the deposit was laterally sectioned, ground, polished and etched according to the standard metallographic sample preparation of Titanium. The samples were observed under the Olympus optical microscope (Olympus BX51M) to study the macrostructure and the microstructure using low and high magnifications. The microhardness of the cross sections was measured using the MH-3 Vickers Microhardness indenter developed by Metkon with a load of 500 g and a dwell time of 15 seconds. The space between the indentations was maintained at 15μm.

Microstructural Characterization

The microstructure of the substrate material is shown in Figure 4a, which is characterized by lamellar structure of alpha and beta phases. The lighter parts are depicted by the alpha phases while the darker parts are the beta phases. The morphology of the powder is spherical gas atomized (see Figure 4b) with the Gaussian distribution shown in Figure 4c. A typical cross section of the deposited layer according

Figure 3. (a) Rofin laser system at CSIR Pretoria, South Africa, (b) powder delivery hoppers, (c) experimental set-up

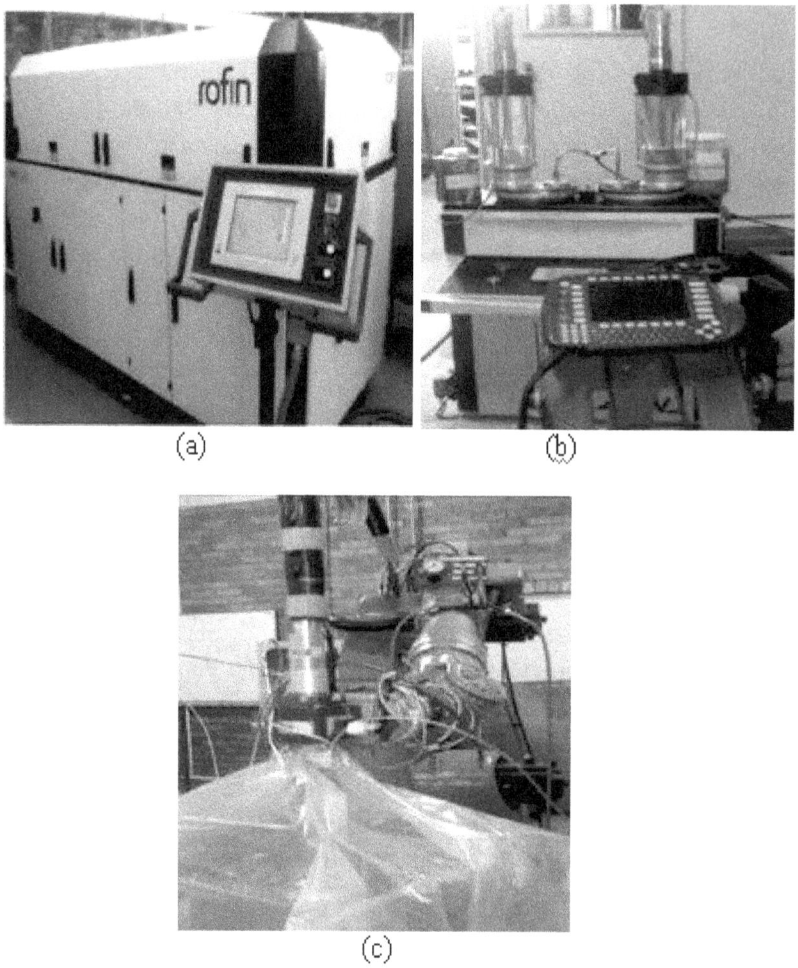

to (Mahamood et al., 2013) is shown in Figure 4d, displaying the Deposit Zone (DZ), the Fusion Zone (FZ) and the Heat Affected Zone (HAZ). The Deposit Zone contains the fully melted deposited powder, the consistency seen in the microstructure shows that the powder is fully melted (see Figure 4e). The microstructure of the DZ is characterised by columnar grain structure with a mixture of alpha and prior beta grains. Martensites are seen in this region. The Fusion Zone (FZ) consists of the mixture of the substrate and the powder that are melted and mixed to produce bonding of the deposited material with the substrate material. The Heat Affected Zone (HAZ) consists of the substrate material only. The heat transferred from the deposit and the heat affected zone was absorbed by the rest of the substrate. The region closest to the fusion zone received more heat that causes microstructural changes in the area and it is reffered to as the heat affected zone. The microstructure in this region is characterised by globular grains structure.

This globular grains resulted from the grain growth of the alpha transformed beta and prior beta grains. The directional solidification taking place in the DZ causes the grains in this zone to grow epitaxially on

Figure 4. (a) The microstructure of the substrate, (b) the micrograph of the Ti6Al4V powder, (c) particle size analysis of the Ti6Al4V powder, (d) a typical cross section of deposit showing different zones (Mahamood et al., 2013), (e) the deposit zone of the sample at laser power 0.8 kW, (f) microstructure of sample at 0.8kW laser power, (g) microstructure of sample at 1.6 kW laser power

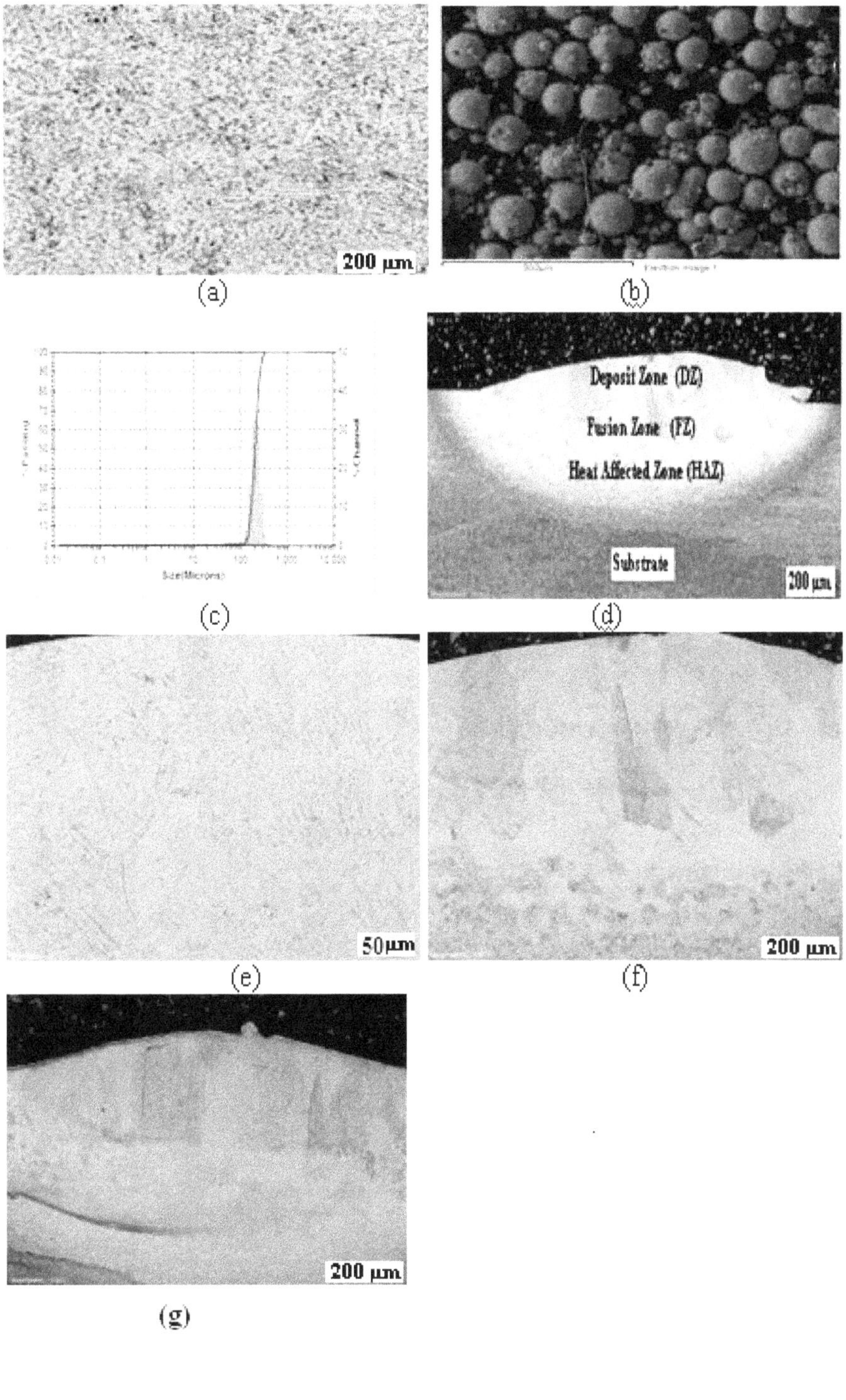

the globular grains in the HAZ. This is because the substrate material is serving as a heat sink thereby enhancing the columnar grain growth whose direction is opposite to the direction of heat flow which is towards the substrate. The effect of varying the laser power on the deposit zone can be seen in Figure 4f and 4g where the number of columnar grains reduces as the laser power increases. This is because at high laser power, the melt pool stays longer on the substrate causing enough time for the grain growth in the HAZ which results in fewer globular grains in the heat affected zone. The columnar grains in the DZ grows epitaxially on these globular grains in the HAZ, which explains why the number of columnar grain structure reduces as the laser power increases. The fusion zone is chacterised by continuous columnar prior β grains structures showing the bonding is metallurgically sound (see the upper part of Figure 5). The microstructures in the fusion zone is exploded in the lower part of Figure 5 showing various alpha grains.

Microhardness

The microhardness profiles of the four samples were taken to be able to see the effect of varying the laser power on the resulting microhardness of the samples. The micrograph of an indented sample at laser power of 2.4 kW is presented in Figure 6a. The average Vickers hardness value of each of the samples was determined and plotted using the best fit line to appreciate the relationship of the average Vickers hardness value to the laser power (see Figure 6b). The average Vickers hardness value is found to increase with an increase in the laser power. This result is consistent with the microstructures observed in the samples, especially the martensite, which changes from fine to coarse microstructure as the laser power increases.

Figure 5. Microstructure of fusion zone of sample at 0.8 kW (Mahamood et al., 2013)

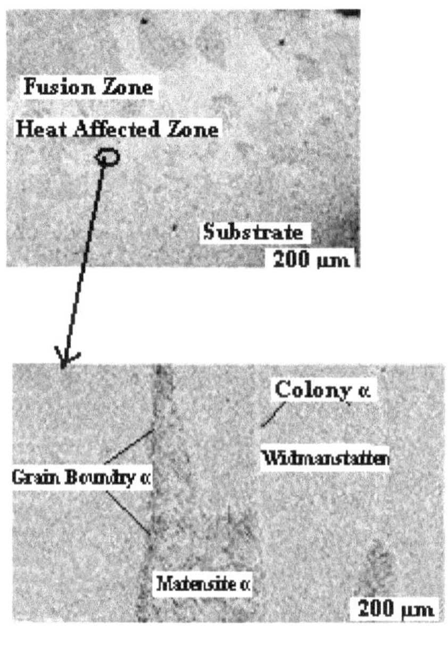

Figure 6. (a) Indented sample at laser power 2.4 kW, (b) the plot of the average microhardness value of the samples at 0.8, 1.6, 2.4 and 3.2 kW

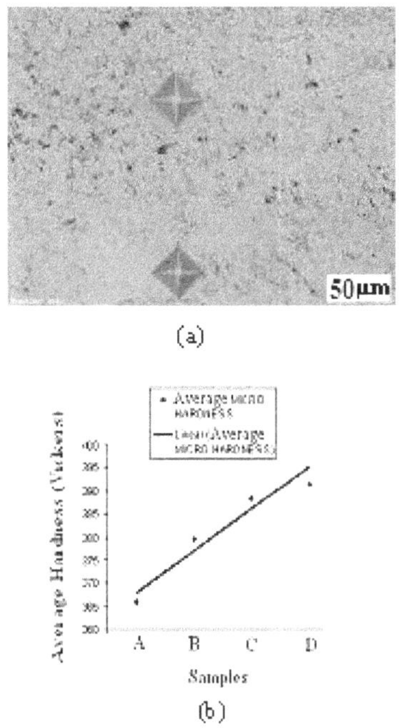

CHARACTERISATION OF LASER DEPOSITED TI6AL4V/TIC COMPOSITE

Introduction

The flexibility offered by the LMD makes it possible to produce composite materials. To improve surface properties of components, Metal Matrix Composites (MMC) of metal and ceramic are often deposited on the surface of the bulk material as reported in the literature (Rosso, 2006). In this section, a laser deposited Ti6Al4V/TiC composite is studied and characterized. The study was conducted using Ti6Al4V and TiC powders delivered into the melt pool by powder feeder with two hoppers carrying each of the powder and delivering them into the melt pool simultaneously. The composite was characterized using the Scanning Electron Microscopy (SEM), Microhardness tester and wear tester to study the microstructure, microhardness, and the wear resistance properties. The results are presented and discussed in detail.

Experimental Procedure

The materials used are the Ti6Al4V plate and powder described in previous section. In addition, Titanium Carbide (TiC) powder of particle size below 60 µm and 99.5% purity was also used. The experimental set-up is the same as the one described earlier. A set of trial runs were performed until a fully dense deposit was obtained. The powders were placed separately in each hopper, and the setting of each hopper was maintained at equal value of 1.44 g/min to give rise to a composite of 50W% TiC / 50W% Ti6Al4V ratio. the optimized process parameters are as follows: The gas flow rate was set at 1 l/min for each of the powder, the laser power was set at 2.4 kW and the scanning speed was set at 0.005 m/s. Multiple tracks were deposited with 50% overlap. After the deposition process, the sample for the microstructural examination was cut laterally to reveal the cross section. The cut samples were mounted, ground, polished and etched according to standard metallographic preparation of titanium. The sample was observed under the Optical Microscope (OM) and the Tescan Scanning Electron Microscopy (SEM) equipped with oxford energy dispersion spectrometry. The microhardness profiling was conducted using an MH-3 Vickers microhardness indenter with a load of 500g and a dwell time of 15 seconds. The wear test was conducted using a Pin on disk tribometer (by Cert) with tungsten carbide ball of diameter 10 mm and a load of 20 N at a sliding distance of 2000 m. The tests were conducted in dry air and the coefficient of friction was obtained for all the samples.

RESULTS and DISCUSSION

Microstructural and Microhardness Analysis

The SEM micrograph of the TiC powder is shown in Figure 7a. The TiC powder particle is characterized by irregular shape produced by ball milling process. The microstructure of the surface of the Ti6Al4V substrate and the Ti6Al4V powder were presented in Figures 4a and 4b respectively. The micrograph of the composite in Figure 7b shows the cross section of the composite. The intermediate zone shows where the composite bond with the substrate material and the bonding is metallurgically sound due to the choice of the processing parameters.

Microhardness Profiling

The plot of the microhardness profiles of the substrate and the composite are shown in Figure 7c.

The average microhardness value of the substrate is about 300 HV while that of the deposited composite is about 400 HV, this shows that the deposit has improved the hardness of the surface of the bulk material.

Wear Test

To further study the property and the integrity of the resulting surface, wear test was conducted with tungsten carbide ball. The SEM micrograph of the wear track of the substrate is shown in Figure 8a. It was observed from Figure 8a that there are ploughing grooves on the wear track with some of the debris welded to the wear track. The wear mechanism as is seen in Figure 8a is a combination of abrasion, adhesion and plastic deformation. At the beginning of the wear action, the wear is generated by the rubbing action of the tungsten carbide ball and the substrate. As the rubbing continues, some of the debris removed from the surface got welded to the surface of the substrate while some are loosen. The wear mechanism then changes from two-body to three body wear mechanism as a result of the loosen debris which further aggravated the wear process thereby causing the characteristic parallel grooves that is typical of abrasive wear mechanism. The wear behaviour of the substrate material is different from the behaviour observed in the deposited TiC/Ti6Al4V composite. Figure 8b shows the SEM micrograph of the surface of the composite. It was found that there is no obvious adhesion of the debris on the wear tracks as shown in Figure 8b because the wear debris from the composite are formed from the hard particle of the Unmelted Carbide UMC particles of the TiC powder. The three-body wear mechanism that

Figure 7. (a) The SEM micrograph of the TiC powder, (b) the optical micrograph of the composite, (c) the plot of microhardness of the substrate and the composite

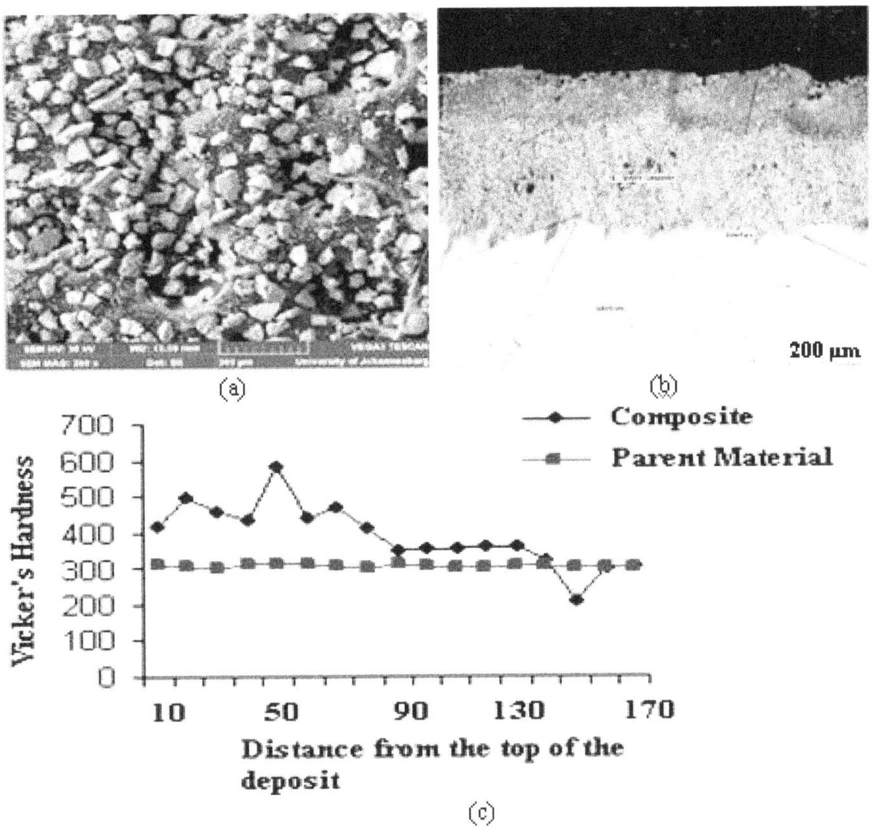

occurred with the TiC/Ti6Al4V composite is such that the loose debris rubs on the surface containing unmelted carbide and this continuous rubbing resulted in the grinding of the debris thereby reducing the debris sizes to finer powder particle forming lubricant and inhibiting further wear action. That is why the wear resistance performance of the surface deposited with the composite is improved compared to the substrate.

The coefficient of friction is obtained for a sliding distance of 2000 m. The plot of the coefficient of the composite and that of the substrate are shown in Figure 8c.

The graph (Figure 8c) shows that, the composite surface has a better wear resistance performance compared to the substrate and that the composite deposit is capable of improving the sliding wear property of the Ti6Al4V surface.

FUTURE RESEARCH DIRECTIONS

The future research will be focused on surface modification laser metal deposited using functionally graded material. Functionally graded material (FGM) is an advanced material that is capable of withstanding harsh working condition. Traditional composite will fail when subjected to extreme temperature

Figure 8. (a) The SEM micrograph of the wear track of the substrate, (b) the wear track of the composite surface, (c) the plot of coefficient of friction against sliding distance

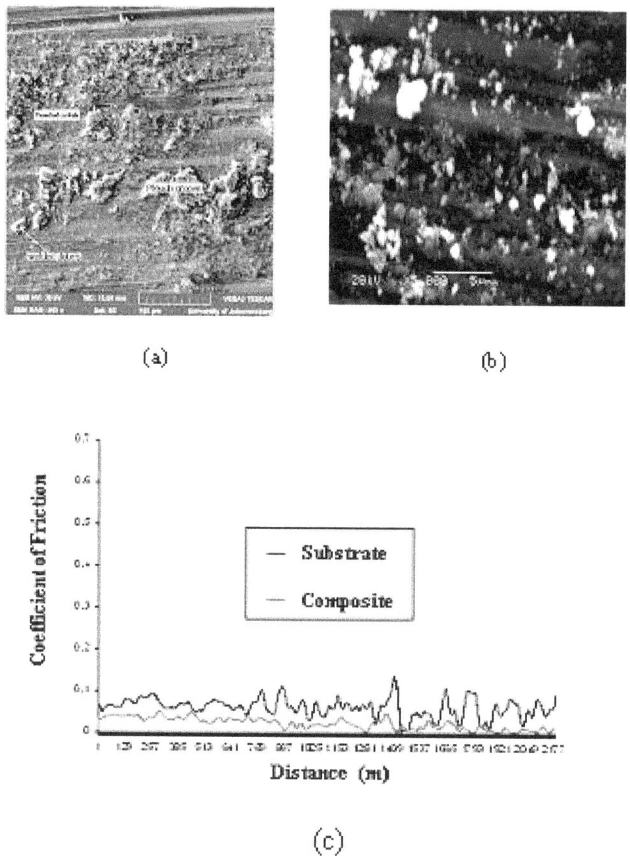

(a) (b)

(c)

due to mismatch in the thermal properties of the composite materials. This failure is basically due to the sharp interface between the composite materials. The functionally graded material replaced the sharp interface with gradual interface. The shape interface in the traditional composite is as a result of sudden change in one material to another. This sharp interface is a high stress concentration zone where failure is initiated. The functionally graded material on the other hand change gradually from one material to the next, thereby eradicating the sharp interface. Laser metal deposition of FGM for surface modification of metals is an interesting research area and the future research will be focused on this area.

CONCLUSION

In this chapter, an overview of laser surface engineering has been presented. Laser additive manufacturing process which is the most popular of all the additive manufacturing processes, because of its ability to repair high valued parts and its ability to produce composite and functionally graded parts and surfaces. LMD is used in the production of complex parts in one single build that are difficult to produce using the traditional manufacturing process. LMD reduces weight due to elimination of assembly; reduces manufacturing time and the overall production cost. The processing parameters affecting the properties of laser metal deposited parts were described

In conclusion, the background of the laser metal deposition process has been discussed, a study on the effect of laser power on the property of laser deposited Ti6Al4V and material characterization of laser deposited Ti6Al4V/TiC composite was also presented. The study reveals that the property of the surface produced with the LMD process is better when compared to the substrate material. The LMD process is a promising technology for manufacturing, repairing and for surface modification of parts for the aerospace industry, defense and as well as many other industries.

ACKNOWLEDGMENT

This work was supported by the Rental Pool Grant of the National Laser Centre - Council of Scientific and Industrial Research (NLC-CSIR), Pretoria South Africa and The Schlumberger Foundation Faculty for the Future (FFTF).

REFERENCES

Akinlabi, E. T., Mahamood, R. M., Shukla, M., & Pityana, S. (2012). *Effect of scanning speed on material efficiency of laser metal deposited Ti6Al4V*. Paper presented at the International Conference on Aerospace, Mechanical, Automotive and Materials Engineering. Paris, France.

Bang-Zong, L., & Jion-Jin, Z. (2005). Research on surface characteristics of non-traditional finishing. *Journal of Zhejiang*, *6A*(10), 1152–1157. doi:10.1631/jzus.2005.A1152

Bergan, P. (2000). Implementation of laser repair processes for navy aluminum components. In *Proceeding of Diminishing Manufacturing Sources and Material Shortages Conference* (DMSMS). Retrieved from http://smapcenter.uah.edu/smap-center/conferences/dmsms2k/papers/decamp.pdf

Boboulos, M. A. (2010). *CAD-CAM & rapid prototyping application evaluation*. PhD & Ventus Publishing Aps. Retrieved from www.bookboom.com

Cai, L. F., Zhang, Y. Z., & Shi, L. K. (2007). Microstructure and formation mechanism of titanium matrix composites coating on Ti-6Al-4V by laser cladding. *Rare Metals*, *26*(4), 342–346. doi:10.1016/S1001-0521(07)60226-5

Gasser, A., Backes, G., Kelbassa, I., Weisheit, A., & Wissenbach, K. (2010). Laser additive manufacturing laser metal deposition (LMD) and selective laser melting (SLM) in turbo-engine applications. *Laser Material Processing*, *2*, 58–63.

Grylls, R. (2003). Laser engineered net shapes. *Advanced Materials and Processes*, *161*(6), 46–57.

Hedges, M., & Keicher, D. M. (2002). Laser engineered net shaping - Technology and applications. In *Proceedings of 3rd National Conference on Rapid Prototyping, Rapid Tooling, and Rapid Manufacturing*, (pp. 17-23). Academic Press.

Kumar, S., & Stucker, B. E. (2005). Development of a Co-Cr-Mo to tantalum transition using LENS for orthopaedic applications. In *Proceedings of 16th Solid Freeform Fabrication Symposium*. University of Texas.

Kuvella, A., & Dahotre, N. B. (2005). Review paper: surface modification for bio implant: The role of laser surface engineering. *Journal of Biomaterials Applications*, *20*(1), 5–50. doi:10.1177/0885328205052974 PMID:15972362

Lekala, M. B., van der Merwe, J. W., & Pityana, S. L. (2012). Laser surface alloying of 316L stainless steel with Ru and Ni mixtures. *International Journal of Corrosion*. doi:10.1155/2012/162425

Liu, J., Dahmen, M., Ventzke, V., Kashaev, N., & Poprawe, R. (2013). The effect of heat treatment on crack control and grain refinement in laser beam welded β-solidifying TiAl-based alloy. *Intermetallics*, *40*, 65–70. doi:10.1016/j.intermet.2013.04.007

Liu, W., & DuPont, J. N. (2003). Fabrication of functionally graded TiC/Ti composites by laser engineered net shaping. *Scripta Materialia*, *48*, 1337–1342. doi:10.1016/S1359-6462(03)00020-4

Mabhali, L., Pityana, S., & Sacks, N. (2010). Laser alloying of Al with mixed Ni, Ti and SiC powders. *Journal of Laser Applications*, *22*(4), 121–126. doi:10.2351/1.3516482

Mahamood, R. M., Akinlabi, E. T., Shukla, M., & Pityana, S. (2012). Evolving properties of laser metal deposited Ti6Al4V. In *Proceedings of International Conference of Rapid Product Development Association of South Africa* (RAPDASA). RAPDASA.

Mahamood, R. M., Akinlabi, E. T., Shukla, M., & Pityana, S. (2013). *Laser metal deposition of Ti6Al4V: A study on the effect of laser power on microstructure and microhardness.* Paper presented at the 2013 International Multiconference of Engineering and Computer Science (IMECS 2013). New York, NY.

Meng, Q., Liu, Y., Yang, H., Shariat, B. S., & Nam, T. (2012). Functionally graded NiTi strips prepared by laser surface anneal. *Acta Materialia*, *60*, 1658–1668. doi:10.1016/j.actamat.2011.11.052

Michael, K.H., Leung, H.C., & Man, J.K., & Yu. (2007). Theoretical and experimental studies on laser transformation hardening of steel by customized beam. *International Journal of Heat and Mass Transfer*, *50*(23–24), 4600–4606.

Nontraditional Manufacturing. (n.d.). Retrieved from http://www.turnxon.com/articles.html

Obielodan, J. O. (2010). *Fabrication of multi-material structures using ultrasonic consolidation and laser-engineered net shaping*. (PhD Thesis). Utah State University, Logan, UT.

Paschotta, R. (2008). *Encyclopaedia of laser physics and technology*. Hoboken, NJ: Wiley.

Peters, M., Kempner, J., Ward, C. H., & Liens, C. (2003). Titanium alloys for aerospace applications. *Advanced Engineering Materials*, *5*, 419–427. doi:10.1002/adem.200310095

Richter, E., Orban, K. H., & Nowotny, S. (2004). Laser cladding of the titanium alloy Ti6242 to restore damaged blades. In *Proceedings of 23rd International Congress on Applications of Lasers and Electro-Optics*. Retrieved from www.ccl.fraunhofer.org/download/cladding_titanium.pdf

Rosso, M. (2006). Ceramic and metal matrix composites: Routes and properties. *Journal of Materials Processing Technology*, *175*(1–3), 364–375. doi:10.1016/j.jmatprotec.2005.04.038

Scott, J., Gupta, N., Wember, C., Newsom, S., Wohlers, T., & Caffrey, T. (2012). Additive manufacturing: Status and opportunity. *Science and Technology Policy Institute*. Retrieved from https://www.ida.org/stpi/occasionalpapers/.../AM3D_33012_Final.pdf

Senthilkumaran, K., Pandey, P. M., & Rao, P. V. M. (2009). Influence of building strategies on the accuracy of parts in selective laser sintering. *Materials & Design*, *30*(8), 2946–2954. doi:10.1016/j.matdes.2009.01.009

Shukla, M., Mahamood, R. M., Akinlabi, E. T., & Pityana, S. (2012). *Laser metal deposition of Ti6Al4V: A study on the effect of laser power on microstructure and microhardness*. Paper presented at the International Conference on Aerospace, Mechanical, Automotive and Materials Engineering. Paris, France.

Sommer, C. (2000). *Non-traditional machining handbook*. Retrieved from http://books.google.com.ng/books?id=p3NoQgAACAAJ

Steen, W. M., & Mazumder, J. (2010). Laser surface treatment. *Laser Material Processing*, 295-347.

Wang, Z. M., & Ezugwu, E. O. (1997). Titanium alloys and their machinability a review. *Journal of Materials Processing Technology*, *68*, 262–270. doi:10.1016/S0924-0136(96)00030-1

Wu, G. Q., Huang, Z., Li, Z. F., & Ruan, Z. J. (2003). TEM observations of grain refinement of laser surface melted γ-TiAl based alloy. *Materials Letters*, *57*(24–25), 3810–3814. doi:10.1016/S0167-577X(03)00183-6

Yan, X., & Gu, P. (1996). A review of rapid prototyping technologies and systems. *Computer Aided Design*, *28*(4), 307–318. doi:10.1016/0010-4485(95)00035-6

Yao, Y. L., Cheng, G. J., Rajurkar, K. P., Kovacevic, R., Feiner, S., & Zhang, W. (2005). Combined research and curriculum development of nontraditional manufacturing. *European Journal of Engineering Education*, *30*(3), 367–376. doi:10.1080/03043790500114581

Zhang, H., Pan, Y., & He, Y. Z. (2011). Grain refinement and boundary misorientation transition by annealing in the laser rapid solidified 6FeNiCoCrAlTiSi multicomponent ferrous alloy coating. *Surface and Coatings Technology*, *205*(16), 4068–4072. doi:10.1016/j.surfcoat.2011.02.054

Zhang, Y., Huang, C., & Vilar, R. (2011). Microstructure and properties of laser direct deposited CuNi17Al3Fe1.5Cr alloy. *International Journal of Minerals. Metallurgy and Materials*, *18*(3), 325–328. doi:10.100712613-011-0442-9

ADDITIONAL READING

Chemical Machining, Metals Handbook, vol. 3.

Kalpakjian, S., & Schmid, S. R. (2013). *Manufacturing Engineering and Technology* (7th ed.). Prentice Hall.

Machining, 8th ed. (pp. 240–248). American Society for Metals.

Timings, R. L., & Wilkinson, S. P. (2000). Manufacturing [Longman.]. *Technology (Elmsford, N.Y.), 2*,

KEY TERMS AND DEFINITIONS

3-D Printing: The fabrication of objects through the deposition of a material using a print head, nozzle, or other printer technology. 3-D printing is often used synonymously with additive manufacturing.

Additive Manufacturing: Additive manufacturing is termed the official industrial standard according to ASTM F2792. It is defined as the process of joining materials to make objects from 3D model data, in layer wise manner, as opposed to subtractive manufacturing methodologies such as turning (http://wohlersassociates.com/additive-manufacturing.html).

CAD: Computer-Aided Design, according to ASTM standard means the use of computers for the design of real or virtual objects.

Coefficient of Friction (COF): Coefficient of friction is the ratio of the force that maintains contact between two surfaces in sliding contact and the frictional force that resists the motion of the object.

Composite: Composite is made up of distinct types of materials.

FGM: Functionally graded material is may be defined as the variation in materials, composition or structure gradually over the entire volume of the bulk material, resulting in corresponding changes in the properties of the material. The functionally graded materials can be tailored or designed for specific function and applications.

Laser Beam Diameter: The diameter of the laser beam measured at the focus point on the material being processed.

Laser Metal Deposition (LMD): An additive manufacturing process in which metal is applied layer by layer on an existing metal base called substrate. The laser beam that is focused on the substrate generates a molten pool of metal on the surface into which one or more metal powders is delivered through a nozzle or nozzles. Upon solidification of the melt pool, the powder added to the melts bonds with the base material to form a new material layer. The process is repeated until a 3D part is created.

Nontraditional Manufacturing Process: Non-traditional manufacturing processes is defined as a group of processes that are used to shape or remove excess material by various techniques such as thermal, electrical or chemical energy or a combinations of these processes. This is different from the conventional methods that involve the use of sharp cutting tools to shape or remove excess materials such as turning, drilling, shaping and milling. Extremely hard and brittle materials that are difficult to machine by traditional machining processes can be successfully machined using Nontraditional machining processes.

Scanning Speed: Scanning speed determines the time with which the laser beam interact with the processed materials. This is the distance travelled by the laser per unit time.

This research was previously published in Surface Engineering Techniques and Applications edited by Loredana Santo and J. Paulo Davim, pages 146-176, copyright year 2014 by Engineering Science Reference (an imprint of IGI Global).

Chapter 10
Enhancement of Surface Integrity of Titanium Alloy With Copper by Means of Laser Metal Deposition Process

Mutiu F. Erinosho
University of Johannesburg, South Africa

Esther T. Akinlabi
University of Johannesburg, South Africa

Sisa Pityana
National Laser Centre, South Africa

ABSTRACT

The laser metal deposition process possesses the combination of metallic powder and laser beam respectively. However, these combinations create an adhesive bonding that permanently solidifies the laser-enhanced-deposited powders. Titanium alloys (Ti6Al4V) Grade 5 have been regarded as the most used alloys for the aerospace applications, due to their lightweight properties and marine application due to their excellent corrosion resistance. The improvements in the surface integrity of the alloy have been achieved successively with the addition of Cu through the use of Ytterbium laser system powered at maximum of 2000 Watts. The motivation for this research work can be attributed to the dilapidation of the surface of titanium alloy, when exposed to marine or sea water for a longer period of time. This chapter provides the surface modification of titanium alloy with the addition of percentage range of Cu within its lattices; and the results obtained from the characterizations conducted on the laser deposited Ti6Al4V/Cu alloys have been improved.

DOI: 10.4018/978-1-5225-9624-0.ch010

1. INTRODUCTION

So many research works have been conducted on titanium and its alloys both presently and in the past. These works have facilitated the researchers to discover the best method on how to improve on the production of the alloys, in order to enhance their functionality in service. Ti6Al4V alloy is the most applicable among the titanium alloys; since it exhibits a combination of mechanical, physical and corrosion-resistance properties which have made it desirable in the aerospace, chemical industries, energy and automotive industrial services. These alloys are also applied in the developing biomedical applications, due to their excellent biocompatibility among metallic materials (Moiseyev, 2006; Lutjering & Williams, 2007).

The enhancements in the mechanical properties of titanium alloys have mostly been achieved through the addition of alloying compounds (Sen et al., 2010; Gogia et al., 1992; Okazaki et al., 1993; Tian & Nemoto, 1997).

The alloying additions in titanium can be divided into three different classes. The α-stabilizers - an example is Aluminium which impart solid solution strengthening to titanium; the neutral additions, such as tin and zirconium, also contribute to solid solution strengthening; and finally, the β-stabilizers, such as vanadium, molybdenum, niobium, iron, copper, chromium and manganese, serve to introduce the β-phase in an otherwise α-phase microstructure (Leyens & Peters, 2003).

A desired microstructure could also be obtained through thermo-mechanical processing, like the basket weave microstructure achieved from heat treatments in the β-phase field. This is found to offer better creep resistance than an equiaxed α/β phase microstructure (Mishra et al., 2005). The addition of copper to titanium alloys influences the mechanical properties through age- hardening (Lutjering & Weissman, 1970). A beneficial effect of precipitation strengthening has been utilized in Ti-2.5Cu (in weight percent (wt.%)) over commercially pure titanium (Donachie 2000). The work was later extended to titanium alloys containing a variety of other alloying additions with minor composition modification. An experiment was also conducted by substituting vanadium for copper with the same 4 wt.% of vanadium; and this yielded the same results. Copper (Cu), a β-stabilizer, exerts the same influence on the β transus temperature as vanadium; and the direction on the β transus of Ti6Al4V alloy was not affected (Bania et al., 1993). Consequently, an attempt was made by Gollapudi et al., (2011) to improve the compressive strength, as well as the hardness of titanium alloys, by utilizing the precipitation-hardening technique in an experiment that was conducted to ascertain the performance of Ti-6Al-1.5V-2.5Cu in comparison to a standard titanium alloy Ti6Al4V; however the ductility value was decreased marginally. A small amount of less than 2 weight percent of Cu was found to improve the resistance of Ti-48Al-2Cr-2Nb to oxidation. It was also believed that the presence of Cu reduces the concentration of chromium in the scale, thereby enhancing the formation of a more- coherent and protective alumina layer on the surface of the sample (Dang et al., 2001). However, Cu has the tendency to be a useful alloying element for Ti-48Al-2Cr-2Nb.

Titanium and its alloys are very expensive light metals; and recently, there has been a renewed interest in titanium powder metallurgy as a cost-operational way of fabricating components from these expensive metals. The problems faced by these alloys have been the issue of biofouling in the marine industries. Titanium and its alloys have been used as the major components for marine use. There has been the clogging of the sea debris to the marine structures, thereby destroying the surfaces - as a result of the bacterial and virus attacks such as barnacles, fungi, bacteria, and marine debris. Diverse numbers of projects have been carried out to provide solutions to improve and enhance the surface reliability of titanium alloys, and to reduce the problems by increasing the effectiveness and reducing the costs of the device's units or products, and to prolong their service terms.

Ti6Al4V alloy is known to be good for marine environments due to its excellent resistance to corrosion; and it has been used as the primary alloy. Copper, on the other hand, is among the elements that stabilizes the β-phase of Ti6Al4V alloy; it has been used in small quantities with the primary alloy to form the modified Ti6Al4V/Cu alloys through the laser deposition process.

1.1. Titanium and Its Alloys

Today, the most widely used method for Titanium (Ti) processing is known as the "Kroll process". It is rarely found in high concentrations; and it is never found in a pure state; and the difficulty in processing the metal makes it very expensive; thus, it is only produced only in a batch process. Ti is classified as a non-ferrous and light metal. Figure 1 shows the representation of the light and the heavy metals.

The metals vary substantially in weight; and Lithium has the lowest density of 0.5 g/cm^3 while Osmium and Iridium are the heaviest metals with a density of 22.5 g/cm^3. The separation point density between the light and heavy metals is 5 g/cm^3; therefore, Ti is the heaviest light metal with a density of 4.51 g/cm^3.

1.2. Metallurgy of Titanium

The physical and chemical behaviour of titanium and its alloys, as well as their intermetallic properties comprise the monarchies in which material science is applied to their production.

1.3. Crystal Structure of Titanium and Its Alloys

Titanium can crystallize in various crystal structures; and each modification is only stable within a particular temperature ranges. Ti, as well as the majority of the titanium alloys, crystallizes at low temperatures in a revised perfectly hexagonal close packed (HCP) structure; and this property applies to alpha titanium (α-Ti). At high temperatures, the body centered cubic (BCC) structure is stable, and is referred to as beta titanium (β-Ti). The atomic unit cells of the HCP α-Ti and the BCC β-Ti are schematically shown in Figures 2 (a) and (b) (Leyens & Peters, 2003).

Figure 1.

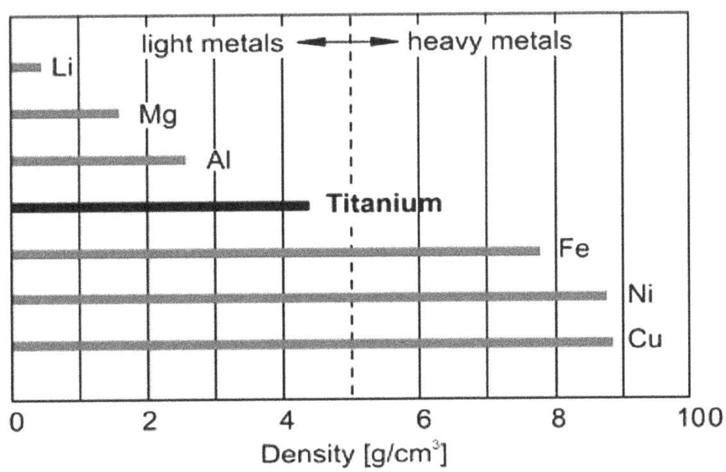

Figure 2.

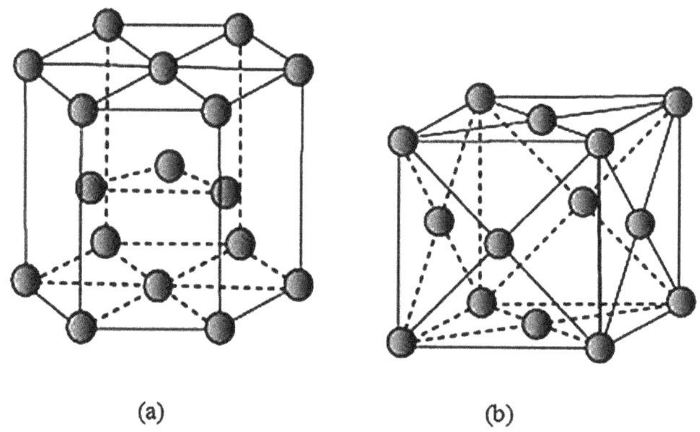

From the HCP and BCC crystal structures, the ease of plastic deformation increases from the HCP lattice to the BCC and to the face centered cubic (FCC) lattice. The slip system of HCP structure is three; while that of BCC is twelve (12). According to the Von Mises' criterion, at least five independent slip systems are required to produce homogeneous plastic deformation of metals. This phenomenon makes it extremely difficult for HCP α-Ti to deform.

1.4. Classification of Titanium Alloys

The alloying elements of Ti are classified as α-stabilizers, β-stabilizers, and neutral:

- **α-stabilizers:** These are the stabilizing elements that spread the α-phase field to a higher temperature; such elements include Aluminium, Carbon, Oxygen and Nitrogen. The α-stabilizing elements are segmented into the β-isomorphous and β-eutectic elements. The β-isomorphous elements are soluble in Ti such as Molybdenum, Vanadium and Tantalum. The β-eutectic elements lead to the formation of intermetallic compounds such as Silicon, Iron, Nickel, Chromium, Copper, Manganese and Hydrogen.
- **β-stabilizers:** These are the elements that shift the β-phase field to lower temperatures. A small amount is added to the α-phase to form near- α alloys.
- **Neutral Elements:** These titanium alloys have no influence on α/β phase boundary such as Tin and Zirconium.

1.5. Properties of Titanium Alloys

The properties of metals, like the titanium alloys, are fundamentally based on the metallic bonding of the atoms in the crystal lattice. The free valence electrons in the lattice result in the standard metallic properties; and these can be improved with the integration of impurities into the crystal lattice (Leyens & Peters, 2003). The high specific strength and excellent corrosion resistance of titanium alloys clarifies their special usage in the aerospace sector (Leyens & Peters, 2003), the chemical industry, medical engineering (Sobiecki et al., 2002) and the leisure sector.

The alloying elements such Aluminium, Vanadium, Copper, Iron, Chromium Tin and Silicon have different influences on the properties of titanium. Similarly, these alloying elements decrease the ductility and the toughness when applied to titanium alloys. Even more so, Aluminium, Zirconium and Molybdenum increase the heat resistance properties (Maslenkov & Maslenkova, 1991).

1.6. Mechanical Properties of the Titanium Alloys

The three processes stated below are assumed to improve the mechanical properties of titanium alloys and other materials. These are alloying, processing and the production of composite materials.

Principle of Alloying

The principle of alloying highlights the increase in the strength of materials. This alloy allows the generation of intermetallic structures, such as density, elastic modulus, the coefficient of thermal expansion; and these properties control the corrosion resistance and oxidation of the material (Leyens & Peters, 2003).

Processing Techniques

The processing techniques permit the properties of the materials to be balanced. Depending on the specific property profile required for the final application, different microstructures can be generated for titanium alloys - by means of thermo-mechanical treatment to optimize for strength, such as the solid solution strengthening, the dispersion strengthening, grain boundary strengthening, texture hardening, ductility, toughness, super plasticity and stress corrosion, to mention a few (Leyens & Peters, 2003).

Production of Composite Materials

This is regarded as a recent option among the mechanical properties of titanium alloys; and it has gained importance in determining the ultra-high strength of materials. Different materials are combined to create a new composite with improved properties. Titanium alloys and Aluminides have been strengthened with fibers to become metal-matrix composites. These ultra-high strength values were better performed by titanium matrix composites (Leyens & Peters, 2003).

Titanium and titanium alloys have found limited use in the mechanical engineering applications because of their poor tribological properties such as poor abrasive wear resistance, poor fretting behaviour and their high coefficient of friction. This poor fretting behaviour of titanium alloys can be improved and enhanced by applying different surface treatments and coatings (Boyer et al., 1994).

According to Fu et al., (1998), there are four main mechanisms that can be used for improving the tribological behaviour of titanium alloys. The first is to increase the surface roughness; the second is to decrease the coefficient of friction; the third is to increase the hardness; and lastly, to induce a compressive residual stress. The problem of friction is related to the crystal structure and the reactivity of titanium; and these problems could be basically overcome by changing the nature of the surface by using surface engineering technologies as different thermochemical treatments; hence, the mechanical properties of the new surface can be improved by a hard compound of titanium (Fu et al., 1998).

1.7. Specific Strength of Titanium Alloys

The strength of titanium and its alloys has been greatly employed in many studies, in which different alloying elements have been added to titanium alloys, in order to improve their mechanical properties.

Cu was known to be among the β-eutectic of the α-stabilizing elements. It has been added to commercially pure titanium, in order to improve its mechanical properties. As highlighted by Murray, Ti-Cu exhibited precipitation strengthening; and with a decrease in temperature, the solid solubility of Cu in Ti was reduced and an intermetallic compound Ti_2Cu was precipitated (Murray, 1992). Figure 3 shows the Ti-Cu binary phase diagram in different temperature ranges.

Various intermetallic phases are illustrated with different atomic and weight percentages of Cu at specified temperatures. The decreasing solubility of the solid with temperature is a characteristic of precipitation hardening; and hence, the addition of Cu to titanium alloys allows manipulation of the mechanical properties through age-hardening (Lutjering and Weissman, 1970). Kikuchi et al., (2003) studied the evaluation of the mechanical properties of cast Ti-Cu alloys with the hope of developing an alloy for dental casting with better mechanical properties than the unalloyed commercially pure titanium

Figure 3.

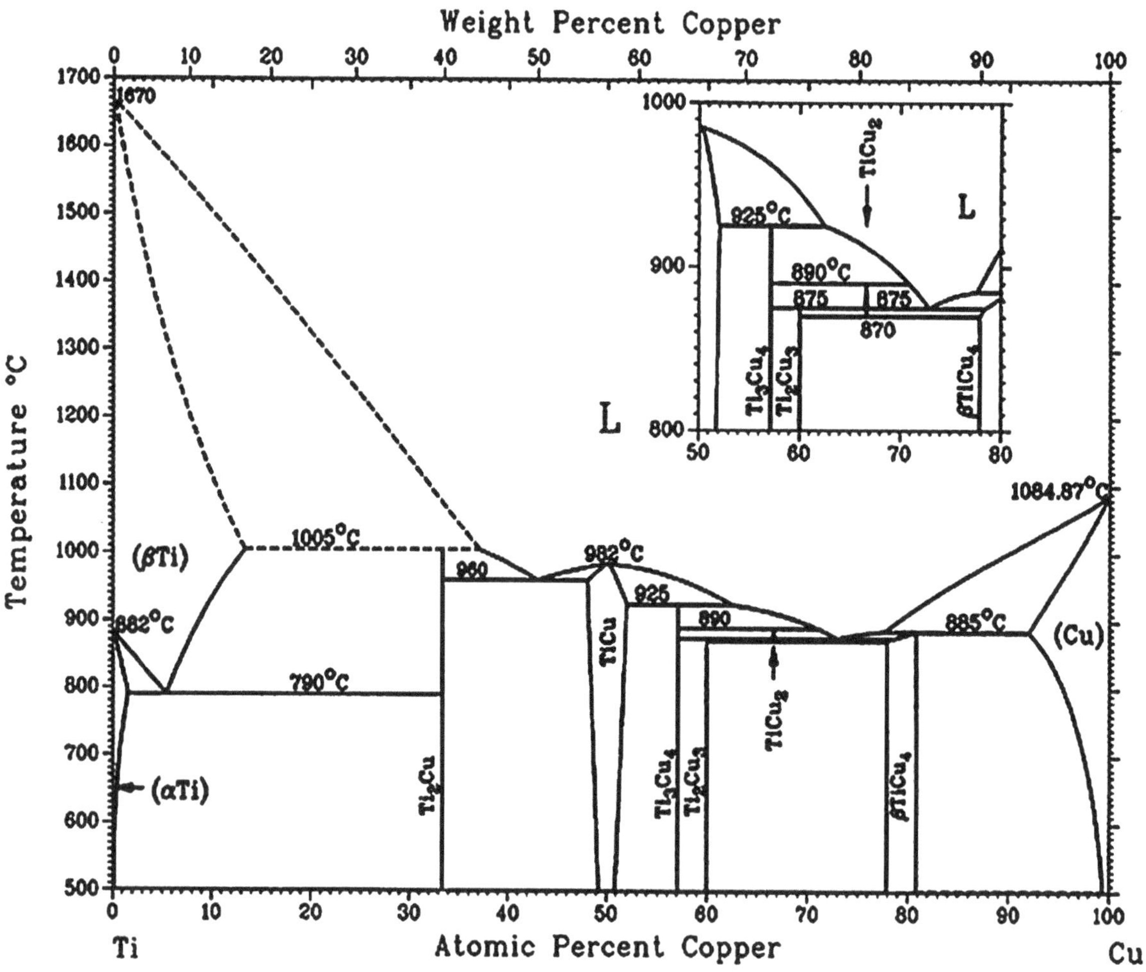

(CP Ti). The Ti-Cu alloys with five different mass percentages (m %) concentrations of Cu (0.5, 1.0, 2.0, 5.0 and 10.0[INSERT FIGURE 001]m %) were prepared in an argon-arc melting furnace. They reported that, the mean tensile strengths of all the cast Ti-Cu alloys were significantly higher than the cast CP Ti. The cast CP Ti became stronger by alloying with Cu; and there was an increment in the value of the tensile strength and the yield strength over that of the CP Ti (Kikuchi et al., 2003).

An attempt was made by Gollapudi et al., (2011) to improve the strength of titanium alloys (Ti6Al4V) by the addition of Cu in wt.%; and by utilizing the precipitation hardening techniques. In their investigation, Ti-6Al-1.5V-2.5Cu alloy was compared with Ti6Al4V alloy to ascertain the performance of the alloys. Both the alloys were melted using a Double Vacuum Arc Melting Technique; and this was followed by forging and rolling in α-β regime. Subsequently, the alloys were heat treated at 1010 °C and then water quenched. The alloys were aged at 500 °C and the ageing features were studied methodically through hardness and tensile testing techniques. The yield strength of Ti-6Al-1.5V-2.5Cu alloy following peak ageing was found to be 1059 MPa which was 70 MPa higher than Ti6Al4V alloy under related conditions (Gollapudi et al., 2011).

1.8. Physical Properties of Titanium and Its Alloys

According to Key to Metal, (Accessed 2013), the arrangement of electrons is responsible for the unique physical properties of titanium. Titanium has two electrons in its third shell, and two electrons in the fourth shell. A crystal structure may be referred to as a physically homogeneous solid in which the atoms are arranged in a repeating pattern. The pattern arrangement is a structure that determines the physical behaviour of a metal. Examples are thermal conductivity, linear coefficient of expansion, electrical conductivity and resistivity Magnetic Properties.

1.9. Chemical Properties of Titanium and Its Alloys

According to Williams Gregor (1791), the chemical behaviour of titanium alloy shows many similarities with those of silicon and zirconium. Titanium alloy belongs to the first transition group; its chemistry in aqueous solution, particularly in the lower oxidation states, has some similarities with those of chromium and vanadium. Titanium is a transition light metal with a white-silvery-metallic colour; and it possesses a strong, lustrous appearance, and excellent corrosion-resistant properties. Pure titanium is not soluble in water; but it is soluble in concentrated acids. This metal forms a passive and protective oxide coating, which leads to its corrosion resistance when exposed to elevated temperatures in air; and it also resists tarnishing at room temperatures.

Corrosion Behaviour of Titanium and Its Alloys

Ti6Al4V spontaneously forms a stable and continuous oxide film upon exposure to oxygen in air or water, which is important in its excellent corrosion resistance behaviour. Its corrosion behaviour occurs spontaneously in aqueous solutions, such as seawater, oxidizing acids, chlorides in the presence of water, rocket propellants and alkalis (Technical Data Sheet, 2013). Wong et al., (2012) revealed that the corrosion potentials of the laser-alloyed specimens decrease as the Ti content increases. The improvement in the corrosion resistance of the alloy was attributed to the presence of Ti in the intermetallic and metallic phases that create the protective oxide. The protective surface film of Ti and its alloys to most environments causes their excellent resistance to corrosion; and it consists basically of TiO_2. Ti is chemically very reactive; and the thin oxide film continues to passivate the base metal, as long as its integrity can be maintained (Wong et al., 2012). Mostly, they occur in oxidizing environments, such as nitric acid solutions or salt solutions, including sulfates, chlorides and hypochlorides; but under reducing conditions, or in the presence of fluoride ions, the protective nature of the oxide film diminishes; and hence, this condition of Ti is not corrosion resistant (Lutjering & Williams, 2007). Ti6Al4V is also susceptible to general corrosion in the presence of reducing acids or dry chlorine gas, and gaseous or cathodic hydrogen. This hydrogen can diffuse into the metal, thereby forming brittle hydrides (Technical Data Sheet, 2013). Titanium alloys are generally resistant to stress corrosion cracking (ASM Handbook, 2005).

1.10. Microstructures of Titanium Alloys

It is well known that the microstructure of titanium alloys has a substantial influence on the properties of titanium alloys. The size, the texture (both fine and coarse microstructures) and the arrangement (both lamellar and equiaxed microstructures) of the α-phases and β-phases describe the microstructure of the conventional titanium alloys (Leyens & Peters, 2003).

Lamellar Microstructure

The lamellar microstructure is generated upon cooling from the β-phase field at the temperatures above the β-transus temperature, approximately 900 °C. The α grains nucleated at the grain boundaries, and then grew into the prior β grain at the temperature below the transus temperature. The (α+β) lamellar microstructure for the Ti6Al4V alloy upon cooling is shown in Figure 4 (a). The β-phase transforms into martensite, with high cooling rates from temperatures above the martensitic start temperature (MST) and through the two-phase field. At temperatures below MST, the β volume fraction is no longer transformed to martensite, and decreases further (Leyens & Peters, 2003).

The morphology and thickness of α lamella are also very important factors; since they influence the mechanical properties of Ti6Al4V alloy (Jones et al., 2008 and Jackson et al., 2009). The lamellae are either fine or coarse, depending on the cooling rate. The slow cooling from the β-phase field results in a pure lamellar microstructures; but it becomes coarser with faster cooling rates. Figure 4 (b) presents the fine lamellae with fine needle-like microstructure after furnace cooling. The hardening effect on the strength and hardness observed for titanium alloys on martensitic transformation is only moderate. The chemical compositions of the α and β phases change in the two-phase field (α+β) with decreasing temperature under equilibrium conditions; thus, at that low temperature, vanadium strongly enriches β and stabilizes this phase (Leyens & Peters, 2003).

Equiaxed Microstructures

The equiaxed microstructure occurs as a result of a recrystallization process. Firstly, the alloy has to be highly deformed in the (α+β) field to introduce the cold work effect into the material, then upon subsequent solution heat treatment in the two-phase fields, a recrystallized and a fine equiaxed microstructure is generated. Figure 4 (c) represents a fine microstructure after recrystallization. Due to prolonged annealing, a coarse and equiaxed microstructure can be formed. Figure 4 (d) shows the coarse microstructure

Figure 4.

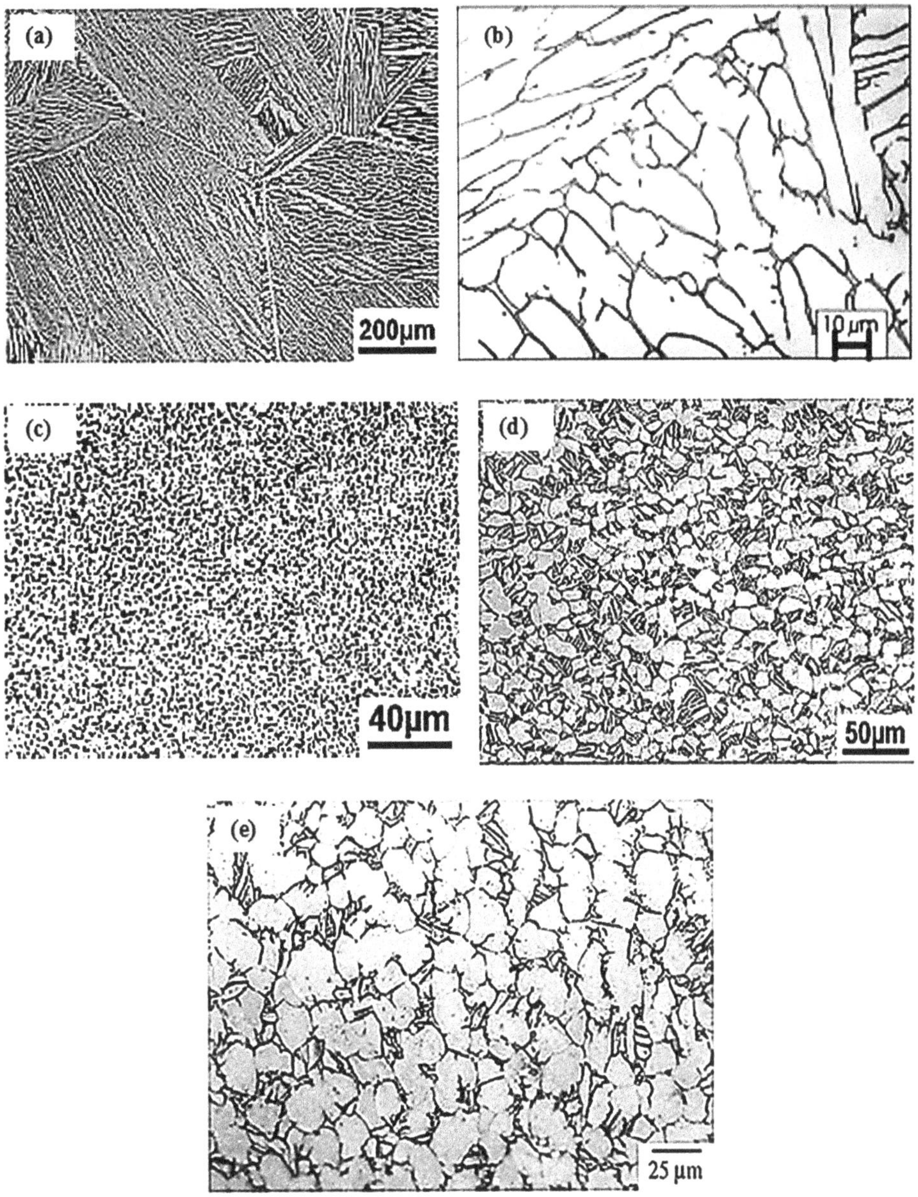

after recrystallization. These microstructures have high ductility, fatigue strength, and are preferred for superplastic deformation (Leyens and Peters, 2003). Kong et al., (2011) indicated that, with different cooling rates, the morphology of the α-phase can appear as equiaxed, acicular, or plate-like; and the different features in the microstructures of the α-phase can lead to different mechanical properties.

Bimodal Microstructures

This could be referred to as the combination of lamellar and equiaxed microstructures. At a temperature below the β-transus temperature, the solution heat treatment of Ti6Al4V results in bimodal microstructures. These microstructures combine the advantages of lamellar and equiaxed structures; and they thus exhibit a well-balanced property profile. A typical example of a bimodal microstructure is shown in Figure 4 (e).

In general, the different microstructures are generated by thermo mechanical treatments. These are considered as a complex sequence of solution heat treatment, deformation, recrystallization, and annealing for stress relief. The fine-scale microstructures increase the strength, ductility, retard crack nucleation, and serve as bedrock for ultra-plastic deformation (Leyens & Peters, 2003).

1.11. Copper And Copper Alloys

Copper (Cu) is a reddish coloured metal with a face-centered cubic crystalline structure. Due to its band structure, it reflects red and orange light; and it absorbs other frequencies in the visible spectrum. It is malleable, ductile, and an extremely a good conductor of both heat and electricity. Cu has a low chemical reactivity and a greenish surface film coating called patina, which is largely formed on the surface in moist air to protect the metal from further attack (Lenntech, 1998-2013.

High Conductivity of Electrolytic Copper

The high conductivity of copper can be seen in the form of wires, tubes, extrusions, bars and sheets. It has excellent ductility. The 99.90% electrolytic tough pitch copper CW004A is the main grade of Cu, which is used for electrical applications, such as building wire, motor windings, cables, and bus-bars.

Engineering Aspects of Non-Electrolytic Copper

Engineering copper, grade CW024A is a non-electrical copper; and it is useful for engineering application. The CW024A grade possesses the following properties, which make them a standard material for engineering purposes:

Due to its thermal conductivity, this grade of copper is used for components where rapid heat transfer is essential. The thermal conductivity of Cu is 394 W/mK which is about twice that of aluminium, and thirty times that of stainless steel. Examples include heat sinks, vehicle radiators, and heat exchangers.

1.12. Titanium and Its Alloy-Copper Modifications

Titanium alloy has been reinforced with 2.5% wt of Cu at an elevated temperature of up to 350 °C, which combines the formability and weldability of unalloyed titanium with improved mechanical properties. In the annealed condition, Ti-2.5Cu alloy is very easy to forge without cracking in the (α+β) field at a preheating temperature of 800-850 °C; and is has good fatigue properties of 3 ratio 5 of the static tensile strength. Over the range of 150-320 °C, aged Ti-2.5Cu alloy was also found to have more creep-resistance than IMI Titanium 317 and the hardest grade of commercially pure Ti at all temperatures. There was an improvement in strength, and no evidence of ductility reduction (Key to Metal, 1999-2010).

Zhang et al., (2013) were able to ball- mill Ti powder with 10 wt.% of Cu powder for 3 to 6 hrs and hot- sintered under a pressure between 15 to 30 MPa and at a temperature between 850 to 1050 °C for antibacterial purposes and applications. Their result showed that Ti-Cu has a strong antibacterial property - both on the surface - and in the alloying composites. The addition of Cu provided the entire alloy with an increase in mechanical properties. They also suggested that the antibacterial property might be related to the release of Cu ions (Zhang et al., 2013).

Cu is well known as a strong β-stabilizing element, and its atomic migration into Ti lattice results in the formation of β-Ti during cooling; and it could travel a longer distance in the Ti lattice than other elements; and open more crystallographic structure of the β matrix (Ghosh & Chauerjee, 2003).

In the research work of Xia et al., (2013), a gas-atomized Ti-48Al-2Cr-2Nb powder was sintered with 2 wt.% Cu powder. The small addition of Cu powder on the sintering densification of gamma Ti-48Al-2Cr-2Nb resulted in an improvement in the mechanical properties and they also reported that Cu and Cr enrich the hexagonal close-packed structure of Ti.

Sakuma et al., (2012) studied the effect of several copper contents on the super elasticity characteristics in Ti-Ni alloy wires. Observations were carried out on Ti-50Ni, Ti-45Ni-5Cu, Ti-40Ni-10Cu and Ti-37Ni-13Cu. Their results revealed that during the loading, the degradation of the strain energy and residual strain increases when the Cu contents decrease.

Much research work has been conducted on Cu-based Ti alloy; and this is widely used for electrical purposes to boost their properties. In this present research work, titanium alloy has been used as the base composite; and the addition of Cu was introduced to improve the mechanical properties of the alloy.

Kano et al., (2013) presented the analysis of their invention and patent work on the production of Ti6Al4V powder with the addition of 1 to 10 wt.% of Cu to improve on the densification. The production was channelled through a Cold Isostatic Pressing and a subsequent Hot Isostatic Pressing process. Their research procedures involved material hydrogenation and dehydrogenation processes to produce the Ti6Al4V powder. Electrolytic Cu powders not more than 45 μm were mixed in a V-type mixing machine at different weight percentages, of 1 wt.%, 3 wt.%, 5 wt.%, 8 wt.% and 10 wt.%, respectively with Ti6Al4V powder. The mixtures were passed through the Hot Isostatic Pressing process for 1 hr under a hydrostatic pressure of 100 MPa and a temperature of 900 °C. The results of the sintered capsule showed a density of not less than 99% and the hardness values increase, as the Cu content increases from 1 wt.% to 10 wt.%.

Stranak et al., (2011) examined the deposition of thin Ti-Cu films on a Ti6Al4V substrate using three different method of magnetron sputtering: direct current, dual, and dual high-power impulse magnetron sputtering. Their study revealed that a larger amount of Cu was released within the first day from the dual high-power impulse magnetron sputtering compared to other methods used. They also envisaged that the Cu released from the Ti-Cu film produced the antimicrobial effect that killed the Staphylococcus epidermis and Staphylococcus aureus bacteria used in the test.

1.13. Laser Technology

The term Laser is an acronym for "Light Amplification by Stimulated Emission of Radiation". The laser device produces intense beams of low divergence light of an electromagnetic radiation ranging from 1 nm to 1000μm in wavelength, 400 to 700 nm in visible spectrum, and 200 to 400 nm for ultraviolet light. Compared with other source of light and photon energy, the wavelength of laser light is of a pure monochromatic type; and it coheres with light of other wave-lengths.

The laser technology process is vibrant to the field of engineering, and in all walks of life; since it can travel a greater distance and be focused to a small bright spot that exceeds the brightness of the sun. The light particle that is called a photon exhibits a particle-like and wave-like property; and it has energy, which can be determined from Bohr's model equation (Introduction to laser technology, Accessed, 2013).

1.14. Types of Laser System

There are different types of lasers in the laser system technology and all the lasers perform dissimilar functions, based on the level of their functionality and purpose.

1.14.1. Discharge Gas Type Lasers

In principle, the gas discharge laser involves a container attached with mirrors at its end, filled with gas and discharged through a jet. Due to their complex nature, the container configurations, the parameters of gas discharge, the gas mixture, and the reaction of the discharged container with the laser optics must be taken into consideration, especially for an inversion. The gas discharge types are explained in the subsections below.

Helium-Neon Lasers

The helium-neon laser (HNL) is a small laser with a beam quality of pure single transverse mode usually less than 1.05 and having an operating life of equal to or greater than 50,000 hrs. The laser was second in the discovery of lasers; and today, the laser is the first in terms of volume usage and applications. In the configuration set-up, helium constitutes the major gas mixture; while neon is the actual lasing medium. The HNL operates on a low current-high voltage principle, with a wavelength of 633 nm for red colour, 543 nm for green colour, 1523 nm for infrared, 594 nm for yellow colour, and 612 nm for orange colour. The output power value of the laser depends on the size of the laser cylinder and the wavelength.

Noble-Gas Ion Lasers

The noble-gas ion lasers are mixed-gas lasers that combine the formation of argon and krypton in their configuration. They operate on a high current-, low voltage-, low pressure principle, and at a high intensity with a wavelength of 488 nm for the blue colour, 514 nm for the green colour, 568 nm for the yellow colour, 647 nm for the red colour and 752 nm for the near infra-red. A cooling mechanism is required to remove the heat generated at the head of the laser system. The lasers are of two classes. These are: low power air-cooled lasers, and high power water-cooled lasers. The two classes have the same features; but the distinguishing limiting factors are the depletion of the cathode and the consumption of gases.

Carbon Dioxide Lasers

The carbon dioxide (CO_2) lasers operate at a pressure, which is equivalent to 10% of the atmospheric pressure. The beam generated from the CO_2 laser is a mixture of helium and nitrogen. The helium possesses the highest percentage of the mixture due to its high thermal conductivity and the cooling rate (Industrial Laser Processes, Accessed, 2013).

The laser is applicable for material processing, due to its very high power-efficiency relationship. The standard wavelength of this laser is 10.6.., and the power output is between the value of less than 1 W and greater than 10 kW. The CO_2 laser could be grouped as a continuous-wave (CW) laser and a pulsed laser. In the CW laser, the operating power is equal to its average power; while in the pulsed laser, the average power leads to a peak power; and it is therefore controlled by an on-and-off duty cycle (Introduction to laser technology, Accessed, 2013).

Helium-Cadmium Lasers

The operation of helium-cadmium (HC) lasers is similar to that of the helium-neon laser; but the cadmium metal is the lasing medium; and it becomes solid at room temperature. The lasers are economical and have an excellent wavelength of 442 nm for violet. In application, they are used for 3D holographic and stereo-lithography features. The operation of HC lasers is accomplished by a process of electrophoresis.

Excimer Lasers

The word 'excimer' is the termed based on the molecular complexity of two stable atoms in an electronic excitation state. This is a type of gas-discharged laser that encompasses the halogen family as XeFl, with a wavelength of 351 nm, XeCl with a wavelength of 308 nm, KrCl with a wavelength of 222 nm, and KrF with a wavelength of 248 nm. All these compounds are used in eye-surgery operations, photolithography and micro-machining applications. With the same beam quality, the excimer output beam diameter can be spot-focused to approximately 40 times smaller than the CO_2 laser beam. The construction of this type of laser is similar to the CO_2 laser; but the major dissimilarity is the corrosiveness of the gas.

Semi-Conductor Diode Lasers

The advantage of diode lasers lies in the fact that it has been generated by the recombination of injected holes and electrons in which electricity is transformed to light. The merit features of the diode laser over other lasers are the small volume enclosure, low power consumption, high speed transmission, and the storage of information compact disc. A double hetero-structure diode laser is constructed with a feedback mechanism, using two mirrors to cause the circulation of light repeatedly and to build up the beam by a stimulated emission of radiation.

1.14.2. Solid State Lasers

Neodymium Doped Yttrium Aluminium Garnet

The Neodymium Doped Yttrium Aluminium Garnet (Nd: YAG) with a compound of $Y_3Al_5O_{12}$ is one of the solid-state lasers that are optically pumped via a laser diode, a flash tube, or a continuous gas lamp in which the light is emitted at a wavelength of 1064 nm in an excitation state. The lasing medium is the Neodymium, a rare earth element with a valence of 3 embedded in the Yttrium Aluminium Garnet (YAG) crystal. The Nd: YAG operation could be a pulsed mode operation, which involves a medium power generation from 20 W to 1800 W; and it is used for deep drilling and welding operations, or a continuous-wave mode with a high power output from 500 W to 5000 W; and it has the advantage of using the flexibility of a robot for welding or cutting operations (Industrial Laser Processes, Accessed, 2013).

Fiber Lasers

In this type of laser, energy is introduced via a pump diode; and a laser beam is generated with a wavelength of 1,064µm and a very small focal diameter with respect to the intensity, which is equivalent to 100 times higher than that of the CO_2 laser. It is emitted via similar average power. They are applicable in the marking and engraving of both plastic and metallic materials (Trotec Laser, 2013).

1.15. Dissimilarities between Co_2 Lasers and the Nd: Yag Laser

There are many distinctions between the CO_2 laser and the Nd: YAG laser; and these are based on the functionality of the materials and the process properties. The wavelength of Nd: YAG laser is shorter and it is best fitted with metallic materials; and it produces more elaborated graphics in the same machine set-up; but the wavelength is not easily adapted for other non-metallic substances, such as plastics, timbers, acrylic, rubbers etc.; whereas the wavelength of CO_2 laser is longer and it is better suited for most of the non-metallic organic materials, as cited above; but it is not easily absorbed by most metals (Laser technology, Accessed 2014). In application, the Nd: YAG laser beams have an advantage over the CO_2 laser in focusing on a very much smaller spot diameter, with higher-power density. However, the higher power and focusing property of CO_2 lasers makes them applicable for deep penetration welding and thick metal cutting. The Nd: YAG lasers are gaining acceptance, due to the increment in the quality of the beam and power proficiency (Laser technology, Accessed 2014).

1.16. Powder Metallurgy

Powder metallurgy (PM) is an advanced manufacturing technology used to produce a wide range of finished and semi-finished engineering components. PM can be used for processing all types of metal and alloys in the form of powder and furnished into a final or near-net shape by using different processes (Global Market Review, Accessed 2014).

1.17. Additive Manufacturing

Additive manufacturing (AM) is a three-dimensional printing technique that is referred to as a layer-by-layer technique of producing three-dimensional (3D) objects directly from a model. AM builds a finished material from the initial production stage in successive layers. Applications are found in industries, such as the aerospace, dentistry, sport, auto parts etc.

Additive-layer manufacturing can be used to attach additional part features; and it can be used in the repair of parts (Sears, 2011).

1.18. Classifications of Additive Manufacturing

There are different categories of AM, depending on the material selection and the design requirement of the output product.

Powder Bed

This class of AM involves the thin spreading of powder in layers; and a laser is allowed to scan the spread powders on the bed. The heated powder is fused together, thereby forming a solid geometry. Laser sintering is a typical process of the powder bed type. An example of the material type used includes thermoplastic. Selective laser melting (SLM) and electron beam melting (EBM) are other processes for processing ferrous and non-ferrous materials (Advanced Manufacturing CRC Limited, Accessed 2015). The EBM is the bed type that uses metal powder; and it is melted by an electron beam in a vacuum to form a layer-by-layer 3D object.

Material-Deposition Process

This requires the heating of materials through a nozzle connection; and it flows through a definite path onto a substrate in a layered manner. The 3D product is obtained by depositing the materials on the prior material deposit in layers; and the laser beam is applied to melt the materials. The blown or spray process and wire extrusion are classic examples of the material-deposition process.

3D Printing

This is a three-dimensional printing process that involves layer-by-layer building of an object onto a platform from CAD files. The object is printed exactly the way it appears on the CAM data. The kind of materials in this category include thermoset, graphite etc.

1.19. Laser Metal Deposition

Laser Metal Deposition (LMD) is referred to as the process of using a laser beam to produce a melt pool on a metal substrate; and the powder is then deposited into the melt pool created via a nozzle and fusion bonded to the substrate to form a new material layer or composite. Both the laser beam and the powder come out through the nozzle attached to the robotic arm of the laser system.

1.20. Mechanical and Metallurgical Bonding in the LMD Process

The LMD process produces an adhesive bonding that permanently attaches the deposited powder material from the jets' blow to the melt pool produced by the laser on the parent or base material. Against the force of gravity, a new layer of metal or composite is formed. The process parameters and the powder composition must be chosen to have a homogeneous distribution in the layer and improved properties. There are significant limitations to some mechanical and metallurgical bonding in the LMD processes; since they do not hold very well and require replacement or repair, due to the parameters used; while some require more heat to melt both the composite and the base metal (Laser Solutions for Manufacturing, 2013).

1.21. Laser Metal Deposition of Titanium Alloy Composites

Mahamood et al., (2013) laser deposited Ti6Al4V alloy (titanium grade 5) powders on Ti6Al4V substrate using the Nd: YAG laser machine and varied the laser powers from 0.8 kW to 3.0 kW; whilst other parameters were kept constant. They revealed that the microstructures of the Ti6Al4V composites were observed to be columnar in nature, and showed a fine and coarse globular alpha phase.

Vrancken et al., (2012) presented the analysis on the mechanical properties and the microstructure of a heat treated Ti6Al4V samples. A YAG fiber laser of wavelength 1070 nm, a laser power of 250 W, scan speed of 0.16 m/sec, and a hatch space of 30μm, were used for their experiment. They discovered that before the heat treatment process, the alloying samples show a full acicular and martensitic microstructure. At a lower magnification, a long columnar grain is formed; and this is regarded as prior β grains. After the heat treatment, the microstructure of the laser deposited material was found to have a fine martensitic structure and a needle-like structure of α and β in the mixture.

In the research work for producing medical parts, (Vandenbroucke & Kruth, 2007), reported a selectively laser melted Ti6Al4V alloy on a titanium base plate, and a Co-Cr-Mo composite on a steel base plate. In their investigation, the energy density produced for Ti6Al4V alloy was 195 J/mm^3 and that of Co-Cr-Mo was 85 J/mm^3. The microhardness and the macrohardness test conducted on Ti6Al4V composite showed that hardness increases with an increase in energy density, which is due to the minimal pores established in the deposited samples and the rapid cooling of the melt pool.

Song et al., (2012) were able to analyse the effect of processing parameters on the selective laser melted Ti6Al4V alloy on a stainless steel plate for the production of titanium alloy parts. They made some single tracks of the deposited materials, using the YLR-100-SM single mode CW Ytterbium fiber laser with a maximum power of 120 W. Three single tracks were considered, with each track having a laser power and scanning speed of 120 W and 0.2 m/sec, 110 W and 0.4 m/sec and 110 W and 1.2 m/sec, respectively. They stated that a crack occurred in the first parameters used, together with partial melting in the second parameter set; while the third set of parameters gave the continuous melting of the final deposited sample.

Yu et al., (2012) performed a similar operation by laser depositing Ti6Al4V powder on a titanium substrate; and they varied the laser power between 380 W and 570 W; while the other process parameters were kept constant: the scanning speed at 1000 mm/min, powder feed rate at 1.43 g/min, spot diameter at 1.2 mm, gas flow rate at 10.5 l/min and layer thickness at 0.5 mm. They revealed that the lowest laser power of 380 W leads to poor bonding and pore defects. However, higher heat generated results in poor oxidation; and this was controlled by a monitoring device to minimize the heat.

Yasa and Kruth, (2011) made an analysis on the laser re-melting on a selective laser melting parts of AISI 316L stainless steel powder; and they maintained that almost all the parts with low scanning speed and high laser power, produced a higher percentage of porosity; and the density was improved after laser re-melting. Thijs et al., (2010) reported the rate of cooling with varying scanning speed, and indicated that a decrease in the size of the α-martensitic structure would lead to an increase in the Vickers microhardness value. Similarly, a higher scanning speed resulted in some porosity, which was an implication of an unmelted powder region within the composite. The literature has similarly reported the use of Ti6Al4V alloy and other metal alloys suitable for biomedical implants for both humans and animals. Okazaki et al., (2005) in their research study examined the formation of a new bone using various metal implants, such as 316L stainless steel, Co-Cr-Mo casting alloy, Ti6Al4V and Ti-15Zr-4Nb-4Ta alloys. They were implanted into the rat femur and tibia for close to 12 months; and they discovered the presence of foreign-body giant cells and inflammatory cells. In the implant of 316L stainless steel and Co-Cr-Mo alloy, osteocytes were observed in the lamellar bone tissue; and these resulted in the formation of capsulated fibrous connective tissue. The bone formation rates of the alloy implants after a month of implantation were markedly high.

Reig et al., (2013) investigated the microstructure and mechanical behaviour of pores during Ti6Al4V powder sintering, and revealed that in order to improve the behaviour of Ti6Al4V for bone application; stiffness must be reduced, in order to generate a porous structure. The stiffness of the developed porous material was varied from 40% to 88% Ti6Al4V after sintering.

The production of nanocrystalline Al, Al-10% Cu, and Al-10% Cu with 5% Ti alloys was conducted for corrosion analysis in the natural sea water using the mechanical alloying method. The presence of Cu and Al-10% Cu alloy were reported to reduce the corrosion rate of Aluminium. The addition of 5% Ti to the alloys produced the best passivation to the surface of Al with the lowest corrosion rate (Sherif et al., 2014).

2. EXPERIMENTAL PROCEDURES AND SETUP

The research work was conducted using the laser techniques at the National Laser Centre of Council of Scientific Industrial Research (NLC-CSIR), Pretoria, South Africa. The LMD of the composites was accomplished on the Ytterbium Laser System equipment (YLS-2000-TR). The system runs at a maximum power of 2000 W and uses a Kuka robot for its task.

Figure 5 shows a typical schematic view of a robotic laser with the nozzle attached to the laser head; and it also shows the deposition process.

Figure 5.

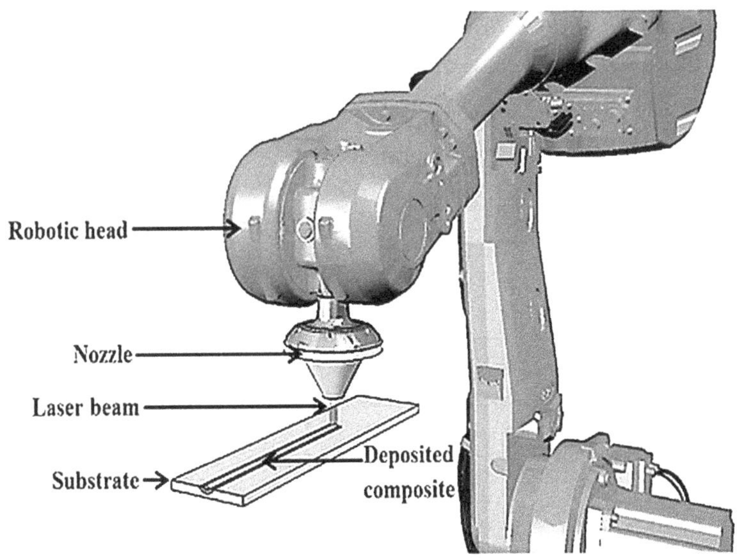

The laser is incorporated with dual-core fiber optic cables of 400 microns to allow the laser beam to be delivered on the substrate via the robot-delivery device. The laser system produces a wavelength, which radiates at 1.047μm with a better efficiency output. The deposition operation is completely protected and shielded with an argon gas, to prevent the deposited composites from oxygen contamination.

2.1. Materials and Methodology

A 99.6% square plate made of titanium alloy grade 5 with a volumetric dimension of 102 X 102 X 7.54 mm^3 was used as the substrate. It was supplied by TITANIUM METAL SUPPLY, 12215 Kirkham Road, Poway, United States. The substrate was grit blasted in other to fascinate the surface. After the grit blasting, the substrate was sanitized with acetone and dehydrated. The two powders used for this experiment are Ti6Al4V and Cu powders and were fed from two different hoppers and flow out through the nozzle.

The powders used for this project work are Ti6Al4V alloy powder and Cu powder. The Ti6Al4V alloy powder was supplied by F.J. Brodmann and Co., L.L.C., Louisiana, TLS Technik GmbH, South Africa. The Cu powder was supplied by the INDUSTRIAL ANALYTICAL (Pty) LIMITED, South Africa.

Table 1 shows the chemical composition of the Ti6Al4V alloy powder; and Table 2 depicts the chemical composition and the mesh-size analysis of the Cu powder.

The powders enter the cylinder and come out through three different hoses connected to the nozzle at an angle of 120° to each other. The nozzle has four jet holes, in which three of them coaxially are for the powders; while the fourth one at the center of the nozzle is for the laser beam.

Table 3 illustrates the experimental matrix used for the laser deposition process. The samples are designated from A to G at varying scanning speed between 0.3 m/min and 1.5 m/min respectively.

Other parameters such as the laser power, powder flow rates and gas flow rates for Ti6Al4V alloy and Cu are all kept constant. The beam diameter of 4 mm and a standoff distance of 12 mm were used throughout the experimental setup.

Table 1. Chemical composition of the Ti6Al4V powder

Element	Al	V	Fe	C	N_2	H_2	O_2	Ti
Wt.%	6.30	3.90	0.17	0.008	0.005	0.005	0.140	Balance

Table 2. Analysis of Cu powder

Powder	Percentage (%)	Hydrogen Loss (%)	Apparent Density (g/cm^3)	Mesh Analysis	Mesh %
Copper powder	99.83	0.31	5.1	+100 mesh -100+140 mesh -140+200 mesh -200+325 mesh -325 mesh	0.3 18.3 21.5 25.7 34.1

Table 3. Experimental matrix

Sample Designation	A	B	C	D	E	F	G
Scanning speed (m/min)	0.3	0.5	0.7	0.9	1.1	1.3	1.5
Additional Constants: Laser power = 1200 Watts; Powder flow rate (Ti6Al4V) = 4.175 g/min; Powder flow rate (Cu) = 0.32 g/min; Gas flow rate (Ti6Al4V) = 3 l/min; Gas flow rate (Cu) = 1 l/min							

2.2. Microstructure

Prior to optical microscopy observation, the Kroll's reagent was prepared with 100 ml H_2O, 2-3 ml HF and 4-6 ml HNO_3. This was prepared according to Struers application note of metallurgical preparation of titanium. The samples were etched for 10-15 seconds, sprinkled with acetone, rinsed under clean running water and dried off. The microstructures of all the etched samples were observed under the BX51M Olympus optical microscope. A mirror-like image on the surface of the samples was observed with no scratches; and this serves as an indication of a good grinding and polishing operation. All the samples were prepared for metallurgical characterizations, according to the ASTM standard (E3-11 ASTM Standard).

2.3. Microhardness

The microhardness profiling was performed on a Vickers hardness tester named Zwick/Roell. The hardness was performed laterally on each sample from the top of the deposit to the substrate thereby making eight indentations on the sectioned surface of the deposited Ti6Al4V/Cu composites to the substrate. A load of 500 grams and a dwell time of 15 seconds were used throughout the hardness test according E384 ASTM standard [11].

3. RESULTS AND DISCUSSION

3.1. Microstructural Evaluation

The micrographs of the laser deposited Ti6Al4V/Cu alloys are discussed in this section. Figures 6 (a) to (h) show the micrographs and the microstructures of the deposited Ti6Al4V/Cu alloys.

The volume of the deposited alloys decreases as the scanning speed increases. This in turns reduced the grain sizes with respect to the increment in the speed of scan. Figures 6 (a) and (b) show the macrograph and microstructure of sample A deposited at laser power of 1200 Watts and scanning speed of 0.3 m/min. Both the α, β and (α+β) phases were observed in the microstructure. Figures 6 (c) and (d) depict the macrograph and microstructure of sample B deposited at laser power of 1200 Watts and scanning speed of 0.5 m/min. α- Ti lamella was found to decrease in length prior the β-phase as the scanning speed increases. Globular microstructures were also formed after the fusion zone. Both the α- Ti lamella and the acicular structures were found breaking into the closer grain boundaries of the (α+β) grains. The gradual disappearance of the α-acicular could be attributed to the decrease in the energy density as the scanning speed increases Erinosho et al., 2015. Widmanstettan structures were observed in the

Figure 6.

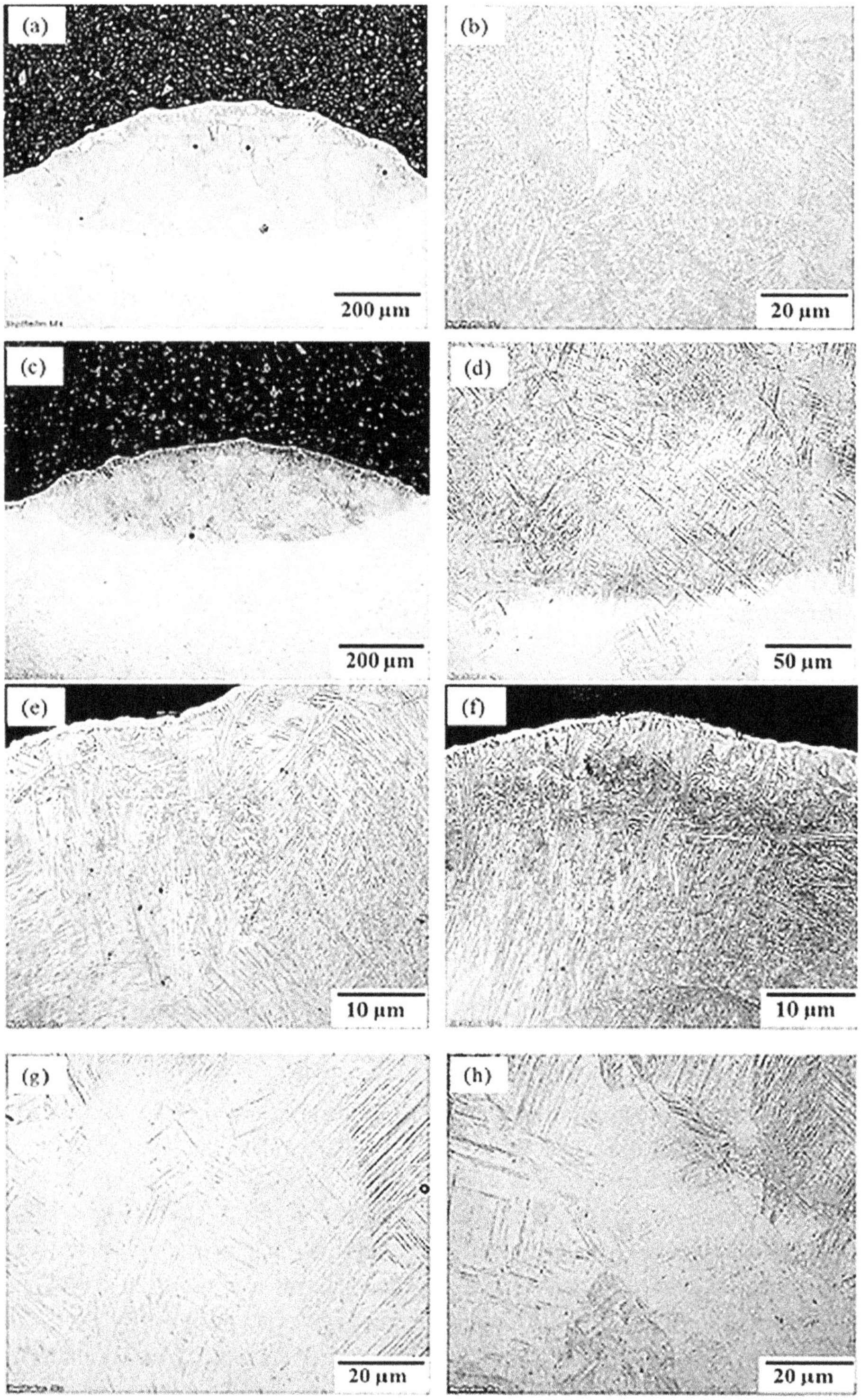

samples (Erinosho et al., 2014). The martensitic structure in the α-phase region of the modified alloy was likewise found to decrease as the scanning speed increases. As the time of deposition decreases, the magnitude of the laser energy induced into the parent material is also reduced.

3.2. Microhardness Analyses

The hardness of the parent material (substrate) was first conducted on the both the surface and the cross section. The average hardness HV of the surface of the substrate is HV 316; while that of the cross section is HV 342. Figure 7 represents the histogram of the HV values for both the surface and the cross section of the substrate.

The average HV of the cross section is about 8% higher than that of the HV on the surface of the substrate. Their occurrence was due to the fact that the bulk sample is a rolled material and the grain boundaries were elongated and compacted in between the rolls applying forces. In other words, the cross section suffers the magnitude of the load; since it was condensed in between the roll surfaces.

The microhardness evaluation of the Ti6Al4V/Cu alloy samples was illustrated in Figure 8. Seven indentations were made on the deposited alloys.

From the histogram plot of the hardness values, Erinosho et al., 2015, it can be inferred that the hardness decreases as the scanning speed increases. The deposit at low scanning speed creates enough room for indentations to be done with the bulk of deposit as compared with the deposit at high scanning speed. Sample A deposited with a laser power of 1200 W and a scanning speed of 0.3 m/min shows the highest hardness value of HV 541±20 as compared to the hardness values of other samples. The hardness of sample A is about 42% higher than that of the hardness of the cross section of the parent material and about 37% greater than the average hardness indented on the surface of the substrate. The indentations made on different phases of the deposited Ti6Al4V/Cu alloys initiated variations in the microhardness values. However, the presence of Cu accommodates strain around the β phase thereby causing its stability.

Figure 7.

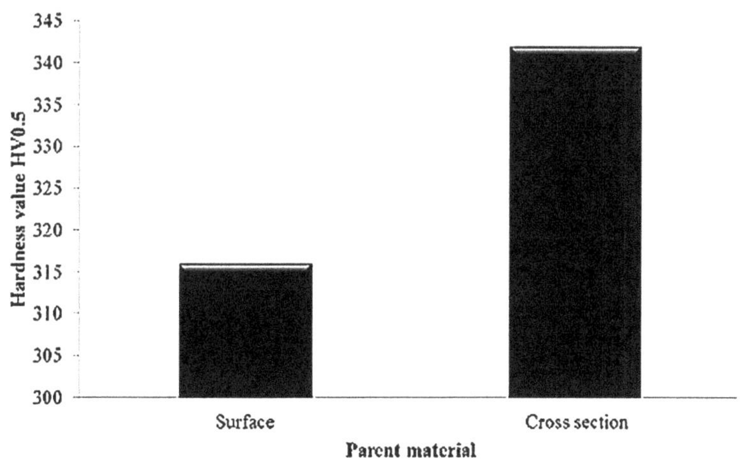

Figure 8.

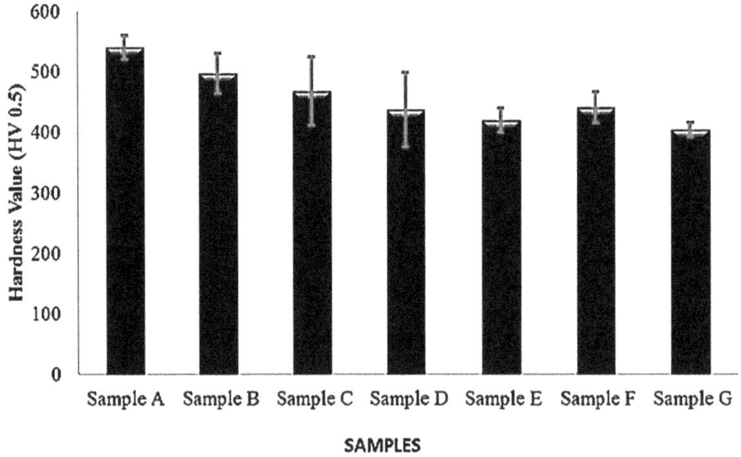

4. CONCLUSION

Today, the combination of different alloys has been a major phenomenon or route for the improvement of their physical and mechanical properties. The deposition of titanium alloy and copper powders is a great accomplishment; and the results obtained have really enhanced the alloy. The gaps between the α/β phases have been modified with the aid of the copper stabilizer added since it can accommodate strain around the β phases. The 3 weight percent of copper has a greater impact in the hardness values obtained as compared with that of the parent material. However, the future direction of this modified TiAl4V/Cu alloy is streamlined towards the marine modules in order to enhance their life span during service.

REFERENCES

Advanced Manufacturing Crc Limited. (n.d.). *Additive Manufacturing Categories Processes and Materials*. Retrieved from http://amcrc.com.au/wp-content/uploads/2013/03/additive-manufacturing-categories-processes.pdf

ASTM E3-11. (2011). *Standard Guide for Preparation of Metallographic Specimens*. ASTM International.

ASTM E384-11e1. (2011). *Standard Test Method for Knoop and Vickers Hardness of Materials*. ASTM International Book of Standards.

Bania, P. J., Eylon, D., Boyer, R. R., & Koss, D. A. (Eds.). (1993). *Beta Titanium Alloys in the 1990's*. Minerals, Metals and Materials Society.

Boyer, R., Welsch, G., & Collings, E. W. (1994). Materials Properties Handbook: Titanium Alloys. ASM International.

Dang, B., Fergus, J. W., Gale, W. F., & Zhou, T. (2001). Effect of Copper on the Oxidation Behaviour of Ti-48Al-2Cr-2Nb. *Oxidation of Metals*, *56*(1), 15–32. doi:10.1023/A:1010387218164

Donachie, M. J. (2000). *Titanium: A Technical Guide*. ASM International.

Erinosho, M. F., Akinlabi, E. T., & Pityana, S. (2014). Laser Metal Deposition of Ti6Al4V/Cu Composite: A Study of the Effect of Laser Power on the Evolving Properties. *World Congress of Engineering (WCE)*, London, UK.

Erinosho, M. F., Akinlabi, E. T., & Pityana, S. (2015). Influence of Scanning Speed and Energy Density on the Evolving Properties of Laser Deposited Ti6Al4V/Cu Composites. In *Proceedings of the World Congress on Engineering*.

Fu, Y., Loh, N. L., Batchelor, A. W., Liu, D., Zhu, X., He, J., & Xu, K. (1998). Improvement in fretting wear and fatigue resistance of Ti–6Al–4V by application of several surface treatments and coatings. *Surface and Coatings Technology*, *106*(2-3), 193–197. doi:10.1016/S0257-8972(98)00528-3

Ghosh, M., & Chauerjee, S. (2003). Diffusion bonded transition joints of titanium to stainless steel with improved properties. *Materials Science and Engineering A*, *358*(1-2), 152–158. doi:10.1016/S0921-5093(03)00298-3

Global Market Review. (n.d.). Retrieved from http://www.ipmd.net/shop/powder_metallurgy

Gogia, A. K., Nandy, T. K., Muraleedharan, K., & Banerjee, D. (1992). The effect of heat treatment and niobium content on the room temperature tensile properties and microstructure of Ti3Al-Nb alloys. *Materials Science and Engineering A*, *159*(1), 73–86. doi:10.1016/0921-5093(92)90400-U

Gollapudi, S., Sarkar, R., Chintababu, U., Sankarasubramanian, R., Nandy, T. K., & Gogia, A. K. (2011). Microstructure and mechanical properties of a copper containing three phase titanium alloy. *Materials Science and Engineering A*, *528*(22-23), 6794–6803. doi:10.1016/j.msea.2011.05.080

Industrial Laser Processes. (n.d.). Laser Systems Product Group of AMT— the Association for Manufacturing Technology. Retrieved from file:///C:/Users/USER/Downloads/IndustrialLaserProcesses_0109_TECH1.pdf

Introduction to Laser Technology. (n.d.). Retrieved from www.mellesgriot.com

Jackson, M., Jones, N. G., Dye, D., & Dashwood, R. J. (2009). Effect of initial microstructure on plastic flow behaviour during isothermal forging of Ti-10V-2Fe-3Al. *Materials Science and Engineering A*, *501*(1-2), 248–254. doi:10.1016/j.msea.2008.09.071

Jones, N. G., Dashwood, R. J., Dye, D., & Jackson, M. (2008). Thermomechanical processing of Ti-5Al-5Mo-5V-3Cr. *Materials Science and Engineering A*, *490*(1-2), 369–377. doi:10.1016/j.msea.2008.01.055

Kano, O., Takatori, H., & Sugawara, S. (2013). *Titanium Alloy Complex Powder containing copper powder, chromium powder or Iron Powder Titanium Alloy material consisting of this powder, and process for production thereof*. Patent Application Publication No: US 2013/0071284 A1.

Key to Metal. (n.d.). *Properties and Applications of Titanium - 2.5% Copper Alloy*. Retrieved from www.keytometals.com

Key to Metal. (2013). *Physical Properties of Titanium and its alloy*. Retrieved from www.keytometals.com/article122

Kikuchi, M., Takada, Y., Kiyosue, S., Yoda, M., Woldu, M., Cai, Z., ... Okabe, T. (2003). Mechanical Properties and Microstructures of Cast Ti-Cu alloys. *Dental Materials*, *19*(3), 174–181. doi:10.1016/S0109-5641(02)00027-1 PMID:12628428

Kong, F. T., Chen, Y. Y., & Zhang, D. L. (2011). Interfacial microstructure and shear strength of Ti-6Al-4V/TiAl laminate composite sheet fabricated by hot packed rolling. *Materials & Design*, *32*(6), 3167–3172. doi:10.1016/j.matdes.2011.02.052

Laser Solutions for Manufacturing. (2013). *Industrial. Penn Well Copyright*. Retrieved from http://www.industrial-lasers.com/articles/print/volume-250/issue-6/features/laser-metaldeposition.html

Lenntech, B. V. (2013). *Water Treatment Solution*. LENNTECH. Retrieved from www.lenntech.com/periodic/element/Ti

Leyens, C., & Peters, M. (Eds.). (2003). Titanium and Titanium Alloys: Fundamentals and Applications. Wiley-VCH Verlag GmbH & Co. KGaA. doi:10.1002/3527602119

Lutjering, G., & Weissman, S. (1970). Mechanical Properties and Structure of Age-Hardened Ti-Cu Alloys. *Metallurgical Transactions*, *1*(6), 1641–1649. doi:10.1007/BF02642011

Lutjering, G., & Williams, J. C. (2007). *Titanium, Engineering Materials and processes* (2nd ed.). Springer.

Mahamood, R. M., Akinlabi, E. T., Shukla, M., & Pityana, S. (2013). Laser Metal Deposition of Ti6Al4V: A Study on the Effect of Laser Power on Microstructure and Microhardness. In *Proceedings of the International MultiConference of Engineers and Computer Scientists 2013*.

Maslenkov, S. B., & Maslenkova, E. A. (1991). *Steels and Alloys at High Temperatures*. Moscow: Metallurgy Press.

Mishra, H., Ghosal, P., Nandy, T. K., & Sagar, P. K. (2005). Influence of Fe and Ni on creep of near α-Ti alloy IMI834. Material Science and Engineering, A399(1-2), 222-231.

Moiseyev, V. N. (2006). *Titanium alloys: Russian aircraft and aerospace applications*. CRC Press Taylor & Froes Group.

Murray, J. L. (1990). Binary Alloy Phase Diagrams (2nd ed.). ASM International.

Murray, J. L. (1992). ASM Handbook of Alloy Phase Diagrams. ASM International.

Okazaki, Y., Gotoh, E., Nishimori, M., Katsuda, S., Manabe, T., & Kobayashi, K. (2005). Osteocompatibility of Stainless Steel, Co-Cr-Mo, Ti-6Al-4V and Ti-15Zr-4Nb-4Ta Alloy Implants in Rat Bone Tissue. *Materials Transactions, 46*(7), 1610-1617.

Okazaki, Y., Ito, Y., Ito, A., & Tateishi, T. (1993). Effect of alloying elements on mechanical properties of titanium alloys for medical implants. *Material Transaction, 34*, 1217-1222. 10.2320/matertrans1989.34.1217

Reig, L., Toja, C., Busquets, D. J., & Amigó, V. (2013). Microstructure and Mechanical Behavior of Porous Ti-6Al-4V Processed by Spherical Powder Sintering. *Materials (Basel)*, *6*(10), 4868–4878. doi:10.3390/ma6104868

Sakuma, T., Hosogi, M., Okabe, N., Iwata, U., & Okita, K. (2002). Effect of Copper content and Super elasticity Characteristics in Ti-Ni and Ti-Ni-Cu alloy wires. *Materials Transactions*, *43*(5), 828–833. doi:10.2320/matertrans.43.828

Sears, J. (2011). *The development of titanium applications via additive manufacturing*. Retrieved from http://www.ipmd.net/articles/001349.html#sthash.vJ8WLvtI.dpuf

Sen, I., Gopinath, K., Datta, R., & Ramamurty, U. (2010). Fatigue in Ti-6Al-4V-B alloys. *Acta Materialia*, *58*(20), 6799–6809. doi:10.1016/j.actamat.2010.09.008

Sherif, E. M., Ammar, H. R., & Khalilb, K. A. (2014). Effects of copper and titanium on the corrosion behaviour of newly fabricated nanocrystalline aluminum in natural seawater. *Applied Surface Science*, *301*, 142–148. doi:10.1016/j.apsusc.2014.02.019

Sobiecki, J. R., Wierzchon, T., & Rudnicki, J. (2001). The influence of glow discharge nitriding, oxynitriding and carbonitriding on surface modification of Ti–1Al–1Mn titanium alloy. *Vacuum*, *64*(1), 41–46. doi:10.1016/S0042-207X(01)00373-6

Song, B., Dong, S., Zhang, B., Liao, H., & Coddet, C. (2012). Effects of processing parameters on microstructure and mechanical property of selective laser melted Ti6Al4V. *Materials & Design*, *35*, 120–125. doi:10.1016/j.matdes.2011.09.051

Stranak, V., Wulff, H., Rebl, H., Zietz, C., Arndt, K., Bogdanowicz, R., ... Hippler, R. (2011). Deposition of thin titanium–copper films with antimicrobial effect by advanced magnetron sputtering methods. *Materials Science and Engineering C*, *31*(7), 1512–1519. doi:10.1016/j.msec.2011.06.009

Technical Data Sheet. (2013). *Carpenter. Titanium alloy*. Dynamet Holdings Inc.

Technology, L. (2005). *CO2 and YAG laser system specifications & applications*. Accessed from http://www.ritapad.com/lasertechnologies.htm

Thijs, L., Verhaeghe, F., Craeghs, T., Van Humbeeck, J., & Kruth, J. (2010). A Study of the Microstructural Evolution During Selective Laser Melting of Ti6Al4V. *Acta Materialia*, *58*(9), 3303–3312. doi:10.1016/j.actamat.2010.02.004

Tian, W. H., & Nemoto, M. (1997). Effect of carbon addition on the microstructures and mechanical properties of γ-TiAl alloys. Intermetallics, 5(3), 237-244.

Vandenbroucke, B., & Kruth, J. (2007). Selective laser melting of biocompatible metals for rapid manufacturing of medical parts. Emerald Group Publishing Limited. DOI doi:10.1108/13552540710776142

Vrancken, B., Thijs, L., Kruth, J., & Van Humbeeck, J. (2012). Heat treatment of Ti6Al4V produced by Selective Laser Melting: Microstructure and mechanical properties. *Journal of Alloys and Compounds*, *541*, 177–185. doi:10.1016/j.jallcom.2012.07.022

Wong, P. K., Kwok, C. T., Man, H. C., & Cheng, F. T. (2012). Corrosion Behavior of Laser-alloyed Copper with Titanium fabricated by high power diode laser. *Corrosion Science*, *57*, 228–240. doi:10.1016/j.corsci.2011.12.013

Xia, Y., Luo, S. D., Wu, X., Schaffer, G. B., & Qian, M. (2013). The sintering densification microstructure and mechanical properties of gamma Ti-48Al-2Cr-2Nb alloy with a small addition of copper. *Materials Science and Engineering A*, *559*, 293–300. doi:10.1016/j.msea.2012.08.100

Yasa, E., & Kruth, J. (2011). Application of Laser re-melting on Selective Laser melting parts. Advances in Production Engineering and Management, 6, 259-270.

Yu, J., Rombouts, M., Maes, G., & Motmans, F. (2012). Material properties of Ti6Al4V parts produced by laser metal deposition. *Physics Procedia*, *39*, 416–424. doi:10.1016/j.phpro.2012.10.056

Zhang, E., Li, F., Wang, H., Liu, J., Wang, C., Li, M., & Yang, K. (2013). A new antibacterial titanium-copper sintered alloy: Preparation and antibacterial property. *Materials Science and Engineering C*. Retrieved from www.elsevier.com/locate/msec

KEY TERMS AND DEFINITIONS

3D Printing: This is a three-dimensional printing process that involves layer-by-layer building of an object.

Bimodal Structure: The structure having two type of arrangement.

Biofouling: This is the fouling of marine structures caused by microorganisms.

Laser Metal Deposition: This refers to as the process of forming a solid sample after the deposition of metallic powder with a beam of laser.

Linear Coefficient of Expansion: This is referred to as the increase in length of a material due to heating to temperatures below its melting point.

Powder Metallurgy: This involves the compaction of heated powder below the melting point and solidifies to give the required shape and size.

This research was previously published in Advanced Manufacturing Techniques Using Laser Material Processing edited by Esther Titilayo Akinlabi, Rasheedat Modupe Mahamood, and Stephen Akinwale Akinlabi, pages 60-91, copyright year 2016 by Engineering Science Reference (an imprint of IGI Global).

Chapter 11
Trend and Development in Laser Surface Modification for Enhanced Materials Properties

Muhammed Olawale Hakeem Amuda
University of Lagos, Nigeria & University of Johannesburg, South Africa

Esther Titilayo Akinlabi
University of Johannesburg, South Africa

ABSTRACT

This article presents a process review of the commonly available laser surface modification techniques for surface property enhancement. This is reinforced with the specific case treatment of research trends in relation to commonly treated materials. The progression from simple surface modification to the production of components with multifunctional characteristics known as functionally graded material is discussed in combination with emerging research focus on the computational simulation of laser surface modification for optimization of process dynamics.

1. INTRODUCTION

Many solid materials possess adequate bulk mechanical properties which commend them for a number of applications but this is not usually the case with their surface properties. In most cases, the bulk material lacks good surface characteristics for effective performance for the timescale over which it is presumed fit-for-purpose. Additionally, literature (Krauss, 1992) indicates that surfaces of materials are subjected to greater stresses and more direct environmental impact than the interior; hence, when such aggressive stresses reached a material's resistance limit, surface initiated fracture, fatigue, wear and corrosion failures occur. Therefore, surfaces of materials are usually twitched to make them robust to the environment in which they will be used in order to derive maximum benefit. The process of treating the surface of materials to improve their surface functionalities and making them robust to their environment is referred to as surface modification or surface engineering. The process involves treatment of the surface or near-surface regions of a material to permit the surface to perform functions that are distinct from those demanded

DOI: 10.4018/978-1-5225-9624-0.ch011

from the bulk materials (ASM International, 2001; Cotell and Sprague, 1994). Surface modification has a chequered history from the advent of civilization to the present age and it manifests in many forms; but irrespective of the form, however, it involves changing the composition, crystal structure, texture, chemistry and microstructure of the substrate of the bulk material up to certain depth towards creating new features and properties in the surface (Burakowski and Wierzchon, 1999). Some of the benefits of surface modification include improved corrosion and oxidation resistance, improved wear resistance, reduced frictional energy loss, improved fatigue resistance, enhanced electrical/electronic properties, thermal insulation, size restoration, biomedical functionalization, and improving aesthetics (Ansari et al., 2014). Because the motivation for surface modification is very wide so is the spread of the process very wide as well. At one end of the wide spectrum, the depth of modified surface could be very thin between 0.001-1.0 mm and at the other end, overlayer surface depth in the range 1- 20 mm are typical (Krauss, 1992). These modification depths are, of course, process specific and thus, each process can only optimise within a specific length scale range. There are several presentations of the process such as ion implantation, nitriding, aluminising, physical vapour deposition (PVD), chemical vapour deposition (CVD), anodising, laser processing, thermal spraying, cold spraying, and liquid deposition methods. The possible range of modification depth in these processes is shown in Figure 1 with ion implantation providing the smallest depth while weld overlay could be in the tens of a millimetre.

Among these several processes, laser surface modification particularly laser deposition has the capacity to provide across the spectrum range (nano to millimetre) of modified surface depth not possible through other processes. Arising from these possibilities which have resulted in a wide range of improved surface properties in treated materials, the growth in laser surface modification process has been exponential in the last three decades; and new application areas are equally emerging (Baker 2010). The process is one of the strong driving forces advancing additive manufacturing particularly in laser engineered net shaping (LENS) manufacturing systems (Gu et al., 2012).

The attraction in laser technology for surface modification is driven by the ability to precisely control the spot onto which the beam is delivered to achieve the desired response. In specifics, laser beam has the ability to precisely deposit a large amount of energy into a material over a short time scale in a spatially confined region near the surface of the material. This permits the control of local surface properties relative to the bulk materials and other regions on the surface (Brown and Arnold, 2010). The degree of distortion and size variations in laser modified layers is minimal compared to such other processes such as weld overlays or conversion coating. The change in composition and structural fluctuation is equally less compared to thermal surface treatment. Furthermore, it provides excellent interfacial bonding between the top layer and the substrate layer resulting in a strongly adherent surface modified layer. Ultimately, it provides possibility for the formation of novel surface alloys not possible with the other processes owing to the non-equilibrium characteristics of the process (Kwok et al., 2000). Thus, the many possibilities of materials response from the effect of the incident laser energy, the interaction time scale and other laser parameters that can result in changes spanning multiple length scales, from the atomic to the macroscale, have been responsible for the growth in laser surface technology.

There exist extensive literature on several aspects of laser principles, laser surface modification and applications which are not integrated but are far and wide apart (Dahotre, 1998; Steen, 2003; Ion, 2005). It has been very difficult and near impossible to aggregate these wide publications on laser and its use for surface modification into a simple, quick but detailed reference guide containing recent developments in the process. Therefore, this article attempts to provide in a single piece, basic generic information on laser surface modification reinforced with several reported works on some materials. While it is not the focus of this article to provide an exhaustive treatment of laser surface modification in all its forms and application across the material spectra, yet, due treatment is accorded to such processes as laser surface alloying, laser surface melting, laser cladding, laser surface sintering and laser surface composite coating. This is preceded by a review of laser beam, its characteristics and laser-material interaction in relation to surface modification. The centric discussion on laser surface treatment techniques is further illuminated with a new paradigm such as the deposition of functionally graded composite coatings and the use of computational simulation and statistical optimization technique to gain further insight into laser surface modification.

2. THE LASER BEAM

Laser composes of light beams propagating mainly in one direction and it represents an acronym for *light amplification by stimulated emission of radiation* (Laser). Unlike other light sources (collimated light beams, light emitting diodes, hot objects), lasers exhibit unique distinguishing features including coherency, temporal nature, monochromatic wavelength, highly concentrated and an irradiance beam profile (Kirkham and Roundy, 2014). The monochromatic nature of a laser beam indicates that it is essentially a single narrow wavelength with virtually no light at wavelengths outside the central peak. The temporal characteristic of a laser beam make it to vary from a continuous wave (CW) to an extremely short pulse (femtosecond) which can provide very high power densities. Its coherency provides capacity for it to travel in a narrow beam with a small and well defined divergence or spread. This offers flexibility in permitting a user to specify exactly the area for illumination by the laser beam. This capability equally makes focusing to a very small and intense spot in a highly concentrated area possible. And it is this ability for focusing that makes the laser beam useful for many applications in physics, chemistry, the medical industry, and industrial sectors most especially in materials processing. Also, the beam irradiance pattern known as the "beam profile" influences the energy density, the concentration and the collimation of the lights; which determines the application of the laser beam.

Laser beam can be generated from the three physical states of matter: gases, liquid and solids and as such are available in many classifications and can be classified into five broad grouping (gas lasers, liquid lasers, solid lasers, semi-conductor lasers and free electron lasers). Baker (2010) reported that the various classes of lasers are grouped based on the active medium, average power, wavelength, and mode of operation. The details of the classification as provided by Ion (2005) in a chart, based on power output against wavelength is shown in Figure 2, and vary from excimer, Nd: YAG, diode through to CO_2 lasers. Fibre laser is however not included in the classification which produces a better quality beam with a wavelength of about 1.06 μm in the near infrared region of the electromagnetic spectrum.

But it represents the trend in laser materials processing. The figure shows that a wide range of laser beams are available for materials processing across the spectrum. Dahotre (2002) reported that these various beams can be deployed in heating, melting and vaporizing materials depending on the relationship between the power density of the laser beam and its interaction with the material (see Figure 3). Most surface modification treatments in materials using lasers fall within the regions labelled heating and melting in Figure 3. However, it must be noted that the characteristics of the beam in terms of its profile greatly influence the outcome of any laser surface modification process. If a proper beam profile is not ensured, then, the expected outcome of the modification may not be achieved.

2.1 Laser Beam Profile and Power Density

The beam profile describes the distribution and form of the spatial intensity of the laser beam. The nature and character of this profile affects the energy density, its concentration, the collimation of the laser lights and the propagation of the beam through space. Specifically, in terms of energy density, the profile determines whether the energy density is highly focused on a localised spot area or it is evenly spread across a given area. Kirkham and Roundy (2014) reported that there are many varieties of laser beam depending on the laser source. Excimer laser exhibit a different beam profile from HeNe laser beam or CO_2 laser beam or a Cr:LiSAF beam. But generally, beam profile can either approximate Gaussian or non-Gaussian distribution. The non-Gaussian is equally referred to as flat-top beam profile. The Gaussian beam provides the highest concentration of focused light, whereas, in the non-Gaussian, uniform energy is distributed across a given area. These broad idealized profiles are shown in Figure 4 though there are deviations from these two broad groupings. The character of the beam influences the energy density delivered to a surface during laser process which in turn affects the response from the material.

Most laser processes particularly those of surface modification are conditioned to work with the Gaussian profile because this is considered as being able to deliver maximum energy density to the surface. Laser surface modification is premised on the beam being focused on specific spot of the substrate material as it traverse the surface to generate maximum material response and also minimizing physical distortion in the material unlike the flat-top profile. The beam profile do degrade over time delivering out of spectrum energy density to the material and this has been reported to have caused severe distortion in the process for which it was applied (Kirkham and Roundy, 2014). This could generate a large error margin in experiment resulting in misleading outcomes. Thus, it is absolutely necessary to consistently ascertain the correct beam profile before the commencement of any laser process. Incidentally, there are several methods for controlling and maintain the beam profile for consistency.

The energy density (E) of a laser beam measures the quantum of energy per unit area (J/mm^2) delivered, and is related to the laser power (q), the scanning velocity (ν) and the radius r_b of the laser beam; as presented mathematically in Equation (1). It is evident from the equation that the energy density is influenced by the character of the laser beam. For a Gaussian beam has a smaller radius than a flat-top profile and this explains why a Gaussian profile delivers higher energy density than a flat-top beam. In a real-time laser surface modification process not all incident energy density is converted for surface modification. It is only the fraction that

$$E = \left(\frac{q}{r_b \nu} \right) \tag{1}$$

is absorbed that does the work and this is referred to as the input energy density. The amount of energy density absorbed by the material's substrate controls the depth of the melt pool and hence the volume of the molten alloy (Baker, 2010). Irrespective of the energy density delivered from any laser beam for surface modification, the process is characterized by high heating and cooling rates in the range 10^4 to 10^{10} Ks^{-1}, thermal gradients between 10^5 and 10^8 K/m and solidification velocities may be up to 30 ms^{-1} (Baker, 2010). And these are influenced by the laser-material interaction dynamics. The laser beam may either be continuous wave or pulsed wave type. In the continuous wave type, the laser produces a continuous output beam overtime, whereas in the pulse wave type, the laser output occurs in pulses of some duration at some repetition rates usually in the range of picosecond to nanosecond. Both types of laser waveform are used for surface modification (Abboud et al., 2007). However, the depth of surface modified layer in both wave forms are influenced by the laser beam wavelength, temporal pulse power (pulse length, peak power and pulse shape), repetition rate, beam energy distribution and beam geometry in terms of the focal spot size and depth of focus (Abboud et al., 2007). For instance, in pulsed laser modified surface, the properties are controlled by pulse energy, pulse width, frequency and scan rate. A pulsed mode in laser surface modification provides temporal limitation in energy coupling into the target which results in a very limited depth of heat conduction into it resulting in reduced heating of the work piece. Such laser irradiation mode reduces incidences of defects such as porosity, bubbles or depressions which occur readily a continuous wave laser (Pinkerton and Li, 2003). Notwithstanding this, pulsed laser beam is more suited for surface texturing or ablation.

2.2 Laser-Material Interaction

The material response during surface modification is greatly dependent on the interaction between the material substrate and the laser beam. Laser-material interaction phenomena are very complex in relation to the thermo-physical properties of the substrate. The full description of the several forms of laser-material coupling is beyond the scope of a single book chapter as this but the interaction is influenced by a combination of optical (reflectivity/absorptivity), thermal and electrical properties of materials which may produce such effect as temperature rise, melting, vaporization, cooling and solidification. Detailed literature treatment of the laser-material interaction is available in many texts on laser materials processing (Gladush and Smurov, 2011; Steen and Mazumder, 2010). Notwithstanding, it is suffice to note that the main laser-solid interaction process is the excitation of electrons from their equilibrium states to some excited states by absorption of photons which manifests in a particular type of response from the material. The processes occurring when the laser beam hits a material depend on the amount of deposited laser energy. This energy in relation to its spatial and temporal distribution determines the type of surface modification that occurs. The penetration depth of the laser beam depends on the interaction time which is determined by the cross section of the beam and the scan rate. Thus, the unique interaction of laser beam with materials over a given timescale can lead to permanent changes in the material's surface composition, chemistry, crystal structure and morphology resulting in improved surface properties that are not readily accomplished through other means. This unique laser-material interaction has been utilized in various forms to provide a wide range of laser surface modification techniques such as laser texturing, selective sintering, cladding, surface melting, etc. (Weng et al., 2014; Lawrence and Waugh, 2014).

3. LASER SURFACE MODIFICATION PROCESSES

The quality of laser surface modification is influenced by three significant parameters and these are laser power, beam size and scan rate. These parameters determine the extent of structural and composition changes in the substrate. For instance, an increasing laser power increases absorptivity in the substrate, melt depth, widens the heat affected zone and provided for a uniform mixing of the melt pool but it also increases tendency to surface rippling and crack formation. A small beam size provides high power density that increases melt depth but decreases melt width whereas, an increasing scanning speed implies less interaction between the beam and the material resulting in narrowing melt width and shallow melt depth. It equally reduces fluid flow which affects the homogenization of the melted zone, changes the shape of the melted zone from circular to elliptical though with reduced tendency for surface rippling and crack formation. The high cooling rate associated with increasing scanning rate reduces the dendritic arm spacing (Akgun and Inal, 1994). This wide range in the possibility of combination of the key laser processing parameters provides opportunity for manipulating the laser process for specific surface modification. Therefore, there are many variants of laser surface modification for altering the surface composition, chemistry, crystal structure, texture and microstructure; and these are treated in this section. The discussion traverses the principles, merits and demerits and a brief critique of researches conducted in each of the processes.

3.1 Laser Surface Ablation and Texturing

Laser ablation represents a process for removing material from a substrate (metals, inorganic insulator/ semiconductors and organic materials) by direct absorption of laser energy resulting in a surface with different morphology. In laser ablation, the chemical composition and crystal structure of the substrate material is not affected. The process is usually conducted with pulse laser beam although the use of the continuous wave beam had equally been reported (Brown and Arnold, 2010). Surface ablation occurs at a particular energy density (fluences) above the melting point resulting in the formation of plumes; and it depends on so many variables including absorption mechanism, material properties, microstructure and morphology, the presence of defects, beam wavelength and pulse duration. The threshold of energy density for ablation in metals is between 10 and 100 kJ/m^2, for inorganic insulators- 5 and 20 kJ/m^2, and for organic materials-1 and 10 kJ/m^2. However, in the case of multiple pulses, the ablation thresholds may be lower than the stated ranges. Literature indicates that materials responses during ablation equally involves a combination of other phenomena such as surface melting and thermally activated processes which may result in cumulative changes in the materials surface texture, morphology and chemistry not ordinarily intended (Brown and Arnold, 2010). Ablation finds application in several areas including improving surface roughness for better adhesion, tribological control, optical and electrical properties and biomedical functionalization of implant. Frerichs et al. (1995) reported that laser-induced ablation of some polymeric materials as a pretreatment step prior to metallization improved their surface roughness for adhesion for the subsequent metallization process. Recently, Rytlewski and Żenkiewicz (2013) deployed laser ablation for the formation of catalytically active surfaces that are fully prepared for direct electroless metallization in thermoplastic and thermosetting polymer composites. Ho et al. (2015) deployed laser ablation to modify the surface of AZ31B magnesium alloy for improved biowettability by inducing alteration in the grain size, surface roughness, surface energy and surface composition. Such effort resulted in increased protein absorption which assisted in cell/osteoblast growth. Similar process was earlier applied in generating craters in silicon and gold target material to modify their surface morphology (Besner et al., 2005).

Another application of laser ablation in the last two decades is the production of patterned microstructure in materials which is equally known as laser surface texturing (LST). In LST, specific textures are introduced onto the substrate surface at specific repetitive positions in the materials using pulsed laser beam. The functionalities of the textured pattern depend on the size, depth and area of the pattern. There are many options of such patterned texture which may include micropores, grooves, circular dimple-like depressions. These patterned textures can improve load capacity, wear rates, lubrication lifetime, wetting characteristics, reduces friction in tribological mechanical systems (Vilhena et al., 2009). Additionally, LST has been extended to texturing of multicrystalline silicon for solar cell. This has eliminated surface processing dependence on grains crystallographic orientations unlike the chemical texturing methods (Dobrzański and Drygała, 2008). Surface bio-functionalization using femtosecond pulsed laser multiscale surface texturing was achieved in titanium and its alloys for dental and orthopaedic implants (Cunha, 2015). And in the recent past, LST has been used to create super-hydrophobic surfaces (high wettability surfaces with $\theta_c > 150^o$) for biological scaffolds, microfluidics, self-cleaning surfaces, and lab-on-chip devices (Brown and Arnold, 2010). Invariably, the application window for LST appears inexhaustive and it is most likely that this will continue to grow into the next decade and beyond. A broad review of the state of the art in LST is available in the literature.

3.2 Laser Surface Melting and Remelting

Laser surface melting (LSM) involves irradiating the surface of the substrate with laser beam to a regulated depth without modifying the surface layer composition. The melting and solidification is very rapid achieving a very high quench rate which permits the development of non-equilibrium structure. Since there is no change in surface composition, the melting facilitates the redistribution of dissolved alloying elements. A major characteristics the LSM technique is the rapid solidification which can generate hardening through the introduction of crystalline defects such as vacancies and dislocations. Though, residual stresses are developed which result in the distortion of a work-piece, but this is readily overcome by applying low powered surface heating procedure after the laser melting process (Baker and Arnold, 2010).

During LSM, the near surface region rapidly reaches the melting point creating a liquid/solid interface which start to move through the alloy while diffusion of dissolved of elements begins in the liquid phase. At the termination of the laser pulse, the surface remained below the vaporization temperature (this is the key difference between laser surface ablation and surface melting, vaporization temperature is not reached in laser melting but in laser ablation) but the maximum melt depth has been attained and inter-diffusion continues; though the re-solidified interface velocity is momentarily zero, and later rapidly increases. The interface moves back to the surface from the region of maximum melt depth. Inter-diffusion continues in the liquid, but the re-solidified metal behind the liquid/solid interface cools so rapidly that solid state diffusion may be negligible. At the completion of re-solidification, a surface alloy has now been created. The melted zone is usually relatively small of about 50 to 1000 microns permitting very high quench rates in the range of 10^3 to 10^6 Ks^{-1} resulting in non-equilibrium martensitic microstructures (Baker, 2010). LSM has been applied to both ferrous (cast iron, plain carbon steels, stainless steels and alloyed steels) and non-ferrous (aluminum alloys, titanium alloys and super alloys of the nickel based series) metals either in inert or nitrogenous environments (Benyounis et al., 2005; Olakanmi et al., 2015; Li et al., 2016; Engeli et al., 2016). In the case of ferrous metals particularly carbon steels, surface melting cause carbon redistribution and/or the dissolution of precipitates resulting in the formation of martensitic or homogenized structure which increased surface hardness many fold translating to better wear resistance. In titanium alloys, LSM produces fine basket weave microstructure resulting in higher hardness and better wear and corrosion resistances (Akgun and Inal, 1994). Microstructures in LSM processes approximate the conventional solidification structure which is essentially dendritic. An example is illustrated in Figure 5 in the case of surface melted nodular cast iron and Ti-6Al-4 V titanium alloy. LSM is currently applied in rapid prototyping, rapid tooling and rapid manufacturing of high grade materials including steels, titanium, aluminum and nickel based superalloys for different range of fixtures such as turbine blades, propellers, nozzles, dental and human implants (Zhang and Attar, 2015; Yasa and Kruth, 2011; Gebhardt et al., 2010).

Laser surface remelting is closely related to LSM except that in surface remelting a smaller localized region is focused with intense reheating followed by rapid solidification to generate extremely fined and tailored microstructure. It is also in another instance referred to as selective laser melting (SLM). It is mainly applied to remove surface defects and improve integrity such as homogeneity and adhesion particularly in overlay coating for improved corrosion resistance. SLM is similar to selective laser sintering (SLS) except in the way the powder is processed. In SLS, only the surface of the powder is melted to produce a fused compact while in the SLM, the powder particles are completely melted and a high density part is produced (Mahamood et al., 2014). It is strictly not a surface modification process but an integral part of additive manufacturing. In some instances, laser surface remelting is loosely classified as LSM. Recent advancement in this area is toward selective laser melting of metal powder mixture in additive manufacturing (Engeli et al., 2016; Olakanmi et al., 2015); and laser glazing for the formation of amorphous structure on substrate surface. In laser glazing, laser beam at power densities in the range of 10^5 to 10^7 W/cm^2 is used to surface melt the substrate for about 10^{-4} to 10^{-7} seconds. The very intense power densities and short interaction time results in very high cooling rates exceeding 10^5 K/s which suppresses the usual nucleation and crystallisation processes that accompany solidification. Rather, amorphous glassy surface layers are produced which have been found to assist greatly in closing cracking pores particularly in thermal barrier coatings (Ahmadi-Pidani et al., 2013). Laser glazing is commonly conducted on the Nd-YAG or Excimer pulsed laser beams.

3.3 Laser Surface Alloying

Laser surface alloying (LSA) uses focused laser beams to melt coatings and a portion of the underlying substrate. In LSA, alloying elements (metallic or non-metallic powder) are either deposited on the substrate surface, or fed through a hopper delivery system or in a gaseous environment and then irradiated by a high-energy laser beam. This causes dissolution of the powders or diffusion of the alloying gas into the laser generated melt pool and chemical reaction take place between the additives and the melted substrate. In a very short period of time, cooling and solidification form a thickness of 0.1-0.5 mm of a new alloy layer (Tian et al., 2005). Through this, the composition, chemistry, crystal structure and microstructure structure of the surface layer is altered; and this has been extended to a wide range of materials. LSA could be conducted in gaseous medium, solid medium or solid + gaseous medium (Baker, 2010). The diverse choice of alloying materials that can be incorporated by LSA permits tailoring of surface properties to impart better wear, corrosion and oxidation properties to mitigate against any degradation mode. This is distinct from the use of laser processing for the formation of a metal matrix composite layer on the surface via incorporation of ceramic powders because in the case of composite layer, the injected ceramic particles do not dissolve in the molten substrate layer; rather it remain embedded in the matrix to produce the composite layer (Thawari et al., 2003).

Though, laser surface processing requires the protection of the molten pool from oxidation and other contaminations using inert gases such as argon or helium, these gases can be replaced with nitrogen to form nitride surface with superior surface chemistry and structure. Extensive literature abound on laser gas nitriding of titanium alloys, stainless steels, and aluminum alloys (Baker, 2010). Laser nitriding with a 100% nitrogen atmospheres in the early works produced a thin 5-10μm surface layer of titanium nitride (TiN) with surface hardness close to ~ 1000-2000Hv and improved corrosion resistance together with a lower coefficient of friction and wear resistance. However, cracking was often a problem in such surface modified titanium. This was resolved by either using diluted nitrogen atmosphere or by preheating prior to nitriding but at the expense of a decrease surface hardness and melt depth (Baker, 2010). Controlling parameters in laser gas nitriding process are energy density, beam diameter, scanning speed, nitrogen pressure or flow rate and these must be regulated to avoid development of cracks in the nitride surface. Bianco et al. (1995) investigated the potential use of CO_2 as alloying gas and reported better abrasive wear resistance in CO_2-laser alloyed titanium than TiN surface layer produced through laser nitriding. The attraction in laser gas alloying is that it affords more precise control of the concentration of the alloying element in the molten pool compared to powder injection. Additionally, laser gas alloying can be undertaken more easily on complex shapes without the feeding difficulties associated with powders or wires.

LSA has also been performed with the addition of different powders through either injection or preplacement in slurry form; though, this is sometimes technically referred to as cladding. Both cases produce partial dissolution of the powders which may provide a strong bond with the matrix and confers significant wear resistance to the substrate (Baker, 2010). The powders may be introduced singly or in mixes and all metallic or non-metallic powders are possible but for powder injection system, the particle size must be within the range 45-120 μm for good flowability. Particle sizes outside this range are only suitable through the preplacement technique. However, in some applications where service conditions require improved surface strength through dispersion hardening and /or improved corrosion properties, then a complete dissolution of the particles during laser processing is preferable. And this is possible using laser powers of ~3kW, when the particle size is less than 10μm.With SiC particles, both of these techniques provide an opportunity for the precipitation, in a fine state, of new phases such as Ti5Si3 and TiC. Thawari et al. (2003) conducted LSA investigation on medium carbon steel with $SiC_{(p)}$ using high power CO_2 continuous wave laser in nitrogenous environment at multiple scans. They established that LSA of $SiC_{(p)}$ can be used to produce hard and wear resistant layers on medium carbon steel but these are significantly influenced by the processing variables particularly the energy density. The microstructure of the laser alloyed layer is influenced by repetitive scanning producing refined microstructure with increasing scanning (see Figure 6). Comparative analysis of laser gas and powder alloying, however, showed that the hardness of the powder alloyed surface is lower than that of the nitrided surface (Baker, 2005). A wide range of powder mix composition has been reported in the literature (Tian et al, 2005; Baker, 2010).

LSA of metals combining gaseous atmospheres and powders have equally been studied particularly mixtures of nitrogen with SiC. Dilute nitrogen atmospheres combined with powder alloying have been found to produce crack- free surfaces which have additional hardness relative to the titanium parent alloy and the powder alloying alone (Baker, 2010).

3.4 Laser Surface Cladding

Laser cladding for surface modification involves the deposition of a new layer of material on a substrate by laser fusion of blown powders or pre-placed powder coatings or wire. This process is equally referred to as laser metal deposition (LMD) if the extrinsic materials being injected are solely metals (Mahamood et al., 2014). Multiple layers can be deposited to form shapes with complex geometry (Zhong and Liu, 2010; Tian et al., 2005; Weng et al., 2014)). Defocused or rastered beam laser beam is used to deliver laser energy for the melting of the cladding material which upon solidification forms strong metallurgical bond with the substrate. It is a widely used process for improving the surface and near-surface properties of a new metal part or resurfacing a worn used component thus permitting restoration of original dimension through machining (Gabler, 2014). Laser cladded materials are noted for improved wear, corrosion, oxidation, erosion and fatigue properties but do exhibit high residual stress due to the rapid cooling associated with the process (Tian et al., 2005). Laser cladding, compared to other cladding techniques such as thermal spray and arc welding, provides high deposition rates, low dilution of the substrate, high cooling rates and low distortion. The process is applicable to both similar and dissimilar material including those that are difficult to clad via conventional processes. Figure 7 is the schematic illustration of laser cladding process by filling powder materials from a synchronous feeding system, through powder preplacement method and wire feeding. The successful deposition of clad materials onto the substrate by laser process is influenced by laser power (specifically the energy density), beam size, laser scanning rate or specimen motion velocity. Precise choice of the melting energy density is critical to avoid excessive dilution. But the most critical parameter is the laser scanning velocity because this determines the dwell time of the laser beam on the molten pool which influence the growing process of the crystals (Emamian et al., 2011; Qian et al., 1997). Therefore, it is necessary to ensure a good control of the process parameters in order to achieve desired properties in laser cladding process. In assuring this, extensive investigations have been conducted on the effect of laser cladding process parameters on the microstructure and properties of coatings of different materials (Shah et al., 2014; Mahamood et al., 2013; Emamian et al., 2011; Luo et al., 2011; Thawari et al., 2003; Kwok et al., 2000). Most of these investigations established that laser parameters played significant roles in the evolving microstructure, dilution rates and the properties of the various laser cladded substrates.

Other than laser processing parameters, the quality of the coating material equally influence the success or otherwise of the cladding process. Broadly, the physicochemical properties of both the coating and substrate materials should be considered; and through this, compatible coating materials would be selected to achieve coatings of desired service conditions. The extensive range of experiments and analysis available on laser cladding has enabled a deeper understanding of microstructure-composition-synthesis-processing relationships resulting in optimization of both the process parameters and the material system. The cladding materials range from single material to multiple material systems comprising single or multiple metal alloys (Majumdar et al., 2009; Meng et al., 2005), single or multiple ceramics (Sun and Lei, 2007; Lin et al., 2012; Selamat et al., 2003) or combination of ceramics and metal alloys as metal matrix composite (Ochonogor et al., 2012; Zhang et al., 2010). The challenge with single or multiple ceramic cladding, however, is the presence of defects particularly at certain location between the reinforced phase and the matrix. Furthermore, the dispersion of the ceramics in the melt may be poor most especially in coatings formed through the preplaced technique. These challenges have been attributed to the brittle nature of ceramics (Weng et al, 2014). Defects in cladding coatings have been minimized through the addition of self-fluxing elements like boron and silicon which prevents the development of pores and entrapment of inclusions (Tobar et al., 2006; Chen et al, 2005). The injection of some rare earth elements or their oxides during laser cladding has also been widely reported in the literature to refine the microstructure of surface coatings (Stanford et al., 2008; Choi et al., 2004). The rare earth elements are able to accomplish this because they readily accumulate on the interface crystalline phases being surface active elements and reduce the critical nucleation energy. They also cause a drag on the movement of the grain boundaries. This is further aided by the unmelted rare earth oxides acting as nuclei to promote nucleation rate (Weng et al., 2014). The extent of research in laser surface cladding suggests that there are still opportunities for further exploration of multiple material cladding.

3.5 Effect and Challenges of Laser Surface Modification on Substrate Properties

The modification of substrate chemistry, microstructure and composition by laser irradiation bears correlation to the properties of the substrate. Extensive research in this area reviewed in detail by Weng et al. (2014), Baker (2010) and Tian et al. (2005) identified the properties affected by laser irradiation to include hardness and residual stress, surface roughness, wear, corrosion, oxidation and erosion, fatigue and biocompatibility. The hardness of laser modified surface increased manifolds relative to the untreated substrate due to the formation of hard dendritic microstructure, quasispeherical particles or needles in a solid solution strengthened matrix. The hardness (Hv) is controlled by the volume fraction of the hard precipitated force which is related to the details of the secondary arm spacing. The secondary arm spacing is itself controlled by the cooling rate. Hardness values in the range 1600-2000Hv have been reported in laser modified surface. Strongly related to the hardness value is the residual stress characteristics in surface modified layer which do alternate between tensile (+ value) and compressive (-ve value) across the modified layer track. The residual stress is either measured parallel to the laser tack or perpendicularly to it. However, the variation in the residual stress value is more pronounced in the direction parallel to the laser track than in the perpendicular direction (Baker, 2010). The residual stress is equally influenced by the number of tracks deposited on the surface modified layer with multiple tracks inducing greater residual stress than single laser track which probably accounts for the greater tendency to crack in multiple tracks than in single track (Tian et al., 2005).

Laser irradiation of surfaces induces rippling effect due to the rapid heating and cooling associated with the process. The characteristics of the rippling are used to establish the surface finish in terms of roughness and waviness. The level of surface roughness obtained from a laser surface modification process depends on laser processing parameters, powder concentration and size, details of the carrier gas flow, track overlap ratio and the environment. The waviness associated with surface roughness is a function of the conventional flow of the melt surface which is equally influenced by the track overlap ratio. Generally, laser surface modification produces a smoother surface (Ra $<10\mu m$) than the as-ground condition or shot peening, but powder preplacement technique generally produce a poor quality surface finish (Baker, 2010). The process equally reduces the coefficient of friction and lowers the tendency for materials transfer and adhesive wear resulting in improved wear resistance. The use of ceramic materials and some other oxide materials such as $SiC_{(p)}$ and ZrO_2 has resulted in improved erosion and oxidation resistances in treated material particularly titanium and titanium alloys (Baker, 2010; Tian et al., 2005). Laser deposited titanium aluminide intermetallic particles also improved the cavitation resistance in Ti6Al4V alloy (Cárcel et al., 2014). Laser deposited layer comprising $TiAl_3$ +TiAl on commercially pure titanium alloy provided excellent oxidation resistance at 1000°C. Similarly, laser synthesised TiN/B/Si/Ni complex coating on titanium alloys was found to be about four times more oxidation resistant than the untreated alloy after 70hrs at 750°C (Baker, 2010). Laser cladding of hydroxyapatite with some metallic alloys such as Ti6Al4V (Tlotleng et al., 2014), NiTi (Yang et al., 2013), SS316L (Khandelwal et al., 2013) produced excellent combination biocompatibility and mechanical properties with strong bonding to the substrate. In the case of the titanium alloy, this was attributed to the formation of calcium titanates and titanium phosphides at the interface (Weng et al., 2014).

In spite of the tremendous success achieved with laser irradiation for surface modification and structural property improvement; it, notwithstanding, has some challenges. These challenges include high dilution, existence of cracks and pores particularly in surface modified layer deposited with coating materials exhibiting wide disparity in physicochemical properties (Weng et al., 2014). This is the situation in the cladding of metal substrate with ceramic coating materials. The very high cooling rate associated with laser irradiation is responsible for the existence of cracks in laser surface modification. This is due to the high level of residual stress resulting from the solidification strain. Pores are facilitated by the presence of residual shielding gas or the gas in-situ during laser processing which has not enough time to escape from the molten pool. Literature suggests that ensuring physicochemical compatibility between coating and substrate materials can minimize the problem of cracking. Furthermore, it is necessary to use appropriate combination of process parameters; and this is best achieved through screening experimentation. Then preheating of substrate material reduces the cooling rate and this in turn relaxes the residual stress which assists in reducing the tendency to cracking (Weng et al., 2014). Mridha and Baker (1994) reported that laser surface alloying especially nitriding in a dilute environment (nitrogen-argon, nitrogen-helium mixed gases) produces crack free surface modified layer.

There is also the difficulty in achieving reproducibility of modified layer in laser surface modification which is a common problem in metal-ceramic coating system (Tian et al., 2005). Surface modification for biocompatibility using hydroxyapatite presents its own challenge in terms of changes in the constituents, lack of adequate density and poor bonding strength between the coatings and substrate. But, Katto et al. (2002) resolved this challenge by using a hybrid laser ablation-assisted laser deposition method. The laser ablation was used to improve the adhesion between the coatings and the substrate by creating a buffer interlayer which significantly improved the crystalline structure and mechanical characteristics of the coating material.

4. EMERGING TRENDS IN LASER SURFACE MODIFICATION RESEARCH

Though, laser surface modification has been extended to wide areas as indicated in Figure 8 to improve both the surface and sub-surface properties of materials yet researchers are continually exploring ways to further expand its horizon while addressing the inherent challenges associated with the process. Therefore, emerging research efforts in laser surface modification are focused on four broad areas and these are: formation of functionally graded material coating, development of amorphous coating, optimization of process parameters and integration of other techniques to laser surface modification with a view to addressing many in-process challenges in stand-alone conventional laser surface modification process.

The existence of a sharp distinct interface between the substrate and the coating has severally been identified as major point of weakness due mainly to different thermal expansion coefficient which set up residual stresses at the interface ultimately resulting in cracking (Pei and De Hosson, 2000). Recent studies (Weng et al, 2014; Sobczak and Drenchev, 2013) have shown that functionally graded coatings produced by a one-step laser process provided a feasible solution for avoiding the interfacial problems that are common in laser coatings. This provides a gradual change in composition, microstructure and properties rather than the sudden change in single layer coating. The research interests in this area include evaluating material mix suitable for gradient coating, the interplay of process parameters and material property. Also, there is the attraction to extend the gradient coating to multiple material with capacity for self-lubricating property. There are equally research openings in the modelling and fracture mechanics of these functionally graded coatings.

Laser amorphous cladding coating has recently been identified for improving both wear and physicochemical properties due to the glassy structure of the coatings. Such coatings have been successfully formed on some steels such as Ni–Fe–B–Si–Nb, Fe–Co–B–Si–Nb, Fe–Cr– Mo–Y–B–C amorphous phases (Weng et al., 2014). Recently, research is equally being focused on amorphous/nanocrystalline phases in laser cladding of Fe3Al–B4C–TiN mixed powders on TA2 alloy substrate (Li et al., 2013). There is therefore research motivation for laser amorphous cladding coating of some commonly used biomedical materials like titanium alloys and stainless steels for improved surface properties.

Laser surface modification is influenced by many factors including laser power, beam scanning speed and size, shielding environment overlap between successive laser tracks, materials thermal properties, heat transfer phenomena, and convection-induced mixing in the molten pool such that experimental characterization alone cannot provide full elucidation of the process. Computational modelling via numerical simulation based on finite element modelling or computational fluid dynamics is being investigated to obtain better information on the dynamics of the process (Lei et al., 2012; Earl et al., 2015). Furthermore, optimization of the combination of process parameters are being attempted using various statistical schemes (Vora and Dahotre, 2013); and this exploration is expected to continue into the nearest future because it offers potential for establishing optimum range of parameters for defect free surface modification process. Optimum range of parameters established through such statistical optimization has the incentive of minimizing wastage and reducing cost in experimentation.

Weng et al. (2014) identified another emerging research area in laser surface modification in which laser irradiation is assisted with some other techniques such as electromagnetic agitation or vibration to improve the microstructure and properties by facilitating diffusion and other reactions in the molten state. Electromagnetic excitation can also act as a source of heating which is beneficial in improving wettability between metals and ceramics while vibration can assist in reducing residual stress (Liu et al., 2012; Wu, 2000). In another area, laser melting is being combined with cold spray technology to develop a hybrid system known as laser assisted cold spray (LACS) technology for materials' deposition to improve both adhesion and cohesion strength than possible with conventional thermal spray technology (Tlotleng et al., 2014). LACS was developed to widen the scope of materials that can be successfully deposited through cold spray technology and at the same time eliminating the need for gas heating since this would be provided by the laser beam. Though, the technology is new but it has been successfully applied for the deposition of hydroxyapatite coating on titanium alloy for biomedical application (Tlotleng et al., 2014). It has equally been applied for the deposition of aluminum cold-sprayed coatings substrates onto aluminum alloy substrates and Ni-20Cr coatings onto Inconel718 substrates (Christoulis et al., 2012). Research direction currently and in the near future would be focused on investigating the suitability of the process for the deposition of hard facing materials and composite coatings which are currently processed via laser cladding. The process is equally being explored for the deposition on thin sections which is difficult with current processes due to likelihood of distortion arising from high solidification shrinkage.

5. CONCLUSION

Laser beam because of its unique characteristics is widely used in surface modification of different materials for improved functionalities. Though it is available in various classes yet each class do have specific area(s) where it is appropriately suited based mainly on the wavelength, the available power and the profile. For laser surface modification, both the continuous and pulse waves are applicable. Pulsed laser beam is mainly used for surface modification without injection of external elements like laser ablation, laser melting or remelting, whereas, the continuous wave whether CO_2 or, Nd:YAG or fibre based finds application in laser surface alloying and cladding. The success of laser surface modification is influenced by the combination of process parameters and material characteristics. In spite of the success achieved in laser surface modification, the process has some challenges which deserve consideration. In the face of this, there are new areas of research in laser surface modification with a view to achieving better understanding of the process. These new areas include development of gradient coating, amorphous coating, computational simulation and optimization and hybrid laser surface modification process.

REFERENCES

Abboud, J. H., Benyounis, K. Y., Olabi, A. G., & Hashmi, M. S. J. (2007). Laser surface treatments of iron based substrates for automotive application. *Journal of Materials Processing Technology*, *182*(1-3), 427–431. doi:10.1016/j.jmatprotec.2006.08.026

Ahmad-Pidani, R., Shoja-Razavi, R., Mozafarinia, R., & Jamali, H. (2013). Laser surface modification of plasma sprayed CYSZ thermal barrier coating. *Ceramics International*, *39*(3), 2473–2480. doi:10.1016/j.ceramint.2012.09.005

Akgun, O. V., & Inal, O. T. (1994). Laser surface modification of Ti-6Al-4V alloy. *Journal of Materials Science, Vl.*, *29*(5), 1159–1168. doi:10.1007/BF00975058

Ansari, S. A., Satar, R., Panda, D. S., Zaidi, S. K., Chibber, S., Khan, M. J., . . . Alqahtani, M. H. (2014). Surface engineering of multifunctional nanocomposites for biomedical applications: a brief update. Iranian Journal of Biotechnology, 12(1).

Baker, T. N. (2010). Laser surface modification of Ti alloys. In *Surface Engineering of Light Alloys - Aluminium, Magnesium and Titanium Alloys* (pp. 398–443). Woodhead Publications. doi:10.1533/9781845699451.2.398

Benyounis, K. Y., Fakron, O. M. A., Abboud, J. H., Olabi, A. B., & Hashmi, M. J. S. (2005). Surface melting of nodular cast iron by Nd-YAG laser and TIG. *Journal of Materials Processing Technology*, *170*(1-2), 127–132. doi:10.1016/j.jmatprotec.2005.04.108

Besner, S., Degorce, J.-Y., Kabashin, A. Y., & Meunier, M. (2005). Surface modifications during femtosecond laser ablation in vacuum, air and water. In *Proceedings of SPIE International Conference on Photonic Applications in Astronomy, Biomedicine, Imaging, Materials Processing, and Education*, (Vol. 5578, pp. 554-558).

Bianco, M., Tosto, S., Pierdominici, F., & Casaccia, C. R. E. (1995). CO2 gaseous alloying of Ti6Al4V by means of a pulsed laser. In T. S. Sudarshan & M. Jeandin (Eds.), *Surface Modification Technologies V111* (pp. 320–326). Institute of Metals.

Brown, M. S., & Arnold, C. B. (2010). Interaction and application to multiscale surface modification. In K. Sugioka, M. Meunier, & A. Piqué (Eds.), *Laser Precision Microfabrication* (pp. 91–120). Springer-Verlag. doi:10.1007/978-3-642-10523-4_4

Burakowski, T., & Wierzchon, T. (1999). *Development of surface engineering. In Surface engineering of metals: Principles, equipment and technologies* (pp. 7–40). Boca Raton, FL: CRC Press.

Cárcel, B., Serrano, A., Zambrano, J., Amigo, V., & Cárcel, A. C. (2014). Laser cladding of TiAl intermetallic alloy on Ti6Al4V: Process optimization and properties. *Physics Proceedia*, *56*, 284–293. doi:10.1016/j.phpro.2014.08.173

Chen, H., Xu, C., Qu, J., Hutchings, I. M., Shipway, P. H., & Liu, J. (2005). Sliding wear behaviour of laser clad coatings based upon a nickel-based self-fluxing alloy co-deposited with conventional and nanostructured tungsten carbide-cobalt hardmetals. *Wear*, *259*(7-12), 801–806. doi:10.1016/j.wear.2005.02.066

Choi, J. O., Kim, J. Y., Choi, C. O., Kim, J. K., & Rohatgi, P. K. (2004). Effect of rare earth element on microstructure formation and mechanical properties of thin wall ductile iron castings. *Materials Science and Engineering A*, *383*(2), 323–333. doi:10.1016/j.msea.2004.04.060

Christoulis, D. K., Jeandin, M., Irrissou, E., Legoux, J.-G., & Knapp, W. (2012). In D. C. Dumitras (Ed.), *Laser-Assisted Cold Spray (LACS) in Nd:YAG* (pp. 59–96). Rijeka, Croatia: InTech Publishers. Retrieved from http://www.intechopen.com/books/nd-yag-laser/laser-assisted-cold-spray-lacs

Cotell, C. M., & Sprague, J. A. (1994). ASM Handbook: *Preface, Surface Engineering* (vol. 5). ASM International.

Cunha, A. (2015). *Multiscale femtosecond laser surface texturing of titanium and titanium alloys for dental and orthopaedic implants*. (Doctoral thesis). University de Bordeaux, Bordeaux, France.

Dahotre, N. B. (1998). *Lasers in surface engineering, Surface Engineering Series* (Vol. 1). Materials Park, OH: ASM International.

Dahotre, N. B. (2002). Laser surface engineering. *Advanced Materials and Processes*, *160*(7), 35–39.

Dobrzański, L. A., & Drygała, A. (2008). Surface texturing of multicrystalline silicon solar cells. *Journal of Achievement in Materials and Manufacturing Engineering*, *31*(1), 77–82.

Earl, C., Castrejón-Pita, J. R., Hilton, P. A., & O'Neill, W. (2015). (Article in Press). The dynamics of surface modification. *Journal of Manufacturing Processes*. doi:10.1016/j.jmapro.2015.10.002

Emamian, A., Corbin, S. F., & Khajepour, A. (2011). The influence of combined laser parameters on in situ formed TiC morphology during laser cladding. *Surface and Coatings Technology*, *206*(1), 124–131. doi:10.1016/j.surfcoat.2011.06.062

Engeli, R., Etter, T., Hövel, S., & Wegener, K. (2016). Processability of different IN738LC powder batches by selective laser melting. *Journal of Materials Processing Technology*, *229*, 484–491. doi:10.1016/j.jmatprotec.2015.09.046

Frerichs, H., Stricker, J., Wesner, D. A., & Kreutz, E. W. (1995). Laser-induced surface modification and metallization of polymers. *Applied Surface Science*, *86*(1-4), 405–410. doi:10.1016/0169-4332(94)00431-5

Gabler, F. (2014). Broad range of lasers make material processing market dynamic. *Industrial Photonics*, *1*(1), 20–21.

Gebhardt, A., Schmidt, F.-M., Hötter, J.-S., Sokalla, W., & Sokalla, P. (2010). Additive manufacturing by selective laser melting: The realizer desktop machine and its application for the dental industry. *Physics Proceedia*, *5*, 543–549. doi:10.1016/j.phpro.2010.08.082

Gladush, G. G., & Smurov, I. (2011). *Physics of laser materials processing: Theory and experiments* (Vol. 146). Springer Series in Materials Science. doi:10.1007/978-3-642-19831-1

Gu, D. D., Meiners, W., Wissenbach, K., & Poprawe, R. (2012). Laser additive manufacturing of metallic components: Materials, processes and mechanisms. *International Materials Reviews*, *57*(3), 133–164. doi:10.1179/1743280411Y.0000000014

Ho, Y. H., Vora, H. D., & Dahotre, N. B. (2015). Laser surface modification of AZ31B Mg alloy for biowettability. *Journal of Biomaterials Applications*, *29*(7), 915–928. doi:10.1177/0885328214551156 PMID:25201909

International, A. S. M. (2001). Introduction to surface engineering for corrosion and wear resistance. In *Surface engineering for corrosion and wear resistance* (#06835G). Retrieved from www.asminternational.org

Ion, J. C. (2005). *Laser processing of engineering materials: Principles, procedures and industrial applications*. Oxford, UK: Elsevier Butterworths-Heinemann.

Katto, M., Nakamura, M., Tanaka, T., & Nakayama, T. (2002). Hydroxyapatite coatings deposited laser-assisted laser ablation method. *Applied Surface Science*, *197-198*, 768–771. doi:10.1016/S0169-4332(02)00411-7

Khandelwal, H., Singh, G., Agrawal, K., Prakash, S., & Agarwal, R. D. (2013). Characterization of hydroxyapatite coating by pulse laser deposition technique on stainless steel 316 L by varying laser energy. *Applied Surface Science*, *265*, 30–35. doi:10.1016/j.apsusc.2012.10.072

Kirkham, K. D., & Roundy, C. B. (2014). Current technology of laser beam profile measurements. In F. M. Dickey (Ed.), *Laser beam shaping: theory and techniques* (2nd ed.; pp. 464–524). London: CRC Press.

Krauss, G. (1992). Advanced surface modification of steels. *Journal of Heat Treating*, *9*(2), 81–89. doi:10.1007/BF02833144

Kwok, C. T., Cheng, F. T., & Man, H. C. (2000). Laser surface modification of UNS 31603 stainless steel. Part I: Microstructures and corrosion characteristics. *Materials Science and Engineering A*, *290*(1-2), 55–73. doi:10.1016/S0921-5093(00)00929-1

Lawrence, J., & Waugh, D. G. (2014). *Laser surface engineering: processes and application* (Vol. 65). London: Woodhead Publishing Series in Electronic and Optical Materials.

Lei, Y.W., Sun, R.L., Tang, Y., & Niu, W. (2012). Numerical simulation of temperature distribution and TiC growth kinetics for high power laser clad TiC/NiCrBSiC composite coatings. *Optical Laser Technology, 44*, 1141–1147.

Li, Z., Li, J., Liu, J., Liu, D., & Wang, H. (2016). Structure and formation mechanism of α/α interface in laser melting deposited α + β titanium alloy. *Journal of Alloys and Compounds*, *657*, 278–285. doi:10.1016/j.jallcom.2015.09.236

Lin, Y. C., Lin, Y. C., & Chen, Y. C. (2012). Evolution of the microstructure and tribological performance of Ti–6Al–4V cladding with TiN powder. *Materials & Design*, *36*, 584–589. doi:10.1016/j.matdes.2011.12.007

Liu, A., Li, B., Sui, Y., & Guo, J. (2012). Wettability of titanium on calcia-stabilized zirconia in electromagnetic field. *Materials Letters*, *89*, 299–301. doi:10.1016/j.matlet.2012.08.099

Luo, F., Yao, J. H., Hu, X. X., & Chai, G. Z. (2011). Effect of laser power on the cladding temperature field and the heat affected zone. *Journal of Iron and Steel Research International*, *18*(1), 73–78. doi:10.1016/S1006-706X(11)60014-9

Mahamood, R., Akinlabi, E. T., Shukla, M., & Pityana, S. (2013). Scanning velocity influence on microstructure, microhardness and wear resistance performance of laser deposited Ti6Al4V/TiC composite. *Materials & Design*, *50*, 656–666. doi:10.1016/j.matdes.2013.03.049

Mahamood, R., Akinlabi, E. T., Shukla, M., & Pityana, S. (2014). Improving surface integrity using laser metal deposition process. In L. Santo & J. P. Davim (Eds.), *Surface Engineering Techniques and Applications: Research Advancements* (pp. 146–176). IGI Global. doi:10.4018/978-1-4666-5141-8.ch005

Mahamood, R., Shukla, M., & Pityana, S. (2014). Laser additive manufacturing in surface modification of metal. In Surface engineering techniques and application: Research advancements. doi:10.4018/978-1-4666-5141-8.ch007

Majumdar, J. D., Manna, I., Kumar, A., Bhargava, P., & Nath, A. K. (2009). Direct laser cladding of Co on Ti–6Al–4V with a compositionally graded interface. *Journal of Materials Processing Technology*, *209*(5), 2237–2243. doi:10.1016/j.jmatprotec.2008.05.017

Meng, Q., Geng, L., & Ni, D. (2005). Laser cladding NiCoCrAlY coating on Ti–6Al–4V. *Materials Letters*, *59*(22), 2774–2777. doi:10.1016/j.matlet.2005.03.056

Mridha, S., & Baker, T. N. (1994). Crack-free hard surfaces produced by laser nitriding of commercial purity titanium. *Materials Science and Engineering A*, *188*(12), 229–239. doi:10.1016/0921-5093(94)90376-X

Ochonogor, O. F., Meacock, C., Abdulwahab, M., Pityana, S., & Popoola, A. P. I. (2012). Effects of Ti and TiC ceramic powder on laser-cladded Ti–6Al–4V in situ intermetallic composite. *Applied Surface Science*, *263*, 591–596. doi:10.1016/j.apsusc.2012.09.114

Olakanmi, E. O., Cochrane, R. F., & Dalgarmo, K. W. (2015). A review on selective laser sintering/melting (SLS/SLM) of aluminium alloy powders: Processing, microstructure, and properties. *Progress in Materials Science*, *74*, 401–477. doi:10.1016/j.pmatsci.2015.03.002

Pei, Y. T., & De Hosson, J. Th. M. (2000). Functionally graded materials produced by laser cladding. *Acta Materialia*, *48*(10), 2617–2626. doi:10.1016/S1359-6454(00)00065-3

Pinkerton, A. J., & Li, L. (2003). An investigation of the effect of pulse frequency in laser multiple-layer cladding of stainless steel. *Applied Surface Science*, *208-209*, 405–410. doi:10.1016/S0169-4332(02)01420-4

Qian, M., Lim, L. C., Chen, Z. D., & Chen, W. I. (1997). Parametric studies of laser cladding processes. *Journal of Materials Processing Technology*, *63*(1-3), 590–593. doi:10.1016/S0924-0136(96)02689-1

Roundy, C. B. (1998). Laser beam quality characterization. In *Proceedings Second GR-I International Conference on New Laser Technologies and Applications*. 10.1117/12.316569

Rytlewski, P., & Żenkiewicz, M. (2013). Applications of lasers in metallization of thermoplastic and thermosetting polymers. *Journal of Achievements in Materials and Manufacturing Engineering*, *57*(2), 59–67.

Selamat, M. S., Watson, L. M., & Baker, T. N. (2003). XRD and XPS studies on surface MMC layer of SiC reinforced Ti–6Al–4V alloy. *Journal of Materials Processing Technology*, *142*(3), 725–737. doi:10.1016/S0924-0136(03)00814-8

Shah, K., Haq, I. U., Khan, A., Shah, S. A., Khan, M., & Pinkerton, A. J. (2014). Parametric study of development of Inconel-steel functionally graded materials by laser direct metal deposition. *Materials & Design*, *54*, 531–538. doi:10.1016/j.matdes.2013.08.079

Sobczak, J. J., & Drenchev, L. (2013). Metallic functionally graded materials: A specific class of advanced composites. *Journal of Materials Science and Technology*, *29*(4), 297–316. doi:10.1016/j.jmst.2013.02.006

Stanford, N., Atwell, D., Beer, A., Davies, C., & Barnett, M. R. (2008). Effect of microalloying with rare-earth elements on the texture of extruded magnesium-based alloys. *Scripta Materialia*, *59*(7), 772–775. doi:10.1016/j.scriptamat.2008.06.008

Steen, W. M. (2003). *Laser materials processing* (3rd ed.). London: Springer. doi:10.1007/978-1-4471-3752-8

Steen, W. M., & Mazumder, J. (2010). *Laser materials processing* (4th ed.). London: Springer Berlin Heidelberg. doi:10.1007/978-1-84996-062-5

Sun, R. L., & Lei, Y. W. (2007). Microstructure and wear resistance of laser clad layer of TiN on TC4 alloy. *Journal of Tianjin Polytechnic University*, *26*, 57–59.

Thawari, G., Sundarararjan, G., & Joshi, S. V. (2003). Laser surface alloying of medium carbon steel with $SiC_{(p)}$. *Thin Solid Films*, *423*(1), 41–53. doi:10.1016/S0040-6090(02)00974-4

Tian, Y. S., Chen, C. Z., Li, S. T., & Huo, Q. H. (2005). Research progress on laser surface modification of titanium alloys. *Applied Surface Science*, *242*(1-2), 177–184. doi:10.1016/j.apsusc.2004.08.011

Tlotleng, M., Akinlabi, E. T., Shukla, M., & Pityana, S. (2014). Microstructures, hardness and bioactivity of hydroxyapatite coatings deposited by direct laser melting process. *Materials Science and Engineering C*, *43*, 189–198. doi:10.1016/j.msec.2014.06.032 PMID:25175204

Tlotleng, M., Akinlabi, E. T., Shukla, M., & Pityana, S. (2014). Application of laser assisted cold spraying process for materials deposition. In L. Santo & J. P. Davim (Eds.), *Surface Engineering Techniques and Applications: Research Advancements* (pp. 177–221). IGI Global. doi:10.4018/978-1-4666-5141-8.ch006

Tobar, M. J., Alvarez, C., Amado, J. M., Rodriguez, G., & Yanez, A. (2006). Morphology and characterization of laser clad composite NiCrBSi–WC coatings on stainless steel. *Surface and Coatings Technology*, *200*(22-23), 6313–6317. doi:10.1016/j.surfcoat.2005.11.093

Vilhena, L. M., Sedlaček, M., Podgornik, B., Vižintin, J., Babnik, A., & Možina, J. (2009). Surface texturing by pulsed Nd:YAG laser. *Tribology International*, *42*(10), 1496–1504. doi:10.1016/j.triboint.2009.06.003

Vora, H. D., & Dahotre, N. B. (2013). Laser surface heat treatment and modification. *Advanced Materials and Processes*, *171*(11), 45–47.

Vora, H. D., & Dahotre, N. B. (2013). Laser surface heat treatment and modification. *Advanced Materials and Processes*, *171*(11), 45–47.

Weng, F., Chen, C., & Yu, H. (2014). Research status of laser cladding on titanium and its alloys: A review. *Materials & Design*, *48*, 412–425. doi:10.1016/j.matdes.2014.01.077

Wu, W. (2000). Influence of vibration frequency on solidification of weldments. *Scripta Materialia*, *42*(7), 661–665. doi:10.1016/S1359-6462(99)00416-9

Yang, S., Li, W. H., & Man, H. C. (2013). Laser cladding of HA/Ti composite on NiTi alloy. *Surface Engineering*, *29*(6), 409–431. doi:10.1179/1743294413Y.0000000115

Yasa, E., & Kruth, J. (2011). Application of laser remelting on selective laser melting parts. *Advances in Production Engineering and Management*, *6*(4), 259–270.

Zhang, K. M., Zou, J., Li, J., Yu, Z. S., & Wang, H. P. (2010). Surface modification of TC4 Ti alloy by laser cladding with TiC+Ti powders. *Transactions of Nonferrous Metal Society of China*, *20*(11), 2192–2197. doi:10.1016/S1003-6326(09)60441-6

Zhang, L.-C., & Attar, H. (2015). Selective laser melting of titanium alloys and titanium matrix composites for biomedical applications: A Review. *Advanced Engineering Materials*. doi:10.1002/adem.201500419

This research was previously published in Advanced Manufacturing Techniques Using Laser Material Processing edited by Esther Titilayo Akinlabi, Rasheedat Modupe Mahamood, and Stephen Akinwale Akinlabi, pages 92-120, copyright year 2016 by Engineering Science Reference (an imprint of IGI Global).

APPENDIX

Figure 1. Classification and typical surface depth of various surface modification techniques (ASM International, 2001)

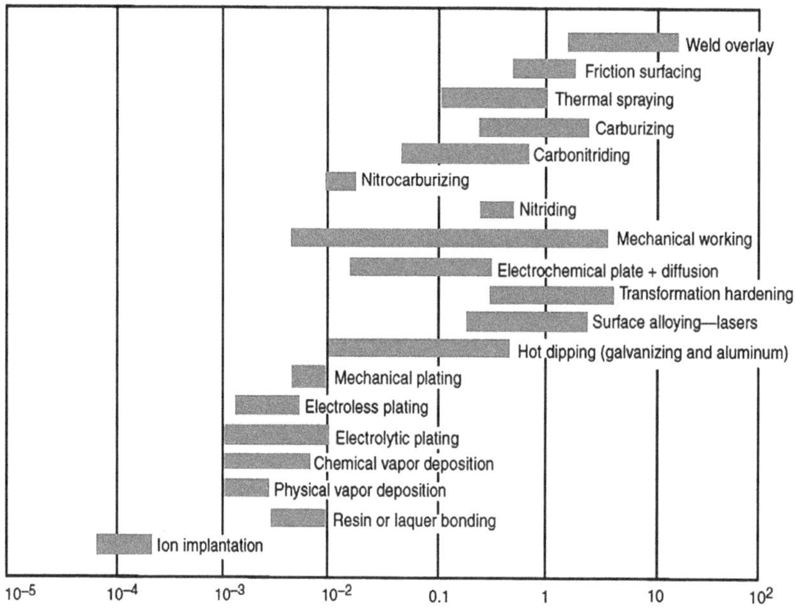

Figure 2. A chart of commercial lasers in terms of average power and wavelength on a background of application windows (Baker, 2010)

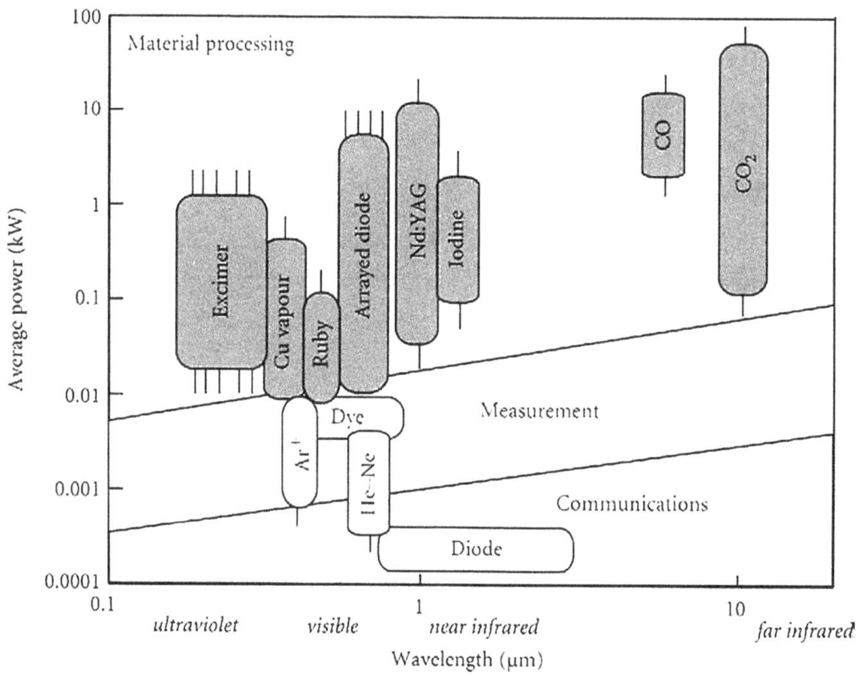

Figure 3. Map of laser processes as a function of power density per unit interaction time (adapted from Dahotre, 2002)

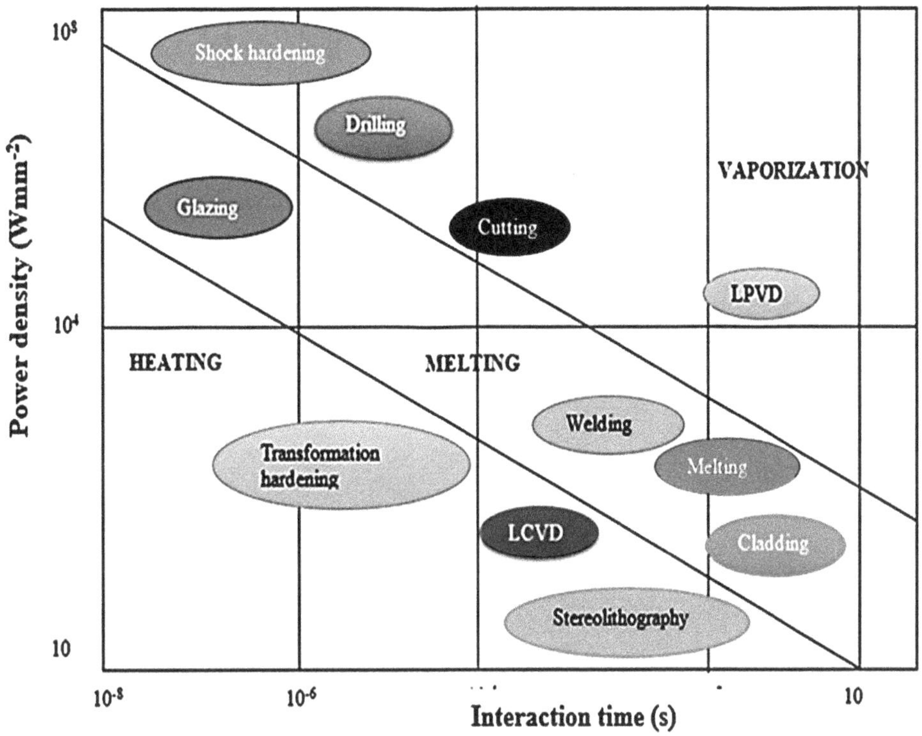

Figure 4. Common laser beam profiles: (a) Gaussian profile for highly concentrated energy distribution and (b) flat-top beam for uniform energy distribution (Roundy, 1998)

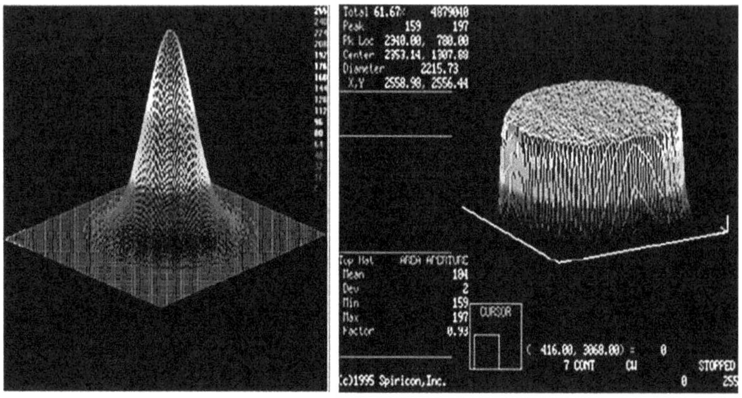

(a) (b)

Figure 5. Dendritic microstructure in laser surface melted: (a) cast iron and (b) Ti-6Al-4V titanium alloy (Benyounis et al., 2005; Akgun and Inal, 1994)

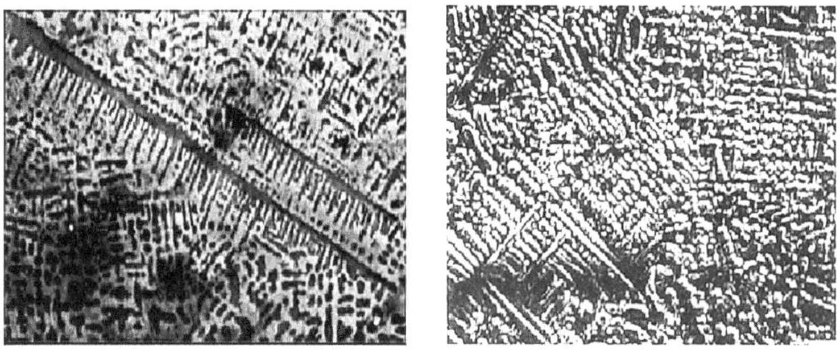

(a) (b)

Figure 6. Cross-sectional micrograph of laser alloyed layers revealing the influence of repetitive scanning: (a) single scan, (b) two-fold scan, and (c) four-fold scan (Thawari et al., 2003)

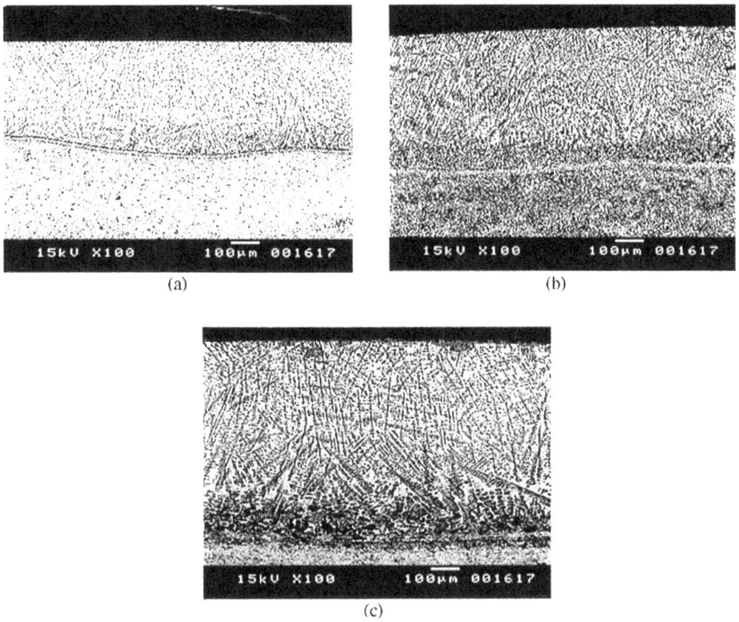

Figure 7. Schematic illustration of common laser cladding processes: (a) powder synchronous feeding system, (b) powder preplacement method, and (c) wire-feeding method (Mahamood et al., 2013; Gabler, 2014)

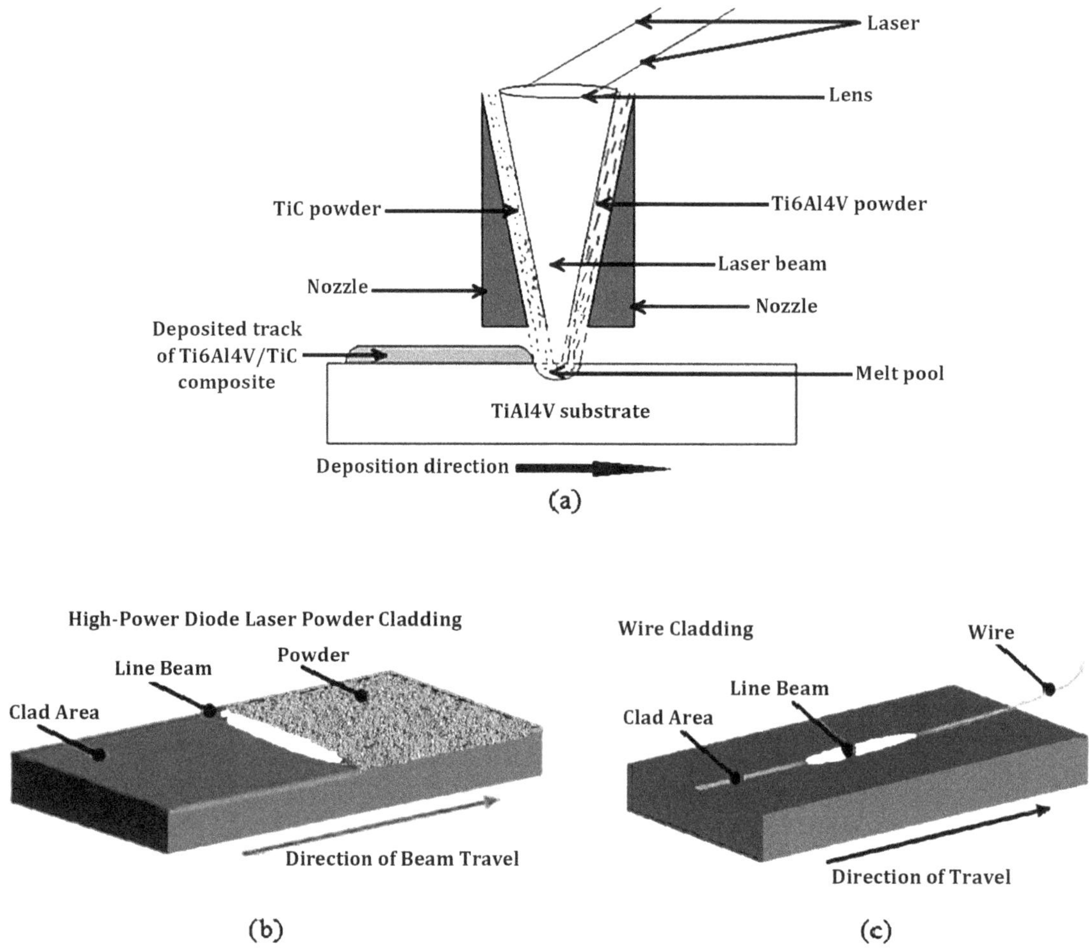

Figure 8. Application window for laser surface modification technique (Vora and Dahotre, 2013)

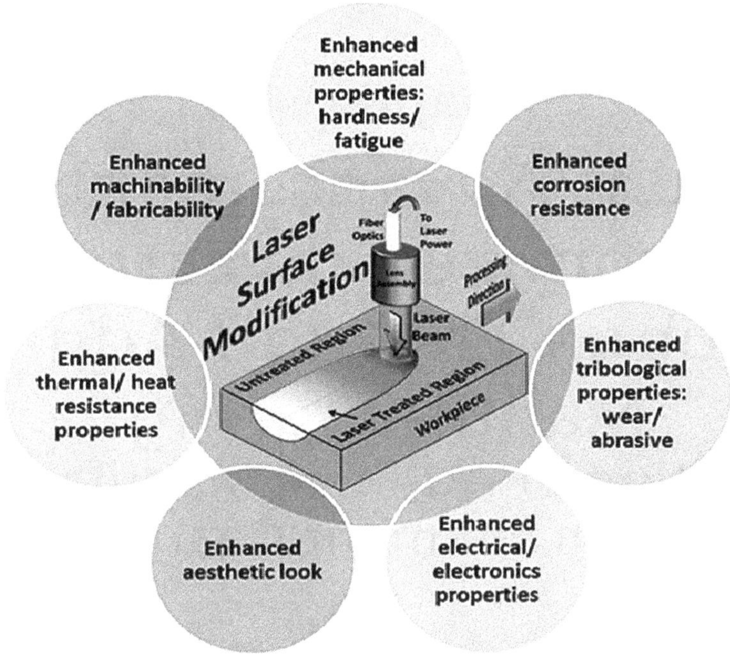

Chapter 12
Recent Advancements in Customized Investment Castings Through Additive Manufacturing:
Implication of Additive Manufacturing in Investment Casting

Sunpreet Singh
Lovely Professional University, India

Chander Prakash
Lovely Professional University, India

M. Uthayakumar
Kalasalingam University, India

ABSTRACT

Conventional investment casting (IC) has suffered from numerous limitations such as rigidity of the process, longer production cycles, higher tooling cost, and waste during different manufacturing stages. With the invent of additive manufacturing (AM) technologies, it is now possible to overcome the aforesaid issues along with additional benefits in terms of comparatively better quality characteristics of the resulting castings. The collaboration of AM and IC provided numerous avenues, specifically in biomedical, aerospace, and automobile sectors. AM technologies supported the IC process both in direct and indirect ways where these systems can be used for both job and mass production applications, respectively. In the chapter, the author will try to discuss the assistance of AM process to IC in detail. Each and every step to be followed will be supported with the practical findings, either by the contributing author or published somewhere else. Moreover, some of the case studies will be discussed in detail to highlight the practical importance of the duo.

DOI: 10.4018/978-1-5225-9624-0.ch012

INTRODUCTION

The casting industry is a vital segment of the manufacturing sector which produces intricate parts, blends, internal features and varying thickness with excellent surface finish with negligible metallurgical limitations, (Pal et al., 2002). In this today's manufacturing era, we rely on conventional investment casting (IC) process for hundreds and thousands of objects that we need for various end user applications. This method of casting is one of the most ever-present manufacturing processes we are utilizing, since 300 BC, before our civilization. The ancient expertise in casting resided among several cultures such as Mesopotamians, Egyptians, Chinese, Thai and Indians (Greer, 2008). However, the basic principles of IC can be traced back to 5000 BC when the early-men employed this method to produce rudimentary tools (Taylor, 1983), followed by jewellery and artistic products (Träger & Bührig-Polaczek, 2002), and the development of aerospace and subsequently engineering components during the 2nd world war (Barnett, 1988). The IC process, often known as lost-wax or precision casting process, is one of the manufacturing processes that have capability to produce high accuracy, intricate shapes and fine impression with tight geometrical tolerance. This is an alternative approach to produce parts that are hard to machine (Singh & Singh, 2016). IC process has numerous applications that includes agricultural equipments, automobile components, aircraft engines, air frames, fuel systems, computer hardware, electrical equipment, electronic hardware and radar, making jewellery, statues and art castings, dentistry and dental tools, prosthetics, guns and armament, hand tools, machine tool components etc. (Pattnaik et al., 2014). The working procedure of conventional IC process is given in Figure 1. The first step of IC process: (a) the injection of molten wax (at about 150°C) inside the precisely machined metallic die (usually split one) with help of an injection machine; (b) patterns are then safely ejected out of the metal cavity; (c) patterns let for strengthening in air conditioned environment so that the geometrical dimensions get fixed, afterwards, the pattern is assembled with another wax assembly that consisted of pouring sprue, runner and gate as integral parts; (d) this tree like structure is dipped into clay and stucco coatings are performed repeatedly to get the required mould is attained (e); (f) de-waxing is carried out to create a hollow cavity; (g) baked to increase hot strength; (h) molten metal is poured into the cavity and part is separated from the tree once solidify and (i) finally the refractory layers are removed on pneumatic vibration system (Singh & Singh, 2013, 2016). The key requirements of an investment casting mould are (Jones & Yuan, 2013): sufficient green (unfired) strength, sufficient fired strength, sufficiently weak to prevent hot-tearing, high thermal shock resistance, high chemical stability, and low reactivity with the metals, sufficient mould permeability and thermal conductivity, low thermal expansion to limit dimensional changes.

Problems associated with ceramic shell materials have been exacerbated following the introduction of the Environmental Protection Act (Jones & Yuan, 2013), wherein fix degree of emissions allowed and this has boosted the use of water-based shells within industry. Moreover, improvements in the quality, reduction in manufacturing costs and explores new markets for the process are the hot topics amongst current research area. Along with this, numerous tools and software are currently being used for the optimization of the mechanical and physical properties at various stages of the casting. The troubles like: long lead time and high tooling costs for low-volume IC production runs is still present in the conventional process. Wax, as pattern material, is suspect of shrinkage that affects the dimensional and surface features of the casting, and various researches has been carried out to control the shrinkage rate (Taşcıoğlu & Akar, 2003; 2007). However, the critical barrier in prototype development and production time could be easily reduced with the introduction of additive manufacturing (AM) technologies. The AM technologies allowing the designer to work more comfortably with the geometrical complexities,

Figure 1. Pictorial description of IC process
(Singh & Singh, 2013).

control over the dimensional tolerances and also to minimize the involved production times (Bidanda & Bártolo, 2007; Kumar & Kruth, 2010; Kruth et al., 2007).

Although traditional machining processes can be used for the development of desired pattern but in case of unusual shape or specific internal features, the machining becomes more difficult. AM allows the production of models directly from engineering designs. AM systems are limited only by the size of the model and not by its complexity (Gibson et al., 2010).

ADDITIVE MANUFACTURING TECHNOLOGIES

The AM technologies (refer Figure 2), often known as three dimensional printing (3DP) or rapid prototyping technology (RPT) or solid freeform fabrication (FFF), is a technology in which materials are correctly printed layer upon layer, converting digital 3D models into solid objects (Wendel et al., 2008; Chia & Wu, 2015; Tseng et al., 2014; Murphy & Atala, 2014). This is an emerging technology in manufacturing industry and this able to turn digital data into physical parts for different applications. In recent years, there has been a dramatic progress in 3D-printing technologies due to investments by governments and companies around the world. This emerging manufacturing technique comprises mainly of fused deposition modelling (FDM), selective laser sintering (SLS), laminated object manufacturing (LOM), and

Figure 2. Schematic for the AM process chain

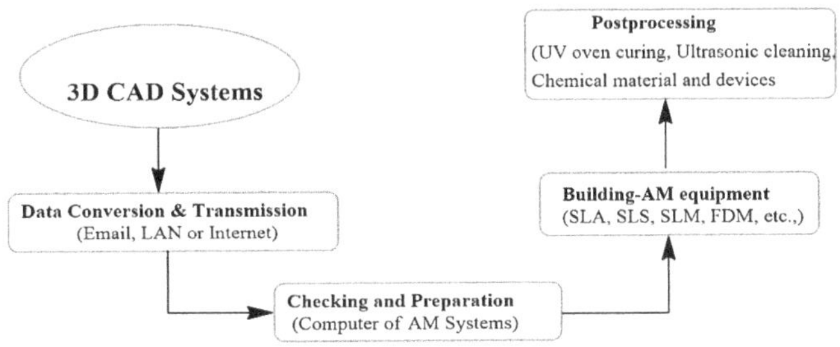

stereo-lithography (SLA). Depending on these manufacturing technologies, a 3D form of the material is created by precisely "printing" it layer by layer. The tangible parts could be created from the designed parts and it acts as a platform for an individual to get their requisite design. The three dimensional (3D) model can be drawn on different software systems, transmitted to AM building equipment and printed into a physical object (Figure 2). Using AM technologies, a rapid pattern substituted for conventional IC wax pattern can be easily prepared, with sustainable mechanical properties, as a wide range of feedstock materials are available that can be processed through different types of AM technologies to assist IC process. Acrylonitrile-butadiene-styrene (ABS), high density polythene (HDPE), low density polythene (LDPE), polyamide, poly-lactic-acid, rubber, resins, wax, photo curable polymers & resins, polymer composites, etc. (Cheah et al., 2015; Guo & Leu, 2013), are majorly used for rapid fabrication of sacrificial patterns. The ceramic moulds are generally baked in the furnace to evaporate these patterns for sufficient time duration, therefore chance of formation residuals are high and these can affect the chemical composition of the casting. The case may become worst when developing a biomedical device. According to Wang and Miranda 2010, carbon residues are mainly resulted due to incomplete combustion of the pattern material (Rosochowski & Matuszak, 2000). Sivadasan et al., 2012 found that in black tar and ash formed after flame burning of the moulds was negligible and can easily be removed by passing compressed air in the mould cavity (Chhabra & Singh, 2011). Further, it would be better to fabricate AM patterns at low density and appropriate design as thermal expansion during burn-out step can induce stresses in the mould and lead to breakage (Omar et al., 2012; Harun et al., 2013). Apart from all the limitations, AM systems are highly beneficial for low and medium volume production runs (Hague et al., 2004). Steps involved in creating a physical prototype with most of these technologies is same and includes: (1) create a CAD model of the design, (2) convert the CAD model to STL format, (3) slice the STL file into thin cross-sectional layers, (4) construct the model one layer atop another and (5) clean and finish the model.

AM technologies have been mainly used for (Madrazo et al., 2009; Dedoussis et al., 2008; Dedoussis et al., 2004; Lee et al., 2004):

- Physical verification of CAD model
- Creating models without regard to draft angles, parting lines, etc.
- Concept presentations and design reviews
- Direct tooling
- Rapid tooling such as investment casting and silicone moulding
- Reducing time-to-market

- Creating anatomical models
- Producing relief models for geographical applications
- Creating 3D portraits

Additive manufacturing technologies can either assist IC process through a direct route or an indirect route (Bártolo & Gibson, 2011), as discussed below:

Direct Route of IC Pattern Making

In the direct route, the rapid patterns are directly used in IC process with minor or negligible intermediate processing. Many technologies can be employed here as given below:

Fused Deposition Modelling

The FDM, an extrusion based AM technology that, utilizes granular or pelletized feedstock with a screw-type extruder. The workhorse materials are usually the amorphous thermoplastic polymer filaments with a diameter of about 1.5-3mm (Turner et al., 2014). In the feed system, the filament is pushed through the system using a pinch roller mechanism like that illustrated in Figure 3(a). A stepper motor is connected to one of the rollers providing energy to move the filament through the system (Agarwala et al., 1996). One or both of the rollers may have a grooved surface like a gear to create sufficient friction. The pressure on the filament, between the rollers, is typically sufficient to slightly deform the filament, but these are designed to avoid crushing the filament. In FDM, the absorption of moisture by the thermoplastic filament raises significant problems when printing parts in an extrusion-based process. The material when extruded from the heated head the moisture vaporize and often lead to morphological changes in the material, blockages (Halidi & Abdullah, 2012). In FDM, layer thickness of 200-500pmm and resolution levels for the positioning of the nozzle of 25pm in x, y plane in line dimensions are achieved based on nozzle diameter (Turner et al., 2014). This method attempts to develop the structures with sharp overhangs or long, unsupported sections due to tough fabrication procedures. However, the extruded filaments have poor material strength during the extrusion process and thus result in drooping or total collapse of the unsupported segment. Filler materials used to support cavities, overhanging and lifted structures could be removed post printing have been developed to eliminate this issue (Chia & Wu, 2015). This extrusion type machine is a specific example to demonstrate the accelerated progress in the development toward a wide availability of 3D printers at an affordable cost; these include servo drives, precision rails, and various electronics items such as computers and software which deal with difficult geometric data.

Stereo-Lithography

The SLA system (refer Figure 3(b)) uses an ultraviolet laser beam to cure a photosensitive monomer resin layer upon layer in order to build prototypes having complex shapes. In the SLA process, a tessellated STL file is first sliced into very thin cross sections, as in case of other instruments. This technology uses a computer-controlled laser to cure a photosensitive epoxy or acrylate resin layer by layer to create a 3D part (Chockalingam et al., 2006). The prototype is built on a platform positioned just under the exterior of fluid epoxy or acrylate resin enclosed in a vat. A low-power highly focused UV laser beam traces

Figure 3. Schematic of FDM (a) (Turner et al., 2014), SLA (b) (Chockalingam et al., 2006), SLA (Kruth et al., 2005) and LOM (Kruth et al., 2003).

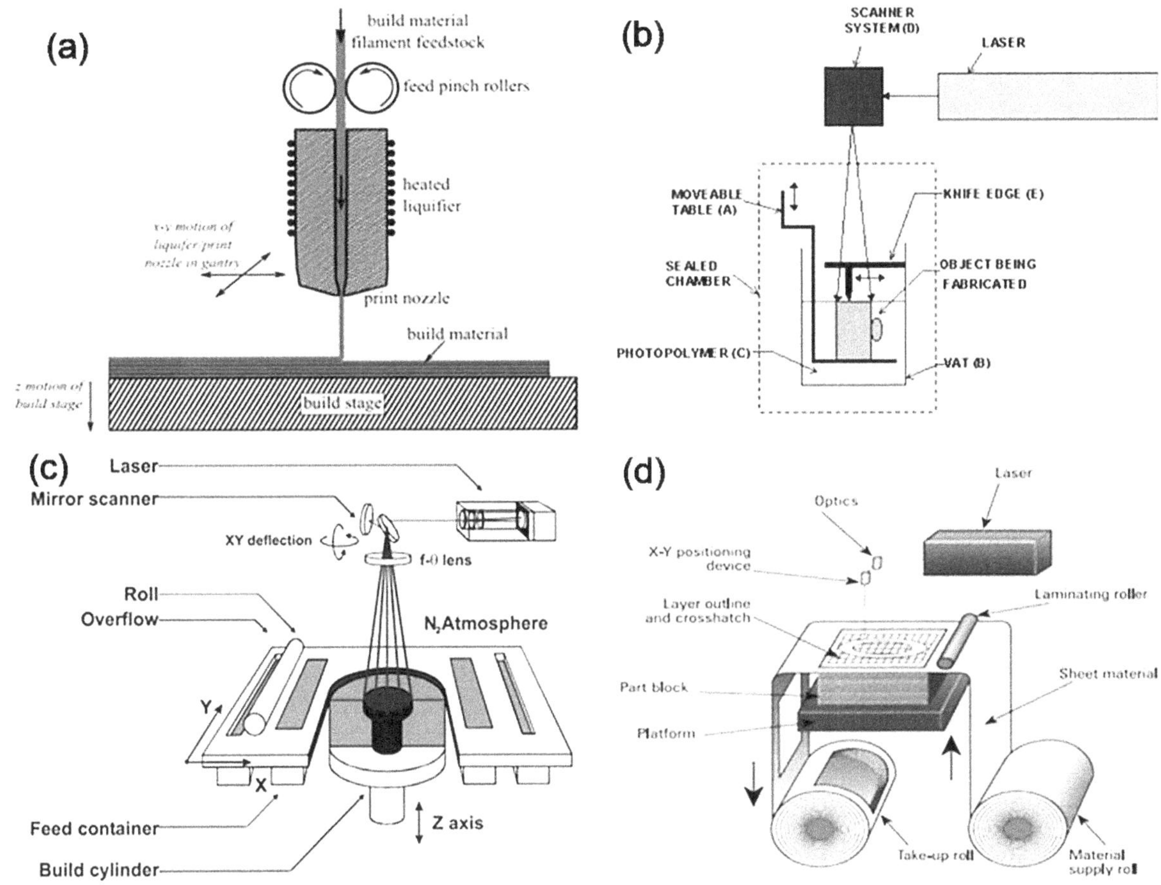

out the first layer, selectively solidifying the resin while leaving the excess area liquid. The elevator incrementally lowers the platform into the vat by a distance equal to the layer thickness value chosen. The photo-curable resin is relatively viscous and does not permit the liquid to instantly coat the whole of the upper surface of the cured part in a consistent way upon moving elevator downside. A recoating mechanism is therefore required to facilitate this process, and a wiper arm traverses over its surface to speedily level the surplus viscous material. The process is continued till the final layer is built. The solid part is removed from the vat and rinsed to remove excess liquid. In the post-curing process, the part is placed in a UV oven to complete the curing after nearly 97% of liquid polymer gets solidified during the build process.

Selective Laser Sintering

The SLS is another type of AM process that allows the manufacturing of both polymer and metallic based workhorse materials (both in powder form) to generate complex 3D parts (Kruth et al., 2005). In this system, the consolidation is obtained by processing the selected areas lying within the geometry of the parts by using highly localized laser beam. A beam deflection system (generally galvano mirrors) scans individual layers as per the cross section, calculated from the CAD model. Figure 3(c) shows a schematic description of SLS setup. There is a wide range of commercially available SLS machines that basically differ in the way the powder is deposited (i.e. roller or scraper), the atmosphere (argon or nitrogen) and in the laser type (CO_2 laser, lamp or diode pumped Nd:YAG laser, disk or fibre laser) (Kruth et al., 2003). The printing is based on hardening or sintering of powders on the platform in terms of successive layers. SLS is one of the common examples of powder bed fusion techniques. These techniques are used to develop industrial prototyping suitable to print polymers, ceramics, metals and combination of these materials into multifaceted and distinctive geometric shapes (Turner & Gold, 2015). The different forms of the powder such as metal or ceramic and thermoplastic polymer are sintered or hardened using a laser and each of the constructed platforms is lowered (up to a pre-defined level) and an additional powder layer is applied and sintered. There is no waste in this manufacturing process, as the un-sintered powder used for building support structures can be re-used for next printing. Appropriate ink materials for SLS have to be a powder ranging from 10 to 100pm, with a reasonable melting point. An additional property is that the particles must have excellent particle flow rate in the bed surface, which is essential to reduce electrostatic forces and surface fictionalisation (Vermeulen et al., 2013). Usually, SLS devices are bulky, sluggish, costly, and necessitate an enormous quantity of material. But the capability to deal with multiple materials in the single bed has been confirmed as most effective in various manufacturing industries. This technique follows similar principles as SLS except for the use of the liquid binding solution for binding of particles within each layer than the use of melting particles along with laser beam in SLS. This technique is widely utilized to design devices with applications in different areas of rapid tooling (Yang et al., 2009).

Laminated Object Manufacturing

The LOM (refer Figure 3(d)) was ever defined in a narrow sense in AM classification. A generalized LOM may be defined here in a wider sense that includes classical LOM, functionally graded materials and coatings on substrate material. LOM has some common features in structures and thermo-elastic responses. It needs high manufacturing temperature and/or uses more than one material, which result in thermal residual phenomenon and reduce workpiece quality (Mekonnen et al., 2016). Effective control of thermal residual phenomenon is a main goal to improve LOM techniques. It allows obtaining desired shapes by sticking successive sheets of a certain rolled material. Each layer is cut with a tool, allows removing the undesired material (Shu & Wang, 2017). The layer is cut with CO2 laser beam to a cross-section determined by CAD file and it cuts the excess material (Das et al., 2003). The excess material provides support for subsequent layers. Originally, LOM was used to make laminated paper models (Park et al., 2000); however now functional parts out of metals, ceramics and polymers are also build using it (Marchelli et al., 2011; Chiu et al., 2006). Literature demonstrated that LOM can successfully produce functional parts and patterns, for its dimensional accuracy (Yi et al., 2004), and low surface average roughness (Ahn et al., 2012). The device, its positioning control systems and dispensing glue system are similar to a traditional printer. When building is completed, a compact layered block is formed.

Indirect Route of IC Pattern Making

In case of indirect route of AM assisted IC process, the master pattern prepared from the printer is used in replication processes, for example silicon moulding, from where number of wax replicas can be generated for IC process. Figure 4 shows the classification of indirect route of pattern replicating process. This collaboration of AM and silicon moulding further speed up the process and cuts the cost, significantly.

Silicon Rubber Tooling

In this process, a liquid silicone and hardener mixture is cast/mould around AM pattern contained in a box, as shown in Figure 5. Initially master patter in hanged inside a plastic, wooden or metallic split box. The runner and gating channels are incorporated by embedding ABS or Perspex rods into the liquid silicone or by cutting the channels in the cured silicone block. Upon curing, the rods are removed to form through channels and the pattern is subsequently removed to form the mould cavity by cutting along the parting line (Cheah et al., 2005). Silicone rubber tooling allows quick production of inexpensive multiple moulds for small and large parts with good part cosmetics (Smith et al., 1999). One mould box can be used for the production of patterns 100-300; however the maximum reported are 1000 wax patterns (Vickers, 2001). Zhang et al. (1999) produced ice patterns at Tsinghua University for silicone moulding of wax pattern production.

Epoxy Tooling

In this process, the AM pattern is surrounded in clay up to the pre-determined parting line and integral runner and gating channels are created by attaching ABS rods on the pattern.

Then the assembly is spray coated with release agent before liquid resin or powder mixture is cast around the pattern. The system is left for curing and then the pattern is separated from the hardened mould half, cleaned and re-coated with release agent in preparation for casting the second mould half (Cheah et al., 2005). Holes form the alignment and locking features for the mould whereas matching pegs form on the second mould half occupying the empty volume of these holes during casting. Epoxy resin mould has been successfully utilized for both plastic and wax injection moulding (Warner 1993). The metal filled epoxy tooling increases the durability and heat transfer characteristics of the mould, for example, PolySteel moulds are reportedly much stronger than aluminium resin tooling.

Figure 4. Classification of indirect pattern making

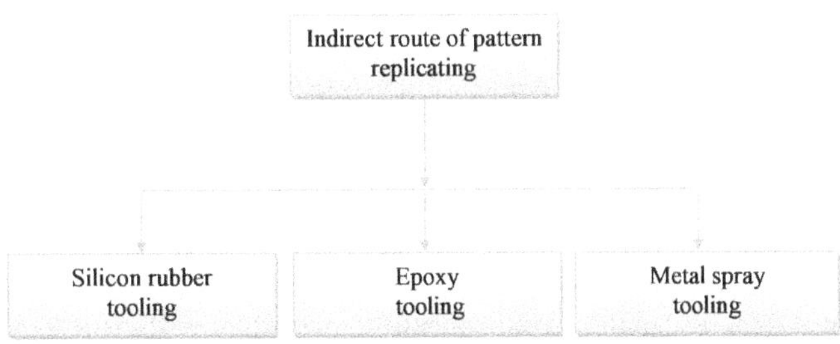

Figure 5. Description of silicon moulding process
(image courtesy: Fracktal Works).

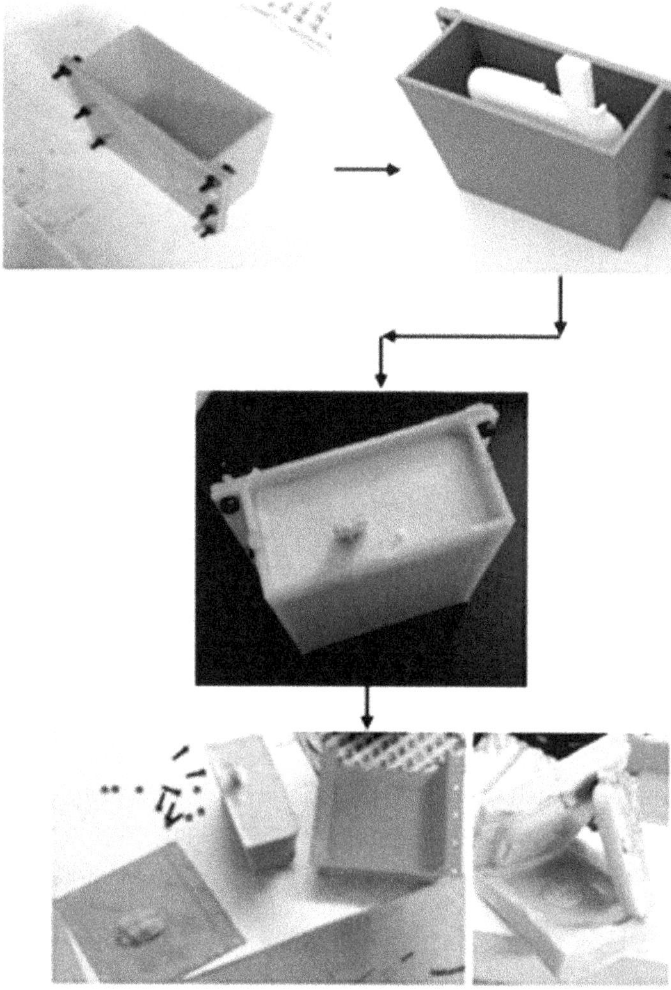

Metal Spray Tooling

To create a spray metal mould, the tiny droplets of tin-zinc or steel melt are sprayed onto AM pattern using an arc spray process (Chua et al., 1999). Further to strengthen the shell, a solid backing is cast around the shell using pure epoxy, metal-filled epoxy or low-melt alloy backfill materials (Cheah et al., 2005). With a proper choice of backfill material and the incorporation of cooling channels, a spray metal mould exhibits good injection cycle times. This process offers numerous advantageous such as low costs, high tolerances and good tooling life of approximately 10,000-100,000 injections (Heine et al., 1955). Spray metal tools have been used in many applications and various plastics have been molded including polypropylene, ABS, polystyrene and difficult process materials such as reinforced nylon and polycarbonate. Spray metal tooling processes were developed by Ford's Spray-Form (Chalmers, 2001) and INEEL rapid solidification process tooling (Knights, 2001).

Figure 6. Methods of improving surface finish of AM parts.

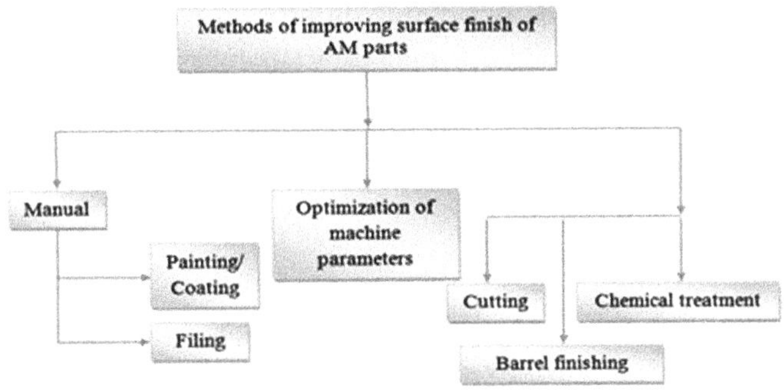

TREATMENTS OF AM PATTERN SURFACES

Poor surface quality of finally produced AM based IC patterns is one of the most critical barrier restricting the commercialization of AM systems within foundries. All the different technologies are working on a same principle of layer manufacturing; hence it is difficult to avoid the stair-case stepping (Novakova-Marcincinova, 2012). The final casted product will occupy the same surface and geometrical features as of sacrificial pattern, thereby previous study highlighted the use of post processing of the AM patterns prior to casting. There are many methods that can be employed for improving the surface topography as given in Figure 7. Table 1 show the various research attempts in order to improve the surface finish of AM parts.

Table 1. Various attempts to improve surface finish of AM parts.

S. No.	AM System	Material	Method Adopted	Ref.
1	FDM	ABS	Chemical vapor treatment	Singh et al., 2017; Singh & Singh, 2017
2			Barrel finishing	Boschetto & Bottini, 2015; Boschetto et al., 2016
3			Parametric optimization	Vijay et al., 2011; Lan et al., 1997; Alexander et al., 1998
4			Hot cutter machining	Pham et al., 1999
5	SLA	Polyethylene wax emulsion	Micro-ultrahigh pressure atomizing coating	Yang et al., 2017
6		E-shell and SI500	Process optimization-meniscus approach	Kunjan et al., 2016
7		CIBA TOOL 5530	Process optimization	Wood, 2016
8		Acrylats, epoxies and vinylethers	New layerless fabrication strategy	Onuh & Hon, 1998
10	SLA	Polyamide 2200	Parametric optimization	Bacchewar et al., 2007
11		Ciba Geigy XB5143	Mass burnishing media finishing	Delfs et al., 2015
12		Silicone, polyurethane, Vinyl-acryl polymers and divers	Vibratory grinder	Schmid et al., 2009
13		Polyamide 6 and polyamide 12	Process parameter optimization	Arjun et al., 2012
14	LOM	Poly(vinyl chloride)	Machine parameter regulation	Pilipović et al., 2011

It has been found from the literature review that the required level of surface finish is achievable through several different attempts including: optimizing process parameters (such as part orientation, built orientation and layer thickness), machining operation (like: CNC machining), chemical machining and barrel finishing laser surface finishing operations.

CASE STUDIES

In this section, a brief discussion of few interesting studies of the various AM technologies assisted IC process has been made in order to outline the feasibility of the combined route for attaining better quality characteristics in-terms of the finally obtained component.

FDM Assisted IC

FDM-ABS Patterns for Investment Casting Process (Harun et al., 2009)

The authors established the physical and collapsibility characteristics of H shaped FDM pattern produced as hollow and solid as shown in Figure 7. They have evaluated the various properties such as roughness, geometrical and distortion occurred and established these characteristics. The surface roughness was recorded highest at the interface of support and model material as compared to other surfaces, and highest roughness value of 32.13μm. Other surfaces of both solid and hollow patterns were having a constant trend between 17.00μm to 19.00μm. The lowest surface roughness was recorded as 6.17μm in case of hollow patterns. In case of dimensional accuracy, the highest and least percentage of deviation found at location 22 and 2, respectively, as 1.23% and 1.14%. Distortion results refer Figure 8, showed that hollow construction distorted 33.11% higher solid patterns. Hollow specimens recorded lower deviation value than solid. It is suggested that the higher percentage of deviations occurred on solid patterns as compared to hollow pattern. From the graph plotted in Figure 9, 90% of the patterns were observed to be burned-off between 300°C and 400°C. However, the complete burn-off occurred at 500°C and ash free burn was visible above 550°C. At 300°C, 400°C, 450°C and 500°C, some cracks were observed on the shell of the solid pattern at the burning temperatures of due to expansion.

Overall, it has been found that hollow patterns were proved as viable in-terms of dimensional accuracy, mould cleanliness, pattern collapsibility and no mould cracking at all temperatures.

Figure 7. H shaped FDM pattern; (a) solid and (b) hollow

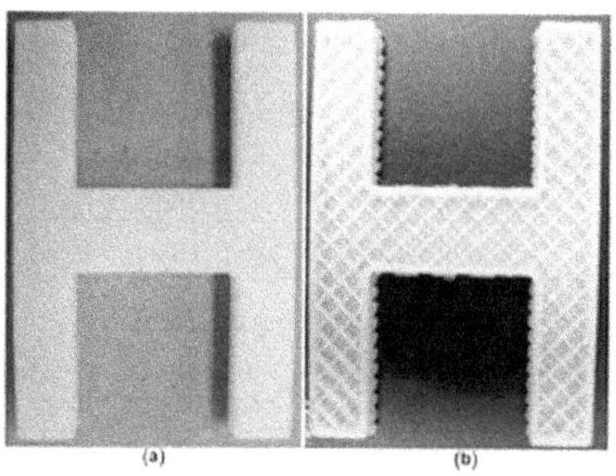

Figure 8. Distortion plot

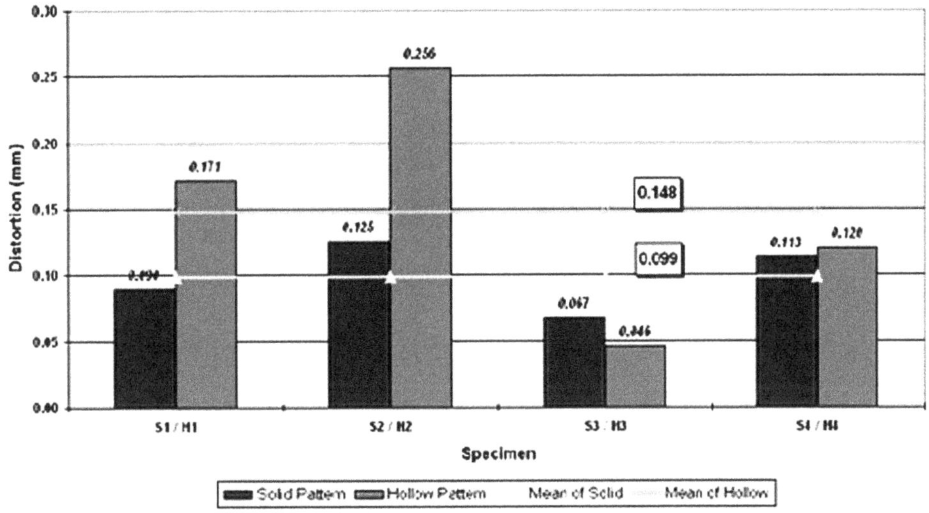

Figure 9. Collapsibility analysis of pattern

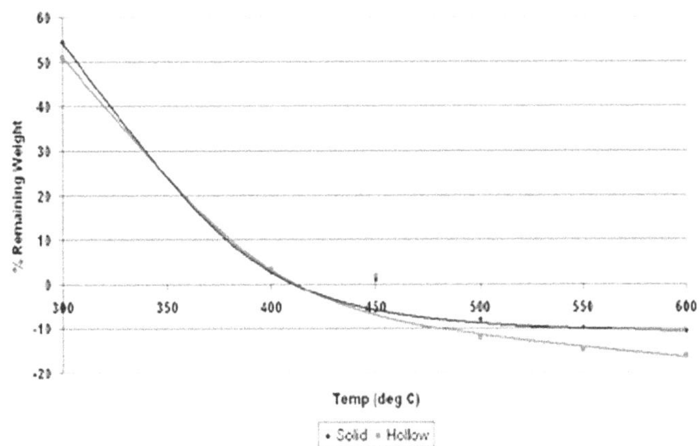

Surface Finish of Biomedical Implant Fabricated by Rapid Investment Casting (Singh et al., 2017)

In this study, the authors have verified the feasibility of four stage combined processes that includes FDM, chemical smoothing, vacuum casting and IC for the development of customized implant with appropriate surface and dimensional accuracy. They have first conducted the pilot study first and proceeded towards IC with the best settings as given in Table 2. Table 3 shows the third stage, i.e. IC stage, as per Taguchi L9 array and Figure 10 shows the S/N response. Table 2 Input parameters and their level.

From Figure 10, it has been found that accuracy decreased within crease in mould wall thickness and was least at 9 layers. In case of 7 layers, accuracy was highest because of less thickness of mould wall that resulted into higher rate of heat transfer and thereby shrinkage occurs. The tolerance grades for

Table 2. Input parameters and their level

Stage	Process	Parameter	Level 1	Level 2	Level 3	Optimum Setting
1	FDM	Orientation (degree)	0	90	-	90
		Part density	Low	Medium	High	High
	Chemical smoothing	Number of cycle	3	4	5	5
		Cycle time (sec)	4	6	8	4
2	VC	PU material	6130	PX223	-	PX223
		% Hardener	90	95	100	100
		Curing temperature (degree)	70	80	90	70
		Curing time (sec)	45	55	65	45

Table 3. Control log of experiment

S. No.	Dry Time (min)	Number of Layers
1	330	7
2	330	8
3	330	9
4	360	7
5	360	8
6	360	9
7	390	7
8	390	8
9	390	9

Figure 10. S/N response to dimensional accuracy (a) and surface roughness (b)

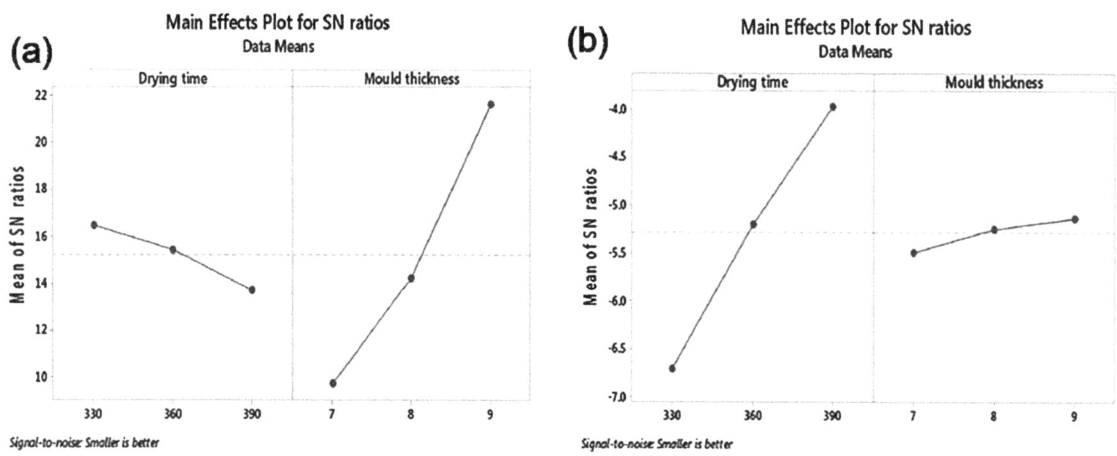

the dimensions obtained were found to be acceptable as per the ISO standard UNI EN 20286-I (1995). Moreover, this combined route is statistically controlled as the resulting Cpk value is 1.44, means non-confirming parts per million are just 20.27.

Hardness of Al–Al_2O_3 Composite Prepared Using Alternative Reinforced FDM Pattern (Singh & Singh, 2016)

Here in this work, FDM assisted IC route was explored for the development of Al-Al2O3 based functionally graded material. Initially, alternative reinforced FDM filaments were fabricated on single screw extruder (refer Figure 11a) by using commercial available Al2O3 powder, Al and nylon-6 granules as filament ingredients. Then cubical patterns (shown in Figure 11b and 11c) of volume 17576mm3, 27000mm3 and 39304mm3 were prepared with FDM at low density, high density and solid density. Barrel finishing was also performed on the resulting patterns in order to improve their surface finish and used for IC process.

Taguchi method used to optimize input parameters without increasing the manufacturing. It has been found that increasing Al2O3 content in FDM filament have increased the hardness value of the casted specimens.

Figure 11. Extrusion of filament (a), prototyping of pattern (b) and reinforced prototypes (c)

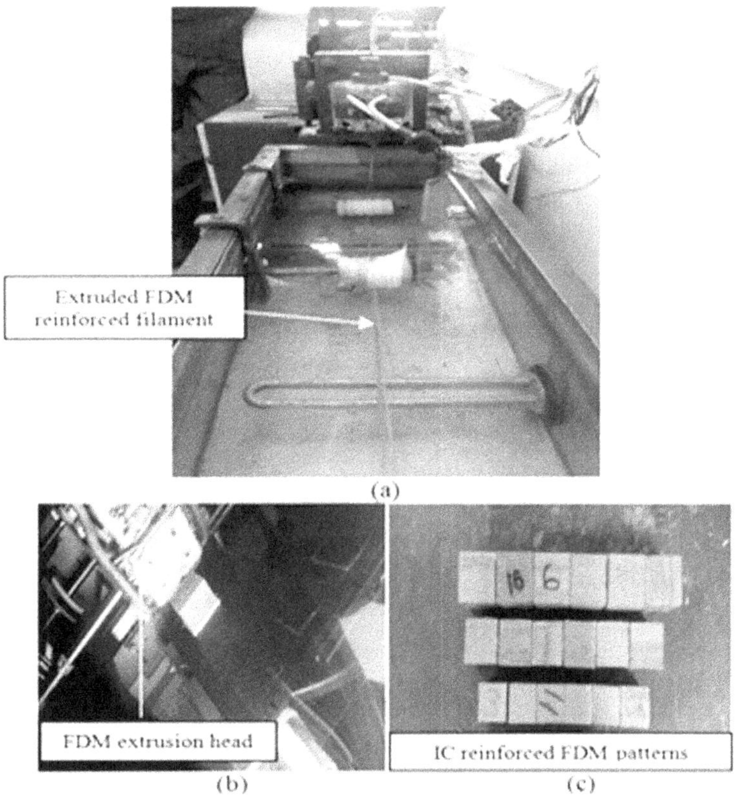

SLA Assisted IC

Investment Casting of Hollow Turbine Blade Based on Stereo-Lithography (Wu et al., 2009)

The major purpose of this research study was to develop a new process of preparing integral ceramic molds for IC process by using SLA. Aqueous gel casting process was utilized to fill the resin pattern with low viscosity ceramic slurry. At last, IC of hollow turbine blade was obtained by vacuum drying, pyrolyzing and sintering. The process was outlined as stable due to guaranteed accuracy but difficult of measuring the inner dimensions and the relative position accuracy of cores and shell are reportedly required to test further. Figure 12 shows the finally produced hollow turbine blade. This novel approach enhanced the adaptability of SLA prototype in the manufacture of integral ceramic mold for IC process.

Ceramic Shell Cracking in Stereo-Lithography-Based Rapid Casting Parts (Chen et al., 2011)

Similar to above, Chen et al. (2011) simplified a thermo-mechanical model for reducing the chances of cracks in ceramic mould prepared with using SLA patterns. In their work, the distribution rule of circumferential stress on the contact boundary between SL patterns and ceramic shell was theoretically derived using displacement method to predict dangerous area of freeform-surface parts. Then the variation rule of circumferential stress at the predicted dangerous area with changing temperature was revealed by applying transient thermo-mechanical finite element analysis on a turbine blade cross section. Circumferential stress prior to 70°C was occurred majorly due to thermal expansion-induced interaction, however relatively low stress and high strength in early stage did not ensure the preventing of shell cracking in the entire pyrolyzing process. By adding high-temperature polymers, strength at higher temperature can be increased to prevent shell cracking.

Figure 12. Developed hollow turbine blade

Lattice Structure for SL Investment Casting Patterns (Norouzi et al., 2009)

This study presented importance of octagonal lattice structure (refer Figure 13a) of SLA patterns for IC process in comparison of former hexagonal structure. The application of 3D-coupled thermal-mechanical FE analysis, magnitudes of hoop stresses on ceramic shell during burnout can be computed. Additionally, practical results obtained by the real-time strain gauging (see Figure 13b) of the cylindrical models during the heating process up to 1008°C.

Numerical results for new octagonal structure indicate that, there was 62% reduction in maximum hoop stresses exerted on the ceramic shell as well as reduction of both axial and hoop strain (near to 50%). Furthermore, 49% of internal mass expressed the probability of dimensional accuracy improvement for the octagonal structure.

LOM Assisted IC

Rapid Tooling and Patternmaking in Foundry Industry (Mueller & Kochan, 1999)

The LOM process in rapid tooling and pattern making is especially advantageous due to the robustness, wood-like properties and their low material costs. Explicit application examples in sand casting, investment casting and ceramics processing show how a reduction of necessary process steps and cycle times can be achieved by the application of LOM models. The axle bracket (refer Figure 14) was a typical pressure die casting part used in the automotive industry. To reduce the effort for the wax injection tool a LOM mold was used instead of a machined metal tool.

SLS Assisted IC

Selective Laser-Sintered of Zirconia Molds (Harlan et al., 2001)

The authors presented an approach of making zirconia and low temperature based copolymer moulds, with SLS, to create titanium casting molds. In this process, SLS shapes were infiltrated with colloidal zirconium and subsequently fired to pyrolyze the copolymer and crystallize the colloidal particles. The resulting mold, consisted of cubic and monoclinic zirconia phases, increased infiltrated weight, improved

Figure 13. New proposed octagonal structure (a), strain measuring systems (b).

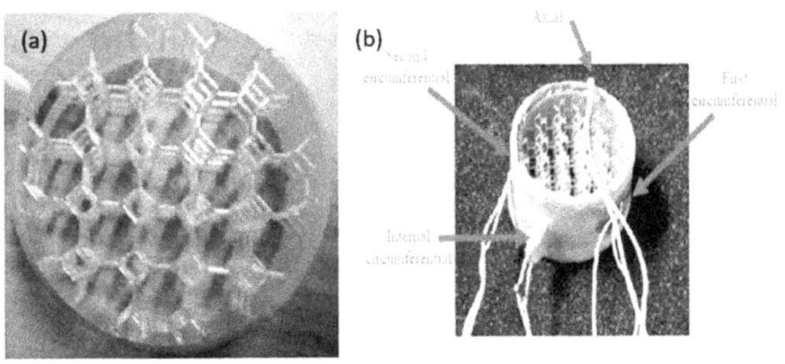

Figure 14. Axle bracket of automobile

flexural strength and density, and reduced roughness of the mold material. The process resulted into flexural strength of 5MPa, average density close to 60% shrinkage occurred (during the first firing cycle) to 13%. Metallographic examination of titanium casting revealed a narrow alpha case and an interior consisting of acicular. This work demonstrated the feasibility of producing tailor titanium parts with complex geometries directly from the scanned data without using any wax tooling or pattern.

Quality and Success Rate of the Patterns (Dotchev & Soe, 2006)

The main objective of their work was to analyze all stages of the CastForme polystyrene (CF) pattern fabrication through laser sintering, to identify their inferior quality and to find out the scope for quality improvement and failures reduction. This paper described manufacturing of patterns for shell or flask investment casting. The process involved data preparation, laser sintering fabrication of "green" part, cleaning and wax infiltration. All process stages were equally important for successful project completion in terms of pattern quality and delivery time. The paper proposed a new approach for wax infiltration. For this the green parts were heater at about 70-80°C and the plate and part was immersed into a vat of molten wax. The infiltration process was observed through a window in the oven door and as soon as all part surfaces was covered with wax; the plate was lifted in order to drain any extra wax from the part. The preheating and infiltration time varied from 30 min to 1 hr as per the geometrical and volumetric requirements.

CONCLUSION

In the light of literature reviewed in this chapter with respect to various innovations and research efforts made with different types of AM technologies, following outcomes can be drawn:

The combination of AM and IC is a mature process that is continuously advancing to achieve the goal of digitization of the foundry workshops in order to eliminate the clutters of injection tools, metallic dies as well as to strengthen their economy by producing superior quality castings in a robust and flexible manner. The outcomes of the literature are encouraging commercialization of the AM solutions in investment and sand casting technologies. It has been found that the utmost contributions of AM in foundry industry is the facilitation of concurrent engineering progresses in simulating, designing, developing, analysis and then production of any geometrical shape. The AM application, discussed here, would serve the purpose of researchers, industrialist, and academicians to explore the viability of the using established practices in their regular productions as well as to look into the areas where further development is required. The current approaches such as FDM, SLA and SLS have been more on applications in novel product development and specialty construction.

AM technologies are doing well enough in production of rapid patterns as per requirements. However, research efforts are still required to make these technologies suitable for mass production applications, aside of indirect routes. Also it is the demand of the time to integrate these methods into manufacturing oriented approaches that could include core, supports and mold production via sand printing and dies and tools.

In literature, numerous metals and alloys are successfully casted through AM assisted IC technology and there is no restriction over the use of materials if is suitable for casting. Further, this hybrid manufacturing system is highly popular in medical sector hence it is utmost importance to study the counter effects of foreign elements on the biological life and performance of the devices. Along with this, four key tasks should also be addressed in the near future: (i) integrate AM into current metal casting, (ii) develop higher speed milling, turning and CNC approaches for pre or post operations, (iii) print pattern like disposable tooling by using AM and (iv) sustain and change traditional manufacturing with AM.

REFERENCES

Agarwala, M. K., Jamalabad, V. R., Langrana, N. A., Safari, A., Whalen, P. J., & Danforth, S. C. (1996). Structural quality of parts processed by fused deposition. *Rapid Prototyping Journal*, *2*(4), 4–19. doi:10.1108/13552549610732034

Ahn, D., Kweon, J. H., Choi, J., & Lee, S. (2012). Quantification of surface roughness of parts processed by laminated object manufacturing. *Journal of Materials Processing Technology*, *212*(2), 339–346. doi:10.1016/j.jmatprotec.2011.08.013

Alexander, P., Allen, S., & Dutta, D. (1998). Part orientation and build cost determination in layered manufacturing. *Computer Aided Design*, *30*(5), 343–356. doi:10.1016/S0010-4485(97)00083-3

Arjun, N. R., Ganiger, M., & Idris, M. (2012). *Studies on optimization of Selective Laser Sintering process to manufacture Fuel tanks*. Available at http://amsi.org.in

Bacchewar, P. B., Singhal, S. K., & Pandey, P. M. (2007). Statistical modelling and optimization of surface roughness in the selective laser sintering process. *Proceedings of the Institution of Mechanical Engineers. Part B, Journal of Engineering Manufacture*, *221*(1), 35–52. doi:10.1243/09544054JEM670

Barnett, S.O. (n.d.). Investment casting - the multi-process technology. *Foundry Trade Journal International*, 1-11.

Bártolo, P. J., & Gibson, I. (2011). History of stereolithographic processes. In Stereolithography (pp. 37-56). Springer US.

Bidanda, B., & Bártolo, P. J. (Eds.). (2007). *Virtual prototyping & bio manufacturing in medical applications*. Springer Science & Business Media.

Boschetto, A., & Bottini, L. (2015). Roughness prediction in coupled operations of fused deposition modeling and barrel finishing. *Journal of Materials Processing Technology*, *219*, 181–192. doi:10.1016/j.jmatprotec.2014.12.021

Boschetto, A., Bottini, L., & Veniali, F. (2016). Finishing of fused deposition modeling parts by CNC machining. *Robotics and Computer-integrated Manufacturing*, *41*, 92–101. doi:10.1016/j.rcim.2016.03.004

Chalmers, R. E. (2001). Rapid tooling technology from Ford country. *Manufacturing Engineering*, *127*(5), 36.

Cheah, C. M., Chua, C. K., Lee, C. W., Feng, C., & Totong, K. (2015). Rapid prototyping and tooling techniques: A review of applications for rapid investment casting. *International Journal of Advanced Manufacturing Technology*, *25*(3-4), 308–320. doi:10.100700170-003-1840-6

Chen, X., Li, D., Wu, H., Tang, Y., & Zhao, L. (2011). Analysis of ceramic shell cracking in stereolithography-based rapid casting of turbine blade. *International Journal of Advanced Manufacturing Technology*, *55*(5), 447–455. doi:10.100700170-010-3064-x

Chia, H. N., & Wu, B. M. (2015). Recent advances in 3D printing of biomaterials. *Journal of Biological Engineering*, *9*(1), 4. doi:10.118613036-015-0001-4 PMID:25866560

Chiu, Y. Y., Liao, Y. S., & Hou, C. C. (2003). Automatic fabrication for bridged laminated object manufacturing (LOM) process. *Journal of Materials Processing Technology*, *140*(1), 179–184. doi:10.1016/S0924-0136(03)00710-6

Chockalingam, K., Jawahar, N., Ramanathan, K. N., & Banerjee, P. S. (2006). Optimization of stereolithography process parameters for part strength using design of experiments. *International Journal of Advanced Manufacturing Technology*, *29*(1), 79–88. doi:10.100700170-004-2307-0

Chua, C. K., Hong, K. H., & Ho, S. L. (1999). Rapid tooling technology. Part 2. A case study using arc spray metal tooling. *International Journal of Advanced Manufacturing Technology*, *15*(8), 609–614. doi:10.1007001700050109

Das, A., Madras, G., Dasgupta, N., & Umarji, A. M. (2003). Binder removal studies in ceramic thick shapes made by laminated object manufacturing. *Journal of the European Ceramic Society*, *23*(7), 1013–1017. doi:10.1016/S0955-2219(02)00266-2

Dedoussis, V., Canellidis, V., & Mathioudakis, K. (2008). Aerodynamic experimental investigation using stereolithography fabricated test models: The case of a linear compressor blading cascade. *Virtual and Physical Prototyping*, *3*(3), 151–157. doi:10.1080/17452750802120201

Dedoussis, V., & Giannatsis, J. (2004). Stereolithography assisted redesign and optimisation of a dish-washer spraying arm. *Rapid Prototyping Journal*, *10*(4), 255–260. doi:10.1108/13552540410551388

Delfs, P., Li, Z., & Schmid, H. J. (2015). Mass finishing of laser sintered parts. *Proceedings of the 26th Annual International Solid Freeform Fabrication Symposium*, 514-526.

Dotchev, K., & Soe, S. (2006). Rapid manufacturing of patterns for investment casting: Improvement of quality and success rate. *Rapid Prototyping Journal*, *12*(3), 156–164. doi:10.1108/13552540610670735

Gibson, I., Rosen, D. W., & Stucker, B. (2010). *Additive Manufacturing Technologies Rapid Prototyping to Direct Digital Manufacturing*. Springer-Verlag New York.

Greer, S. E. (2009). A comparison of the ancient metal casting materials and processes to modern metal casting materials and processes. Hartford, CT: Rensselaer Polytechnic Institute.

Guo, N., & Leu, M. C. (2013, September 1). Additive manufacturing: Technology, applications and research needs. *Frontiers of Mechanical Engineering*, *8*(3), 215–243. doi:10.100711465-013-0248-8

Hague, R., Mansour, S., & Saleh, N. (2004). Material and design considerations for rapid manufacturing. *International Journal of Production Research*, *42*(22), 4691–4708. doi:10.1080/00207840410001733940

Halidi, S. N., & Abdullah, J. (2012). Moisture effects on the ABS used for fused deposition modeling rapid prototyping machine. In *Humanities, Science and Engineering Research (SHUSER), IEEE Symposium* (pp. 839-843). IEEE. 10.1109/SHUSER.2012.6268999

Harlan, N. R., Bourell, D. L., Beaman, J. J., & Reyes, R. (2001). Titanium castings using laser-scanned data and selective laser-sintered zirconia molds. *Journal of Materials Engineering and Performance*, *10*(4), 410–413. doi:10.1361/105994901770344818

Harun, W.S., Safian, S., & Idris, M.H. (2009). Evaluation of ABS patterns produced from FDM for investment casting process. *Computational Method and Experiments in Materials Characterisation IV*, *1*, 319-28.

Heine, R. W., Loper, C. R., & Rosenthal, P. C. (1955). *Principles of metal casting*. Tata McGraw-Hill Education.

Jones, S., & Yuan, C. (2003). Advances in shell moulding for investment casting. *Journal of Materials Processing Technology*, *135*(2), 258–265. doi:10.1016/S0924-0136(02)00907-X

Knights, M. (2001). Rapid tooling is ready for prime time. *Plastics Technology*, *47*(1), 46–53.

Kruth, J. P., Levy, G., Klocke, F., & Childs, T. H. (2007). Consolidation phenomena in laser and powder-bed based layered manufacturing. *CIRP Annals-Manufacturing Technology*, *56*(2), 730–759. doi:10.1016/j.cirp.2007.10.004

Kruth, J. P., Mercelis, P., Van Vaerenbergh, J., Froyen, L., & Rombouts, M. (2005). Binding mechanisms in selective laser sintering and selective laser melting. *Rapid Prototyping Journal*, *11*(1), 26–36. doi:10.1108/13552540510573365

Kruth, J. P., Wang, X., Laoui, T., & Froyen, L. (2003). Lasers and materials in selective laser sintering. *Assembly Automation*, *23*(4), 357–371. doi:10.1108/01445150310698652

Kumar, S., & Kruth, J. P. (2010). Composites by rapid prototyping technology. *Materials & Design*, *31*(2), 850–856. doi:10.1016/j.matdes.2009.07.045

Kunjan, C., Jawahar, N., Chandrasekhar, U., Praveen, J., & Karthic, M. (2016). Development of Process Model for Optimal Selection of Process Parameters for Geometric Tolerances and Surface Roughness in Stereolithography. *International Journal of Advanced Design and Manufacturing Technology*, *9*(3), 103–107.

Lan, P. T., Chou, S. Y., Chen, L. L., & Gemmill, D. (1997). Determining fabrication orientations for rapid prototyping with stereolithography apparatus. *Computer Aided Design*, *29*(1), 53–62. doi:10.1016/S0010-4485(96)00049-8

Lee, C. W., Chua, C. K., Cheah, C. M., Tan, L. H., & Feng, C. (2004). Rapid investment casting: Direct and indirect approaches via fused deposition modelling. *International Journal of Advanced Manufacturing Technology*, *23*(1-2), 93–101. doi:10.100700170-003-1694-y

Madrazo, I., Zamorano, C., Magallón, E., Valenzuela, T., Ibarra, A., Salgado-Ceballos, H., ... Guízar-Sahagún, G. (2009). Stereolithography in spine pathology: A 2-case report. *Surgical Neurology*, *72*(3), 272–275. doi:10.1016/j.surneu.2008.04.034 PMID:18614210

Mahesh, M., Wong, Y. S., Fuh, J. Y., & Loh, H. T. (2004). Benchmarking for comparative evaluation of RP systems and processes. *Rapid Prototyping Journal*, *10*(2), 123–135. doi:10.1108/13552540410526999

Marchelli, G., Prabhakar, R., Storti, D., & Ganter, M. (2011). The guide to glass 3D printing: Developments, methods, diagnostics and results. *Rapid Prototyping Journal*, *17*(3), 187–194. doi:10.1108/13552541111124761

Mekonnen, B. G., Bright, G., & Walker, A. (2016). *A Study on State of the Art Technology of Laminated Object Manufacturing (LOM). In CAD/CAM, Robotics and Factories of the Future* (pp. 207–216). New Delhi: Springer.

Mueller, B., & Kochan, D. (1999). Laminated object manufacturing for rapid tooling and patternmaking in foundry industry. *Computers in Industry*, *39*(1), 47–53. doi:10.1016/S0166-3615(98)00127-4

Murphy, S. V., & Atala, A. (2014). 3D bioprinting of tissues and organs. *Nature Biotechnology*, *32*(8), 773–785. doi:10.1038/nbt.2958 PMID:25093879

Norouzi, Y., Rahmati, S., & Hojjat, Y. (2009). A novel lattice structure for SL investment casting patterns. *Rapid Prototyping Journal*, *15*(4), 255–263. doi:10.1108/13552540910979776

Novakova-Marcincinova, L. (2012). Application of fused deposition modeling technology in 3D printing rapid prototyping area. *Manuf. and Ind. Engg.*, *11*(4), 35–37.

Omar, M. F., Sharif, S., Ibrahim, M., Hehsan, H., Busari, M. N., & Hafsa, M. N. (2012). Evaluation of direct rapid prototyping pattern for investment casting. *Advanced Materials Research*, *463*, 226–233. doi:10.4028/www.scientific.net/AMR.463-464.226

Onuh, S. O., & Hon, K. K. (1998). Optimising build parameters for improved surface finish in stereolithography. *International Journal of Machine Tools & Manufacture*, *38*(4), 329–342. doi:10.1016/S0890-6955(97)00068-0

Pal, D. K., Ravi, B., & Bhargava, L. S. (2002). E-manufacturing one-off intricate castings using rapid prototyping technology. In *International Conference on e-Manufacturing, Bhopal* (pp. 259-63). American Foundry Society.

Pandey, P. M., Thrimurthulu, K., & Reddy, N. V. (2004). Optimal part deposition orientation in FDM by using a multicriteria genetic algorithm. *International Journal of Production Research*, *42*(19), 4069–4089. doi:10.1080/00207540410001708470

Park, J., Tari, M. J., & Hahn, H. T. (2000). Characterization of the laminated object manufacturing (LOM) process. *Rapid Prototyping Journal*, *6*(1), 36–50. doi:10.1108/13552540010309868

Pattnaik, S., Jha, P. K., & Karunakar, D. B. (2014). A review of rapid prototyping integrated investment casting processes. *Proceedings of the Institution of Mechanical Engineers, Part L: Journal of Materials: Design and Applications, 228*(4), 249-77.

Pham, D. T., Dimov, S. S., & Gault, R. S. (1999). Part orientation in stereolithography. *International Journal of Advanced Manufacturing Technology*, *15*(9), 674–682. doi:10.1007001700050118

Pilipović, A., Raos, P., & Šercer, M. (2011). Experimental testing of quality of polymer parts produced by laminated object manufacturing–LOM. *Tehnicki Vjesnik (Strojarski Fakultet)*, *18*, 253–260.

Rosochowski, A., & Matuszak, A. (2000). Rapid tooling: The state of the art. *Journal of Materials Processing Technology*, *106*(1-3), 191–198. doi:10.1016/S0924-0136(00)00613-0

Schmid, M., Simon, C., & Levy, G. N. (2009). Finishing of SLS-parts for rapid manufacturing (RM)–a comprehensive approach. *Proceedings SFF*, 1-10.

Shu, X., & Wang, R. (2017). Thermal residual solutions of beams, plates and shells due to laminated object manufacturing with gradient cooling. *Composite Structures*, *15*(174), 366–374. doi:10.1016/j.compstruct.2017.04.060

Singh, J., Singh, R., & Singh, H. (2017). Dimensional accuracy and surface finish of biomedical implant fabricated as rapid investment casting for small to medium quantity production. *Journal of Manufacturing Processes*, *25*, 201–211. doi:10.1016/j.jmapro.2016.11.012

Singh, R., & Singh, M. (2017). Surface roughness improvement of cast components in vacuum moulding by intermediate barrel finishing of fused deposition modelling patterns. *Proceedings of the Institution of Mechanical Engineers. Part E, Journal of Process Mechanical Engineering*, *231*(2), 309–316. doi:10.1177/0954408915595576

Singh, R., & Singh, S. (2013). Effect of process parameters on surface hardness, dimensional accuracy and surface roughness of investment cast components. *Journal of Mechanical Science and Technology*, *27*(1), 91–97. doi:10.100712206-012-1218-5

Singh, R., Singh, S., Singh, I. P., Fabbrocino, F., & Fraternali, F. (2017). Investigation for surface finish improvement of FDM parts by vapor smoothing process. *Composites. Part B, Engineering*, *111*, 228–234. doi:10.1016/j.compositesb.2016.11.062

Singh, S., & Singh, R. (2016). Effect of process parameters on micro hardness of Al–Al2O3 composite prepared using an alternative reinforced pattern in fused deposition modelling assisted investment casting. *Robotics and Computer-integrated Manufacturing*, *37*, 162–169. doi:10.1016/j.rcim.2015.09.009

Singh, S., & Singh, R. (2016). Fused deposition modelling based rapid patterns for investment casting applications: A review. *Rapid Prototyping Journal*, *22*(1), 123–143. doi:10.1108/RPJ-02-2014-0017

Singh, S., & Singh, R. (2016). Precision investment casting: A state of art review and future trends. *Proceedings of the Institution of Mechanical Engineers. Part B, Journal of Engineering Manufacture*, *230*(12), 2143–2164. doi:10.1177/0954405415597844

Sivadasan, M., Singh, N.K. & Sood, A.K. (2012). Use of fused deposition modeling process in investment precision casting and risk of using selective laser sintering process. *International Journal of Applied Research in Mechanical Engineering*.

Smith, B. J., St Jean, P., & Duquette, M. L. (1996). A comparison of rapid prototype techniques for investment casting Be-Al. In *Proceedings of Rapid Prototyping and Manufacturing'96* (Vol. 2, pp. 1–11). Dearborn, MI: Conference.

Taşcıoğlu, S., & Akar, N. (2003). A novel alternative to the additives in investment casting pattern wax compositions. *Materials & Design*, *24*(8), 693–698. doi:10.1016/S0261-3069(03)00097-9

Taşcıoğlu, S., & Akar, N. (2007). Conversion of an investment casting sprue wax to a pattern wax by chemical agents. *Materials and Manufacturing Processes*, *18*(5), 753–768. doi:10.1081/AMP-120024973

Taylor, P. R. (1983). An illustrated history of lost wax casting. *Proceedings of the 17th Annual BICTA Conference.*

Tiwary, V., Arunkumar, P., Deshpande, A. S., & Khorate, V. (2015). Studying the effect of chemical treatment and fused deposition modelling process parameters on surface roughness to make acrylonitrile butadiene styrene patterns for investment casting process. *International Journal of Rapid Manufacturing*, *5*(3-4), 276–288. doi:10.1504/IJRAPIDM.2015.074807

Träger, H., & Bührig-Polaczek, A. (2000). *Foundry technology*. Ullmann's Encyclopedia of Industrial Chemistry. doi:10.1002/14356007.a12_035

Tseng, P., Murray, C., Kim, D., & Di Carlo, D. (2004). Research highlights: Printing the future of microfabrication. *Lab on a Chip*, *14*(9), 1491–1495. doi:10.1039/c4lc90023e PMID:24671475

Turner, B., Strong, R., & Gold, S. (2014). A review of melt extrusion additive manufacturing processes: I. Process design and modeling. *Rapid Prototyping Journal*, *20*(3), 192–204. doi:10.1108/RPJ-01-2013-0012

Turner, B. N., & Gold, S. A. (2015). A review of melt extrusion additive manufacturing processes: II. Materials, dimensional accuracy, and surface roughness. *Rapid Prototyping Journal*, *21*(3), 250–261. doi:10.1108/RPJ-02-2013-0017

Vermeulen, M., Claessens, T., Van Der Smissen, B., Van Holsbeke, C. S., De Backer, J. W., Van Ransbeeck, P., & Verdonck, P. (2013). Manufacturing of patient-specific optically accessible airway models by fused deposition modeling. *Rapid Prototyping Journal*, *19*(5), 312–318. doi:10.1108/RPJ-11-2011-0118

Vickers C. (n.d.). *An alternative route to metal components for prototype and low-volume production. Rapid prototyping casebook*. Academic Press.

Vijay, P., Danaiah, P., & Rajesh, K. V. (2011). Critical parameters effecting the rapid prototyping surface finish. *Journal of Mechanical Engineering and Automation*, *1*, 17–20. doi:10.5923/j.jmea.20110101.03

Wang, S., Miranda, A. G., & Shih, C. (2010). A study of investment casting with plastic patterns. *Materials and Manufacturing Processes*, *25*(12), 1482–1488. doi:10.1080/10426914.2010.529585

Warner, M. C. (1993). Rapid prototyping methods to manufacture functional metal and plastic parts. Rapid Prototyping System: Fast Track to Product Realisation, 137-44.

Wendel, B., Rietzel, D., Kühnlein, F., Feulner, R., Hülder, G., & Schmachtenberg, E. (2008). Additive processing of polymers. *Macromolecular Materials and Engineering*, *293*(10), 799–809. doi:10.1002/mame.200800121

Wood, I. (2016). *Development of a layerless additive manufacturing stereolithography machine to improve surface quality and dimensional accuracy* (Doctoral dissertation). University of Ontario Institute of Technology.

Wu, H., Li, D., & Guo, N. (2009). Fabrication of integral ceramic mold for investment casting of hollow turbine blade based on stereolithography. *Rapid Prototyping Journal*, *15*(4), 232–237. doi:10.1108/13552540910979749

Yang, J., Shi, Y., Shen, Q., & Yan, C. (2009). Selective laser sintering of HIPS and investment casting technology. *Journal of Materials Processing Technology*, *209*(4), 1901–1908. doi:10.1016/j.jmatprotec.2008.04.056

Yang, Q., Lu, Z., Zhou, J., Miao, K., & Li, D. (2017). A novel method for improving surface finish of stereolithography apparatus. *International Journal of Advanced Manufacturing Technology*, *393*(5-8), 1537–1544. doi:10.100700170-017-0529-1

Yi, S., Liu, F., Zhang, J., & Xiong, S. (2004). Study of the key technologies of LOM for functional metal parts. *Journal of Materials Processing Technology*, *150*(1), 175–181. doi:10.1016/j.jmatprotec.2004.01.035

Zhang, W., Leu, M. C., Ji, Z., & Yan, Y. (1999). Rapid freezing prototyping with water. *Materials & Design*, *20*(2), 139–145. doi:10.1016/S0261-3069(99)00020-5

This research was previously published in the Handbook of Research on Green Engineering Techniques for Modern Manufacturing edited by M. Uthayakumar, S. Aravind Raj, Tae Jo Ko, S. Thirumalai Kumaran, and J. Paulo Davim, pages 24-48, copyright year 2019 by Engineering Science Reference (an imprint of IGI Global).

Section 4
Prototyping

Chapter 13
Integrated Manufacturing System for Complex Geometries:
Towards Zero Waste in Additive Manufacturing

Divya Zindani
National Institute of Technology Silchar, India

Kaushik Kumar
Birla Institute of Technology, India

ABSTRACT

The chapter proposes an integrated manufacturing system consisting of three main components: digital prototyping, physical prototyping, and lost core technology. The integrated system combines the beneficial aspects of computer-aided design, computer-aided engineering, rapid prototyping, and rapid tooling. The proposed integrated system is an attempt to compress the product development time while saving cost. The system can be efficient in designing of mold, parts with complex ducts and cavities, and carrying out design analysis through optimization and simulations. The system is therefore an attempt to minimize the waste of material that occurs in the development of a product and is therefore an efficient green technology for the manufacturing industries.

INTRODUCTION

The escalating competition worldwide and the globalization have resulted in immense pressure on different manufacturing units to reduce the production time and cost of the product while meeting the quality requirements. The time-to-market the products have been now been reduced to weeks and therefore the new product must be made quickly and cheaply meeting the market requirements. One of the costly and time consuming phases in product development is the manufacturing of moulds for both the

DOI: 10.4018/978-1-5225-9624-0.ch013

development of prototype part and production component manufacturing. The conventional machining approach to produce mould entails long lead times and costs. And therefore the sequential approach of conventional machining doesn't meet the requirement of rapid product development. Thus it has become imperative on the part of manufacturing units to research on to innovate new technologies that can aid in rapid product development.

Additive manufacturing has emerged as potential technology that has the ability to reduce the production development time. Rapid prototyping (RP) was the first of a kind process for creating a 3D object through layer by layer technique and computer-aided design (CAD). The RP process was developed in 1980's for production of models and prototypes and therefore help design engineers to create what they had in mind. The major advantages of the RP processes are the reduction in cost and time as well as the possibility to create any complex shape (Ashley, 1991). The versatile RP processes have been applied by medical doctors, scientists, professors, artists and market researchers (Noorani, 2006; Flowers & Moniz, 2002; Chua et al., 1998) for creation and analysis of models for various studies. For instance doctors create a model for a body part for which the operation is to be performed. They use the model to better plan the procedure. The prototypes made from rapid prototyping (RP) can be used for different evaluations such as visual inspection, ergonomic evaluations etc., thereby leading to comprehensive design analysis earlier in the product development cycle. Since its inception, RP has been used in different industries where both time and precision are of paramount importance. RP is however not used for large scale production.

RP is now evolving towards rapid tooling (RT) which has the capability to be used for commercial scale production. Therefore RT is used effectively for making of moulds for commercial purposes. Further, there are products that are required to be provided with complex cavities, bypass or even to completely close hollow bodies. There are other products in which case the assembly is to be avoided such as in the case of pneumatic and hydraulic parts. For such cases one of the evolving technologies is the Lost Core or Soluble Fusible metal core technology.

Industries are now looking for a feasible option to use the advantages of both the RT and RP. The present chapter proposes an integrated manufacturing system to produce complex shape products that have complex internal cavities.

RAPID PROTOTYPING PROCESSES

The different rapid prototyping processes can be categorized into: powder based, solid based and liquid based.

Liquid Based Processes

Stereolithography (SL)

The SL process was developed by 3D systems, Inc. and is one of the most widely used process of rapid prototyping. SL is a liquid based process wherein the ultraviolet light makes contact with the resin which leads to solidification or curing of photosensitive polymer. The SL process incepts with a 3D CAD model and is translated to STL file. In the STL format, the 3D CAD model is cut into slices and the information of each layer is stored. A platform supports the piece and other overhanging structures. The UV laser then solidifies the selective locations of the resin. The platform supporting the piece is lowered on completion of a layer and the excess material is then drained on completion of the entire product. The drained material then can be reused (Noorani, 2006; Cooper, 2001; Kruth, 1991). Microstereolithography is the improvised version ofstereolithography supporting for higher resolution.

Certain errors are introduced in the final product from SL process. One such error is that of overcuring which occurs mainly on the overhung parts. The overcuring may be attributed to the absence of fusion with the bottom layer. Scanned line shape is another error which is introduced as a result of scanning process. The layer thickness is variable because of the high viscosity of the resin and therefore introduces an error in the border position control. Lack of surface finish is another error of the SL process (Kim et al., 2010).

SL offers to build part with different materials and the process is called multiple material SL (Nagy & Matyasi, 2003).

Fused Deposition Modeling (FDM)

In FDM process a thin filament plastic is fed to a machine. The machine consists of print heads that melts the plastic filament and extrude it. The materials used in the FDM process are acrylonitrile butadiene styrene (ABS), polycarbonate (PC), polyphenylsulfone (PPSF), PC-ISO and PC-ABS blends. The FDM process has numerous advantages such as no requirement of chemical post-processing, less expensive machines and no resins to cure (Noorani, 2006; Cooper, 2011). The major disadvantage is the relatively low resolution on the z axis in comparison to other processes, requirement of finishing processes and relatively high amount of time required to build a part. Some models however allow for two modes of operation: dense mode and sparse mode, which aids in saving of production time but at the cost of mechanical properties (Morvan et al., 2005). Figure 1 delineates the FDM process outline.

Polyjet

The polyjet is a liquid based rapid prototyping process that uses inkjet to produce physical models. A photopolymer is deposited by the x and y axes moving inkjet head. The deposited photopolymer is then cured by the ultraviolet lamps after the deposition of each layer is completed. The overhanging parts are supported by using a gel-type polymer which is then removed after the part is completed. The polyjet process can produce parts of multiple colors (Wholers, 2010; Singh, 2011; Petrovic et al., 2011). Further, the polyjet process produces parts with very high resolution. However, the parts produced by the polyjet process are relatively weaker than that produced by other RP processes.

Powder Based Processes

Selective Laser Sintering (SLS)

SLS is a 3D printing process in which a carbon dioxide laser beam is used to fuse powders. The chamber is heated to the melting point of the material. The carbon dioxide laser beam fuses the powder, lying on a bed, at specific position in accordance with the design. The bed is supported with a piston which lowers on completion of a layer by the amount equivalent to that of the layer thickness. A wide range of materials can be used with the SLS process such as plastics, metals, polymers and combination of

Figure 1. Process layout for fused deposition modeling
(Make Parts Fast, 2018)

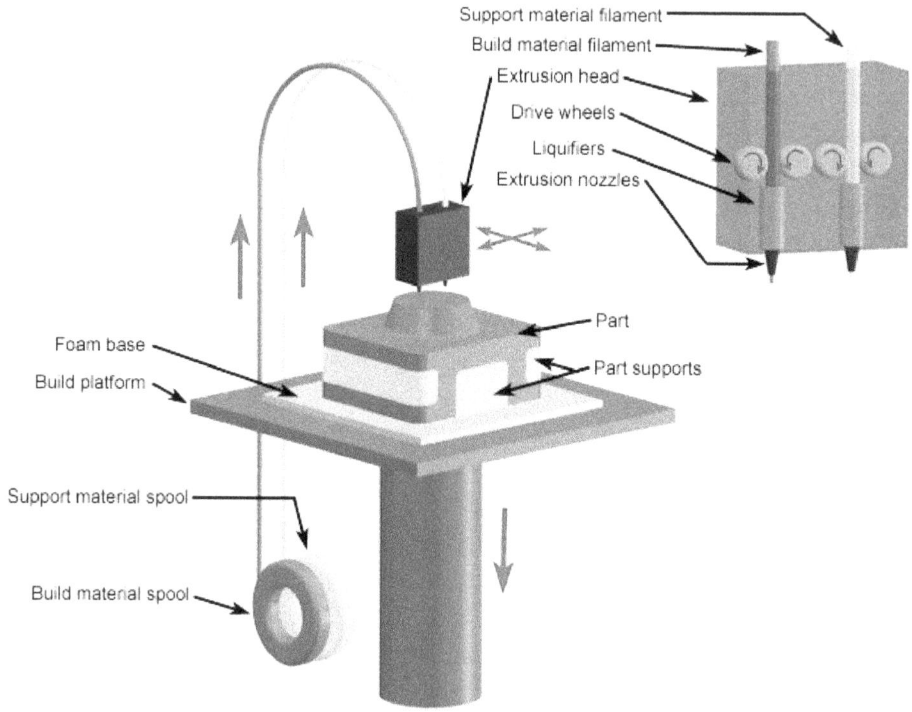

metals, metals and ceramics, metals and polymers (Halloran, 2011; Hwa-Hsing et al., 2011; Salmoria et al., 2011). Another advantage of the SLS process is the recycling of the unused powder. The SLS process however suffers from a number of disadvantages such as the accuracy of the part is determined by the size of the material particles, requirement of inert gas to avoid the oxidation and to ensure that the process is carried out at constant temperature. Process outline of SLS is depicted in Figure 2.

Electron Beam Melting (EBM)

EBM process is similar to the SLS process. A very high voltage electron beam melts powder. The use of electron beam is only the differentiating factor between the EBM and the SLS processes. The voltage of electron beam ranges between 30 to 60 kV. The EBM process takes place in a vacuum chamber. The main advantage of the EBM process production of parts with a wide variety of pre-alloyed metals. Further, the process has find applicability in outer space (Murr et al., 2012; Semetay, 2007). The process is mainly used for producing metal parts which is one of the disadvantages of the EBM process.

Figure 2. Process layout for selective laser sintering
(Make Parts Fast, 2018)

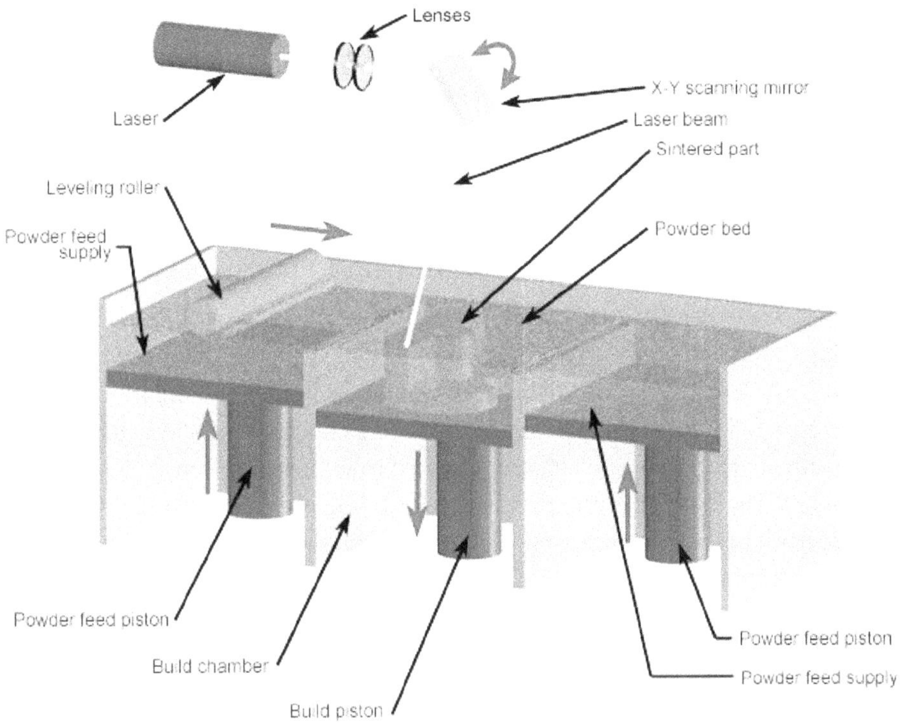

Prometal

This powder-based process is used for the fabrication of tools and dies. The process uses a liquid binder to bind the steel powder. The powder is located on a piston supported bed that lowers when each layer is completed. The excess powder is removed when the required thickness of the layer is achieved. No postprocesssing is required while building a mold. However, finishing operations such as infiltration and sintering are required if a functional part is being built (Cooper, 2001; Kruth, 1991).

Laser Engineering Net Shaping (LENS)

A high powered laser beam is used to melt the metal powder. The metal powder is injected at a specific location. The molten metal material solidifies on cooling. The LENS process takes place in a chamber with argon atmosphere. A wide range of materials and their combinations can be used with the LENS process. For instance materials such as stainless steel, titanium-6, nickel based alloys, aluminium-4 vanadium, copper alloys etc can be used with the LENS process. The LENS process is also suitable for repairing of parts that are impossible or expensive to be repaired by other processes. One of the major disadvantages of the LENS process is the presence of residual stress due to uneven cooling and heating process.

3DP

The data from the CAD drawing is printed with the aid of water-based liquid binder supplied in a jet onto a starch based powder. The starch-based powder is supported on a powder bed and glues to each other with the aid of liquid based blinder. The 3DP process is similar to that of inkjet printing process that is used for two-dimensional printing. The 3DP process is capable of handling a high variety of polymers (Halloran, 2011; Cooper, 2001).

Solid Based Processes

Laminated Object Manufacturing (LOM)

Helisys of Torrance, CA developed the LOM process. The LOM process combines both the subtractive and additive techniques to build a part. The material is fed in the form of sheet that is advanced over a platform with the help of sheet feed mechanism. A heated roller is used to bond the sheets together with the use of heat energy. The outline of the part is then cut with the help of laser. The platform is then lowered by a depth equal to the thickness of the sheet and then the new layer of sheet advances over the previously built layer. The main advantages of the LOM process are the low cost, no deformation, and requirement of no postprocessing and the possibility to build large parts. The disadvantages of the LOM process are wastage of material, difficulty to produce complex cavities and low surface definition. The LOM process can be used for creating models of papers, composite and metals (Noorani, 2006; Cooper, 2001; Vaupotic, 2006). Process outline of laminated object manufacturing is shown in Figure 3.

LOST CORE TECHNOLOGY (LCT)

The Lost core technology requires the core of the part with complex cavities. A low melting point alloy is used for casting the component's interior. A composite or plastic material is used to overlay the casted component. The overlay may be cured in some cases. The low melting alloy is allowed to melt out with the overlay still in place. The melting temperature doesn't leads to any distortion or damage to the fabricated part. The steps involved in making the core consist of the following:

- Casting of the core.
- Inserting the core into mold.
- Performance of the casting process.
- Ejecting the mold and the core.
- Core and mold separation by melting

The alloys used for making of cores have their melting point ranging 50°C to 300°C. it should be ensured that the melting temperature of the alloy is consistent with that of the working temperature of the over molding. The mechanical properties of the alloy are to be given due consideration in cases where the core is heavy or is required to withstand high pressures. The amount of energy consumed by

Figure 3. Process layout for laminated object manufacturing
(Make Parts Fast, 2018)

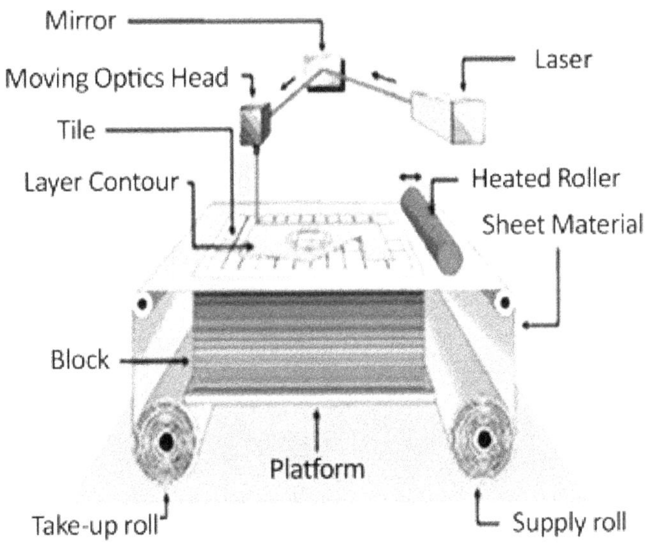

melting is another factor to be given due consideration (Polifke et al., 1998). A wide range of materials can be used for making cores such as wax, Tin-Bismuth alloy, Zn-Al eutectic mixture etc. The lost core technology is suitable for producing complex internal ducts and cavities.

THE INTEGRATED MANUFACTURING SYSTEM

A product development system should consider all aspects related to the product such as product design, its manufacturability and recycling. All these aspects should be considered at the early stage of design cycle so that design engineers can easily make changes during the iterative process of product development. The integrated manufacturing system for producing geometries with complex cavities and ducts consists of three main phases: Digital prototype, Physical prototype and Rapid tooling (lost core technology process). The integrated manufacturing system is shown in the Figure 4 below.

The digital prototyping phase begins with the 3D model using the computer aided design software (CAD). The dimensional as well as the aesthetical aspects of the product is fixed at this stage. The mold and its other components are also designed using the CAD software such as Pro-Engineer, CATIA, SOLIDWORKS, Unigraphics etc. For an already existing component reverse engineering (RE) process can be used for creating the CAD model. The RE process consists of three main steps: digitization of the existing part, extraction of part features and the 3D CAD model. Variety of contact and non-contact digitizers can be used for digitization of the part. Segmentation of the digitized data accomplishes the task of feature extraction. Fitting of the digitized surfaces to the set of data points will accomplish the 3D CAD modeling of the part under consideration. Part digitization can be accomplished with a wide range of systems such as coordinate measuring machine (CMM), laser triangular scanner (LTS), computer tomography (CT), cross-sectional imaging measurement etc.

Figure 4. Process layout for integrated manufacturing system

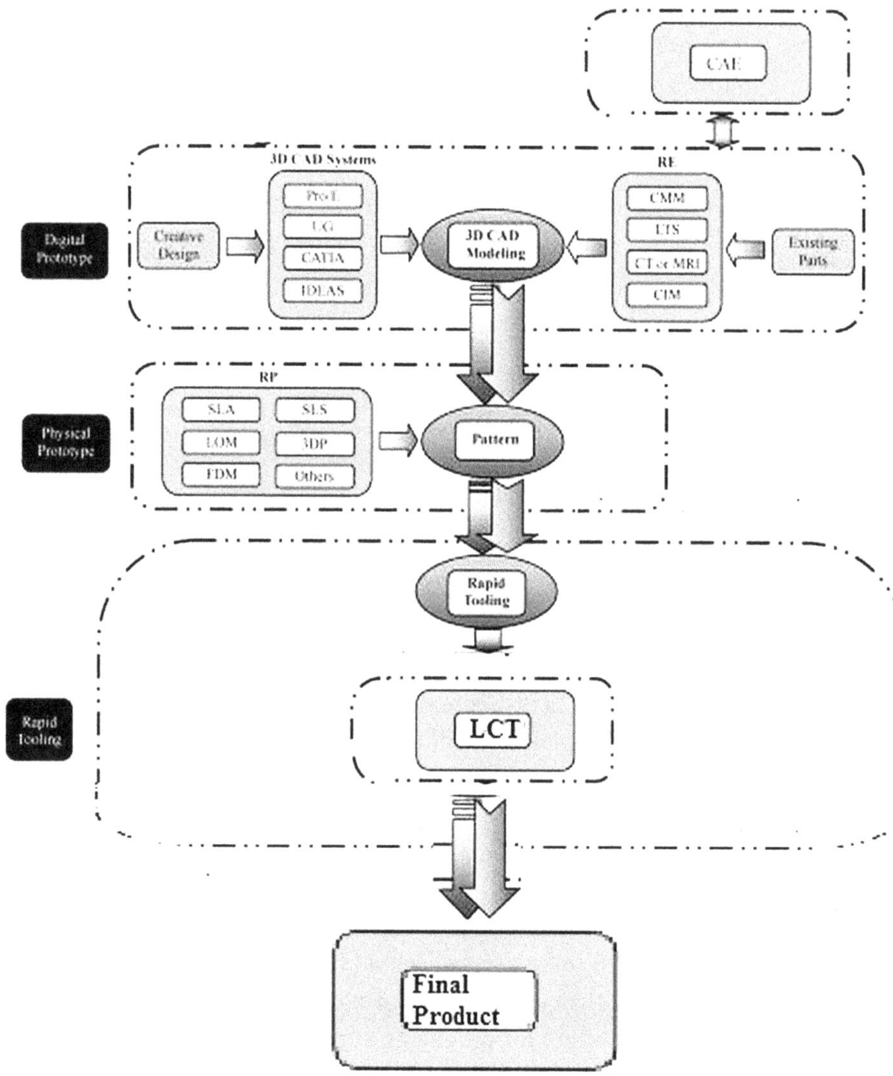

The optimized product is the primary requirement of the manufacturer. To guide through the process of optimization of the product design, a number of computer aided engineering (CAE) systems can be used. Few examples of CAE systems are ANSYS, AUTOFORM, DYNAFORM etc. the 3D CAD model can be transferred to the CAE systems for the optimization analysis. CAE environment can help in predicting the filling pattern, temperature and pressure for the mold. The CAE systems aids in optimizing the part geometry, and the core. The CAE model will help in reduction of shrinkage, warpage and residual stresses.

The digital prototype is now converted to physical prototype using the different rapid prototyping techniques. The 3D CAD model is converted to STL file format which is typically an interfacing system between the 3D CAD model and the RP systems. The RP process creates the physical prototype layer by layer from the STL format of the 3D CAD model. The RP techniques save a lot of time to market the product with improved quality and reduced cost. Over the years, industries have used a wide range of RP techniques such as SL, SLS, FDM, LOM, 3DP etc.

Next the mold is prepared with the physical prototype of the part. The physical prototype is now master pattern for preparing the mold. The master pattern is suspended in a mold frame equipped with the gates and runners. The master pattern is set up on the parting line of the mold casting frame. The casting frame will then be filled with silicon-rubber resin which is properly mixed using computer-controlled equipment. The resin fills the casting frame surrounding the master pattern. The casting is then moved to a heating chamber for curing. The master pattern is then removed from the silicone mold after the hardening is accomplished. The gates and runners are cut off to obtain the final mold cavity.

For the parts with complex cavities and ducts, the lost core technique is then followed. Lost core technology is one of the rapid tooling process. The core prepared from the RP process is used for the fabrication of complex shaped cavities and ducts. The core is prepared using a low melting alloy such as tin-bismuth alloy. The core is placed inside the already prepared silicon mold. The mold is now filled with the desired resin. The mold is then moved to the heating chamber and the casting is removed from the mold after the completion of the hardening process. The obtained casting has core embedded in it. Therefore to remove the core, the casting is then placed in the oil bath. The temperature of the oil bath is similar to the melting point of the core material. The core material melts out leaving behind the complex cavity.

The desired part with the complex cavity is finally obtained after sequential procedural steps.

CONCLUSION

The chapter presents an integrated manufacturing system that optimizes the product design while minimizing the wastage of the material. The system comprises of three different modules: Digital prototyping, Physical prototyping and Rapid tooling. 3D CAD systems are used to establish the digital prototypes and the CAE systems are used to optimize product design. The RP techniques then are used to obtain the master pattern and the mold. The RT in the form of lost core technology is used to produce the complex shaped mold cavity. Thus the proposed system can reap benefits in number of ways such as saving of time, cost and material. The system can thus be used by different manufacturing industries seeking to develop cost effective product with enhanced quality and less time to market the product.

REFERENCES

Ashley, S. (1991). Rapid prototyping systems. *Mechanical Engineering (New York, N.Y.)*, *113*(4), 34.

Chua, C. K., Chou, S. M., Lin, S. C., Eu, K. H., & Lew, K. F. (1998). Rapid prototyping assisted surgery planning. *International Journal of Advanced Manufacturing Technology*, *14*(9), 624–630. doi:10.1007/BF01192281

Cooper, K. (2001). *Rapid prototyping technology: selection and application*. CRC Press. doi:10.1201/9780203910795

Core Tech System Inc. (n.d.). *Engineering Talk From Core Tech Systems*. Core Tech Systems.

Flowers, J., & Moniz, M. (2002). Rapid prototyping in technology education. *Technology Teacher*, *62*(3), 7.

Halloran, J. W., Tomeckova, V., Gentry, S., Das, S., Cilino, P., Yuan, D., ... Alabi, T. R. (2011). Photopolymerization of powder suspensions for shaping ceramics. *Journal of the European Ceramic Society*, *31*(14), 2613–2619. doi:10.1016/j.jeurceramsoc.2010.12.003

Hwa-Hsing, T., Ming-Lu, C., & Hsiao-Chuan, Y. (2011). Slurry based selective laser sintering of polymer-coated ceramic powders to fabricate high strength alumina parts. *Journal of the European Ceramic Society*, *31*(8), 1383–1388. doi:10.1016/j.jeurceramsoc.2011.02.020

Kim, H., Choi, J. W., & Wicker, R. (2010). Scheduling and process planning for multiple material stereolithography. *Rapid Prototyping Journal*, *16*(4), 232–240. doi:10.1108/13552541011049243

Kruth, J. P. (1991). Material incress manufacturing by rapid prototyping techniques. *CIRP Annals-Manufacturing Technology*, *40*(2), 603–614. doi:10.1016/S0007-8506(07)61136-6

Lost Core Injection Moulding. (n.d.). Copyright. Retrieved from http://LostCoreInjectionMoulding.htm

Make Parts Fast. (2018). Retrieved from www.makepartsfast.com

MCP. (n.d.). Retrieved from http://www.mcp-group.de

Morvan, S., Hochsmann, R., & Sakamoto, M. (2005). ProMetal RCT (TM) process for fabrication of complex sand molds and sand cores. *Rapid Prototyping*, *11*(2), 1–7.

Murr, L. E., Gaytan, S. M., Ramirez, D. A., Martinez, E., Hernandez, J., Amato, K. N., ... Wicker, R. B. (2012). Metal fabrication by additive manufacturing using laser and electron beam melting technologies. *Journal of Materials Science and Technology*, *28*(1), 1–14. doi:10.1016/S1005-0302(12)60016-4

Noorani, R. (2006). *Rapid prototyping: principles and applications*. John Wiley & Sons Incorporated.

Petrovic, V., Vicente Haro Gonzalez, J., Jorda Ferrando, O., Delgado Gordillo, J., Ramon Blasco Puchades, J., & Portoles Grinan, L. (2011). Additive layered manufacturing: Sectors of industrial application shown through case studies. *International Journal of Production Research*, *49*(4), 1061–1079. doi:10.1080/00207540903479786

Polifke, M. (1998). *Lost Core Technology On The Way To New Application*. Spe-Antec.

Salmoria, G. V., Paggi, R. A., Lago, A., & Beal, V. E. (2011). Microstructural and mechanical characterization of PA12/MWCNTs nanocomposite manufactured by selective laser sintering. *Polymer Testing*, *30*(6), 611–615. doi:10.1016/j.polymertesting.2011.04.007

Semetay, C. (2007). *Laser Engineered Net Shaping (LENS) modeling using welding simulation concepts*. Lehigh University.

Singh, R. (2011). Process capability study of polyjet printing for plastic components. *Journal of Mechanical Science and Technology*, *25*(4), 1011-1015.

Szilvśi-Nagy, M., & Matyasi, G. Y. (2003). Analysis of STL files. *Mathematical and Computer Modelling*, *38*(7-9), 945–960. doi:10.1016/S0895-7177(03)90079-3

Vaupotic, B., Brezocnik, M., & Balic, J. (2006). Use of PolyJet technology in manufacture of new product. *Journal of Achievements in Materials and Manufacturing Engineering*, *18*(1-2), 319–322.

Wohlers, T. (2010). *Wohlers Report 2010*. Wholers Associates.

This research was previously published in the Handbook of Research on Green Engineering Techniques for Modern Manufacturing edited by M. Uthayakumar, S. Aravind Raj, Tae Jo Ko, S. Thirumalai Kumaran, and J. Paulo Davim, pages 14-23, copyright year 2019 by Engineering Science Reference (an imprint of IGI Global).

Chapter 14
Innovation in Product Design:
IoT Objects Driven New Product Innovation and Prototyping Using 3D Printers

Ravi Ramakrishnan
Amity University, India

Loveleen Gaur
Amity University, India

ABSTRACT

The IoT and 3D printing can become a potent combination when it comes to launching new business initiatives driven by hard-core data and analytics and not really based on human perceptions or limited survey data. Previous empirical research has shown that drivers of new product performance are a mix of strategic, development process, organizational, and market environmental factors. This chapter attempts (1) to understand how introduction of IoT sensors embedded in customer appliances or wearable's sending real time customer information coupled with rapid prototyping using remotely located 3D printers can help address design considerations for new products, and (2) to provide an overview of how using IoT data and 3D printers for new product development and prototyping as an early stage activity can be done without using human imagination of restricted market survey data.

INTRODUCTION

Entering new markets and predicting customer behaviour and analysing his purchase pattern for a market driven product design and development is a risky proposition. The primary advantage of the Internet of Tings (IoT) in such business cases is their ability to create situational awareness and enable applications, human beings and machines to better understand surrounding for timely informed decisions in a dynamic daily changing environment. Sensor generated data related to our physical world can help us understand better and create newer products and services. 3D printing will help convert such data into visible prototypes (Nashiya, 2015).

DOI: 10.4018/978-1-5225-9624-0.ch014

"The worldwide 3D printing industry in expected to grow from \$3.07B in revenue in 2013 to \$12.8B by 2018 and 3D printing technology is destined to transform almost every major industry (Allied Market Research, 2015). 3D printing also called additive manufacturing is a process of making three dimensional solid objects from a digital file using specific raw materials which can vary from plastic to metal. 3D printing starts with providing a virtual design of the object to be printed in a Computer Aided Design (CAD) file format. Also of relevance are 3D scanners which use different technologies to generate a 3D model such as time-of-flight (uses laser light to probe the subject), structured / modulated light (project a pattern of light on the subject and look at the deformation of the pattern on the subject) or volumetric scanning (industrial computed tomography) (David, 2014).

The next step is to convert the 3D model using specialised software by a process called "Slicing" which is dividing a 3D model into hundreds or thousands of horizontal layers.3D printing has several ways of printing and all are additive with differences in the way layers are built to create the final object. These methods range from melting, selective laser sintering (SLS) and fused deposition modelling (FDM) or curing a photo-reactive resin with a UV laser or another similar power source one layer at a time (known as stereo lithography SLA).

The American Society for Testing and Materials (ASTM) developed a set of standards that classify the Additive Manufacturing processes into 7 categories as below (ASTM52900, 2015):

- **Vat Photo Polymerisation:** This uses a container filled with photopolymer resin which is then hardened with UV light source.
- **Material Jetting:** In this material is applied in droplets and overlapped layer by layer to build an object by using a small diameter nozzle.
- **Binder Jetting:** In this two materials are used: powder base material and a liquid binder, powder is spread in equal layers and binder is applied through jet nozzles that "glue" the powder particles in the shape of a programmed 3D object.
- **Material Extrusion:** This uses a technology called fused deposition modelling (FDM) in which the plastic or metal filament is heated and sprayed from a nozzle, which can move bi-directionally, into layers which gets hardened immediately.
- **Powder Bed Fusion:** The technology involves high power laser to fuse small particles of plastic, metal, ceramic or glass powders into a mass of desired 3D shape.
- **Sheet Lamination:** This involves sheets of metal, paper or polymer welded together by ultrasonic waves or adhesive glue.
- **Directed Energy Deposition:** This consists of a multi axis Robotic arm and a nozzle depositing metal powder with an energy source which can be laser, electron beam or plasma arc.

IoT and 3D printing can become a potent combination when it comes to launching new business initiatives driven by hard-core data and analytics and not really based on human perceptions or limited survey data which are generally bound to fail and may not be fully exhaustive or indicative. Previous empirical research has shown that drivers of new product performance are a mix of strategic, development process, organisational and market environmental factors (Geroski, Machin and Van Reenen 1993, Soni, Lilien and Wilson 1993, Capon, Farley and Hoenig 1990). Entering new markets and predicting customer behaviour and analysing consumers' purchase pattern for a market driven product design and development is a risky proposition.

New product development (NPD) in business or engineering refers to the steps required to bring a product from concept or ideation stage to the market. Krishnan (2001) describes new product development as the transformation of a market opportunity into a product available for sale which can be a tangible product or a service. Reducing time for NPD has led to concepts such as concurrent engineering and traditionally has eight stages namely idea generation, idea screening, idea development and testing, business analysis, beta testing, technical implementation, commercialisation and new product pricing.

In the traditional model, design feedback came from only sales figures and consumer complaints—and generally after design and production have already been completed which is too late for any meaningful corrective action or repairs. The IoT can play a major role in idea screening, beta testing, continuous product improvement and provide inputs for pricing by capturing key data elements in the background relating to the usage and efficiency of the product. Similarly, 3D printing can help in idea development and technical implementation. In this chapter we are explaining how 3D printing and IoT can change the traditional model of NPD to an agile iterative model which will involve a launch – data gather – product evolution cycle thereby reducing time.

BACKGROUND

Three-dimensional (3D) printing is a manufacturing method in which objects are made by fusing or depositing materials—such as plastic, metal, ceramics, powders, liquids, or even living cells—in layers to produce a 3D object. 3D printing or additive manufacturing starts with a 3D model which is then recreated into a solid form by superimposing layer by layer keeping in consideration all the complexity and detail without additional processing required by traditional manufacturing which involves injecting material into pre-fabricated moulds. Recent technology advancements have made 3D printers more affordable leading to an explosion in the commercial use of this technology (Kalani & Richard, 2014).

NPD teams are continuously dependant on data for product improvement and not hearing the voice of the consumer cause products to fail. Transforming data to actionable insights is the key to success in NPD scenarios. Embedded sensors in products are generating statistical dumps of data generated by actual consumer usage in a real world scenario and are much more realistic than laboratory experiments in controlled environment. These data will provide all departments of a business such as manufacturing, quality, and logistics a tool to make critical decisions.

Elisha et al. (2016) explained the data delivery and gathering in IoT which is of very high significance for this proposed model of 3D printing and introduced literature-backed solutions for topology, reducing costs and measuring network failure. Their work also covers avoiding inaccuracy due to RFID tag collision and ability to revive from node failures based on research from using heuristic approaches, to implementing localization techniques that contributes to finding the optimal network topology.

Mitzi et al. (1994) suggested that new product success is driven by a mix of strategic, development process, organisational and market environmental factors. Previous empirical research on new product performance has proved that a wide variety of factors can influence the outcomes of new product development activity. Mitzi Montoya-Weiss and Roger Calantone (1994) conducted a comprehensive review of this literature and observed a wide variety of study designs and methodological approaches. They developed quantitative comparisons of the results, which provide a look at the persistent exploratory nature of this research. They report a wide variation in results that are surprisingly nonconvergent.

Sue (2016) showed a future vision where it may be possible to even develop a fully functioning organ or rapidly produce made to order fixtures in the operation theatre or even create custom surgical equipment immediately customized for each patient based on human body anatomy 3D bio-printing offers additional important advantages such as highly precise cell placement and high digital control of speed, resolution, cell concentration, drop volume, and diameter of printed cells. Organ printing takes advantage of 3D printing technology to produce cells, biomaterials, and cell-laden biomaterials individually or in tandem, layer by layer, directly creating 3D tissue-like structures.

Ventola (2014) illustrates that medical applications for 3D printing are rapidly expanding with revolutionary consequences for medical industry. These are broadly in seven categories such as tissue and organ fabrication, creation of prosthetics, implants and anatomical models and pharmaceutical research regarding drug dosage forms, delivery and discovery. Companies that use 3D printing for commercial medical applications have also emerged. These include: Helisys, Ultimateker, and Organovo, a company that uses 3D printing to fabricate living human tissue (Klein, 2013).

Angewandte (2016) showed the positive 3D printing developments in the field of microfluidic devices which are still made with extensive manual labour and which can reduce labour and cost by a large extent along with better safety. As part of this 3D model is developed and the design can be assembled in modules by remote teams, and it's mechanical and fluidic behavior can be simulated using finite-element modeling. The processing is environmentally friendly and economically efficient and is predicted to replace most plastic molding techniques.

Bailey (2012) showed some techniques, methodologies and toll sets where challenges of performance, quality and reliability in electronic industry manufacturing can be achieved by reducing downtime in component malfunction for critical appliances. Inkjet printing can be used in electronics packaging as interconnections between electronic components. Conductive inks and dielectric inks substitute traditional printed circuit board (PCB) with inkjet-printed interconnections. Inorganic inks offer better conductivity as they consist of metal nanoparticles and organic solvent which make the ink printable. This will help reduce breakdown of traditional PCB or electronic components while ensuring higher quality aesthetics.

Ramakrishnan (2016) explained how IoT devices can help track location of the source of origin of information using BLE beacons. This might be relevant for NPD driven by IoT-3D printing as it may be necessary to identify the source of request for changes to the product and such changes may be again subject to a specific customer location, situation or context.

Existing research exists in the areas of new product development focuses on managing subjects and sources of ambiguity or, using IoT based configurable platforms for product life cycle management. However, the role of sensor based inputs in prototyping new products using 3D printers dynamically is still a greenfield area with not much research. The authors have tried to present concepts which are empirically tested in order to show the relation and benefits of combining IoT sensors with 3D printers for dynamic prototyping including new product developments and enhancements or corrections to design.

Figueroa (2013) suggested models for 3D printing of objects without manual intervention in the processing pipeline using 3D sensing devices like the Kinect sensor which can monitor and map gestures and movements thereby working in an augmented reality scenario. In this experimental prototype used for validating the model 3D Kinect camera tracks different poses of a human subject and generates a 3D model using a remote 3D printing of the entire subject similar to teleportation. The study presents initial successful results while also addressing the limitations of 3D printer and the speed of printing complex poses.

MacDonald (2014), emphasized how time to market can be reduced leading to better opportunity realisation for next generation products using 3D printing techniques like stereo lithography and fused deposition modelling in electronic products. The main focus of the study is to show the time saved when prototyping or manufacturing using 3D printing to a few hours as compared to designing on traditional technologies like pre printed circuit boards which may take days.

Hakima (2010) listed the architecture and networking technologies that are required to support the internet of things since traditional data requirements are far different from the IoT data networks which transport massive amount of data and hence need a robust and cost effective network. The difference being in terms of interval of data transfer which is much higher than the traditional networks and second is the volume of data. IoT will have to allow connectivity of a large number of objects of different type thereby presenting challenges of heterogeneity and scalability of communication networks with new functionality requirements of identifying, locating and sensing leading to high level of task automation.

Past research has shown that a product launch represents a major investment in revenue and profit for a company (Bhuiyan, 2011). A successful product launch can yield new markets, provide access to new customers, as well as increase business with existing customers. A product launch which is not based on proper market research or customer feedback may result in huge commercial losses and eroding of brand value.

Huang et al. (2014) proposed a unified framework of targeted marketing using customer preferences where the authors investigated two marketing strategies - the customer oriented and product oriented, introduced marketing models to test the relation between customers and products, and explained that the required parameters of a market value function depends on three information: customer profile, product profile, and transaction data.

Horvath (2001) proposed a method to extract knowledge background during definition of product objects in the form of initiated effect on product and also add the human thought process which is the starting point for initiation of a change in product feature. They conclude with an extended and integrated product definition model. Strategic supply chain management demands collaboration from all participants and understanding of customer and supplier constraints and the method will provide benefits such as radically improved customer responsiveness, enhanced customer service and satisfaction, increased flexibility for changing market conditions, improved customer retention and more effective marketing.

Han et al. (2015) conducted an interesting study on modelling customer preferences, proposed a concept selection method to evaluate future market performance of concepts and to choose the best one of them, and validated the method using field study. IoT introduction can further add on to this model by giving real time insight on customer preferences thereby getting the best model to market in a very fast time.

Ramakrishnan et al. (2016) has made an interesting point wherein machines can communicate with each other and follow a human world like hierarchy of superior subordinate relations taking the case of a smart electricity distribution in residential areas. This will help reduce unbilled energy and prevent wastage of energy. The same concept can be applied in IoT driven innovation wherein even the inputs send by IoT sensors can be validated by other IoT machines before being sent for 3d printing.

3D printed world has the potential to bring localised production of consumer goods or industrial goods from the current manufacturing location to right up-to customer door steps. For example, a broken door handle can be just printed and send to a customer immediately from a remote location by the manufacturer. Medical facilities will 3D print ankle replacements and even tissues while manufacturers can use Nano printing to print paper thin batteries. Add to this the IoT capabilities of detecting the requirement

Figure 1.

continuously and beforehand (without waiting for a break down) and we have a fully automated world with auto replenishment and detection systems across.

The authors are of the opinion that with an integration of 3D printing and IoT we have a technology that will transform the way business is conducted and the speed of responsiveness to customers will increase manifold. The time of fulfillment across extended supply chain will decrease considerably as manufacturing becomes highly customized and real time. Product development will be driven by objects (customer experience managers) communicating with other objects (3D design printers) and product versions will roll out more faster with a very high level of transformation and value addition as compared to the previous versions. This will be in deep contrast to the current slowly evolving product life cycle where each major upgrade is predominantly incremental and not radical different from the previous release. The above literature review shows that the author's views are very much in line with industry trends and have potential to replace human decision making on design and prototyping with intelligent captured data using IoT devices.

MAIN FOCUS OF THE CHAPTER

Intelligent Objects Observers: New Product Development

The tasks relating to new product development are as follows (Figure 1).

1. **Generating and Screening Ideas:** This involves a systematic search for new ideas; the sources can be internal customers, employees, competitors, distributors or suppliers. Customers can be a major source of ideas specifically of the company which already has products in the market. The resultant ideas need to be screened for commercial and technical viability and companies do this using formal review committees or market research.
2. **Testing Concepts:** The screened ideas are now converted into detailed concepts and then tested with customers either physically or symbolically to understand the customers perception of the product. The company can then project these findings to get an idea of the sales market volume.
3. **Analysing Market and Business Strategy:** This stage involves describing and identifying the target market, product positioning and sales and the market share and profit goals for the immediate year. It also covers pricing, distribution and marketing strategies. Finally it involves listing down long term sales, profit goals and marketing mix.
4. **Market Testing:** Post development of the concept to a product and test marketing to a restricted target segment are done prior to a full launch but this stage has risks of sabotage or competitors copying the concept.

5. **Commercializing:** The final step is commercialisation which involves manufacturing, advertising and promotion.

Manufacturing systems will become more accurate, environmental friendly and customer focussed with modern information systems. Sensors embedded inside applications or as standalone machines can be used as mobile or fixed robot to monitor an environment such as office building or factory. Craig et al. (1995) described the concept, design and initial implementation of systems called intelligent observers. The Intelligent Observer (IO) is a mobile robot which moves through an environment (such as an office building or a factory) while independently observing moving targets selected by a human operator. The robot carries cameras which allow it to track objects while keeping track of its own location. It interacts with humans who issues task-level commands, such as indicating a target to track by clicking in a camera image. The user could be located far away from the observer itself, communicating with the robot over a network. As the IO performs its tasks, the system provides real-time visual feedback to the user. These intelligent observers can fuse information from various sensors and using geometric information about the environment reconstructs a 2D or 3D view of the observed scene. Sensors today can measure anything from noise vibrations, electricity consumption at appliance level, proximity, timed usage, heat and pressure, frequency and strength of transmitted and received signals from Bluetooth or Infrared or Wireless access and many more as sensors and actuators are emerging as powerhouses of the IoT wave. The observers can be used for shop floor monitoring, security, or landmark detection.

Existing Reference Cases Worldwide

Since not many organisations have done this kind of dynamic prototyping of products based on inputs from IoT objects on customer preferences and manufacturing bottlenecks, the secondary sources of data including white-papers, scientific journals and feasibility study reports will be referred to on what all types of objects can be readily produced using 3D printing - e.g., Lithium Ion cells or even medical tissues being printed have been possible. We will also research the materials which can be used for 3D printing such as polyester, metals and fibres. Finally using secondary research survey, we will discuss on the kind of information that IoT devices and sensors are able to track in manufacturing products with reference to end consumer behaviour.

This paper will also focus on studies relating to the limitations while keeping a positive view of future trends to suggest solutions which can be implemented either today or maybe over a period of time as more advances are registered in 3D printing and IoT technology.

There are many cases of organisations having implemented IoT to help improve their products or develop new products. The IoT implementation may involve turning products into sensors or integrating them to provide data-based evidence for product engineers and also for better customer service. Data collection with the IoT devices is better than asking a customer to respond to questions relating to service or failures and move towards a proactive approach at a much reduced cost.

Earlier the organization had limited interaction with its customer's up to the point of sale. Organizations relied largely on reporting from field engineers post factor, random survey and product returns to gauge device durability and customer satisfaction. But through new capabilities to analyse the real-time IoT data provided by connected devices, companies can now gain key insights into product performance, consumer trends, and purchasing behaviour.

The first case is of Bosch, a leading healthcare product company. Bosch runs "Health Buddy" an early human response dependant IoT device that monitors the vital signs of patients and records patient inputs. This is targeted at patients with chronic health conditions who can be monitored on a continuous basis from their own homes and data is transmitted to remote medical practitioners. Health Buddy, acquired in 2007 from Health Hero, was a home management platform for CHF, depression, diabetes, hypertension, COPD, or weight management. It consisted of a small box with a screen and four buttons. Health Buddy would ask a patient survey questions daily, as well as prompt them to take their weight or blood pressure via connected devices, and send the data to a care provider. However, Bosch has officially closed its US subsidiary Robert Bosch Healthcare, reducing the scope its healthcare operations to a 50-person team based in Germany. This showed that the attempt was not very successful and also was prone to having human inputs proactively rather than a human independent monitoring. The data from this equipment was planned to be used by remote healthcare providers to develop custom made appliances for weight control which was planned to be done using 3D printers. It also allowed health care providers to track the efficiency of existing medications and thereby provide information to develop new more potent and effective variants of the medicine (Neil, 2011).

Trane, a division of Ingersoll Rand that focuses on industrial air conditioning, has changed both its product development practices and the products delivered based on IoT data and its insights. Trane makes HVAC products for a wide variety of uses, from homes to offices to manufacturing facilities. This segment has seen a major change as per available literature wherein manufacturers like Trane tended to own and control less of the Intellectual Property that went into producing equipment. Suppliers were inventing and controlling innovations. On other hand the purchasing process for HVAC equipment had become highly intermediated. Owners of facilities increasingly relied on consultants for advice about which systems to buy. Trane now addresses a whole new set of questions:

- What is this product's relationship to other building systems? What impact will it have on or from the smart grid?
- What new levels of efficiency and reliability can be achieved, both mechanically and from operational optimizations?
- How will connected products and services change Trane's relationship with customers?

The solutions and insights provided to the above facets by the IoT data shows that Trane is now able to find and defend new sources of value and show value add based on their real-time operating environment versus their initial design. Trane's devices are now not restricted to a turn on and off function during start and end of office hours but can do a lot more things such as matching its operation to the building's actual occupancy load, or a more complicated optimization like switching energy sources. The data allow an intelligent decision to be made at another level and that affects the capabilities needed from the device. Trane is able to address the needs of key influencers in the buying process by creating applications and services that help optimize HVAC systems (IoT World Show, 2016). This shows that IoT has indeed influenced product design using real time data flow and hence added value to the buyer. While as of now Trane is not using 3D printing but Haier in China has already developed the first 3D printed air conditioner (Whiteney, 2015). The authors see a future potential for Trane to integrate the IoT data with the 3D printer in line with what Haier has done in China to design the next generation of HVAC systems.

Coca-Cola uses data from its Freestyle and other vending machines to gain insights on their customers purchasing pattern including where, when, and how they are consuming their products. A rich insight thrown from the connected smart vending machines is spikes in its beverage consumption on college campuses before certain television shows air; this has helped up move towards targeted marketing. Testing of these devices started in around 60 locations in 2009 in US and another 200 locations in 2010 followed by Great Britain 16 locations in London in 2012. The project is a pioneering example of organizations harnessing the power of business mobility to fundamentally transform processes and experiences for consumers. Customers can select a base product and supplement it with additional flavours which provide insights for Coca-Cola to launch new products of the most popular flavours in other locations as a derivate of the customer insight (Wikipedia). Coca-Cola has also pioneered into the use of 3D printing (Ricardo, 2013) in advertisements and for recycling plastic bottles to help protect the environment. As per the authors while currently the two developments i.e. IoT and 3D printing is not being used on combination in future Coca-Cola can integrate the two by adaptations of the Freestyle machine to provide consumers small 3D printed souvenirs and fashionable items (Journey Staff, 2014).

IRhythm, a company that manufactures a patch that can detect heart problems, uses the data flowing from devices to ensure that its products operate as intended and better understand usage patterns. The ZIO Service integrates a discreet patch that continuously monitors and records up to 20,000 minutes worth of patient heartbeats over 14 days, with delivery of a concise, high quality report. With over 400,000+ Patients and 100+ Million Hours of Annotated ECG this product has been very successful and also the data is being used to further fine tune the product for better effectiveness. This data can be used by medical practitioners to understand the efficiency of machines thereby providing more insights than field trials of medicines on real time basis thereby helping with new formulation releases. As per the authors with new developments in the field of medical tissue printing it is now possible to recreate 3D bio printed heart valves (Robert, 2015) and the IRhythm data can be used in conjugation with a 3D printer in the future.

Optomec helps make consumer and industrial devices smart and connected with 3D Printed sensors and antennas for Internet of Things applications. Optomec Aerosol Jet technology is a high volume printing solution for the production of 3D antennas and 3D sensors that are tightly integrated with an underlying product ranging from Smartphones to Industrial Components. With Aerosol Jet systems, customers can print high-precision integrated electronics onto 3D plastic, ceramic and metallic structures at extremely fine resolutions. Aerosol Jet is used to make consumer and industrial products smart and connected so that they can provide feedback to engineers to improve future designs and to service technicians to proactively maintain products based on their condition. This data on product performance, customer behaviors and environmental factors gives manufacturers unprecedented levels of insight, fundamentally changing the way products are developed, maintained and supported. Since life cycle costs of industrial products can be 10 to 15 times higher than the original equipment purchase cost, smart connected devices offer customers and manufacturers significant potential for cost savings and improved product performance. These printers are being used worldwide by companies like Bosch, Air Force Research Laboratory, and Neotech.

Audi's Smart Factory attempts to optimize car production efficiency, with the help of highly specialized equipment that uses the IoT to communicate seamlessly. The factory uses 3D printers to create custom parts as needed, while using 3D projections to match the customized parts to the car chassis. This process also uses drones and other cross-functional electronics, communicating with one another to increase production speed while maintaining a high-quality product. Humans and robots work hand

Figure 2.

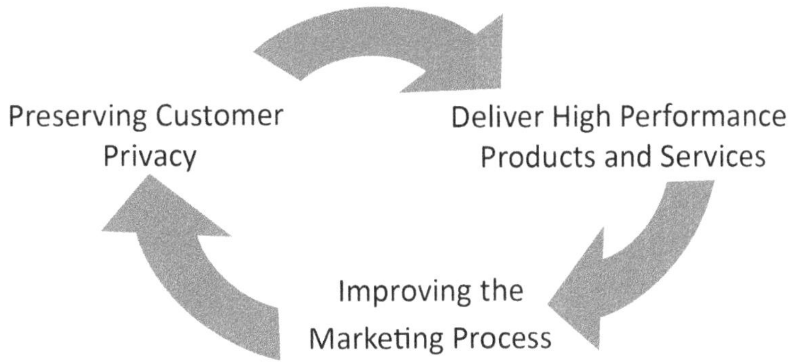

in hand, producing car-body parts from a 3D printer and resulting in cars driving in piloted mode from the assembly line. Employees in the pre-series center combine the virtual and the physical worlds of automobile development extremely precisely. The system allows the projection of virtual 3D components onto the car. In Audi's tooling division, a 3D printer produces complex metal parts, and intelligent tools are able to exactly distribute the high pressure forces to stamp sheet-metal parts that are accurate to one hundredth of a millimeter. And in the assembly process at the Ingolstadt plant, a robot works with the employees on the production line without a protective fence, passing them components at exactly the right speed and in an ergonomically pleasant position (Research and Markets Report, 2015).

Streetline is another example of how a company is using IoT analytics to build new products. The company recently launched Streetline IoT Gateway-to create smart cities and smart campuses. Analysing the data streaming into the Gateway, Streetline realized they could build a successful business by harnessing smart data and advanced analytics to help cities, universities and campuses provide guidance on available parking and parking enforcement. This case illustrates that IoT can help build new data based business models which can be readily 3D prototyped. The Gateway allows Streetline customers to effectively capture data from in-ground sensors and video cameras, for effective decisions. The product has derived value added features such as better parking and ability to decrease congestion, and thereby build a revenue stream for Streetline at the same time which is reflecting positively in the company's top line.

Potential for Innovation

The above study has thrown up the following areas of business relevance for IoT and 3D Printing with respect to new product development as shown in Figure 2.

Deliver High Performance Products and Services

The IoT in combination to 3D Printing can vastly speed up product R&D. By using sensor, diagnostic, and user interaction data from devices, companies get full product transparency. By observing how customers are using their products, businesses can quickly update features or adjust future release models to closely align with customer preferences and needs. They can also forecast where consumer preferences are going in the future. This insight about customer future preferences can eliminate tedious trial and error, and companies can know real time what is successful and what has low acceptance, leading to better product usage. With this data, it will be possible to generate a 3D print design of a new prototype. In future it will be possible that IoT devices directly interact with remote 3D printers to print an evolving and progressive prototype for new products with a reduced role for human product engineers.

Customer service improves with access to this level of detail on product usage, product performance and customer preferences. Proactive notification to customers affected by an outage or affected device can achieve very high levels of customer service. Companies can also then better position their customer support centres to allow more immediate service, reducing costs and appeasing customers simultaneously. This is the driving force behind IoT captured customer usage and preference data and 3D printers generating immediate services and parts replacement without the traditional waiting time for procuring or sourcing spares against faulty components as is the norm now.

Improving the Marketing Process Using IoT Data and 3D Printed Prototypes

One of the ways IoT focussed new product development companies can move away from a general strategy to a tailor made product for customers is by using 3D printers to create models specific to data generated by a customer and his usage pattern. Advertisers and marketers can reach individual segments of their customer base. So instead of flooding all customers with the same offer, marketing teams can create personalized offers that keep the customer engaged and satisfied with the product. Reduced marketing costs result from core targeted, successful marketing efforts. With data-driven insights from connected devices, companies can know from the start of a campaign how and where to position their new product lines on social media and elsewhere to get the best response from the market of customers and prospects. Philips (Gary, 2016) has launched a connected healthcare devices campaign to monitor lifestyle and use that data for positioning their health care services and products which again can be 3D printed.

Preserving Customer Privacy

Using IoT for product development by capturing buyer's preferences through his actions also raises concerns of protecting customer privacy. These concerns can be reduced by omitting personally identifiable information to limit the identification of individuals Weber (2010), defined privacy in more detail. Privacy includes the concealment of personal information as well as the ability to control what happens with this information, the right to privacy can be considered as either a basic and inalienable human right, or as a personal right or possession. With IoT it may be possible to violate privacy without the knowledge or approval of the subject matter. Privacy data may become the interest of the state as well as private marketers which is a cause of concern.

Traditional product development involved a sequential phase of requirement definition, design or prototyping, building and testing unlike operational performance feedback which is a mainstay of product development in the IoT era. IoT suggests a proactive approach which is very much needed to allow engineers to apply analytics to operational and performance data, deriving meaningful insights. As a result, engineering teams can learn continuously with data and usage, thereby enhancing product performance much more quickly than they can use traditional models.

Figure 3 shows the authors illustration of the three most important aspects to be kept in mind while designing an IoT-3D solution.

A simple case in reference which can be used by an electronics equipment company is to optimize the positioning of its remote control sensor. Traditionally if a sensor was not fitted properly it would

Figure 3.

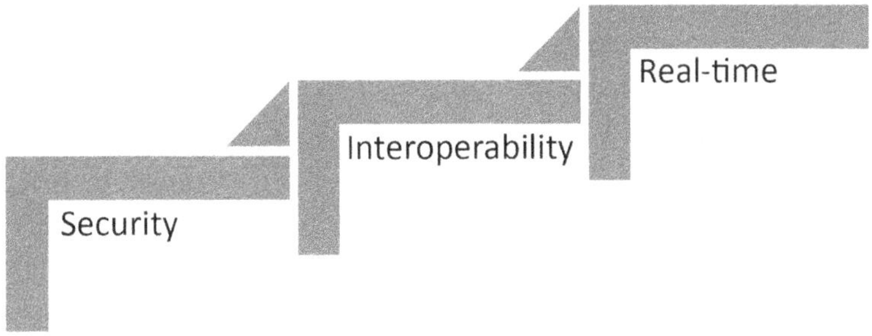

often result in a customer attempting many times in pressing the remote control and ultimately getting through, some customers may complaint or giving a feedback (some may ultimately adjust and change their operating habit) and ultimately result a product recall or design change. However, this would result in a significant loss of goodwill and also it may take a lot of time for such feedback to be consolidated to convince the need for a product change. With an integrated IoT-3D printing technology in place the object - in this case the electronic equipment - would transmit data about this design flaw to a remote 3D printer which will print a revised model to address this design flaw thereby leading to a near real time design change keeping in mind the customer preferences and feedback on existing design limitations.

RESULTS AND ANALYSIS

Taking the case of automobile industry a car with IoT sensors can transmit and monitor information about car usage per day in kilometers and the speed of travel to suggest a lower model or a higher cost effective laced, all this information can be machine generated without needing human discretion or judgment. Add to this the ability to monitor the wear and tear of tyres or engine which can help in passing data related to these thereby allowing remote 3D printers to come up with different designs which can be more economical to both the customer and the manufacturer. This process of additive intelligence powered by data captured by IoT devices and modeled by 3D printers to arrive at near perfect models of accessories or automobiles will add immense value.

Yet another use case can be of consumer electronics wherein an air conditioner can suggest to the owner or the manufacturer what the optimum capacity should be based on sensors monitoring the running of compressor and give a cost benefit analysis of a replacement viz a viz the running cost. Going to a subcomponent part level it may be easier to predict faults and rectify them in design considerations by combining the data of failures and faults generated by IoT embedded chips inside consumer appliances and remote 3D printers which can actually work intelligently on the point of failure data.

The authors have done research on a very common place problem of the apparel industry and the fashion clothes segments in India. India's population specially the millennial is addicted to buying clothes and their buying trends and frequencies have been driven even more by the proliferation of high street shops and malls and many international and home grown brands in the apparels section for men, women and children riding on the back of an improving economic ecosystem where surplus disposable incomes have increased multi folds with cheap circulation of plastic money. However, for merchants and for garment manufacturers getting insights on what really attracts the customers is more a culmination of traditional thought process, ad hoc market research and sales figures. As a retailer it would be worthwhile to stock goods which are more in demand and can convert to sale quicker to avoid having slow moving apparels with limited commercial viability in stocks. Add to it the crowd generated in trying out different outfits and then repackaging the tried garments leading to investment of human resources. The net result could be a very low ratio of garment trials and actual sales. When the garments are repackaged along with it goes the valuable data insight on how many people actually tried but did not like it thereby losing opportunity to get direct customer insight and review. A Bloomberg report of this year (James, 2016) clearly shows that 3D printing of garments is now a near possibility and will trickle down to masses. However, there are limitations of rigidness and material may not be similar to clothes or as comfortable as normal clothes with the current infrastructure. A rich use case which may develop is the ability of IoT driven SmartMirrors equipped with cameras which can analyze and predict customer preferences for garments and at the instant get a 3D printed garment more attuned to his style thereby making an immediate sale.

In another use case for retailers, IoT technology can in instance of accessories such as hardware, jewelry, footwear, eyewear which can be detected using SmartMirrors or intelligent cameras provide real time data to nearby 3D printers and manufacturing designs on demand.

The concept can also be very relevant in the automobile industry where each consumer has a unique driving habit as a result of which the same vehicle driven by multiple people gives different results and efficiencies. Today it will be impossible to imagine a world where we have tailor made and finely attuned models for a group of customers driven by their behavioral and usage patterns along with tailor made pricing but the situation is possible using IoT sensors sending usage information to remote 3D printers which in turn can design and produce custom made offerings. This is being observed now with Connected Cars project. By 2018, one in five cars on the road will be self-aware and able to discern and share information on their mechanical health, global position, and status of their surroundings (Gartner, 2015).

SOLUTIONS AND RECOMMENDATIONS

The authors have done a research on the customer behavior on garment shops to further substantiate our findings and assumptions above. As part of the research over 100 prospective buyers of ready-made garments were observed and the summary of findings is listed below.

- 94 of the respondents did not convert to sale or did not buy the clothes from the retail shop. Of these around 68 tried different clothes while the remaining 26 did not like the clothes design hanging and did not feel it worthwhile to go for trials.
- Of the 68 people who tried around 31 did not find color to their liking and 23 did not find the right size while the others felt the cost and the feature were not in sync in other words overpriced.
- The prospects spend on an average around 45 minutes in the retail shop.
- The average sales person took three minutes to bring the clothes back from the trial room and place it in the correct position and rack.

The authors suggest below how the situation could improve using the IoT and 3D printing (assuming technical feasibility with improving technology).

Table 1. Comparison of the current methods for design and the IoT and 3D printing-driven design

	Current Methods for Design	IoT and 3D Printing-Driven Design
When it is initiated	• Before Product Launch • Based on market feedback • Based on technology evolution	• Real time customer behavior analysis • Real time based on complaints, failures and usage patterns
How it works	• Usually prototyping happens by design engineers and then a pilot launch happens which helps assess market response. • Next design change is more technology driven rather than customer driven and happens less frequently	• Real time design changes possible based on continuous usage data transmitted across networks and with rich analytical engines supporting and suggesting findings. • Human intervention in design is more of commercial and technical validation.
Benefits	• Leverages on human experience which further learns from past mistakes.	• Design is rich interpretation of customer desire and preferences and may lead to perfect design
Limitation	• More reactive than proactive. • Based on majority views • Suffers from customer accepting a feature since no credible alternative possible • Feedback mechanism seldom reaches design team and customers settle for the best of the worst approach	• Each customer may have his own perception and style of operating leading to tailor made recommendations at even individual level. • Less standardization • Commercial feasibility, safety or other unstated customer low priorities may be ignored • Product may have hundreds of evolution cycle.
Application Areas	Across industries: • More real time in case of food items. • Rarely used for automobiles this goes with expert opinions.	Across industries: • Consumer electronics TV/Fridge/White goods. • Automobiles and transportation. • Apparels and clothing.

- IoT devices could capture using embedded tags inside the garments on how many times a particular garment reached the trail room (which implies an interest level). The same tags can also see garments which were never picked up and did not evoke any interest and hence may be categorized as ones whose design was not as per customer preferences. This data could provide insights to the garment manufacturer on what has a higher potential to sell so that similar designs can be launched more.
- IoT devices could also tag using SmartMirrors or interactive kiosks customer preferences for a specific color and size. If the color is not available this data can again signal the manufacturer or the retailer on the color preferences being most sought after or size of garments in demand.
- A faster response would mean lesser time spent in the shop thereby making space for new customers or prospects.

The same approach can be used for other areas including customer electronics, automobile manufacturing, and lighting or air conditioning products. Table 1 compares the current methods for design and the IoT and 3D printing-driven design.

FUTURE TRENDS AND RESEARCH DIRECTIONS

Future trends in IoT devices will be in the direction of device originated monetary transactions, social objects and devices with limited form factors. This will increase the potential to help further in new product development aligned to customer expectations and preferences based on the data gathered on behavior patterns which will be generated by the above IoT devices.

Figure 4.

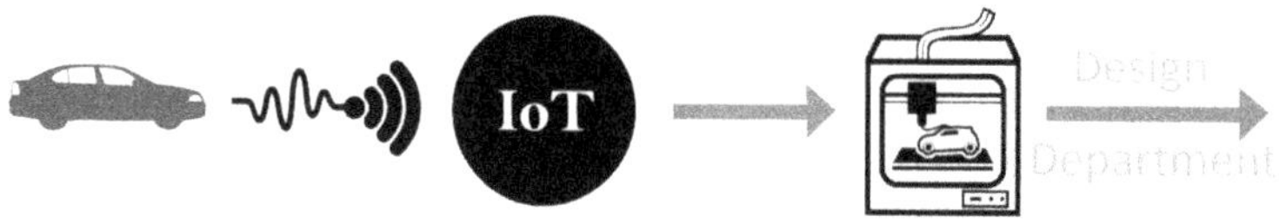

According to Kubler (2015) in traditional product companies, value creation was identified with customer needs and manufacturing engineered devices while today IoT with analytics, subscription, apps, contextual information and cognitive capabilities has become the future. While information and knowledge is the core of the IoT era, it nonetheless remains challenging to perceive and extract the real value of those assets, as information is not as tangible and concrete as physical assets.

The IoT and the data from connected devices are revolutionizing the interaction between companies and customers and adding value. Product managers with the huge data visibility from IoT devices, sensors can create higher performing superior efficiency products directly based on customer stated and unstated needs. Marketers can build better and sustained relationships with their customers powered from insights of data on customer's behavior and also provide more personalized service and products, while keeping data confidential.

There are numerous use cases in different verticals of product engineering where IoT gathered data will be an input for new product design and adoption while also continuously generating newer prototypes using remote 3D printers at Original Equipment Manufacturers premises for product refinement.

Some use cases of future trends in IoT would involve Intelligent Objects monitoring production lines, conveyor belts, automobile engines/tyres/fuel consumption, energy consumption and distribution, medical treatment and efficacy, logistics movement and many other areas sending data and also receiving instructions based on analysis to act and prevent damage. A transformer or an electric meter shutting down to prevent a blackout when a home appliance draws excess current is one such case. These will further contribute to feature enhancements in new product launches.

Figure 4 shows a schematic of the futuristic models where connected cars (with IoT sensors embedded in all components be it tires, engines, clutch plates) will transmit information of an existing customer to a remote 3D printer which will 3D print a metal and rubber or plastic components of the vehicle tailor-made to the customer to provide him more economy and savings. For example, this might be a custom-made clutch plate which activates with the desired force or even a customized seat with flexible height or fabric based on the height of the driver.

In the above futuristic model which seems very much probable it is suggested that all appliances such as vehicles will be fitted with sensors which will send information on driver specific usage such as gear changes or acceleration patterns, this can be transferred to a 3D printer connected to a computer which will create a new design (based on data being processed) and send to the design department for human approval. Figure 5 shows an iterative prototyping improvement model based on the IoT and 3D printing.

Industrial IoT technologies will open up a host of new capabilities to machine designers. They will add significant value and competitive advantage to their OEM products. But designers will have to anticipate how their machines fit into the new object-driven world. That will require the specification of robust hardware, ample processing power, and sufficient memory to fulfill the enhanced software requirements that lie ahead. In addition, machines must work with the latest standards to transport data to higher-level networks for asset-management applications.

Figure 5.

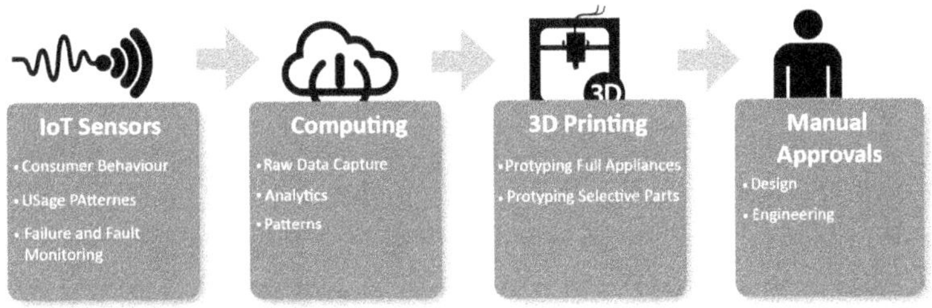

End-user customers will benefit from simpler equipment commissioning at lower cost which is designed and derived from their usage data and their subconscious feedback provided to servers based on their day-to-day usage actions of an appliance, readily prototyped by a remote printer sitting thousands of miles away near the design lab of an equipment manufacturer, and validated at the last stage by human beings primarily from commercial and security viewpoint.

Innovative product development teams are beginning to look at enormous streams of real time machine data into actionable information to improve the product development process by understanding where product innovation is necessary, which features are most used or least user or more desirable, and how to lower their overall cost of ownership.

The area opens up a wide opportunity for engineers and researchers to work on current limitations and disadvantages being faced by consumers due to design faults or short-sightedness of design engineers thereby quantifying the benefits of this technology from the consumer point of view and the manufacturer stand point. The authors see significant benefits in terms of costs and customer conveniences for all stakeholders as faults come down, usage picks up and people more readily adapt to new devices and appliances or newer service offerings.

The inclusion of IoT in devices also increases opportunities and challenges for targeted marketing which focuses on identifying specific group of products with respect to certain customers or vice versa while also developing appropriate solutions for those targeted groups.

CONCLUSION

In this paper, a collection of current papers on data collection and delivery has been reviewed. Although there are many researches in independent areas of IoT and 3D printing with different solutions but there is no research on merging both these latest technologies to touch upon a hereto human resource dominated area of new product design and enhancements. The chapter objectives of understanding how IoT enabled devices could potentially work as messengers of customer feedback to design team and further how the objects can themselves substitute to some extent the human design team using this information has been discussed by the authors giving real life examples along with technology limitations. Further the benefits of such a fusion of IoT with 3D printing in terms of cost, reduce MTBF and constant upgrade and evolution were highlighted. Current New Product developments practices are still a major disconnect from the field situation in terms of customer observation and feedback and are more laboratory based and confined to the product design team. Numerous cases some of which are futuristic have also been discussed and a small study with a limited sample size on how actual benefits can be realized has also been discussed.

While both IoT and 3D printing are still evolving technologies and hence suffer from nascent issues like network congestion, reliability of sensors, capability limitations of 3D printers, limitations of material and human reluctance to handover control to machines, it is more or less evident from early adopters that both concepts of IoT and 3D printing are going to change the way NPD teams handle design evolution in the coming future. Whether we can combine both of these and handover control to SmartObjects and relegate ourself to secondary citizens of the WWW while making objects the "FirstCitizens" in terms of compute power, data transmission volumes and productivity workloads is something that will become clear in days to come.

Finally, the major drivers which will ensure that this technology combination proves successful from new product design aspect will be three folds: Firstly, manufacturer's willingness to compromise on standardization vis-à-vis meeting customer desires, secondly technology evolution pace in both 3D printing and IoT and finally human behavior to accept data-driven object-suggested design over traditional knowledge and insights.

This research was previously published in The Internet of Things in the Modern Business Environment edited by In Lee, pages 189-209, copyright year 2017 by Business Science Reference (an imprint of IGI Global).

REFERENCES

Allied Market Research. (2015). *3D Printing Market by Technology and Material-Global Opportunity Analysis and Industry Forecast 2014-2020*. Electronic Systems and Devices.

ASTM International. (2015). *Standard Terminology for Additive Manufacturing*. Retrieved Sep 09.2013 from http://web.mit.edu/2.810/www/files/readings/AdditiveManufacturingTerminology.pdf

Au, A. U., Huynh, W., Horowitz, L. F., Folch, A., & Angew. (2016). 3D-Printed Microfluidics. Angewandte Chemie International Edition, 55(12), 3862-3881. doi:10.1016/j.wneu.2013.07.009

Bailey, C. (2012). Design, simulation and modelling for nanopackaging applications: current capabilities and future requirements. *12th IEEE Conference on Nanotechnology*. 10.1109/NANO.2012.6322238

Barnaghi, P., Wang, W., Henson, C., & Taylor, K. (2012). Semantics for the Internet of Things: Early Progress and Back to the Future. *International Journal on Semantic Web and Information Systems*, *8*(1), 1–21. doi:10.4018/jswis.2012010101

Becker, C., Gonzalez-Ba, H. H., Latombe, J. C., & Tomasi, C. (1995). An Intelligent Observer. *Fourth International Symposium on Experimental Robotics, ISER'95*.

Bhasin, H. (2016). *New Product Development*. Retrieved July 3, 2016 from http://www.marketing91.com/new-product-development/

Bhuiyan, N. (2011). A framework for successful new product development. *Journal of Industrial Engineering and Management*, *4*(4), 746–770. doi:10.3926/jiem.334

Brun, E. & Steinar, A. (2009), Managing ambiguity in New Product Developments. *Creativity and Innovation Management, 18*(1), 24-34. doi:10.1111/j.1467-8691.2009.00509.x

Capon, N., John, U. F., & Scott, H. (1990). Determinants of Financial Performance: A Meta-Analysis. *Management Science*, *36*(10), 1143–1159. doi:10.1287/mnsc.36.10.1143

Chan, S. L., & Ip, W. H. (2011). A dynamic decision support system to predict the value of customer for new product development. *Elsevier Science Direct*, *52*(1), 178–188.

Colmenar, E., Al-Turjman, F., & Biglarbegian, M. (2014). Data delivery and gathering in IoT applications: An overview. *Local Computer Networks Workshops (LCN Workshops), IEEE 39th Conference*, 790-795.

Craig, B., González-Baños, H., Latombe, J. C., & Tomasi, C. (2005). An Intelligent Observer Human Augmentation. *Lecture Notes in Control and Information Sciences*, *223*(4), 151–160.

Figueroa, N., Dong, H., & Saddik, A. H. (2013, Nov 24). *From Sense to Print: Towards Automatic 3D Printing from 3D Sensing Devices*. Paper presented at IEEE International Conference on Systems, Man, and Cybernetics(SMC). 10.1109/SMC.2013.833

Geroski, P., Steve, M., & Reenen, J. C. (1993). The Profitability of Innovating Firms. *The Rand Journal of Economics*, *24*(2), 198–211. doi:10.2307/2555757

Guo, B., Zhang, D., Wang, Z., Zhiwen, Y., & Xingshe, Z. (2013). Opportunistic IoT: Exploring The Harmonious Interaction between Human and the Internet of Things. *Journal of Network and Computer Applications*. doi:10.1016/j.jnca.2012.12.028

Hakima, C. (2010). *The Internet of Things – Connecting Objects to the Web*. London, UK: Wiley.

Hakima, C. (2013). *The Internet of Things – Connecting Objects*. London, UK: Wiley.

Hausman, K. K., & Horne, R. (2014). *3D Printing for dummies*. Hoboken, NJ: John Wiley & Sons, Incorporation.

Hermans, P. (2014). *How could the Electronics Industry use 3D Printing?* Retrieved Jan 17,2014 from http://www.3dprintingelectronicsconference.com/3d-printing-electronics-conference/how-could-the-electronics-industry-use-3d-printing/

Hongming, C., & Xu, L. D. (2014). IoT based Configurable Information Service Platform for Product Life cycle management. *IEEE Transactions on Industrial Informatics, 10*(2), 1558-1567. doi:10.1109/TII.2014.2306391

Horvath, L. (2001). Collaboration: The key to value creation in supply chain management. *Supply Chain Management: An International Journal*, *6*(5), 205–207. doi:10.1108/EUM0000000006039

IoT Case Studies. (2016). *Heidelberg Pioneers the Smart, Connected Printing Press*. Retrieved Jan 1,2016 from http://www.ptc.com/internet-of-things/use-cases

iRythm Technologies. (2016). Retrieved on Aug 10,2016 from http://www.irhythmtech.com/patients-heart-arrhythmias-afib.php

James, T. (2016). *The Future of Fashion Is 3D Printing Clothes at Home*. Retrieved April 15, 2016 from http://www.bloomberg.com/news/articles/2016-04-15/3D-printing-is-poised-to-bring-haute-couture-into-the-home

Kime, M. B. (2016). *Trane At IoT World*. Retrieved May 4,2016 from http://www.trane.com/commercial/north-america/us/en/about-us/newsroom/press-releases/trane-at-iot-world.html

Klein, G. T., Lu, Y., & Wang, M. Y. (2013). *3D printing and neurosurgery--ready for prime time?* Retrieved Jul 16, 2013 from https://www.ncbi.nlm.nih.gov/pubmed/23871811

Krishnan, V., & Karl, T. U. (2001). Product Development Decisions: A Review of the Literature. *Journal of Management Science*, *47*(1), 1–21. doi:10.1287/mnsc.47.1.1.10668

Kubler, S., Min-Jung, Y., Cassagnes, C., Främling, K. D., & Skilton, M. (2015). *Opportunity to Leverage Information-as-an-Asset in the IoT -- The Road Ahead*. Paper published in Future Internet of Things and Cloud (FiCloud), 3rd International Conference. doi: 10.1109/FiCloud.2015.63

Macdonald, E., Salas, R., Espalin, D., Perez, M., Aguilera, E., Muse, D. & Wicker. (2014). 3D Printing for the Rapid Prototyping of Structural Electronics. *IEEE Access, 2*, 234-242. doi:10.1109/ACCESS.2014.2311810

McCaskill, S. (2015). *Coca-Cola Is Using The IoT To Create New Coke Flavors*. Retrieved from http://www.techweekeurope.co.uk/networks/m2m/coca-cola-using-iot-create-new-coke-flavours-180916#wBezSPZAQXStZszf.99

Montoya-Weiss, M. M., & Calantone, R. (1994). Determinants of New Product Performance: A Review and Meta-Analysis. *Journal of Product Innovation Management*, *11*(5), 397–417. doi:10.1016/0737-6782(94)90029-9

Moses, A. (2012). *LG smartfridge tells you what to buy*. Retrieved January 10, 2012 from http://www.smh.com.au/digital-life/hometech/lg-smart-fridge-tells-youwhat-to-buy-cook-and-eat-20120110-1ps9z.html

Nashiya, S. (2015). *3D Printing + IoT= Future*. Retrieved Sep 01, 2015 from https://www.fingent.com/blog/3d-printing-iot-the-future

OPTOMEC. (2016). *Technology*. Retrieved August 03, 2016, from Optomec.com: http://www.optomec.com/Additive-ManufacturingTechnology/Overview

Osweiler, K. (2013). *IRhythm drives innovation with the internet of things*. Retrieved Dec 4,2013 from https://gigaom.com/2013/12/04/irhythm-drives-innovation-with-the-internet-of-things-2/

Ramakrishnan, R., & Gaur, L. (2016, Jan 21). Smart electricity distribution in residential areas: Internet of Things (IoT) based advanced metering infrastructure and cloud analytics. *International Conference on Internet of Things and Applications (IOTA)*. 10.1109/IOTA.2016.7562693

Ramakrishnan, R., Gaur, L., & Singh, G. (2016). Feasibility and Efficacy of BLE Beacon IoT Devices in Inventory Management at the Shop Floor. *International Journal of Electrical and Computer Engineering, 6*(6). Retrieved from http://iaesjournal.com/online/index.php/IJECE/article/view/10807

Research and Markets. (2016). *Business Report 2016-2022: Intelligent, Digitized & Automated Smart Factories to Transform the Manufacturing Industry - Research and Markets.* Retrieved September 15, 2016 from http://www.researchandmarkets.com/research/s2dt9h/smart_factory

Rex, D., & Stoli, S. (2014). *Application of 3D laser scanning in a Production Environment.* Retrieved Jan 21, 2014 from http://publications.lib.chalmers.se/records/fulltext/202372/202372.pdf

Ricardo, B. (2013). *Coca-Cola shows off the untapped potential for 3D printing in advertising.* Retrieved August 20,2013 from http://venturebeat.com/2013/08/20/coca-cola-3d-printing/

Rishikesha, T. K., & Ganesh, N. P. (1999). *Creating Successful New Products: Challenges for Indian Industry.* Retrieved July 31, 1999 from http://www.iimb.ernet.in/~rishi/epwcsnp.pdf

Soni, P. K., Gary, L. L., & David, T. W. (1993). Industrial Innovation and Firm Performance: A Re-conceptualization and Exploratory Structural Equation Analysis. *International Journal of Research in Marketing, 10*(1), 365–380. doi:10.1016/0167-8116(93)90019-U

Staff, J. (2014). *Transforming 3D Printing Using Recycled Plastic Bottles.* Retrieved Jun 17, 2014 from http://www.coca-colacompany.com/stories/ekocycle-transforming-3d-printing-using-recycled-plastic-bottles

Streetline Inc. (2013). Retrieved Jan 10, 2013 from http://www.prnewswire.com/news-releases/streetline-closes-25m-financing-to-enable-internet-of-things-for-smart-cities-186319301.html

Sue, P. (2016). The Atj Report. *American Journal of Transplantation.* Retrieved Sep 10, 2016 from http://onlinelibrary.wiley.com/doi/10.1111/ajt.2016.16.issue-5/issuetoc

Swedberg, C. (2013). Trane Restocks Supplies at Assembly Plant Via RFID. *RFID Journal.* Retrieved online Jul 23, 2013 from http://www.rfidjournal.com/articles/view?10851

Szczerba, R. J. (2015). *No Donor Required: 5 Body Parts You Can Make With 3-D Printers.* Retrieved Jun 17, 2015 from http://www.forbes.com/sites/robertszczerba/2015/06/17/no-donor-required-5-body-parts-you-can-make-with-3-d-printers-2/#7f2b96094a3b

Ventola, C. L. (2014). *Medical Applications for 3D Printing: Current and Projected Uses.* Retrieved Oct 2014 from https://www.ncbi.nlm.nih.gov/pmc/articles/PMC4189697/

Vermesan, O., Friess, P., Guillemin, P., Gusmeroli, S., Sundmaker, H., Bassi, A., ... Doody, P. (2011). *Internet of Things Strategic Research Roadmap.* Aalborg, Denmark: River Publishers.

Versel, N. (2011). *Bosch Healthcare chief is bullish on Telehealth-enabled care management.* Retrieved September 15, 2011 from http://www.mobihealthnews.com/13245/bosch-healthcare-chief-is-bullish-on-telehealth-enabled-care-management

Weber, J. (2016). *3D Printing is poised to bring haute couture into the home*. Retrieved April 15, 2016 from http://www.bloomberg.com/news/articles/2016-04-15/3d-printing-is-poised-to-bring-haute-couture-into-the-home

Weber, R. H. (2010). Internet of Things – New security and privacy Challenges. *Computer Law & Security Report, 2010*(January). Retrieved from https://www.researchgate.net/profile/Rolf_Weber3/publication/222708179_Internet_of_Things_-_New_security_and_privacy_challenges/links/0c96053cab03fee371000000.pdf

Whitney, H. (2015). *The 'World's First' 3D Printed Air Conditioner Sells in China for $6395*. Retrieved Mar 13, 2015 from https://3dprint.com/50805/3d-printed-air-conditioner

KEY TERMS AND DEFINITIONS

3D Printer: 3D printing or additive manufacturing is a process of making three dimensional solid objects from a digital file. 3D printers are devices which implement this 3D printing model.

IoT: Internet of things is a connected network of smart objects each of which has a capability of uniquely identifying each other, communicating across digital network and intent to broadcast information about itself and surroundings usually aided by sensor technologies.

MTBF: Mean time between failures of a machine is a statistical calculation of the time gap between two successive failures

Smart Mirror: Traditional mirrors which function more than just showing reflective image to showing data, capturing photographs or displaying other content with capability to accept inputs form the user in written and verbal formats.

Chapter 15
Selection of Prototyping Process and Part Orientation for Virtually Manufactured Gears

Divya Zindani
NIT Silchar, India

Kaushik Kumar
Birla Institute of Technology, India

ABSTRACT

This chapter explores transformation of a virtually produced gear to a physical product using rapid prototyping technique. The same can be segregated into two parts. The first being selection of the best possible rapid prototyping technique using dynamic mechanical analysis and the next was optimization of the process parameters of the selected rapid prototyping technique considering tensile strength as the criteria. The results indicated that fused deposition modeling process provides the best solution amongst the techniques studied. Moreover, as orientation of the product in rapid prototyping techniques plays an important role in providing the strength and also accuracy of the product profile, from results it was also ascertained that strength and profile accuracy of the product was optimized with horizontal position.

INTRODUCTION

The most prominent challenge faced by product designers is to be able to present a tangible product. Despite advancements in 3D CAD technologies, such situations usually end with the product designer being asked: "Can you prove it?" Additive manufacturing or in the context of this work, rapid prototyping, has come as a shot in the arm. With the ability to present a tangible representation of the concept in the form of a physical object within short product lifecycles and even shorter development times, this concept is gaining importance and popularity with designers and engineers.

DOI: 10.4018/978-1-5225-9624-0.ch015

In the present work, an arrangement is required to be made to transform a virtual object into reality using Rapid Prototype (RP) technique in order to produce a virtually manufactured gear. The database of the output product i.e. the virtual gear serves as the input files for RP machine and plastic gears are manufactured.

RAPID PROTOTYPING TECHNIQUES

Rapid prototyping or 'Layered manufacturing technology' describes a wide range of techniques wherein 3-dimensional objects are constructed in laminated form. Though diverse, each of these shares the same set of fundamental parameters.

First, the object to be fabricated must be described in terms of an accurate 3-dimensional design representation. This must then be reformatted to describe the object in terms of a number of slices with finite thickness. This "slice information" is then used to fabricate the appropriate number of slices from the desired material. These slices are assembled to form the solid object. In practice, a number of the current technologies combine the slice fabrication and assembly processes by using a previously deposited slice as a template for the deposition of subsequent slices. Hence Rapid prototyping (RP) or layered manufacturing technology got its existence to fill the need in the manufacturing industry for development of representative or functional prototypes of objects normally manufactured in large quantities by tooled processes (Kai and Fai 1997). Models and prototypes manufactured with RP technology not only for visualization purposes but also to build functional parts. It is an important technology as it has potential to reduce the manufacturing lead time of the product up to 30–50% even when the relative part complexity is very high (Massod et al. 2001).

However, many RP techniques are available and the output of these processes also differs. Therefore, it is essential to select the appropriate RP technique for the generation of the gears. The quality of rapid prototyped parts mainly depends on careful selection of process variables. Hence, identifying the RP process parameters that affect the quality of processed parts is of utmost importance. In recent years, eminent researchers such as Masood, S.H. and his team (Massod et al. 2000, Massod et al. 2003, Omar et al. 2015 and Omar et al. 2016) have explored a number of ways to improve the mechanical properties and part quality using various experimental design techniques and concepts.

DIFFERENT RP PROCESSES

Various RP processes have been developed and also being developed in order to improve the realization of a virtual product to the extent possible. In this case as the product to be produced is of a non-metallic material hence out of all the RP techniques available only Stereolithography (SL), Selective Laser Sintering (SLS), Fused Deposition Modeling (FDM) and Laminated Object Manufacturing (LOM) have been studied.

STEREOLITHOGRAPHY

Stereolithography (SL) is a popular RP process in which intricate parts of a plastic monomer are directly built by photo polymerization process with the model constructed using a computer-aided design (CAD) package (Kai and Fai 1997). The process of SL involves; modeling of part with a CAD package to generate 3D solid model; conversion of 3D solid model into standardized triangular language (STL) file format to create volumetric mesh and creation of support structure; slicing of STL format of 3D solid model to provide a series of cross-sectional layers; exporting the sliced model to the computer of stereolithography apparatus (SLA); building the support structure and the part layer by layer (from bottom to top) over a vat of specially designed liquid resin with a helium–cadmium or argon laser, which traces the outline of the planar sections and solidify the resin in SLA; removal of support structure to get the green part; post-curing of green part to undergo final polymerization in a post-curing apparatus (PCA), which is either a controlled furnace or an ultraviolet oven. The dies made through SL process are subjected to high tension due to high injection pressure. Strength is crucial in case of rapid tooling since the parts have to withstand pressures during the test of fitment and when they are used as a die for injection molding.

SELECTIVE LASER SINTERING

Selective laser sintering (SLS) is an additive manufacturing technique that uses a high power laser (for example, a carbon dioxide laser) to fuse small particles of plastic, metal (Direct Metal Laser Sintering), ceramic, or glass powders into a mass representing a desired 3-dimensional object. The laser selectively fuses powdered material by scanning cross-sections generated from a 3-D digital description of the part (for example from a CAD file or scan data) on the surface of a powder bed. After each cross-section is scanned, the powder bed is lowered by one layer thickness, a new layer of material is applied on top, and the process is repeated until the part is completed. Compared to other methods of additive manufacturing, SLS can produce parts from a relatively wide range of commercially available powder materials. These include polymers such as nylon, (neat, glass-filled or with other fillers) or polystyrene, metals including steel, titanium, alloy mixtures, and composites and green sand. The physical process can be full melting, partial melting, or liquid-phase sintering. In many cases large numbers of parts can be packed within the powder bed, allowing very high productivity. Additionally, the selective laser sintering (SLS) process makes parts that have greater stability than SL and selective laser sintering (SLS) parts generally do not lose their shape or post-cure over time.

FUSED DEPOSITION MODELING

Fused deposition modeling (FDM) is an additive manufacturing technology commonly used for modeling, prototyping, and production applications. The technology was developed by S. Scott Crump in the late 1980s and was commercialized in 1990. FDM works on an "additive" principle by laying down material in layers. A plastic filament or metal wire is unwound from a coil and supplies material to an extrusion nozzle which can turn on and off the flow. The nozzle is heated to melt the material and can be moved in both horizontal and vertical directions by a numerically controlled mechanism, directly controlled by a computer-aided manufacturing (CAM) software package. The model or part is produced by extruding small beads of thermoplastic material to form layers as the material hardens immediately after extrusion from the nozzle. The FDM process constructs three-dimensional objects directly from CAD data. A temperature-controlled head extrudes thermoplastic material layer by layer. The FDM process starts with importing an STL file of a model into the preprocessing software. This model is oriented and mathematically sliced into horizontal layers varying from +/- 0.127 to 0.254 mm thickness. A support structure is created where needed, based on the part's position and geometry. After reviewing the path data and generating the tool paths, the data is downloaded to the FDM machine. Several materials are available, with different trade-offs between strength and temperature properties like Acrylonitrile Butadiene Styrene (ABS) polymer, Polycarbonates, Polycaprolactone, Polyphenylsulfones and Waxes. A "water-soluble" material can be used for making temporary supports while manufacturing is in progress, this soluble support material is quickly dissolved with specialized mechanical agitation equipment utilizing a precisely heated sodium hydroxide solution.

LAMINATED OBJECT MANUFACTURING

The first commercial Laminated Object Manufacturing (LOM) system was shipped in 1991. LOM was developed by Helisys of Torrance, CA. The main components of the system are a feed mechanism that advances a sheet over a build platform, a heated roller to apply pressure to bond the sheet to the layer below, and a laser to cut the outline of the part in each sheet layer. Parts are produced by stacking, bonding, and cutting layers of adhesive-coated sheet material on top of the previous one. A laser cuts the outline of the part into each layer. After each cut is completed, the platform lowers by a depth equal to the sheet thickness (typically 0.002 to 0.020 in), and another sheet is advanced on top of the previously deposited layers. The platform then rises slightly and the heated roller applies pressure to bond the new layer. The laser cuts the outline and the process is repeated until the part is completed. After a layer is cut, the extra material remains in place to support the part during the build. The excess material supports overhangs and other weak areas of the part during fabrication. The cross-hatching facilitates removal of the excess material. Once completed, the part has a wood-like texture composed of the material layers. Moisture can be absorbed by the paper, which tends to expand and compromise the dimensional stability. Therefore, most models are sealed with paint or lacquer to block moisture increase.

SELECTION OF APPROPRIATE TECHNIQUE FOR THE PRESENT STUDY

The ability to evaluate and determine the best part building orientation for different rapid prototyping (RP) processes is important for building a satisfactory part/prototype. It is also an essential step towards the identification of the most suitable RP process for a given RP application.

In the present study, primary criteria for the selection of the RP technique is based on the fact that built plastic gears will have the optimum combination of strength, dimensional accuracy and surface finish. It is quite easy to ascertain the accuracy of the RP product on last two aspects, namely, dimensional accuracy and surface finish. Achievable accuracy, surface finish, manufacturing time and building cost are significantly affected by the differences in layering methods of various RP techniques. However, strength analysis is not that straightforward especially when the product is a complicated object like gears. As a result, considerations for the selection of suitable orientation that relates to above-mentioned criteria vary with different RP processes.

Therefore, before proceeding further with the manufacturing of gears by any one of the RP techniques, Dynamic Mechanical Analysis was performed on the test specimens made from three most popular and most widely used processes utilizing non-metallic materials, namely, Stereolithography, Laminated Object Manufacturing, and Fused Deposition Modeling. The output results of the analysis are compared and the best option is selected. Once the RP technique is selected, evaluation of best part building orientation is considered for that particular RP method.

EXPERIMENT SCHEME FOR DYNAMIC MECHANICAL ANALYSIS

The objective of the work was to find out the best possible RP process for the fabrication of virtually manufactured gears. The test specimens are made as shown in Figure 1. Dynamic Mechanical Analysis is performed on the test specimens (dimension 12.5 mm X 6 mm X 3 mm) made from each of the processes.

The samples of stereolithography (SL) are prepared on SLA -250 / 50 machine while the samples of laminated object manufacturing (LOM) are produced by LOM 1015Plus machine. Dimension SST 768 machine is employed to manufacture the samples of fused deposition modeling (FDM). In the dynamic mechanical analysis, the strength of the material produced by RP processes is determined under varying temperatures at which the product may operate during its uses. Thus the sample is subjected to a sinusoidal stress (σ) of varying frequencies. The temperature is varied during the test. The resulting strain (ε) is also sinusoidal with respect to the frequency (Figure 2) and in case of elastic materials, it occurs exactly in phase with the stress. However, when internal molecular motion occurs at the same frequency as the applied stress, the material exhibits viscoelastic characteristics and the strain response lags slightly behind the stress. The phase angle is expressed as an angle δ. The modulus of a viscoelastic material (E*) is a complex quantity,

$$E^{*} = E' + E'' \tag{1}$$

where E' is the storage modulus or in-phase component, and E'' is the loss modulus or out of phase component.

Tangent of the phase angle (δ) can be given by

Figure 1. Dynamic Mechanical Analysis test specimens

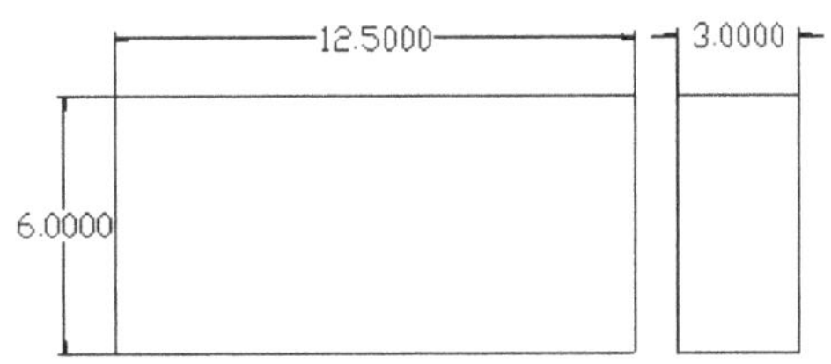

Figure 2. Sketch for storage modulus and loss modulus

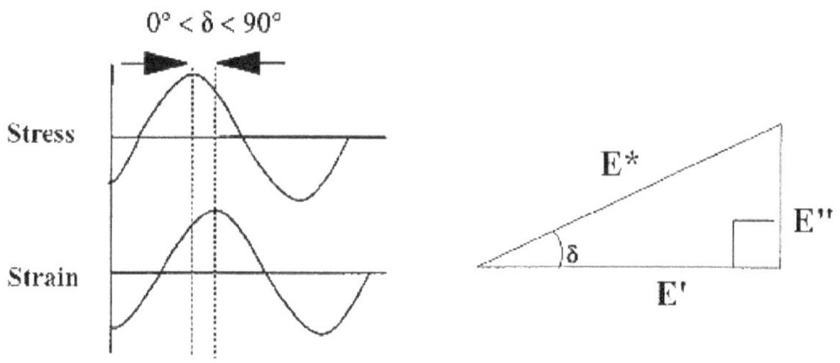

$$\tan\delta = E'' \,/\, E' \tag{2}$$

Since the material used for the present study exhibits viscoelastic behavior to some extent, DMA is carried out with varying frequency range.

EXPERIMENTATION

Dynamic mechanical analysis is performed in 3 point bending mode, the specimens were clamped between strain gauges and subjected to small sinusoidal strain (static strain of 0.2% and dynamic strain of 0.1%) at the frequency of 1 Hz. The measurements were carried out at 20–1200C temperature range with a rate of heating 5^0C/min and the oscillation amplitude is set at 15 µmm. The machine used in this measurement was DMA Q800, TA Instruments, USA. The data was collected and was subsequently analyzed with the help of proprietary software DMA V4.2.

RESULTS AND DISCUSSION

The dynamic mechanical properties of products are significantly dependent upon the type of materials and processing techniques. Polymeric materials exhibit more than one relaxation region or so-called transition over a wide range of temperature during dynamic thermo-mechanical analysis. The variations of dynamic mechanical properties (storage modulus, loss modulus, and tan δ with temperature for different samples prepared by various prototyping processes are graphically presented in (Figures 3, 4 and 5). In Figure 3, a variation of the storage modulus made by SL, LOM, and FDM is measured as a function of temperature. It is observed from the graph that the decrease in storage modulus with temperature occurred in all three cases. However, maximum storage modulus was observed in case of the samples prepared by FDM and the minimum by those prepared by LOM process.

As the basic material used in case of LOM is paper, which has comparatively lower strength, the sample shows minimum storage modulus. Whereas the sample prepared by FDM is thermoplastic in nature and has very high interlayer integrity when it is processed layer by layer. But in the sample prepared by SL, although thermosetting in nature, the interlayer adhesion is poor due to incomplete curing of the interior portion of the sample and results in lower storage modulus compared to material prepared by

Figure 3. Variations of storage modulus with temperature for different samples prepared by FDM, SLA, and LOM

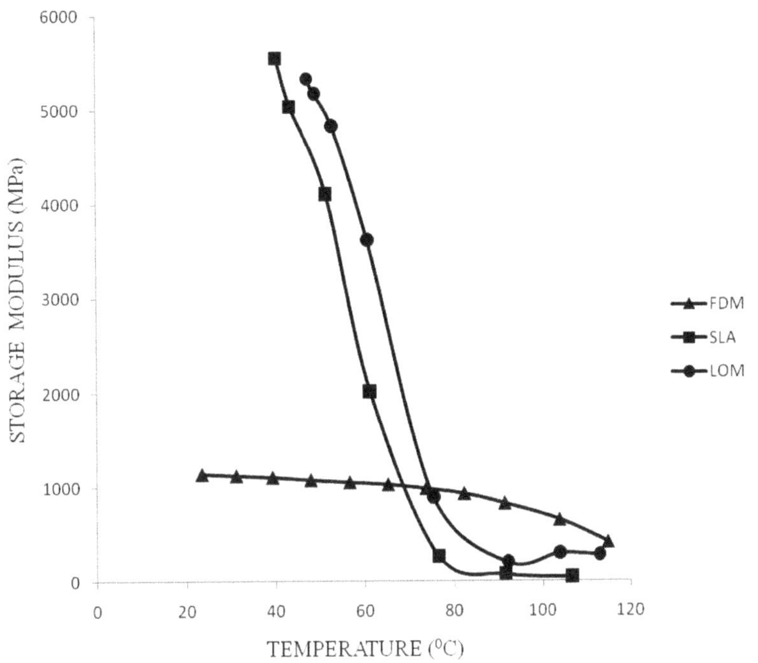

FDM. It can be inferred that the FDM sample gives the maximum storage modulus up to 800C which can be assumed to be the maximum working temperature for gears made from plastic-type materials. The same procedure is followed for the loss modulus of the samples made by three processes as well. If the mechanical energy is not stored elastically, it may be lost and converted to heat through molecular friction (that is viscous dissipation) within the material. The extent to which this happens is reflected by the magnitude of loss modulus, higher the value of loss modulus, higher is the loss of mechanical energy. Figure 4 exhibits the variation of the Loss modulus for SL, LOM, and FDM samples, respectively as a function of temperature. It is revealed that the loss modulus is the minimum for the FDM sample over the range of temperature. So it can be inferred that the FDM sample exhibits more dynamic strength as it has the minimum loss modulus amongst the three processes considered. In fact, its loss modulus may even reach as low as zero around 800C indicating almost elastic behavior at the maximum working temperature of a plastic type of gears. The ratio of loss modulus to storage modulus (E"/ E') is measured as the mechanical loss factor or tan δ. The variation of tan δ as a function of temperature for different formulated samples is shown in Figure 5. As shown in Figure 5, tan δ, which corresponds to the damping properties of the material, is found to be maximum in case of the sample prepared by FDM and minimum in case of the sample generated by SL. As the material used in FDM is in nature of thermoplastic, having a more flexible polymeric backbone, it allows more movement of macromolecule chain with an increase in temperature. Whereas the material used in SL is thermosetting in nature and the rigid cross-linked backbone does not allow larger long chain movement with an increase in temperature. Hence, Fused Deposition Modeling is selected as the process for fabrication of plastic type gears for its better damping character as well as higher modulus even at a higher working temperature (80^0C).

Figure 4. Variations of loss modulus with temperature for different samples prepared by FDM, SLA, and LOM

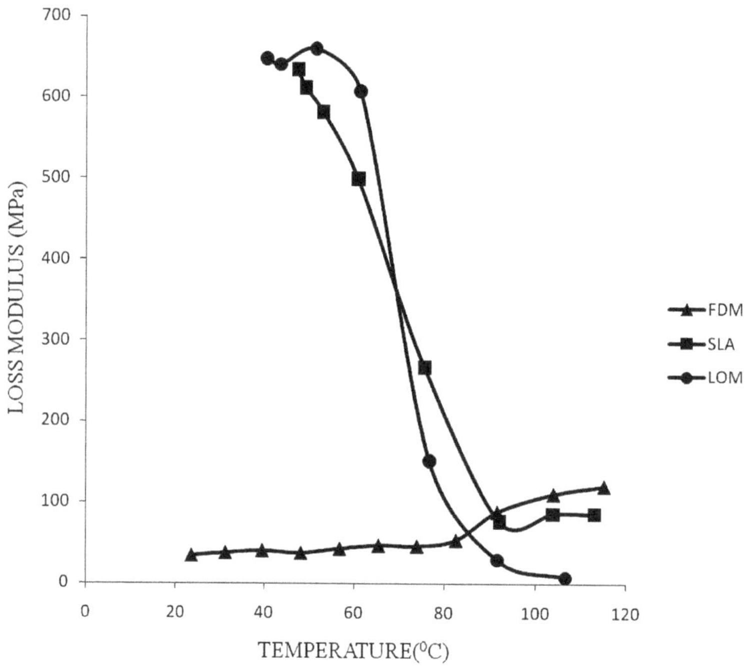

Figure 5. Variations of tan δ with temperature for different samples prepared by FDM, SLA, and LOM

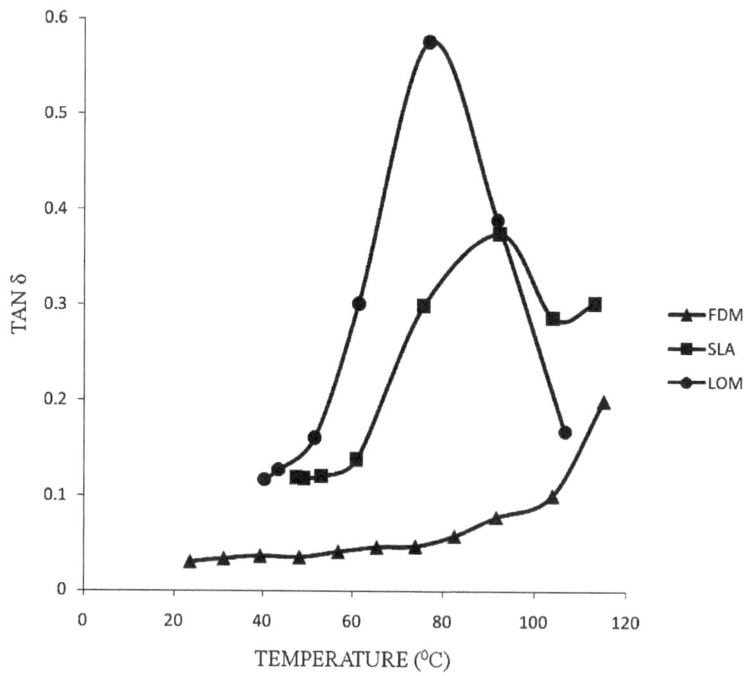

PART DEPOSITION ORIENTATION FOR FDM

So FDM has been identified as the most suitable RP technique for the present study hence now evaluation of best part building orientation is required to be estimated. It is a very important factor as it affects strength, build time, support structure, dimensional accuracy, surface finish and cost of the prototype. A number of process specific parameters and constraints are required to be considered while making this decision. Determining the optimal part deposition orientation is a difficult and time-consuming task as one has to trade-off among various contradicting objectives (Pham and Demov 2001) like part strength, surface finish and build time. There have been several attempts (Alexander et al. 1998, Frank and Fadel 1994, Chang et al. 1995, Lan et al. 1997, Xu et al. 1997, McClurkin and Rosen 1998, Pham et al. 1999, West et al. 2001, Goyal et al. 2000, Mishra and Thirumavalavan 2014, Phatak and Pande 2012, Paul and Anand 2011 etc.) to determine suitable part deposition orientation for different objectives like accuracy, build time, support structure etc. The build time of a prototype can be reduced by depositing it with thicker layers; however, this may lead to the rough surface. Moreover generation of curved surfaces, if not oriented properly, gets deformed due to phenomena are known as Staircase Effect. Staircase Effect is so called because of the step formed between two adjacent layers when the layers are built. Staircase Effect in a build will result not only in very rough surface finish but also inaccurate surface profile. However, deposition using thin layers increases built time drastically. In the present work, average part surface roughness and the build time of the part are considered as two objectives and are minimized by minimizing their weighted sum.

SELECTION OF FDM PROCESS PARAMETERS TO MAXIMIZE STRENGTH OF THE PRODUCT

The FDM machine used for the present experiment is SST 768 with installed proprietary software Catalyst EX. The strength of FDM processed component primarily depends on input process parameters. They are defined as follows:

1. **Orientation:** Part build orientation or orientation refers to the inclination of a part in a build platform with respect to X, Y and Z axes. X and Y-axes are considered parallel to build platform and Z-axis is along the direction of the part build.
2. **Layer Thickness:** It is the thickness of the layer deposited by the nozzle and depends upon the type of nozzle used i.e. the height of each layer of material extruded to produce a part. In the present machine, it may be either 0.010 or .013 inch (0.2540 or 0.3302 mm). The resolution will affect build time and surface finish i.e. 0.010 will be a smoother finish but will take a longer time to build, the default value being 0.2540 mm.
3. **Model Interior:** It establishes the type of fill to be used for the part. It refers to the way in which the part is filled. It controls the density of the part. There are two types of build styles in the machine used.
    a. "Solid" fills the interior part completely. It will build a part with a stronger interior.
    b. "Sparse" minimizes the material volume and build time by leaving gaps. It utilizes a unidirectional raster.

Support style refers to the portions of a model that extends outward to prevent the part from collapsing during the building process. It is of four types.

1. "Basic" is the standard raster tool path support structure. It supports all part features with small support raster curves.
2. "Sparse" reduces the amount of support material volume. It uses less amount of material than basic support.
3. "Surround" fill the surround of small features or small parts. It is used for all parts which are small in nature and requires more support due to the geometry.
4. "Break-away" is similar to "Sparse", but it contains discrete boxes and it is easier to remove than other three types of support structures.

The objective of the study is to determine a combination that may lead to a product with maximum strength with dimensional accuracy maintained within a proper range. Accordingly, a set of experiments is carried out using tensile test specimens produced by FDM having varying parameter setting as mentioned above.

EXPERIMENTATION

In order to make tensile test specimen, a 3D model of the same is developed using CREO software (Figure 6) and is saved as 'stl' file. The dimensions of the test specimen are chosen according to ASTM D 638 (Figure 7).

The 'stl' versions of the model are loaded on the FDM machine and an attempt is made to create eight different RP models on the FDM machine with three different parameters such as orientation, layer thickness, and model interior. Table 1 shows the variation in parameters along with build time, build volume and support volume required for these eight samples. However, these values are given by the software even before the samples are actually made by FDM.

Three different orientations are the obvious choice i.e. horizontal, vertical and standing, however, for standing orientation, due to heavy vibration of the machine and very small base, the samples fell before completion of the product. This implies that this condition should be avoided. So two different

Figure 6. 3D model of tensile test specimen drawn using CREO software

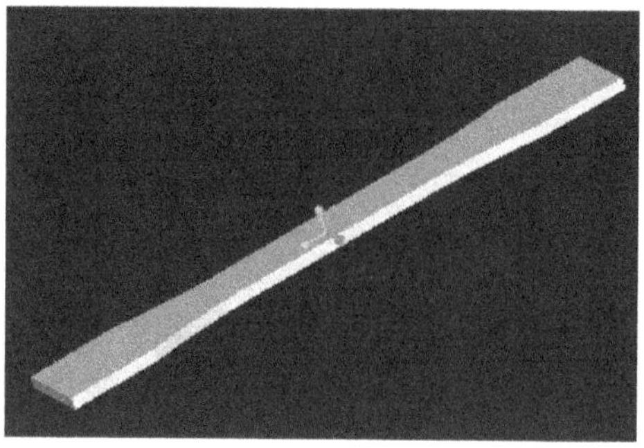

Figure 7. The Tensile model with dimensions (As per ASTM D 638)

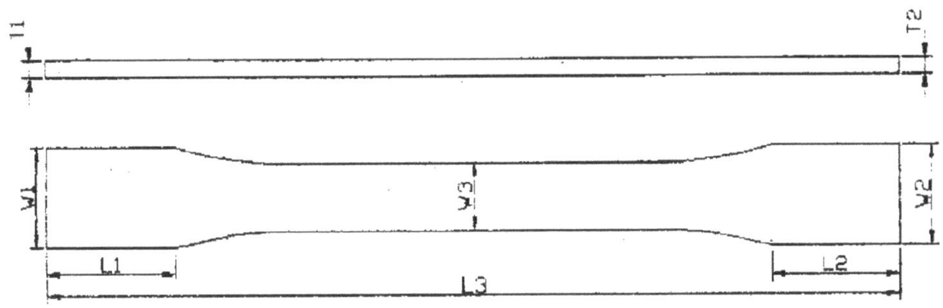

Table 1. The various parameters and their variations used for experimentation

Code No.	Orientation	Layer Thickness	Model Interior	Build Duration (Hrs.)	Model Material Required (cc)	Support Material Required (cc)
H1	Horizontal	0.2540	Solid	2.133333	36.13	12.31
H2	Horizontal	0.2540	Sparse	1.7	26.03	12.31
H3	Horizontal	0.3302	Solid	1.416667	36.07	15.4
H4	Horizontal	0.3302	Sparse	1.4	26.27	15.4
V1	Vertical	0.2540	Solid	3.45	34.55	8.59
V2	Vertical	0.2540	Sparse	2.683333	29.29	8.59
V3	Vertical	0.3302	Solid	2.166667	34.26	10.27
V4	Vertical	0.3302	Sparse	2.616667	34.29	10.27

orientations were considered by the software (Catalyst EX) and are shown in Figure 8. For each of the horizontal and vertical orientations, four samples were created by FDM.

Next, Instron universal-testing machine-Model 3360 was used to find the tensile strength of all these samples at 1.0 mm/min crosshead velocity.

Figure 8. The various orientations (Horizontal and Vertical) as seen by the Catalyst EX software

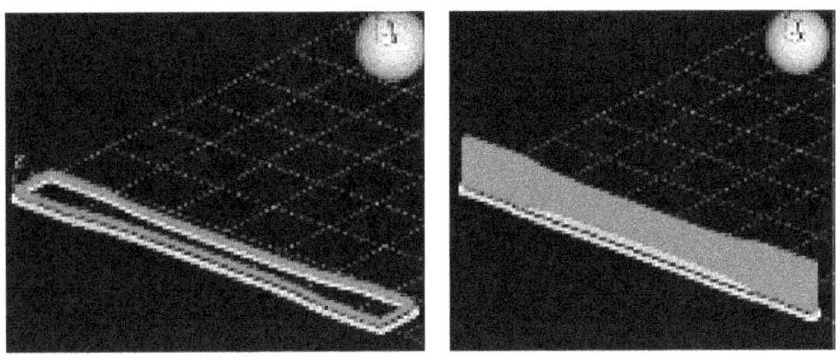

RESULT AND DISCUSSION

Tensile Test Results

The output results of the tensile test are presented in Table 2. The table also includes the dimensional variation of different test specimens.

Figure 9 shows the ultimate tensile stress of sample prepared in both the direction one in horizontal another in a vertical position. Figure 9 also indicates that the in both the cases sample H1 and H3 gives maximum ultimate tensile stress compare to sample H2 and H4 as the model interior is sparse which increases the interlayer gaps and hence the stress decreases. Moreover, it was also observed that the between sample H1 whose layer thickness is 0.2540 and sample H3 having layer thickness 0.3302, H1 shows more stress compare to H3. These results can be attributed to the fact that as layer thickness decreases the interlayer void space also decreases results in more dense structure compare to a material having more layer thickness. Similar kind of results also observed in the sample prepared in the vertical direction. The solid model gives more stress compared to the sparse structure. But the variation in stress

Table 2. The results of the tensile testing along with variation in dimensions of all samples

Sl. No.	Sample Code	Ultimate Stress (MPa)	Ultimate Strain (%)	Modulus (MPa)	Variation in Dimension (%)	
					Max.	Min.
1.	H1	20.32	4.75	635.40	5.63	0.11
2.	H2	13.76	4.25	439.30	5.00	0.11
3.	H3	18.59	5.33	559.80	8.43	0.00
4.	H4	12.76	4.08	418.45	7.81	0.63
5.	V1	17.78	7.17	665.10	7.52	1.69
6.	V2	8.55	40.42	572.80	7.12	1.25
7.	V3	18.94	13.42	665.45	8.80	0.42
8.	V4	15.85	10.42	651.00	6.48	0.95

Figure 9. Ultimate tensile stress of sample prepared in horizontal and vertical position

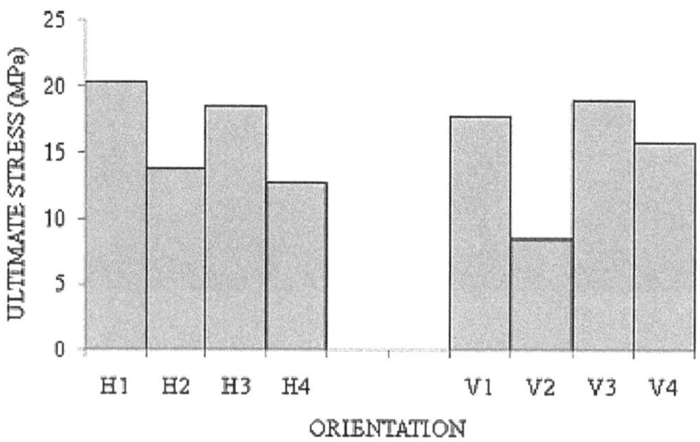

between samples with various layer thicknesses is not very much as the contact area between layers is less in the sample prepared in a vertical position.

Figure 10 shows the variation of ultimate tensile strain for samples prepared in a different orientation. It was observed from data that the vertically oriented sample shows much higher strain compare to sample prepared in a horizontal orientation. As in the tensile strength measurements the test specimen drawn in the direction which is parallel to the layer orientation in case of sample horizontal oriented and perpendicular to the layer orientation in case of sample vertically oriented. Now when the sample is drawn in tensile strength machine the layer which is in machine direction restrict more and do not allow stress transfer rapidly whereas in case of vertical sample since the layer-oriented opposite direction, during tensile drawing the interlayer may slip on over other and stress may transfer to different interface and allows more extension which is shown in graphs.

Figure 11 depicts the tensile modulus of all samples. It is obvious that the solid sample H1 and H3 show higher modulus compare to the sparse sample H2 and H4. But the modulus of H3 sample is found to be lesser than the sample prepared in H1 condition as the layer thickness is higher which increases the interlayer gaps and hence the strength decreases. Similar kinds of results were observed in modulus in case of the sample prepared in a vertical condition. As the area of the layer is less and number of the interface of the layer is high the change in modulus of the sample prepared in vertical condition with different parameter within the small elongation is not a considerable amount. Though solid sample in vertical orientation shows higher modulus compare to sparse sample but differences in value are not considerable.

CONCLUSION

There are two parts in this work. The first being selection of the best possible rapid prototyping technique using dynamic mechanical analysis and the next was optimization of the process parameters of the selected RP technique considering tensile strength as the criteria.

The samples prepared using SL, FDM and LOM, subjected to dynamic mechanical analysis indicated that the storage modulus for liquid-based processes (SL) is less due to the incomplete curing of the resin.

Figure 10. Ultimate tensile strain of sample prepared in horizontal and vertical position

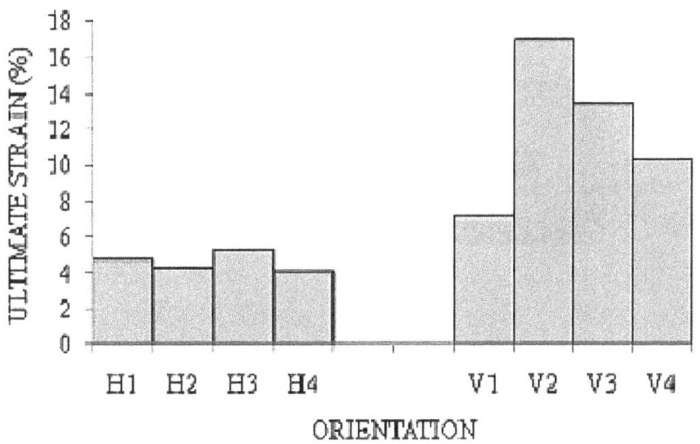

Figure 11. Modulus of sample prepared in horizontal and vertical position

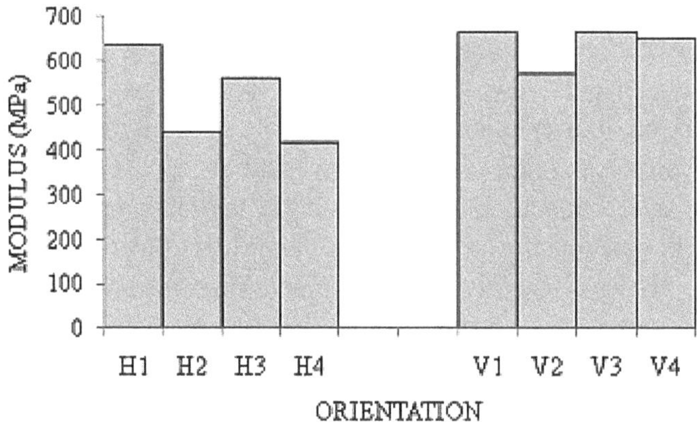

It occurs as the resin remains in contact with the laser for a shorter time and even with post-processing also the curing is never 100% complete. This results in lower storage modulus. In LOM paper are glued together using an adhesive which also shows lower strength. Hence in such situation, FDM, wherein an extruded wire creates the model, is strongest of all these within the specified temperature range.

The orientation of the product in RP techniques plays an important role in providing the strength and also the accuracy of the product profile. Form the results two orientations were found to be suitable H1 and V4. V4 orientation results in highest yield Stress, ultimate strain etc. but the curved profile of tooth of the gear would be improper due to 'staircase effect'. So H1 parameters, although offers lesser strength, is more proper considering the product profile. Therefore H1 condition was chosen for gear generation by FDM and depicted in Figure 12.

FUTURE WORK

The work can be extended for generation of other gears and also evaluation of metal RP for generating metallic products.

Figure 12. Spur gear created using FDM

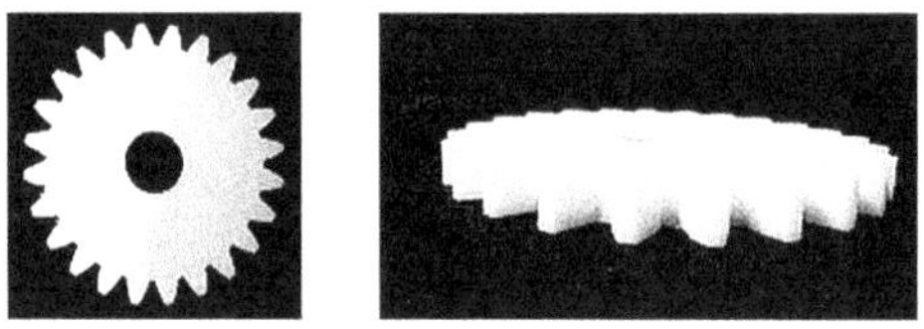

ACKNOWLEDGMENT

The authors sincerely acknowledge the comments and suggestions of the editors and reviewers that have been instrumental for improving and upgrading the paper in its final form.

REFERENCES

Alexander, P., Allen, S., & Dutta, D. (1998). Part orientation and build cost determination in layered manufacturing. *Computer Aided Design*, *30*(5), 343–356. doi:10.1016/S0010-4485(97)00083-3

Cheng, W., Fuh, J. Y. H., Nee, A. Y. C., Wong, Y. S., Loh, H. T., & Miyajawa, T. (1995). Multi-objective optimization of part-building orientation in stereolithography. *Rapid Prototyping Journal*, *1*(4), 12–23. doi:10.1108/13552549510104429

Frank, D., & Fadel, G. (1994). Expert system based selection of the preferred direction of build for rapid prototyping. *Journal of Intelligent Manufacturing*, *6*(5), 334–339.

Goyal, A., Sudha, A., Kumar, A., & Rao, P. V. M. (2000). Optimum part orientation for rapid prototyping based on estimated number of adaptive slices. *Proceedings of International Conference on E-manufacturing*, 247–253.

Kai, C. C., & Fai, L. K. (1997). *Rapid Prototyping, Principles and Applications in Manufacturing. John Wiley & Sons* .

Lan, P.-T., Chou, S.-Y., Chen, L.-L., & Gemmill, D. (1997). Determining fabrication orientation for rapid prototyping with stereolithography apparatus. *Computer Aided Design*, *29*(1), 53–62. doi:10.1016/S0010-4485(96)00049-8

Masood, S. H., Rattanawong, W., & Iovenitti, P. (2000). Part build orientations based on volumetric error in fused deposition modeling. *International Journal of Advanced Manufacturing Technology*, *16*(3), 162–168. doi:10.1007001700050022

Masood, S. H., Rattanawong, W., & Iovenitti, P. (2001). A volumetric approach to part-built orientations in rapid prototyping. *Journal of Materials Processing Technology*, *119*(1-3), 348–353. doi:10.1016/S0924-0136(01)00924-4

Masood, S. H., Rattanawong, W., & Iovenitti, P. (2003). A generic algorithm for part orientation system for complex parts in rapid prototyping. *Journal of Materials Processing Technology*, *139*(1–3), 110–116. doi:10.1016/S0924-0136(03)00190-0

McClurkin, J. E., & Rosen, D. W. (1998). Computer-aided build style decision support for stereolithography. *Rapid Prototyping Journal*, *4*(1), 4–13. doi:10.1108/13552549810197505

Mishra, A. K., & Thirumavalavan, S. (2014). A Study of Part Orientation in Rapid Prototyping. *Middle East Journal of Scientific Research*, *20*(9), 1197–1201.

Mohamed, O. A., Masood, S. H., & Bhowmik, J. L. (2015). Optimization of fused deposition modeling process parameters: A review of current research and future prospects. *Advances in Manufacturing*, *3*(1), 42–53. doi:10.100740436-014-0097-7

Mohamed, O. A., Masood, S. H., & Bhowmik, J. L. (2016). Optimization of fused deposition modeling process parameters for dimensional accuracy using I-optimality criterion. *Measurements*, *81*, 174–196.

Paul, R., & Anand, S. (2011). Optimal Part Orientation in Rapid Manufacturing Process for Achieving Geometric Tolerances. *Journal of Manufacturing Systems*, *30*(4), 214–222. doi:10.1016/j.jmsy.2011.07.010

Pham, D. T., & Demov, S. S. (2001). *Rapid Manufacturing: The Technologies and Applications of Rapid Prototyping and Rapid Tooling*. London: Springer-Verlag London Limited. doi:10.1007/978-1-4471-0703-3

Pham, D. T., Demov, S. S., & Gault, R. S. (1999). Part orientation in stereolithography. *International Journal of Advanced Manufacturing Technology*, *15*(9), 674–682. doi:10.1007001700050118

Phatak, A. M., & Pande, S. S. (2012). Optimum Part Orientation in Rapid Prototyping Using Genetic Algorithm. *Journal of Manufacturing Systems*, *31*(4), 395–402. doi:10.1016/j.jmsy.2012.07.001

West, A. P., Sambu, S. P., & Rosen, D. W. (2001). A process planning method to improve build performance in stereolithography. *Computer Aided Design*, *33*(1), 65–79. doi:10.1016/S0010-4485(00)00064-6

Xu, F., Wong, Y. S., Loh, H. T., Fuh, J. Y. H., & Miyazawa, T. (1997). Optimal orientation with variable slicing in stereolithography. *Rapid Prototyping Journal*, *3*(3), 76–88. doi:10.1108/13552549710185644

KEY TERMS AND DEFINITIONS

Dynamic Mechanical Analysis: A technique to study the viscoelastic nature of materials especially polymers. The applied stress is sinusoidal in nature and is usually accompanied with varying temperature and frequency.

Fused Deposition Modeling: An additive manufacturing process that builds a part from thermoplastic filament by extruding and heating the filament.

Laminated Object Manufacturing: A rapid prototyping technique where a part is built from layers of papers.

Layer Thickness: The thickness of the layer deposited by the nozzle for creating the solid.

Model Interior: The type of fill to be used for the part (i.e., sparse or solid).

Part Orientation: The inclination of the being built with respect to the X, Y, and Z axes of a RP machine.

Rapid Prototyping: The process of making physical prototypes for scale down physical model obtained from CAD data.

Selective Laser Sintering: An additive manufacturing technique in which high-power laser source is used to fuse the powdered material to obtain three dimensional physical prototypes.

Stereolithography: One of the 3D printing techniques of producing prototypes by converting liquid plastics to solid plastics.

Tensile Modulus: The ability of material to withstand change in length when under tension.

Ultimate Tensile Strain: The strain at the point of failure.

Ultimate Tensile Stress: The maximum stress a material can sustain while being stretched before breaking.

This research was previously published in the Handbook of Research on Ergonomics and Product Design edited by Juan Luis Hernández Arellano, Aide Aracely Maldonado Macías, Juan Alberto Castillo Martínez, and Porfirio Peinado Coronado, pages 364-380, copyright year 2018 by Engineering Science Reference (an imprint of IGI Global).

Chapter 16
Complexity Analysis in Additive Manufacturing for the Production of Tissue Engineering Constructs

Kouroush Jenab
Society of Reliability Engineering – Ottawa Chapter, Canada

Philip D. Weinsier
Bowling Green State University – Firelands, USA

ABSTRACT

Additive Manufacturing (AM) is a process of making a Three-Dimensional (3D) solid object of virtually any shape from a digital model that is used for both prototyping and distributed manufacturing with applications in many fields, such as dental and medical industries and biotech (human tissue replacement). AM refers to technologies that create objects through a sequential layering process. AM processes have several primary areas of complexity that may not be measured precisely, due to uncertain situations. Therefore, this chapter reports an analytical model for evaluating process complexity that takes into account uncertain situations and additive manufacturing process technologies. The model is able to rank AM processes based on their relative complexities. An illustrative example for several processes is demonstrated in order to present the application of the model.

1. INTRODUCTION

The fabrication of micro- and nano-structured biodegradable constructs for tissue engineering (TE) is dependent upon the development of techniques and methodologies. Many techniques have been developed for the production of TE such as additive manufacturing (AM). AM is suitable for TE constructs with complex shapes and microstructure because it has a high level of automation and accuracy. In literature pertaining to AM technologies (powder-based and photosensitive-based techniques), four categories have been introduced (Hutmacher et al., 2004):

DOI: 10.4018/978-1-5225-9624-0.ch016

Figure 1. Chronological taxonomy of the AM techniques

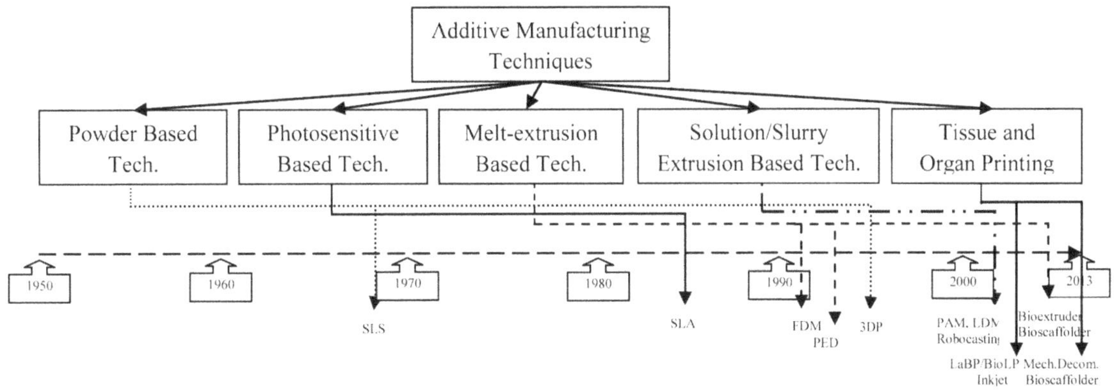

1. stereolithography (SLA)
2. selective laser sintering (SLS)
3. three-dimensional printing (3DP)
4. fused deposition modelling (FDM)

SLA, which is based on laser or ultraviolet (UV) light, is an AM technique that selectively polymerizes layers of a photosensitive polymer. In 1989, selective laser sintering (SLS), which is based on the selective sintering of a polymer, ceramic, or hybrid powder bed, and high-intensity laser beam, was developed by Deckard (1989). Later, another powder-based technology called three-dimensional printing (3DP) was developed that was based on the controlled deposition of a binder material laid on a powder layer using an inkjet head. Since these techniques were not suitable for the rapid manufacturing of prototypes for industry, Crump (1992) proposed a fused deposition modelling (FDM) technique, which was an extrusion of a polymeric filament through a heated nozzle. As an extension, precision extrusion deposition (PED), which was a melt–extrusion AM technique, was introduced. This technique was based on a screw extruder, enabling the processing of polymer pellets for the controlled deposition of a melt filament. In conjunction with melt-extrusion-based technology, Vozzi et al. (2003) studied the micropositioning system with a pressure-activated microsyringe that was well-known as a pressure-assisted microsyringe (PAM). Conversely, low-temperature deposition manufacturing (LDM) was used to exploit a controlled cooling chamber. Xiong et al. (2002) mentioned that LDM allows the deposition process of a polymeric slurry to operate at low temperatures. Finally, the fourth category is the tissue and organ printing technologies such as Laser-assisted BioPrinting (LaBP), and Biological laser printing (BioLP™). As an efficient replacement for a cartridge-based system, the LaBP techniques were introduced by Koch et al. (2012). Also, to print a variety of cell types, Biological laser printing (BioLP™) was developed. Figure 1 shows the chronological taxonomy of the AM techniques.

Complexity of an AM process can be defined through two major aspects: technical design and processes. However, these aspects of the AM technique entail a lot of information that may not be measured precisely because of uncertain situations (Pich et al., 2002; Shafiei-Monfared & Jenab, 2011a/b). As a result, AM project failures are numerous in practice (Tatikonda & Rosenthal, 2000). In other studies, the authors suggest that uncertainty required adapting a management style to the process uncertainty profile, as measured by the dimensions of process size, product structure, and experience with the technology

(Jenab & Liu, 2010; McFarlan, 1981). Also, a product empirical classification method was proposed that used degree of technical uncertainty and complexity of the product (Dvir et al., 1998; Shafiei-Monfared & Jenab, 2010). As complexity measures become an efficient yardstick to manage a group of AM techniques, having a quantitative model to analyze the relative complexity under an uncertainty situation is a must. Therefore, this study proposes a model based on a fuzzy analytic hierarchy process (FAHP) and triangular fuzzy numbers to express the comparative judgments of the AM process analyst.

2. LITERATURE REVIEW

Complexity can be used as a measure for comparative analysis in AM processes. However, quantitative analysis of complexity has not received the scholarly attention and recognition that other seminal concepts, such as flexibility or robustness, have received (Bashir & Thomson, 2001; El-haik & Yang, 1999; Rodriguez-Toro et al., 2003; Smith & Jenks, 2006). Literature pertaining to complexity can be classified as computational, software, manufacturing and operation, projects, EntSLS/SLA/3DPrise, and information systems complexities.

Computational complexity is a branch of computational theory that deals with computational problems such as problem instances, representing problem instances, decision problems as formal languages, function problems, and measuring the size of an instance. In this regard, Chakraborty and Choudhury (1999) studied two computing operations (i.e., addition and multiplication) with different processing times. They proposed a statistical approach based on a weighting system. Dehmer et al., (2006) investigated an algorithm to measure the structural similarity of a generalized graph. They showed that the efficiency of their algorithm was promisingly better than those of the classical approaches used by Kaden (1982). De Reyck and Herroelen (1996) studied the relationship between the hardness of problem instance and the topological structure of its network by measuring complexity. Liu et al. (2007) analyzed the complexity of computing an AU measure within the D–S theory framework. They delineated the conditions that affect computational complexity. However, Huynh and Nakamori (2010) critiqued the work by Liu et al. (2007) and rectified several mistakes in the formulation of their F-algorithm.

Vieira et al. (2008) developed a model based on a q-norm complexity measure that minimized computational complexity in machine learning algorithms. Furthermore, Vladislavleva et al. (2009) presented a Pareto Genetic Programming model that traded off between computing objectives, numerical accuracy, and nonlinearity order. Also, Lathrop et al. (2011) showed the impact of geometry on computability and complexity of nano-scale self-assemblies.

In another study, Grant and Tu (2005) discussed a six-level taxonomy of EntSLS/SLA/3DPrise integration and the ability of EntSLS/SLA/3DPrise Resource Planning (SLS/SLA/3DP) to mitigate the integration complexity. Newell et al. (2005) studied different strategies and complexity factors that are significant in implementing an EntSLS/SLA/3DPrise System (ES). Similarly, Lockett and Brown (2005) explored the complexity issues of emerging vertical application service providers in ES. Also, Acar et al. (2010) proved that the complexity of Acar's Comprehensive Situation Mapping (CSM) is low because of its computational properties. Therefore, using CSM in the field of entrepreneurship can improve the capability for analyzing the opportunities. Kendall et al. (2010) discussed a new process for developing information systems (IS) that took into account complexity of processes. They conducted a survey to analyze the importance and complexity of strategic selection of the IS development approach. Also, Lefley and Sarkis (2005) explored failures in SLS/SLA/3DP adaptation projects. They pointed out that project evaluation and selection complexity would result in such failures. Therefore, a financial appraisal profile model was proposed that not only mitigated the complexity issues but also improved SLS/SLA/3DP project success.

In the field of software, we use metrics that can be classified into product metrics, process metrics, and project metrics. Product metrics describe the characteristics of the software product by size, complexity, design features, performance, and quality level. Kemerer and Porter (1992) proposed a qualitative method for determining complexity of a software product. In an object-oriented system, cyclomatic complexity is used to evaluate the application of an algorithm. A software product with a low cyclomatic complexity may imply that it is not complex (McCabe, 1992). Mahmood and Lai (2008) evaluated the complexity factor, which helped focus on the complex scenarios even though they may not be used often. They showed that complexity measures may be used to classify components based on their inherent characteristics.

Manufacturing complexity means that a crew deals with a system made up of many subsystems with different functions and states, so the effect of functions is difficult to assess (Kauffman, 1993). As a consequence, the cost of a product will be increased. In relation to the engineering design project, Steward (1981) introduced the Design Structure Matrix (DSM) approach to model project tasks. A value in the lower diagonal portion of the matrix represents a precedence (go forward) relationship, whereas the upper diagonal portion of the matrix represents an iteration (go backward) relationship.

Also, complexity in manufacturing is generally considered in relation to technical and process aspects. Ambiguity and product complexity lead to information inadequacy. Ambiguity refers to a lack of awareness of the crew about a system (i.e., subsystem interrelation, geometry, topology, manufacturability, and assembly factors) (Cho et al., 2009; Schrader et al., 1993). Geometry is defined as polygon meshes including coordinate values of vertices of the meshes making up the model. The topology is concerned with redundant references to vertices and edges that are shared by entities. Manufacturability deals with equipment setup time, manufacturing yield, direct supervision, and finishing. Assembly factor takes into account the number of parts, assembly time, and assembly tools that are impacted by complexity (Jenab & Liu, 2010; Kashyap & Sinha, 2011; Rodriguez-Toro et al., 2003; Samy & El Maraghy, 2010). For example, Zhu et al. (2011) presented a dynamic programming model to determine complexity index in sequence planning in mixed-model assembly lines. Furthermore, Wang et al. (2011) developed a multi-objective optimization model to balance between product variety and manufacturing complexity.

This study aims at developing a fuzzy model to rank the additive manufacturing processes for a tissue engineering construct based on their relative complexity. The remainder of the paper is organized as follows. Section 3 is devoted to the method of the complexity model. An illustrative example is demonstrated to present the application of the model in section 4. Finally, section 5 is devoted to the conclusion.

3. METHOD OF FUZZY COMPLEXITY ANALYSIS FOR TISSUE ENGINEERING CONSTRUCTS

Methodological considerations become important when we consider additive manufacturing complexity problems under uncertain situations (Meixner, 2011). This issue is increasingly receiving more attention year after year because of the growth of the application of AM processes in tissue engineering constructs. Therefore, the evaluation process must be improved by robust techniques such as FAHP in order to come close to reality and human cognitive behavior. FAHP, of the fuzzy extension of AHP, is proposed to handle the fuzzy problems that have a hierarchical form. In the FAHP approach, the values in the pairwise comparison matrix are fuzzy numbers where they can be modified according to the product analyst's preference (Sarfaraz et al., Inpress; Schultz et al, 2013; Jitrawichawet et al, 2013; Jenab & Dhillon, 2005).

The process starts by constructing a pairwise comparison matrix for the criteria (C_1 : high accuracy; C_2 : mechanical properties; C_3 : production rate; and, C_4 : material support requirement), which consists of fuzzy numbers instead of single values as in a normal AHP process. For example, if the weight of criterion C_1 relative to criterion C_2 is (l_1, m_1, u_1), the weight of C_2 relative to C_1 is ($\frac{1}{l_1}$, $\frac{1}{m_1}$, $\frac{1}{u_1}$). Then, we calculate the value of fuzzy synthetic extent with respect to each criterion using the following formula:

$$S_i = \sum_{j=1}^{m} M_{gi}^{j} \odot \left[\sum_{i-1}^{n} \sum_{j=1}^{m} M_{gi}^{j}\right]^{-1} \tag{1}$$

where $M_{g_i}^{j}$ is the value of the extent analysis of the *i*th object.

Next, we calculate the weight of each criterion by normalizing the degree of possibility for each S_i to be greater than the others using the following formula:

$$V\left(S_1 \geq S_2\right) = \begin{cases} 1 & \text{if and only if } m_1 \geq m_2 \\ \dfrac{l_1 - u_2}{\left(m_2 - u_2\right) - \left(m_1 - l_1\right)} & \text{Otherwise} \end{cases} \tag{2}$$

where $S_1 = \left(l_1, m_1, u_1\right)$, and $S_2 = \left(l_2, m_2, u_2\right)$

We repeat the last two steps for the AM processes (alternative) with respect to each criterion. Finally, we add up the weight per alternative multiplied by the corresponding criteria to find the final complexity ranking for all alternatives. In the following steps, the model is described in detail:

Step 1: Determine the AM process complexity criteria. In this study, we assume that C_1: high accuracy; C_2: mechanical properties; C_3: production rate; and, C_4: material support requirement.

Step 2: Using Equation (1), calculate the weights of the criteria. The pairwise comparison matrix of the criteria is formed by referring to the tissue engineering product analyst's preference.

Step 3: Using Equation (2), calculate the weights of the alternatives with respect to each criterion.

Step 4: Having the weights of each alternative with respect to each criterion, calculate the final raking number by adding up the weight per alternative multiplied by the corresponding criteria.

Step 5: Rank the alternatives with respect to their final ranking number.

4. CASE STUDY

We explore all AM processes to produce a tissue engineering construct. To select the best process based on manufacturing complexity, we assume the following criteria (Step 1):

C_1: High accuracy
C_2: Mechanical properties
C_3: Production rate
C_4: Material support requirement

Based on the available technologies and the decision criteria, the hierarchical structure of the decision-making process can be constructed as depicted in Figure 2.

As shown in Table 1, the pairwise comparison matrix of the criteria is formed by the product analyst's preference (Step 2).

Then, the value of the fuzzy synthetic extent with respect to each criterion is calculated:

$$S_1 = \left(7,13,16\right) \odot \left(\frac{1}{42.57},\frac{1}{30.62},\frac{1}{18.13}\right) = \left(0.164,0.425,0.883\right)$$

$$S_2 = \left(2.3,4.4,6.75\right) \odot \left(\frac{1}{42.57},\frac{1}{30.62},\frac{1}{18.13}\right) = \left(0.054,0.144,0.372\right)$$

Table 1. Pairwise comparison matrix of the criteria

	C_1	C_2	C_3	C_4
C_1	$(1,1,1)$	$(2,5,6)$	$(3,5,6)$	$(1,2,3)$
C_2	$\left(\frac{1}{6},\frac{1}{5},\frac{1}{2}\right)$	$(1,1,1)$	$(1,3,5)$	$\left(\frac{1}{7},\frac{1}{5},\frac{1}{4}\right)$
C_3	$\left(\frac{1}{6},\frac{1}{5},\frac{1}{3}\right)$	$\left(\frac{1}{5},\frac{1}{3},1\right)$	$(1,1,1)$	$\left(\frac{1}{8},\frac{1}{5},\frac{1}{2}\right)$
C_4	$\left(\frac{1}{3},\frac{1}{2},1\right)$	$(4,5,7)$	$(2,5,8)$	$(1,1,1)$

Figure 2. Hierarchical structure of the relative complexity of the manufacturing process

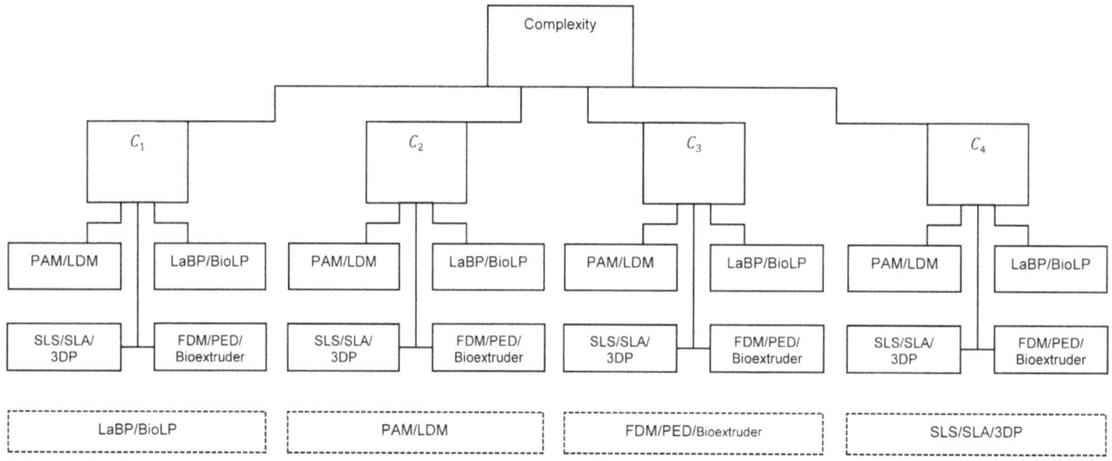

$$S_3 = (1.5, 1.72, 2.82) \odot \left(\frac{1}{42.57}, \frac{1}{30.62}, \frac{1}{18.13} \right) = (0.035, 0.056, 0.175)$$

$$S_4 = (7.32, 11.5, 17) \odot \left(\frac{1}{42.57}, \frac{1}{30.62}, \frac{1}{18.13} \right) = (0.172, 0.376, 0.938)$$

Now, we calculate the weight of each criterion by normalizing the degree of possibility for each S_i to be greater than the others:

$$V(S_1 \geq S_2) = 1$$

$$V(S_1 \geq S_3) = 1$$

$$V(S_1 \geq S_4) = 1$$

$$V(S_2 \geq S_1) = \frac{0.164 - 0.372}{(0.144 - 0.372) - (0.425 - 0.164)} = 0.425$$

$$V(S_2 \geq S_3) = 1$$

$$V\left(S_2 \geq S_4\right) = \frac{0.172 - 0.372}{\left(0.144 - 0.372\right) - \left(0.376 - 0.172\right)} = 0.463$$

$$V\left(S_3 \geq S_1\right) = \frac{0.164 - 0.175}{\left(0.056 - 0.175\right) - \left(0.425 - 0.164\right)} = 0.029$$

$$V\left(S_3 \geq S_2\right) = \frac{0.054 - 0.175}{\left(0.056 - 0.175\right) - \left(0.144 - 0.054\right)} = 0.579$$

$$V\left(S_3 \geq S_4\right) = \frac{0.172 - 0.175}{\left(0.056 - 0.175\right) - \left(0.376 - 0.172\right)} = 0.009$$

$$V\left(S_4 \geq S_1\right) = \frac{0.164 - 0.938}{\left(0.376 - 0.938\right) - \left(0.425 - 0.164\right)} = 0.94$$

$$V\left(S_4 \geq S_2\right) = 1$$

$$V\left(S_4 \geq S_3\right) = 1$$

$$d'\left(C_1\right) = V\left(S_1 \geq S_2, S_3, S_4\right) = \min\left(1,1,1\right) = 1$$

$$d'\left(C_2\right) = V\left(S_2 \geq S_1, S_3, S_4\right) = \min\left(0.425,1,0.463\right) = 0.425$$

$$d'\left(C_3\right) = V\left(S_3 \geq S_1, S_2, S_4\right) = \min\left(0.029,0.579,0.009\right) = 0.009$$

$$d'\left(C_4\right) = V\left(S_4 \geq S_1, S_2, S_3\right) = \min\left(0.94,1,1\right) = 0.94$$

Now we have the weight for each criterion, as shown in Table 2.

We then repeat the same calculation for the alternatives with respect to each criterion (Step 3). Having a pairwise comparison matrix of alternatives (Table 3), Table 4 presents the weights of the alternatives with respect to $\boldsymbol{C_1}$.

$$S_1 = (11,19,26) \odot \left(\frac{1}{45.52}, \frac{1}{31.17}, \frac{1}{17.92} \right)$$
$$= (0.21, 0.61, 1.451)$$

$$S_2 = (2.63, 5.2, 7.82) \odot \left(\frac{1}{45.52}, \frac{1}{31.17}, \frac{1}{17.92} \right)$$
$$= (0.058, 0.167, 0.436)$$

$$S_3 = (1.5, 1.77, 3.2) \odot \left(\frac{1}{45.52}, \frac{1}{31.17}, \frac{1}{17.92} \right)$$
$$= (0.033, 0.057, 0.223)$$

Table 2. Weights of the criteria

	W'	W
C_1	1	0.421
C_2	0.425	0.179
C_3	0.009	0.004
C_4	0.94	0.396

Table 3. Pairwise comparison matrix of the alternatives with respect to C_1

C_1	LaBP/BioLP	PAM/LDM	FDM/PED/BIOEXTRUDER	SLS/SLA/3DP
LaBP/BioLP	$(1,1,1)$	$(3,5,8)$	$(5,8,9)$	$(2,5,8)$
PAM/LDM	$\left(\frac{1}{8},\frac{1}{5},\frac{1}{3}\right)$	$(1,1,1)$	$(1,3,5)$	$\left(\frac{1}{2},1,\frac{3}{2}\right)$
FDM/PED/BIOEXTRUDER	$\left(\frac{1}{9},\frac{1}{8},\frac{1}{5}\right)$	$\left(\frac{1}{5},\frac{1}{3},1\right)$	$(1,1,1)$	$\left(\frac{1}{5},\frac{1}{3},1\right)$
SLS/SLA/3DP	$\left(\frac{1}{8},\frac{1}{5},\frac{1}{2}\right)$	$\left(\frac{2}{3},1,2\right)$	$(1,3,5)$	$(1,1,1)$

$$S_4 = (2.8,5.2,8.5) \odot \left(\frac{1}{45.52}, \frac{1}{31.17}, \frac{1}{17.92} \right)$$
$$= (0.061,0.167,0.474)$$

$$V(S_1 \geq S_2) = 1$$

$$V(S_1 \geq S_2) = 1$$

$$V(S_1 \geq S_4) = 1$$

$$V(S_2 \geq S_1)$$
$$= \frac{0.242 - 0.436}{(0.167 - 0.436) - (0.61 - 0.242)} = 0.305$$

$$V(S_2 \geq S_3) = 1$$

$$V(S_2 \geq S_4) = 1$$

$$V(S_3 \geq S_1)$$
$$= \frac{0.21 - 0.223}{(0.057 - 0.223) - (0.61 - 0.21)} = 0.023$$

$$V(S_3 \geq S_2)$$
$$= \frac{0.058 - 0.223}{(0.057 - 0.223) - (0.167 - 0.058)} = 0.6$$

$$V(S_3 \geq S_4)$$
$$= \frac{0.061 - 0.223}{(0.057 - 0.223) - (0.474 - 0.061)} = 0.596$$

$$V(S_4 \geq S_1)$$
$$= \frac{0.21 - 0.474}{(0.167 - 0.474) - (0.61 - 0.21)} = 0.373$$

Table 4. Weights of the alternatives with respect to C_1

C_1	W'	W
LaBP/BioLP	1	0.588
PAM/LDM	0.305	0.179
FDM/PED/BIOEXTRUDER	0.023	0.014
SLS/SLA/3DP	0.373	0.219

$$V\left(S_4 \geq S_2\right) = 1$$

$$V\left(S_4 \geq S_3\right) = 1$$

$$d'\left(C_1 \mid \text{LaBP / BioLP}\right) = V\left(S_1 \geq S_2, S_3, S_4\right) = \min\left(1,1,1\right) = 1$$

$$d'\left(C_1 \mid \text{PAM / LDM}\right) = V\left(S_2 \geq S_1, S_3, S_4\right) = \min\left(0.305,1,1\right) = 0.305$$

$$d'\left(C_1 \mid \text{FDM / PED / BIOEXTRUDER}\right) = V\left(S_3 \geq S_1, S_2, S_4\right) = \min\left(0.023,0.6,0.596\right) = 0.023$$

$$d'\left(C_1 \mid \text{SLS / SLA / 3DP}\right) = V\left(S_4 \geq S_1, S_2, S_3\right) = \min\left(0.373,1,1\right) = 0.373$$

Similarly, we calculate the weights of the alternatives with respect to $\boldsymbol{C_2}$ as shown in Tables 5 and 6.

$$S_1 = \left(9,13,18\right) \odot \left(\frac{1}{37.75}, \frac{1}{25.1}, \frac{1}{17.28}\right) = \left(0.215,0.518,1.042\right)$$

$$S_2 = \left(4.25,6.32,10.5\right) \odot \left(\frac{1}{37.75}, \frac{1}{25.1}, \frac{1}{17.28}\right) = \left(0.113,0.252,0.608\right)$$

Table 5. Pairwise comparison matrix of the alternatives with respect to C_2

C_2	LaBP/BioLP	PAM/LDM	FDM/PED/ BIOEXTRUDER	SLS/SLA/3DP
LaBP/BioLP	$(1,1,1)$	$(2,3,4)$	$(4,5,8)$	$(2,4,5)$
PAM/LDM	$\left(\frac{1}{4},\frac{1}{3},\frac{1}{2}\right)$	$(1,1,1)$	$(2,3,5)$	$(1,2,4)$
FDM/PED/BIOEXTRUDER	$\left(\frac{1}{8},\frac{1}{5},\frac{1}{4}\right)$	$\left(\frac{1}{5},\frac{1}{3},\frac{1}{2}\right)$	$(1,1,1)$	$\left(\frac{1}{4},\frac{1}{2},1\right)$
SLS/SLA/3DP	$\left(\frac{1}{5},\frac{1}{4},\frac{1}{2}\right)$	$\left(\frac{1}{4},\frac{1}{2},1\right)$	$(1,2,4)$	$(1,1,1)$

Table 6. Weights of the alternatives with respect to C_2

C_2	W'	W
LaBP/BioLP	1	0.52
PAM/LDM	0.596	0.31
FDM/PED/BIOEXTRUDER	0.022	0.011
SLS/SLA/3DP	0.304	0.158

$$S_3 = (1.575,2.02,2.75)\odot\left(\frac{1}{37.75},\frac{1}{25.1},\frac{1}{17.28}\right)$$
$$= (0.042,0.081,0.225)$$

$$S_4 = (2.45,3.75,6.5)\odot\left(\frac{1}{37.75},\frac{1}{25.1},\frac{1}{17.28}\right)$$
$$= (0.065,0.149,0.376)$$

$$V(S_1 \geq S_2) = 1$$

$$V\left(S_1 \geq S_3\right) = 1$$

$$V\left(S_1 \geq S_3\right) = 1$$

$$V\left(S_2 \geq S_1\right) = \frac{0.215 - 0.608}{\left(0.252 - 0.608\right) - \left(0.518 - 0.215\right)} = 0.596$$

$$V\left(S_2 \geq S_3\right) = 1 \quad V\left(S_2 \geq S_4\right) = 1$$

$$V\left(S_3 \geq S_1\right) = \frac{0.215 - 0.225}{\left(0.081 - 0.225\right) - \left(0.518 - 0.215\right)} = 0.022$$

$$V\left(S_3 \geq S_2\right) = \frac{0.113 - 0.225}{\left(0.081 - 0.225\right) - \left(0.252 - 0.113\right)} = 0.4$$

$$V\left(S_3 \geq S_4\right) = \frac{0.065 - 0.225}{\left(0.081 - 0.225\right) - \left(0.149 - 0.065\right)} = 0.702$$

$$V\left(S_4 \geq S_1\right) = \frac{0.215 - 0.376}{\left(0.149 - 0.376\right) - \left(0.518 - 0.215\right)} = 0.304$$

$$V\left(S_4 \geq S_2\right) = \frac{0.113 - 0.376}{\left(0.149 - 0.376\right) - \left(0.252 - 0.113\right)} = 0.719$$

$$V\left(S_4 \geq S_3\right) = 1$$

$$d'\left(C_2 \mid \text{LaBP / BioLP}\right) = V\left(S_1 \geq S_2, S_3, S_4\right) = \min\left(1,1,1\right) = 1$$

$$d'\left(C_2 \mid \text{PAM / LDM}\right) = V\left(S_2 \geq S_1, S_3, S_4\right)$$
$$= \min\left(0.596, 1, 1\right) = 0.596$$

$$d'\left(C_2 \mid \text{FDM / PED / BIOEXTRUDER}\right)$$
$$= V\left(S_3 \geq S_1, S_2, S_4\right)$$
$$= \min\left(0.022, 0.4, 0.702\right) = 0.022$$

$$d'\left(C_2 \mid \text{SLS / SLA / 3DP}\right) = V\left(S_4 \geq S_1, S_2, S_3\right)$$
$$= \min\left(0.304, 0.719, 1\right) = 0.304$$

Also, the weights of alternatives with respect to $\boldsymbol{C_3}$ are calculated and shown in Tables 7 and 8.

$$S_1 = \left(3.2, 5.25, 6.5\right) \odot \left(\frac{1}{35.25}, \frac{1}{27.31}, \frac{1}{17.03}\right)$$
$$= \left(0.091, 0.192, 0.382\right)$$

$$S_2 = \left(9, 15, 19\right) \odot \left(\frac{1}{35.25}, \frac{1}{27.31}, \frac{1}{17.03}\right)$$
$$= \left(0.225, 0.549, 1.116\right)$$

Table 7. Pairwise comparison matrix of the alternatives with respect to $\boldsymbol{C_3}$

C_3	LaBP/BioLP	PAM/LDM	FDM/PED/BIOEXTRUDER	SLS/SLA/3DP
LaBP/BioLP	$\left(1,1,1\right)$	$\left(\frac{1}{5},\frac{1}{4},\frac{1}{2}\right)$	$\left(1,3,4\right)$	$\left(1,1,1\right)$
PAM/LDM	$\left(2,4,5\right)$	$\left(1,1,1\right)$	$\left(4,6,8\right)$	$\left(2,4,5\right)$
FDM/PED/BIOEXTRUDER	$\left(\frac{1}{4},\frac{1}{3},1\right)$	$\left(\frac{1}{8},\frac{1}{6},\frac{1}{4}\right)$	$\left(1,1,1\right)$	$\left(\frac{1}{4},\frac{1}{3},1\right)$
SLS/SLA/3DP	$\left(1,1,1\right)$	$\left(\frac{1}{5},\frac{1}{4},\frac{1}{2}\right)$	$\left(1,3,4\right)$	$\left(1,1,1\right)$

Table 8. Weights of the alternatives with respect to C_3

C_3	W'	W
LaBP/BioLP	0.305	0.188
PAM/LDM	1	0.617
FDM/PED/BIOEXTRUDER	0.01	0.006
SLS/SLA/3DP	0.305	0.188

$$S_3 = \left(1.625, 1.81, 3.25\right) \odot \left(\frac{1}{35.25}, \frac{1}{27.31}, \frac{1}{17.03}\right) = \left(0.046, 0.066, 0.23\right)$$

$$S_4 = \left(3.2, 5.25, 6.5\right) \odot \left(\frac{1}{35.25}, \frac{1}{27.31}, \frac{1}{17.03}\right) = \left(0.091, 0.192, 0.382\right)$$

$$V\left(S_1 \geq S_2\right) = \frac{0.225 - 0.382}{\left(0.192 - 0.382\right) - \left(0.549 - 0.225\right)} = 0.305$$

$$V\left(S_1 \geq S_3\right) = 1$$

$$V\left(S_1 \geq S_4\right) = 1$$

$$V\left(S_2 \geq S_1\right) = 1$$

$$V\left(S_2 \geq S_3\right) = 1 \qquad V\left(S_2 \geq S_4\right) = 1$$

$$V\left(S_3 \geq S_1\right) = \frac{0.091 - 0.23}{\left(0.066 - 0.23\right) - \left(0.192 - 0.091\right)} = 0.525$$

$$V\left(S_3 \geq S_2\right) = \frac{0.225 - 0.23}{\left(0.066 - 0.23\right) - \left(0.549 - 0.225\right)} = 0.01$$

$$V\left(S_3 \geq S_4\right) = \frac{0.091 - 0.23}{\left(0.066 - 0.23\right) - \left(0.192 - 0.091\right)} = 0.525$$

$$V\left(S_4 \geq S_1\right) = 1$$

$$V\left(S_4 \geq S_2\right) = \frac{0.225 - 0.382}{\left(0.192 - 0.382\right) - \left(0.549 - 0.225\right)} = 0.305$$

$$V\left(S_4 \geq S_3\right) = 1$$

$$d'\left(C_3 \mid \text{LaBP / BioLP}\right) = V\left(S_1 \geq S_2, S_3, S_4\right) = \min\left(0.305, 1, 1\right) = 0.305$$

$$d'\left(C_3 \mid \text{PAM / LDM}\right) = V\left(S_2 \geq S_1, S_3, S_4\right) = \min\left(1, 1, 1\right) = 1$$

$$d'\left(C_3 \mid \text{FDM / PED / BIOEXTRUDER}\right) = V\left(S_3 \geq S_1, S_2, S_4\right) = \min\left(0.525, 0.01, 0.525\right) = 0.01$$

$$d'\left(C_3 \mid \text{SLS / SLA / 3DP}\right) = V\left(S_4 \geq S_1, S_2, S_3\right) = \min\left(1, 0.305, 1\right) = 0.305$$

Table 9 presents the pairwise comparison matrix of the alternatives with respect to C_4 used to calculate the weight of the alternatives in Table 10.

$$S_1 = \left(15, 20, 25\right) \odot \left(\frac{1}{47}, \frac{1}{35.34}, \frac{1}{23.97}\right) = \left(0.252, 0.566, 1.043\right)$$

Table 9. Pairwise comparison matrix of the alternatives with respect to C_4

C_4	LaBP/BioLP	PAM/LDM	FDM/PED/BIOEXTRUDER	SLS/SLA/3DP
LaBP/BioLP	$(1,1,1)$	$(4,5,7)$	$(6,8,9)$	$(4,6,8)$
PAM/LDM	$\left(\frac{1}{7},\frac{1}{5},\frac{1}{4}\right)$	$(1,1,1)$	$(3,5,7)$	$(1,3,5)$
FDM/PED/BIOEXTRUDER	$\left(\frac{1}{9},\frac{1}{8},\frac{1}{6}\right)$	$\left(\frac{1}{7},\frac{1}{5},\frac{1}{3}\right)$	$(1,1,1)$	$\left(\frac{1}{4},\frac{1}{3},1\right)$
SLS/SLA/3DP	$\left(\frac{1}{8},\frac{1}{6},\frac{1}{4}\right)$	$\left(\frac{1}{5},\frac{1}{3},1\right)$	$(1,3,4)$	$(1,1,1)$

Table 10. Weights of the alternatives with respect to C_4

C_4	W'	W
LaBP/BioLP	1	0.63
PAM/LDM	0.496	0.313
FDM/PED/BIOEXTRUDER	0.015	0.009
SLS/SLA/3DP	0.076	0.048

$$S_2 = (5.14,9.2,13.25) \odot \left(\frac{1}{47},\frac{1}{35.34},\frac{1}{23.97}\right) = (0.109,0.26,0.553)$$

$$S_3 = (1.5,1.65,2.49) \odot \left(\frac{1}{47},\frac{1}{35.34},\frac{1}{23.97}\right) = (0.032,0.047,0.26)$$

$$S_4 = (2.33,4.5,6.25) \odot \left(\frac{1}{47},\frac{1}{35.34},\frac{1}{23.97}\right) = (0.049,0.127,0.288)$$

$$V(S_1 \geq S_2) = 1$$

$$V\left(S_1 \geq S_3\right) = 1$$

$$V\left(S_1 \geq S_4\right) = 1$$

$$V\left(S_2 \geq S_1\right) = \frac{0.252 - 0.553}{\left(0.26 - 0.553\right) - \left(0.566 - 0.252\right)} = 0.496$$

$$V\left(S_2 \geq S_3\right) = 1 \quad V\left(S_2 \geq S_4\right) = 1$$

$$\mathrm{V}\left(\mathrm{S}_3 \geq \mathrm{S}_1\right) = \frac{0.252 - 0.26}{\left(0.047 - 0.26\right) - \left(0.566 - 0.252\right)} = 0.015$$

$$V\left(S_3 \geq S_2\right) = \frac{0.109 - 0.26}{\left(0.047 - 0.26\right) - \left(0.26 - 0.109\right)} = 0.415$$

$$V\left(S_3 \geq S_4\right) = \frac{0.049 - 0.26}{\left(0.047 - 0.26\right) - \left(0.127 - 0.049\right)} = 0.725$$

$$V\left(S_4 \geq S_1\right) = \frac{0.252 - 0.288}{\left(0.127 - 0.288\right) - \left(0.566 - 0.252\right)} = 0.076$$

$$V\left(S_4 \geq S_2\right) = \frac{0.109 - 0.288}{\left(0.127 - 0.288\right) - \left(0.26 - 0.109\right)} = 0.377$$

$$V\left(S_4 \geq S_3\right) = 1$$

$$d'\left(C_4 \mid \mathrm{LaBP} / \mathrm{BioLP}\right) = V\left(S_1 \geq S_2, S_3, S_4\right) = \min\left(1,1,1\right) = 1$$

$$d'\left(C_4 \mid \text{PAM / LDM}\right) = V\left(S_2 \geq S_1, S_3, S_4\right)$$
$$= \min\left(0.496, 1, 1\right) = 0.496$$

$$d'\left(C_4 \mid \text{FDM / PED / BIOEXTRUDER}\right)$$
$$= V\left(S_3 \geq S_1, S_2, S_4\right)$$
$$= \min\left(0.015, 0.415, 0.725\right) = 0.015$$

$$d'\left(C_4 \mid \text{SLS / SLA / 3DP}\right)$$
$$= V\left(S_4 \geq S_1, S_2, S_3\right)$$
$$= \min\left(0.076, 0.377, 1\right) = 0.076$$

The weights of each alternative with respect to each criterion are summarized in Table 11. By adding the weight per alternative and multiplying by the corresponding criteria (Table 11), we get the final relative complexity ranking for all alternatives (Table 12):

1. LaBP/BioLP
2. PAM/LDM
3. MRPII
4. SLS/SLA/3DP

Table 11. Summary of the alternative weights with respect to the criteria

	LaBP/BioLP	PAM/LDM	FDM/PED/BIOEXTRUDER	SLS/SLA/3DP
C_1	0.588	0.179	0.014	0.219
C_2	0.52	0.31	0.011	0.158
C_3	0.188	0.617	0.006	0.188
C_4	0.63	0.313	0.009	0.048

Table 12. Relative complexity of the alternatives

	LaBP/BioLP	PAM/LDM	FDM/PED/BIOEXTRUDER	SLS/SLA/3DP
Relative complexity	0.591	0.257	0.011	0.14

5. CONCLUSION

The manufacturing processes are based on the building of a solid object from three-dimensional (3D) model data by joining materials, usually layer upon layer to enhance the production of the tissue engineering constructs. This may have an impact on the design and manufacturing of AM products. Therefore, measuring the complexity becomes an important task for management. Recently, literature pertaining to complexity measures shows that some studies aim at quantifying additive manufacturing complexity. However, lack of information and uncertain situations have been obstacles in these studies. In this study, we developed the adapted FAHP model that aims at measuring the relative complexity of the additive manufacturing processes. This relative complexity measure can be used for cost estimation and budgeting. This model takes into account high accuracy, mechanical properties, production rate, and material support requirement. These decision criteria can represent the strength and weakness of AM technologies/processes as follows:

1. Powder-based techniques do not need to have support materials heated during the fabrication process as they have high production rates and can produce large sample sizes.
2. Photosensitivity-based techniques also do not need the support material and are suitable for encapsulating cells.
3. Melt-extrusion-based techniques have good mechanical properties and do not have materials trapped in the scaffold.
4. Solution/slurry extrusion-based techniques can be used for producing hydrogel scaffold encapsulating cells.
5. Tissue and organ printing provide deposition of high cell densities and small volumes with high accuracy; a wide range of cell lines were already printed achieving high cell vitality.

The complexity ranking of these processes can be extracted from the FAHP model. This study shows that FAHP is a good method, because it eliminates inconsistencies due to personal feelings or judgments of the process analyst. Considering different additive manufacturing process technologies, where more than one technology is emphasized, inconsistencies in pairwise comparisons are introduced and factored throughout the model. Without applying FAHP, the overlooked inconsistencies can lead to misunderstanding of the complexity of the AM processes. As a result, we may allocate insufficient budget and resources to the processes. For future studies, a fuzzy artificial intelligent decision support system (DSS) for the evolution of manufacturing complexity should be developed.

ACKNOWLEDGMENT

We would like to thank the reviewers for their time reviewing our paper and providing valuable comments and suggestions for improvement. Their dedication and expert knowledge significantly contributed to this paper.

REFERENCES

Acar, W., Anokhin, S., & Troutt, M. D. (2010). Using decision sciences to enhance entrepreneurial foresight: The comprehensive situation mapping approach. *International Journal of Strategic Decision Sciences*, *1*(1), 81–100. doi:10.4018/jsds.2010103005

Bashir, H., & Thomson, V. (2001). Models for estimating design effort and time. *Design Studies*, *22*(2), 141–155. doi:10.1016/S0142-694X(00)00014-4

Chakraborty, S., & Choudhury, P. P. (1999). A realistic measure for an algorithm's complexity. *Applied Mathematics Letters*, *12*, 113–118. doi:10.1016/S0893-9659(99)00110-X

Cho, S., Alamoudi, R., & Asfour, S. (2009). Interaction-based complexity measure of manufacturing systems using information entropy. *International Journal of Computer Integrated Manufacturing*, *22*(10), 909–922. doi:10.1080/09511920902951393

Crump, S. S. (1992). *Apparatus and method for creating three-dimensional objects* (US Patent No. 5121329). Washington, DC: US Patent Office.

De Reyck, B., & Herroelen, W. (1996). Theory and methodology on the use of the complexity index as a measure of complexity in activity networks. *European Journal of Operational Research*, *91*, 347–366. doi:10.1016/0377-2217(94)00344-0

Deckard, C. R. (1989). *Method and apparatus for producing parts by selective sintering* (US Patent No. 5017753). Washington, DC: US Patent Office.

Dehmer, M., Emmert-Streib, F., & Kilian, J. (2006). A similarity measure for graphs with low computational complexity. *Applied Mathematics and Computation*, *182*, 447–459. doi:10.1016/j.amc.2006.04.006

Dvir, D., Lipovetsky, S., Shenhar, A., & Tishler, A. (1998). In search of project classification: A non-universal approach to project success factors. *Research Policy*, *27*, 915–935. doi:10.1016/S0048-7333(98)00085-7

El-Haik, B., & Yang, K. (1999). The components of complexity in engineering design. *IIE Transactions*, *31*(10), 925–934. doi:10.1080/07408179908969893

Grant, D., & Tu, Q. (2005). Levels of entSLS/SLA/3DPrise integration: Study using case analysis. *International Journal of EntSLS/SLA/3DPrise. Information Systems*, *1*(1), 1–22.

Hutmacher, D. W., Sittinger, M., & Risbud, M. V. (2004). Scaffold-based tissue engineering: Rationale for computer-aided design and solid free-form fabrication systems. *Trends in Biotechnology*, *22*(7), 354–362. doi:10.1016/j.tibtech.2004.05.005 PMID:15245908

Huynh, V. N., & Nakamori, Y. (2010). Notes on reducing algorithm complexity for computing an aggregate uncertainty measure. *IEEE Transactions on Systems, Man, and Cybernetics. Part A, Systems and Humans*, *40*(1), 205–209. doi:10.1109/TSMCA.2009.2030962

Jenab, K., & Dhillon, B. S. (2005). Group-based failure effects analysis (GFEA). *International Journal of Reliability Quality and Safety Engineering*, *12*(4), 291–307. doi:10.1142/S0218539305001835

Jenab, K., & Liu, D. (2010). A graph-based model for manufacturing complexity. *International Journal of Production Research*, *48*(11), 3383–3392. doi:10.1080/00207540902950860

Jitrawichawet, S., Sarfaraz, A., & Jenab, K. (2013). Reliability based foreign supplier selection using fuzzy-AHP. *International Journal of Applied Decision Sciences*, *6*(3), 245–262. doi:10.1504/IJADS.2013.054946

Kaden, F. (1982). Graphmetriken und distanzgraphen. *ZKI-Informationen, Akad. Wiss. DDR*, *2*(82), 1–63.

Kashyap, N., & Sinha, S. (2011). Estimation of job complexity in an industrial organization. *International Journal of Industrial and Systems Engineering*, *7*(1), 26–44. doi:10.1504/IJISE.2011.037241

Kauffman, S. A. (1993). *The origins of order*. Oxford, UK: Oxford University Press.

Kemerer, C. F., & Porter, B. S. (1992). Improving the reliability of function point measurement: An empirical study. *IEEE Transactions on Software Engineering*, *18*(11), 1011–1024. doi:10.1109/32.177370

Kendall, K. E., Kong, S., & Kendall, J. E. (2010). The impact of agile methodologies on the quality of information systems: Factors shaping strategic adoption of agile practices. *International Journal of Strategic Decision Sciences*, *1*(1), 41–56. doi:10.4018/jsds.2010103003

Koch, L., Deiwick, A., Schlie, S., Michael, S., Gruene, M., Coger, V., & ... (2012). Skin tissue generation by laser cell printing. *Biotechnology and Bioengineering*, *109*(7), 1855–1863. doi:10.1002/bit.24455 PMID:22328297

Lathrop, J. I., Lutz, J. H., Patitz, M. J., & Summers, S. M. (2011). Computability and complexity in self-assembly. *Theory of Computing Systems*, *48*, 617–647. doi:10.100700224-010-9252-0

Lefley, F., & Sarkis, J. (2005). Applying the FAP model to the evaluation of strategic information technology projects. *International Journal of EntSLS/SLA/3DPrise. Information Systems*, *1*(2), 69–90.

Liu, C., Grenier, D., Jousselme, A. L., & Bossé, E. (2007). Reducing algorithm complexity for computing an aggregate uncertainty measure. *IEEE Transactions on Systems, Man, and Cybernetics*, *37*(5), 669–679. doi:10.1109/TSMCA.2007.893457

Lockett, N. J., & Brown, D. H. (2005). An SME perspective of vertical application service providers. *International Journal of EntSLS/SLA/3DPrise. Information Systems*, *1*(2), 37–55.

Mahmood, S., & Lai, R. (2008). A complexity measure for UML component-based system specification. *Software, Practice & Experience*, *38*, 117–134. doi:10.1002pe.769

McCabe, T. J. (1992). *Structured testing: A software testing methodology using the cyclomatic complexity metric*. Washington, DC: The National Institute of Standards and Technology.

McFarlan, F. W. (1981). Portfolio approach to information systems. *Harvard Business Review*, *59*(5).

Meixner, O. (2011). *Fuzzy AHP group decision analysis and its application for the evaluation of energy sources*. Retrieved from http://creativedecisions.net/~rozann/0Proceedings/Final_Abstracts/50_Meixner_Fuzzy_AHP_REV_FIN-a.pdf

Newell, S., Cooprider, J. G., David, G., Edelman, L. F., & Logan, T. A. (2005). Analyzing different strategies to entSLS/SLA/3DPrise system adoption: Reengineering-led vs. quick-deployment. *International Journal of EntSLS/SLA/3DPrise. Information Systems*, *1*(2), 1–16.

Pich, M. T., Loch, C. H., & De Meyer, A. (2002). On uncertainty, ambiguity, and complexity in project management. *Management Science*, *48*(8), 1008–1023. doi:10.1287/mnsc.48.8.1008.163

Rodriguez-Toro, C. A., Tate, S. J., Jared, G. E. M., & Swift, K. G. (2003). Complexity metrics for design (simplicity + simplicity = complexity). *Institute of Mechanical Engineers*, *217*(5), 721–725. doi:10.1243/095440503322011461

Samy, S. N., & El Maraghy, H. (2010). A model for measuring products assembly complexity. *International Journal of Computer Integrated Manufacturing*, *23*(11), 1015–1027. doi:10.1080/0951192X.2010.511652

Sarfaraz, A., Jenab, K., & Suwanvaraboon, A. (2014). Using fuzzy AHP for warranty management improvement. *International Journal of Industrial and Systems Engineering*.

Schrader, S., Riggs, W. M., & Smith, R. P. (1993). Choice over uncertainty and ambiguity in technical problem solving. *Journal of Engineering and Technology Management*, *10*, 73–99. doi:10.1016/0923-4748(93)90059-R

Schultz, R., Sarfaraz, A., & Jenab, K. (2013). Analysis of risk and reliability in project delivery methods. *International Journal of Strategic Decision Sciences*, *4*(4).

Shafiei-Monfared, S., & Jenab, K. (2010). Complexity analysis of an operation in demand-based manufacturing. *International Journal of Production Research*, *49*(17), 5303–5315. doi:10.1080/00207543.2010.518993

Shafiei-Monfared, S., & Jenab, K. (2011a). A novel approach for complexity measure analysis in design projects. *Journal of Engineering Design*. doi:10.1080/09544828.2011.554389

Shafiei-Monfared, S., & Jenab, K. (2011b). A fuzzy complexity model for entSLS/SLA/3DPrise maintenance projects. *IEEE Transactions on Engineering Management*. doi:10.1109/TEM.2010.2096560

Smith, J., & Jenks, C. (2006). *Qualitative complexity: Ecology, cognitive processes and the re-emergence of structures in post-humanist social theory*. New York: Routledge.

Steward, D. V. (1981). The design structure system: A method for managing the design of complex systems. *IEEE Transactions on Engineering Management*, *28*(3), 71–74. doi:10.1109/TEM.1981.6448589

Tatikonda, M. V., & Rosenthal, S. R. (2000). Technology novelty, project complexity, and product development execution success. *IEEE Transactions on Engineering Management*, *47*, 74–87. doi:10.1109/17.820727

Vieira, D. A. G., Takahashi, R. H. C., Palade, V., Vasconcelos, J. A., & Caminhas, W. M. (2008). The q-norm complexity measure and the minimum gradient method: A novel approach to the machine learning structural risk minimization problem. *IEEE Transactions on Neural Networks*, *19*(8), 1415–1430. doi:10.1109/TNN.2008.2000442 PMID:18701371

Vladislavleva, E. J., Smits, G. F., & den Hertog, D. (2009). Order of nonlinearity as a complexity measure for models generated by symbolic regression via Pareto genetic programming. *IEEE Transactions on Evolutionary Computation*, *13*(2), 333–349. doi:10.1109/TEVC.2008.926486

Vozzi, G., Flaim, C., Ahluwalia, A., & Bhatia, S. (2003). Fabrication of PLGA scaffolds using soft lithography and microsyringe deposition. *Biomaterials*, *24*(14), 2533–2540. doi:10.1016/S0142-9612(03)00052-8 PMID:12695080

Wang, H., Zhu, X., Wang, H., Hu, S. J., Lin, Z., & Chen, G. (2011). Multi-objective optimization of product variety and manufacturing complexity in mixed-model assembly systems. *Journal of Manufacturing Systems*. doi:10.1016/j.jmsy.2011.03.002

Xiong, Z., Yan, Y., Wang, S., Zhang, R., & Zhang, C. (2002). Fabrication of porous scaffolds for bone tissue engineering via low-temperature deposition. *Scripta Materialia*, *46*(11), 771–776. doi:10.1016/S1359-6462(02)00071-4

Zhu, X., Hu, S. J., Koren, Y., & Huang, N. (2011). A complexity model for sequence planning in mixed-model assembly lines. *Journal of Manufacturing Systems*. doi:10.1016/j.jmsy.2011.07.006

This research was previously published in Analytical Approaches to Strategic Decision-Making edited by Madjid Tavana, pages 240-257, copyright year 2014 by Business Science Reference (an imprint of IGI Global).

Chapter 17
The "Arm" Line of Devices for Neurological Rehabilitation

Alessandro Scano
Institute of Industrial Technologies and Automation, Italy

Andrea Chiavenna
Institute of Industrial Technologies and Automation, Italy

Tito Dinon
Institute of Industrial Technologies and Automation, Italy

Alessio Prini
Institute of Industrial Technologies and Automation, Italy

Giulio Spagnuolo
Institute of Industrial Technologies and Automation, Italy

Matteo Malosio
Institute of Industrial Technologies and Automation, Italy

Lorenzo Molinari Tosatti
Institute of Industrial Technologies and Automation, Italy

ABSTRACT

In the modern scenario of neurological rehabilitation, which requires affordable solutions oriented toward promoting home training, the Institute of Industrial Technologies and Automation (ITIA) of the Italian National Research Council (CNR) developed a line of prototypal devices for the rehabilitation of the upper limb, called "Arm." Arm devices were conceived to promote rehabilitation at affordable prices by capturing all the main features of the state-of-the-art devices. In fact, Arm devices focus on the main features requested by a robot therapist: mechanical adaptation to the patient, ranging from passive motion to high transparency, assist-as-needed and resistive modalities; proper use of sensors for performance monitoring; easy-to-use, modular, and adaptable design. These desirable features are combined with low-cost, additive manufacturing procedures, with the purpose of meeting the requirements coming from research on neuro-motor rehabilitation and motor control and coupling them with the recent breakthrough innovations in design and manufacturing.

DOI: 10.4018/978-1-5225-9624-0.ch017

INTRODUCTION

The use of robotic devices for upper-limb neuro-motor rehabilitation is usual practice in clinical centers. In respect to conventional therapies, robots allow to increase training intensity and help patients to promote their active contribution. Furthermore, robots can act as measurers of patients' performances and adapt their interaction modalities to the emerging needs during the rehabilitation course. Robots like ARMin, MIT Manus, Armeo Spring, Braccio di Ferro, represent the state of the art devices for rehabilitation of the upper-limb and for promoting motor recovery. According to the available assessments and studies in the literature, their efficacy is slightly/moderately higher than the one of conventional therapies. Furthermore, robots are used in research to learn more about physiological and pathological motor control and neuromuscular diseases. Unfortunately, while being the state of the art devices for neuro-motor stimulation and training, such robots are very expensive and not compliant to user-friendly requirements that are needed for semi-autonomous home use. Consequently, they can be used only in clinical environments, under the supervision of medical personnel. Furthermore, sanitary costs related to rehabilitation are increasing and clinical centers can hardly support their burden. The possibility of delocalizing rehabilitation from clinical centers opens the chance for training performed in home environment, with time and costs savings for both the sanitary system and patients. In this scenario, which requires affordable solutions oriented toward promoting home training, the Institute of Industrial Technologies and Automation (ITIA) of the Italian National Research Council (CNR) developed a line of prototypal devices for the rehabilitation of the upper-limb, called *-Arm*. *Arm* devices were conceived to test the possibility of promoting rehabilitation at affordable prices but capturing all the main features of the state of the art devices. In fact, *Arm* devices focus on the main features requested by a robot therapist: mechanical adaptation to the patient, ranging from passive motion to high transparency, assist-as-needed and resistive modalities; proper use of sensors for performance monitoring; easy-to-use, modular and adaptable design. These desirable features are combined with low-cost, additive manufacturing procedures, with the purpose of meeting the requirements coming from research on neuro-motor rehabilitation and motor control and coupling them with the recent breakthrough innovations in design and manufacturing. *Arm* devices cover both clinical and home-oriented training and are designed for adaptation to patients with different motor impairment.

The *Arm* prototypes are:

- **LINarm**: linear device, freely orientable in space, suitable for functional movements. It features a variable stiffness actuation, allowing to adapt the mechanical behavior of the device to patients' needs. Functional Electrical Stimulation, simple Virtual Environments and a Patient Model, gathering data from integrated sensors and modulating the level of assistance, are integrated in the set-up. The LINarm++ Echord++ Project ended in October 2016 and guided the development of a second, more refined prototype, enhancing the original concept.
- **PLANarm**: planar device, freely orientable in space, suitable for planar functional movements. The state of the art planar robots used in literature for motor control and motor learning research inspired PLANarm. It features a variable stiffness actuation, allowing adapting the mechanical behavior of the device depending on patients' needs.
- **DUALarm**: Low-Cost device for bimanual rehabilitation, exploiting the capability of the less affected limb to provide rehabilitation to the more affected limb. DUALarm is completely realized in 3D printing technology and aims at being an easy-to-use, low-cost, open-source project.

Currently, reaching movements can be trained, but the device is conceived to be suitable for training of other functional gestures.

- **LIGHTarm**: Exoskeleton for the rehabilitation of the upper-limb, designed in two versions: LIGHTarm, not actuated, and conceived to support the weight of the impaired limb. The mechanical design includes high backdrivability, focusing on shoulder rhythm and elbow singular configurations.
- **VIRTUALarm**: Kinect One-based platform for motor monitoring, including body and limb tracking and a biomechanical evaluation of the performance in relation to databases of healthy subjects. Assessments include range of motion, motion dynamics, effort, motor control indexes, body segments barycenter tracking.

In this chapter, the main mechanical and design features of the devices will be described, along with the preliminary trials executed with the devices on healthy people and neurological patients. Instrumental assessments during the use of the devices, related to kinematics, EMG and human-robot interaction, will be discussed. Finally, a discussion will describe potential, appealing scenarios of the use of *Arm* devices in clinical and home environment.

BACKGROUND

Neurological Diseases and Stroke: Epidemiology Neurological Diseases are one of the main sources of disability, especially in Western countries. A wide variety of pathologies and symptoms can lead to partial or complete disability, which influences people life up to preventing self-autonomy. Stroke is one of the most noticeable sources of motor impairment. The World Health Organization (WHO) defines stroke as the clinical syndrome of rapid onset of focal (or global) cerebral deficit, lasting more than 24 hours with no apparent cause other than a vascular one (Hsieh, Chen, Li, & Lai, 2015). According to the recently published Global Burden of Disease 2010 Study (GBD 2010), stroke is the second leading cause of death globally and the third leading cause of premature death and disability as measured in DALY (Daily-affected living years). It is estimated that interventions to reduce mortality from stroke and ischemic heart disease by 10% are projected to reduce economic burden by an average of USD $25 billion annually with a threefold return on investment (Hsieh et al., 2015). Ischemic heart disease, sharing many risk factors of stroke, is the leading cause of both death and premature death and disability. The GBD 2010 clusters DALY for cerebrovascular disease in the category "Cardiovascular and circulatory diseases." Within the combined burden of neurologic disorders and cerebrovascular disease as measured in the GBD 2010, hemorrhagic stroke (35.7%) and ischemic stroke (22.4%) contribute the largest proportions of DALY (from GBD 2010 Results by Cause 1990–2010) (Chin & Vora, 2014).

Robots and Devices for Rehabilitation in Clinical Environment: An Overview

The Schema Theory of motor learning (Schmidt, 1975) (Sherwood & Lee, 2003) provides a framework for understanding the way human Central Nervous System (CNS) produces voluntary movement. Such issue has been addressed with the generalized motor program (GMP) concept proposed by Schmidt. In his theory, Schmidt argued that motor output is achieved through neural representations of the movements, called motor programs (Muratori, Lamberg, Quinn, & Duff, 2014). Motor programs are not related to specific tasks; rather, each motor program accounts for a large class of similar actions. Such framework not only is compliant with storage limit capabilities connected to human motor memory, but also explains that new skills can be performed by drawing from already stored similar classes of movements. Furthermore, Schmidt argues that stored rules of each GMP rely on invariant features that the program specifies (Muratori et al., 2014). These invariants help define classes of movement and help our CNS in minimizing the amount of information that must be stored.

The use of robotic devices for upper-limb neuro-motor rehabilitation is usual practice in clinical centers. It can be seen as the use of robotic therapists that help to reshape GMPs that have been lost due to stroke and, consequently, lists of motor commands that the CNS specifies to produce purposeful movement. Consequently, the robot should act as an intelligent care-giver, by triggering assistance and promoting the process of motor learning and have a biomimetic behavior (Habib, 2011). Assist-as-needed control systems have been developed in order to minimize the intervention of the robot that, according to such logics, intervenes only in case of failure in executing the motor task (Morasso et al., 2009). Thus, the requirements for rehabilitation robots include high compliance and softness: a little exerted force should guarantee motion. Backdriveability is requested to allow patients to interact and experience the results of their muscular activation patterns (Maciejasz, Eschweiler, Gerlach-Hahn, Jansen-Troy, & Leonhardt, 2014). Force fields created by the robot can be used to assist the execution of the motor task, or to resist patients' capabilities to provide challenging exercises. Furthermore, robots can act as objective measurers of patients' performances. Such capability allows adapting their interaction modalities to the emerging needs during the rehabilitation course.

Furthermore, in respect to conventional therapies, robots allow to increase training intensity and help patients to promote their active contribution. Resilience can be monitored and the therapy can be made more repeatable.

Two basic approaches to robotic therapy for the upper-limb can be identified. The first one includes end-effector robots, the second one exoskeletons. End-effector robots require patients to hold the handle and displace the robot. Some state-of-the-art examples are the ARMin, MIT Manus, Braccio di Ferro. They include a handle that can be displaced in a planar workspace, together with simple games or virtual environments that address patients' efforts towards specific tasks and targets. During the interaction, the robot, by providing haptic feedbacks and force fields, can ease the execution or making it more challenging. Exoskeletons, instead, mimic the kinematics of the limb and wrap the affected limb, such as Armeo Spring. They are designed to support the weight of the limb and execute 3D movements, usually in interaction with games or virtual environments that ask to perform specific tasks. According to the available assessments and studies in the literature, robots efficacy is slightly/moderately higher than the one of conventional therapies (Lo et al., 2010). Furthermore, robots are used in research to learn more about physiological and pathological motor control and neuromuscular diseases.

The Emerging Need for Home-Oriented, Cost-Effective Solutions

Unfortunately, while being the state of the art devices for neuro-motor stimulation and training, previously mentioned robots are very expensive and not compliant to user-friendly requirements that are needed for semi-autonomous home use. Consequently, they can be used only in clinical environments, under the supervision of medical personnel. Furthermore, sanitary costs related to rehabilitation are increasing and clinical centers can hardly support their burden. The possibility of delocalizing rehabilitation from clinical centers opens the chance for training performed in home environment, with time and costs savings for both the sanitary system and patients. The neuro-motor rehabilitation process involves training exercises whose intensity and quality, along with the amount of damage, are related to functional recovery. Conventional approaches to therapy, robotic or supported by virtual interactive environments are standard practice in the clinic as well as the possibility of surgical and pharmacological support. In the United States, the costs related to the provision of these therapies are expected to double by 2030, going to affect 4% of the population. There are also scientific evidence that the benefits related to rehabilitation are preserved over time if the partially restored features are constantly kept under workout and used continuously in daily life. In Western countries, it is therefore necessary to identify new strategies to promote the rehabilitation process as an intensive and low-cost. The previous conditions indicate as a natural consequence the transfer of neurological therapies at the household level. To the benefits listed above, are added objective factors of decrease of the crowding of hospital centers, the possibility to make daily motor training without having to reach the reference structure, with savings of time not only for the health system but also for the patient.

The ability to run training within the family context, generally more comfortable for the patient, is an added value. It should however be noted that the delivery of treatments at the clinical level takes advantage of: 1) technologies that represent the state for training oriented neuro-motor recovery, 2) equipment and sensors for the assessment of therapy, and 3) medical and physiotherapy that can support the patient. These factors may not be present at home; their absence poses considerable challenges for the effective implementation of motor training, although the home solution seems to be the way of the natural and preferred out of the problems highlighted above.

The first requirement to be analyzed is the absence of a direct supervision to therapy. For this reason, in addition to adequate patient and family education on the proper execution of the training, rehabilitation is based on computer applications, which require the presence of virtual assistants, equipped with artificial intelligence, which direct the activities and show the patient how to rehabilitate, just as would a physical therapist at the clinic structure.

The type of rehabilitation activities proposed at the household level can occur in two main types of approaches. The first is based on the use of motor learning strategies centered on the teaching of specific functional motor tasks related to activities of daily living (coaching); during the training, the patient focuses on the repetitive execution of one or more specific gestures, maximizing the intensity of the workout in order to regain that specific functionality (Cameirão et al., 2016). The second approach, generally associated with the presence of virtual environments, is otherwise to promote physical activity in conjunction with the execution of fun and motivating activities. According to the Cochrane review of Laver (K. E. Laver, George, Thomas, Deutsch, & Crotty, 2015; K. Laver, George, Thomas, Deutsch, & Crotty, 2012), the adoption of virtual environments to support the rehabilitation does not offer substantial improvements compared to conventional therapy and may present cognitively complex challenges and require coordination of multiple motor districts. However, if provided with appropriate logical and appropriate to the patient's ability, the approach related to gaming can enhance motivation by offering a variety of activities and engaging, and more regular training and long-lasting. An element that promotes this aspect is the possible presence of the storyboard, which is a narrative progression motor experience, more emotionally engaging. Functional involvement is also the promotion of appropriately balanced logic, enabling the patient to co-operate or "compete on equal terms" with other people.

Applications should guide the patient in therapy and adapt to his motor skills, by calibrating the difficulty of the exercise in terms of gestures (stimulating the execution of specific motor tasks) and in terms of required skills (within the same gesture stimulating higher execution speed or accuracy).

Scoring systems and performance evaluation are useful for both the monitoring of the training distance, for immediate feedback to patients who may be aware of the trend of their motor performance. These applications are based on low-cost technologies, including optical tracking sensors or accelerometers. These sensors make it possible to track the movement of the body and then bind to the patient's motor gestures or an interaction with a virtual environment or performance evaluation. Some examples of this type of approach are Neuro@Home, and VirtualRehab Jintronix, rehabilitative platforms that use Kinect sensors, derived from industry of video games, for the integration of exercises related to functional motors objectives drawn directly from clinical staff.

The domestic therapy can also make use of electromechanical devices dedicated to functional recovery. These devices act as a support to the movement, replicating the core functionality of the clinical counterparts, with less complexity and extremely lower price solutions. One of the state of the art device for home rehabilitation is Tailwind, a commercial device that allows you to perform bimanual movements of the upper limbs and rehabilitate them to motion.

THE "ARM" LINE OF DEVICES

In this scenario, which requires affordable solutions oriented toward promoting home training, the Institute of Industrial Technologies and Automation (ITIA) of the Italian National Research Council (CNR) developed a line of prototypal devices for the rehabilitation of the upper-limb, called *"Arm"*. *Arm* devices were conceived to test the possibility of promoting rehabilitation at affordable prices but capturing all the main features of the state of the art devices. In fact, *Arm* devices focus on the main features requested by a robot therapist: mechanical adaptation to the patient, ranging from passive motion to high transparency, assist-as-needed and resistive modalities; proper use of sensors for performance monitoring; easy-to-use, modular and adaptable design. These desirable features are combined with low-cost, additive manufacturing procedures, with the purpose of meeting the requirements coming from research on neuro-motor rehabilitation and motor control and coupling them with the recent breakthrough innovations in design and manufacturing. *Arm* devices cover both clinical and home-oriented training and are designed for adaptation to patients with different motor impairment.

Variable Stiffness Actuation

From the invention of the MIT-Manus (Krebs et al., 1999), force feedbacks and force-based controls are considered important features for neurorehabilitation devices (Marchal-Crespo & Reinkensmeyer, 2009), enabling them to react to the actual force exchanged with the patient, modifying, in most cases, the stiffness of the device used by the patient. Within this context, Variable Stiffness Actuators (VSA) (Vanderborght et al., 2013) are considered a promising technology, thanks to their inherent adjustment of the mechanical stiffness, force estimation and robustness to external perturbations and model errors in physical human-robot interacting scenarios (Jafari, 2014). In particular, this type of actuation has been adopted to realize some prototypes specifically dedicated to neurorehabilitation (Vitiello et al., 2013; Zhang, Guo, Pang, Gao, & Guo, 2014). Among the VSA, it is worth to mention the Agonist-Antagonist VSA (AAVSA), made up of two Series Elastic Actuators (SEA), arranged in parallel and exerting opposite forces. The principle of operation of AAVSAs are at the basis of two VSA devices described hereafter, LINarm and PLANarm.

Figure 1. Architecture of an Agonist-Antagonist Variable Stiffness Actuator: two non-linear Series Elastic Actuators arranged in parallel control the position and the stiffness of a mobile mass m

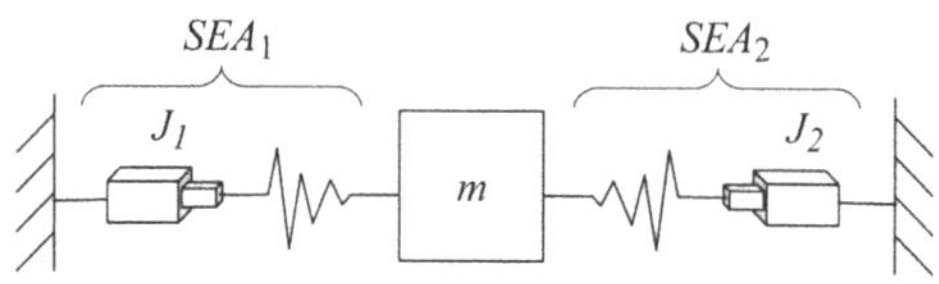

LINarm

The upper limb is an incredible adaptive organ that allows performing high-level interactive movements and is capable of accomplishing numerous functional tasks in an infinitive number of kinematic solutions. It is a sophisticated multi-joint system designed for multitasking functioning and capable of performing from heavy jobs to tasks requiring fine movements, not to mention exploration and manipulation of objects (Rau, Disselhorst-Klug, & Schmidt, 2000). Certainly, not all movements may be trained and this variability makes the selection of a reduced number of them necessary, a selection that, preferably, should be done based on key determinants of the movement control.

As reported above, two functional gesture only are sufficient to study shoulder and elbow compound movement that are frequently used in Activities of Daily Living (ADLs): 1) the Reaching Movement against gravity and 2) the Hand-to- Mouth Movement (Caimmi et al., 2015), Figure 2. These two gestures are representative of important compound movements like shoulder flexion (against gravity) and elbow extension, and, elbow flexion (against gravity) and shoulder flexion.

Figure 2. The 3D Reaching Movement (RM) and the Hand-to-Mouth Movement (HtMM)

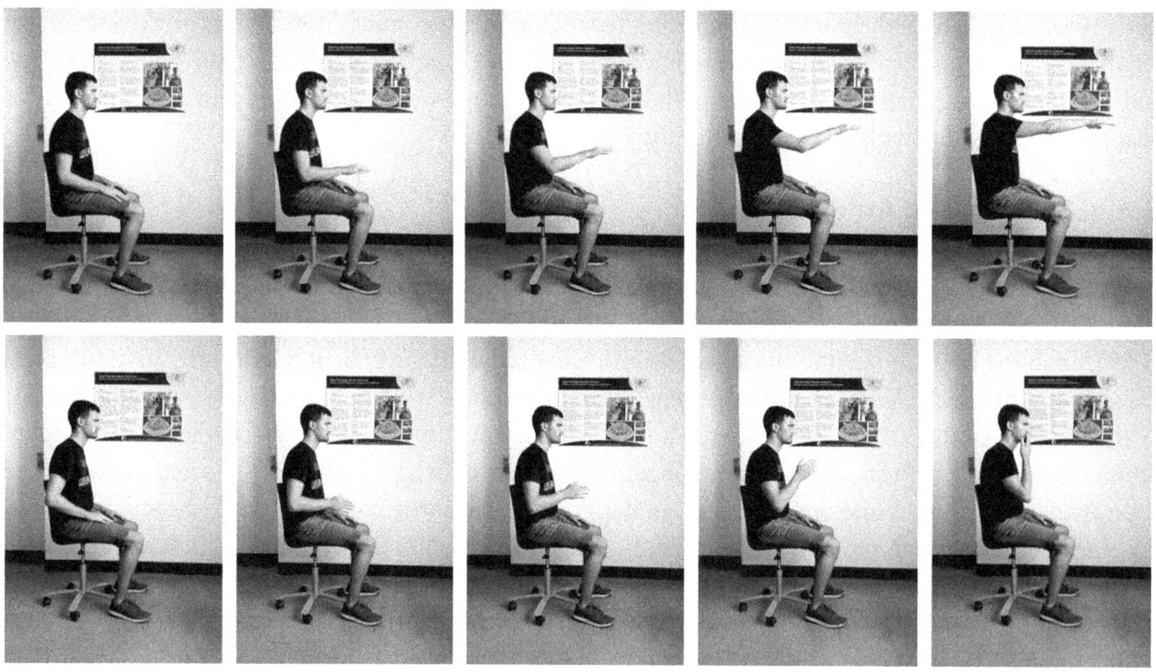

Considering the relevance of RM and HtMM and promising results obtained so far, a solution to enable patients to be trained by a robotic device with one these two movements is considered beneficial and constituted the origin of the LINarm device development. Furthermore, for both RM and HtMM, linear movement represents a good approximation. For these reasons, the development of a modular solution enabling differently oriented movements, approximating them as linear, is considered a good compromise and strategy to be included in a low-cost device for home rehabilitation.

Focusing on the mechatronic actuation schemes of rehabilitation devices, rigid mechanical actuations coupled with force sensors or back-drivable transmissions are typically employed to infer the interaction force between the device and the patient. However, in the last years, the effectiveness of adding mechanical compliance to the actuation system is being explored.

LINarm is a low-cost VSA robotic device for the rehabilitation of the upper limb. It is characterized by a more compact and optimized design, and embeds a novel VSA architecture exploiting a wire or bend-based mechanical transmission. The peculiar novel VSA scheme, enabling a compact and light design, is illustrated in Figure 3 embedding the principle of VSA actuation (G Spagnuolo, Malosio, Dinon, Tosatti, & Legnani, 2017a).

As depicted in the Figure 4 LINarm is characterized by two motors (Pololu), two motor-driven drums which wrap the wires that are connected with two non-linear spring embedded in the handle. The two motors permit to control the stiffness and the position of the handle. The handle slides on a linear guide composed by two rounds bar contained by two mechanical end-stroke.

Figure 3. VSA scheme of LINarm

Figure 4. LINarm mechanical design: Render and demonstrator

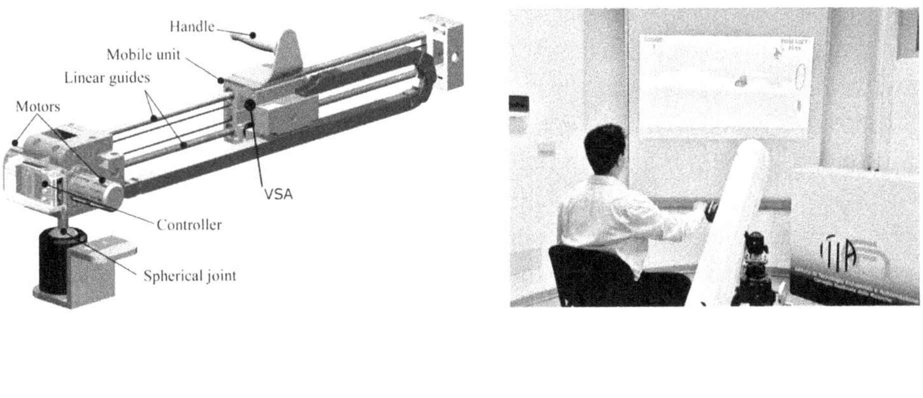

Figure 5. LINarm position control loop

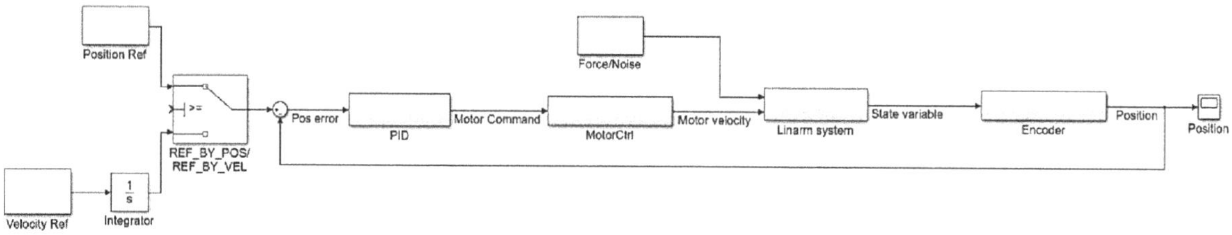

The controller is composed of an Arduino DUE board and two Pololu driver required to control the motors. These are low-cost components, chosen in order to contain the overall cost of the device, but are in any case sufficient to ensure the proper functioning of the device. The controller is also connected to the handle encoder and to two mechanical switch that represent the electrical end-stroke. The controller software developed in C++ is loaded in the Arduino board.

The control software must be able to control the robot in different modalities that represent the state of art of the control strategies in rehabilitation. For this kind of device we can divide the control strategies in two main group. The assistive control group and the challenge-based control group. In the first one, the robot help the patient to reach a pre-determinate target. In the latter, a generic resistance can be added to the movement made by the patient. Gravity in the anti-gravity movement can be considered as a resistance and for this reason LINarm is equipped with a gravity compensation algorithm. The principal control strategies provided in the LINarm robot are:

- **Passive Control**: In this control strategy the patient is guided along the path to reach the target. With passive control the patient could remain totally passive and still reach the expected target. Anyway, studies have shown that also in this case the patient tends to actively participate in the exercise. At mechatronic level this kind of control is characterized by a position control closed on the motors position feedback. A scheme of the real-time passive control is depicted in Figure 5.
- **Admittance Control**: This is a challenged-based control strategy. In this kind of control the interaction force between the patient and the robot handle is converted in a velocity command. So in this kind of control strategy the velocity depends proportionally to the force that the patient is able to apply to the handle. The admittance control is described by the formula:

$$v = a \cdot F$$

Where v represent the velocity command, F represent all the external force while a is the admittance value. It easy to understand that a higher admittance result in a lower perceived resistance during the robot movement. As said previously during the anti-gravity movement the gravity can be felt as a resistance. It is possible to add a virtual force, able to reduce or eliminate the gravity force component. In this control strategy, no help is given to the patient in order to achieve a possible target. The target can be shown to the patient using a virtual reality or other type of visual input.

Assistance control: as said in the passive control the great part of the work is done by the robot and the patient simply follow the robot during along the path. During this kind of exercise the correct movement is assured by the robot but a negative effect can be represented by the slacking hypothesis. In this case, due to the excessive assistance provided by the robot, the patient stop to actively follow the robot during its movement. On the other hand a pure admittance control in patient with severe motor disease can bring to incomplete exercise due to difficulties faced by the patient in reaching the target. A hybrid control strategies can be considered, and this strategy is represent by the assistance control. In this control paradigm, a virtual force could be added to the admittance input force in order to reach the target:

$$v = a \cdot (F_e + F_t)$$

The above formula represent the concept behind the assistance control. The F_t represent the assistance force, while F_e represent all the external force, including interaction force. F_t force can be varied according the particular needs.

Patients is helped by a force linearly proportional to the target distance. The value k_t is representative of this proportionality and could be considered as the stiffness of a virtual spring connected to the handle and to the target.

All modalities can be selected by a PC through a serial communication with the LINarm controller. It’s also possible access to the LINarm robot data (like motor and handle position, interaction force ecc.) and command all the device functions. This characteristic has been exploited in the LINarm++ platform in order to expand LINarm functionalities by integrating additional modules and augment application scenarios and applicable clinical techniques.

The LINarm++ (Matteo Malosio et al., 2016) rehabilitation platform is a multisensory and multimodal system made up of the following set of optional modules interacting with the patient:

- LINarm device, a low-cost unobtrusive sensory system for measuring the patient’s physical activity and his physiological state, in order to obtain and constantly update a comprehensive state of the patient;
- An easily wearable FES system, allowing selective and effective stimulations of upper-limb muscles;
- Virtual reality (VR) in order to represent target exercise. Different level of difficult can be selected during the exercise.

The *patient model*, a software that starting from the kinematic, physiological and VR data is able to select the best VR difficult level, and LINarm and FES assistance level.

These modules are virtually connected by different software communication channel realized exploiting the ROS framework functionalities. Hardware and software modularity meets the need of satisfying the requirements and budget availability of different clinical centers. Its affordability and semi-autonomy in targeting patient’s needs can enable more than one patient to be treated simultaneously, increasing labor productivity and improving the overall services for the patients. These same characteristics facilitate its use in a domestic environment, giving the patients the chance of being treated directly at home with challenging and personalized exercises, exploiting device adaptability to conditions and improvements of each specific patient. The platform features self-adaptation of rehabilitation parameters during its functioning and includes auto-tuning procedures during the setup, as the one to identify the muscles to be stimulated by the NMES system, allowing a rough positioning of the electrode pseudo matrixes on the arm, without requiring deep medical knowledge. The system aims at satisfying the increasingly

Figure 6. Simplified representations of the two main submechanisms embedded in PLANarm. Continuous lines refer to the two horizontal antagonist actuators, dotted lines refer to the two vertical antagonist actuators. (Left) The tendons subsystem includes two orthogonal pairs of antagonistic non-linear series elastic actuators. The direction of the tendons is defined by four carriages c_{ij} which embeds deflecting pulleys. (Right) The carriages subsytem depicts the principle of operation of the tendon-based system which guarantees that the tendons are constantly orthogonal if no external force is applied to m.

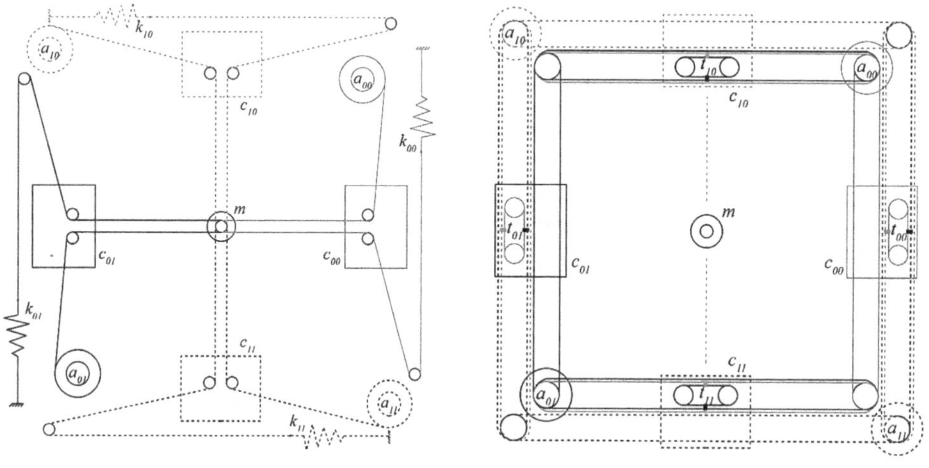

larger request of more and more effective and affordable technologies complementary to traditional rehabilitation techniques, to face population ageing and the related increase of neurological diseases. The first LINarm++ prototype is currently available, technical and its clinical assessment is planned in the near future.

PLANarm

Despite the existence of one-DOF end-effector devices and exoskeletons embedding VSAs (Vitiello et al., 2013; Zhang et al., 2014), no end-effector VSA planar device has been so far developed and exploited for rehabilitation purposes, even if different planar-based force-feedback devices has been developed. Only one general-purpose prototype featuring planar movements and embedding a VSA architecture was developed so far, at authors' knowledge. It is a cable-driven device arranged on a triangular framework, featuring three non-linear SEAs (Zhou, Jun, & Krovi, 2014). Because of the presence of only three tendons, its capability of adjusting the stiffness of the load along different directions is reduced. Each tendon is, moreover, controlled independently by two actuators, for a total requirement of six actuators.

In order to take a step forward with respect to this previously existing prototype, PLANarm is a planar device (Di, Industriale, & Informazione, 2015) embedding a novel architecture conceived with the aim of favoring the isotropy in terms of stiffness and decoupling of dofs, by an orthogonal two-dof planar tendon-driven VSA actuation scheme. Its peculiar architecture allows to tune, in parallel, both the equilibrium position and the stiffness of a mobile platform along two orthogonal directions, in order to favor isotropy and to tune stiffness ellipsis along two independent axis (G. Legnani, Tosi, Fassi, Giberti, & Cinquemani, 2010; Giovanni Legnani, Casalo, Righettini, & Zappa, 1996; Molinari-Tosatti, Fassi, & Legnani, 2003).

The kinematic architecture of the submodules making up PLANarm is depicted in Fig. 6. The orthogonality of agonist-antagonist tendons connected to the handle is guaranteed by a proper mechanical transmission. For a deeper understanding of geometrical, analytical and numerical aspects refer to (M Malosio, Corbetta, & Ramirez, 2016).

The handle is free to slide on the rectangular ferromagnetic surface of the device. The bilateral constraint is guaranteed by a set of rollers and a magnet, which generates an attractive force between the handle and the surface, guaranteeing that the handle remains vertical, even if an external force is applied to it (Figure 7).

The prototype (Figure 8) is actuated by four rotational DC motors Pololu 30:1 Metal Gearmotor 37Dx68Lmm with an embedded encoder, assembled at the corners of the PLANarm (a_{ij} in 6). The non-linear elastic elements have been designed taking inspiration of the principle of operation described in (G Spagnuolo, Malosio, Dinon, Tosatti, & Legnani, 2017b), designing proper cam profiles. The electronics includes an Arduino Due micro-controller, in charge of executing low-level and real-time control algorithms to directly control the motors, and two motor drivers to power the four actuators. The Arduino

Figure 7. Design of the PLANarm prototype which embeds a two-dof planar tendon-driven VSA actuation scheme and cam-based non-linear springs

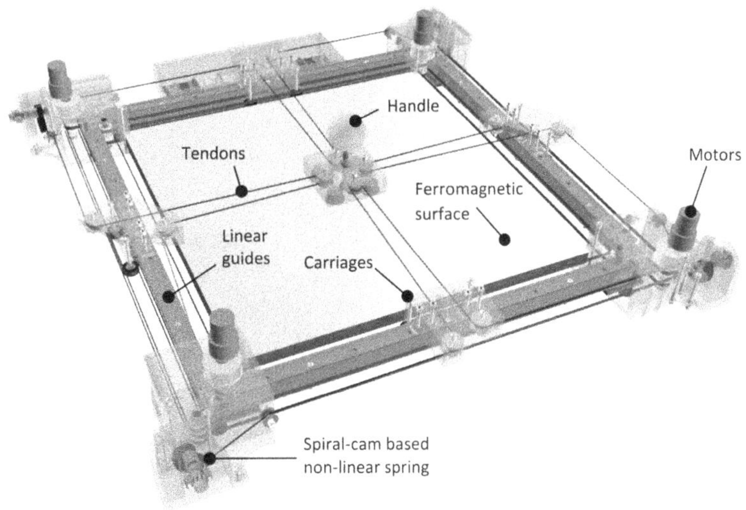

Figure 8. The PLANarm prototype and handle

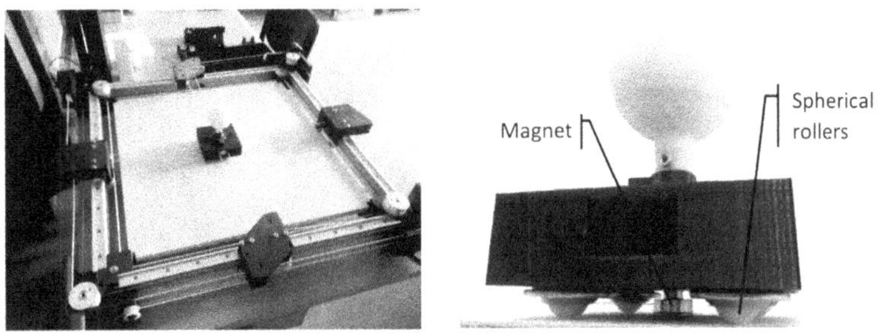

board is connected to a computer through a micro-USB cable to communicate with a PC, in charge of executing the high-level software application, interfacing the user and the device, with graphical and numerical feedback, providing required information to correctly command the control algorithms at low level. The control structure and algorithms executable by PLANarm are very similar to the ones illustrated for the LINarm device, applied to planar movements instead merely linear.

DUALarm

The global requirement of technologies able to facilitate the execution of neurorehabilitation therapies at patients' home and the diffusion of low-cost and home-oriented rehabilitation devices are the basis and source of inspiration of DUALarm (Figure 9), a low-cost, open-source and 3Dprintable device for bilateral upper-limb rehabilitation, whose core mechanism could be exploited in different installation configurations (Figure 10).

Bilateral UL training is a relatively new form of stroke rehabilitation. It is based on the premise that, during simultaneous movements, the non-paretic UL may support the movement of the paretic UL at brain level. The interest in this therapy, which is becoming more and more common, arose partly accidentally and partly from insights, gained from the motor control literature. In this literature, rhythmic inter-limb-coordination studies investigated extensively coupling and interaction effects between the two ULs in healthy subjects (Van Delden, Peper, Kwakkel, & Beek, 2012). It was well proved that human beings do show a basic tendency towards in-phase (i.e. symmetrical movements) or anti-phase

Figure 9. The DUALarm prototype

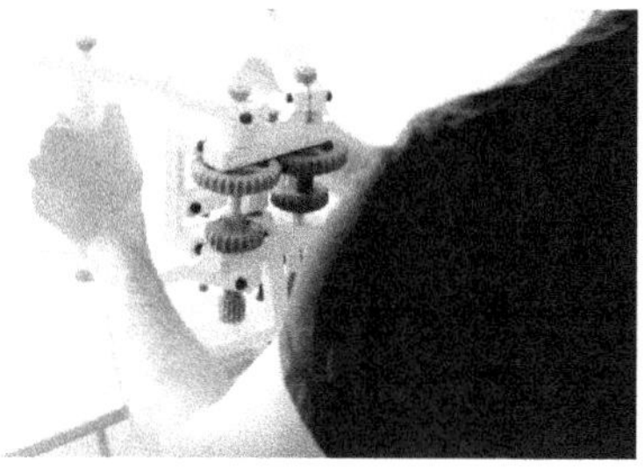

Figure 10. DUALarm designed set-ups

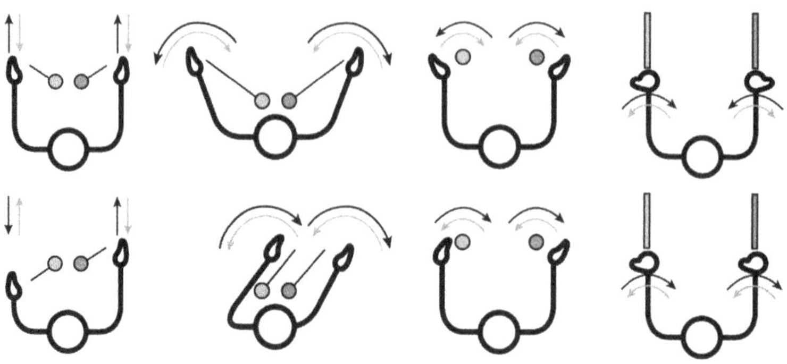

(i.e. alternating movements) coordination, with a prevalent 1:1 frequency locking mode for UL bilateral movements (Swinnen, 2002).The tendency towards these patterns reflects the coupling between the ULs. In bilateral UL training, this coupling is exploited using interactions between both sides of the central nervous system through intact connecting structures, such as the corpus callosum (Cauraugh & Summers, 2005) In this work, the authors are presenting the current version of DUALarm a low-cost, open-source and 3D-printable rehabilitation device based on a geared mechanism (Fig. 11) enabling in-phase and anti-phase movements and developed to support reaching rehabilitation movements of the UL. DUALarm has been designed in order to take the most of 3Dprinting technology, limiting as much as possible the number of commercial components. The version presented in Fig. 11 is an almost-completely 3D-printed plastic (PLA) device, equipped by two handles (one for each side) held by the subject and standing upon a regular table, with an optional clamp to fix it. The core mechanism is made up of: a properly supported gearbox, a mode-selection mechanism and a signal acquisition system. The pivotal component of the device is the mechanism of the gearbox, based on a plastic structure and containing five wheel-gears and three aluminum shafts (10 mm in diameter). Referring to Fig. 11, three groups of gears are recognizable: the fixed (hereafter f) group rotating about *af* and made up of gears *gfu, gfm, gfl* stationary constrained to an aluminum shaft, the inversion gear *gim* rotating about *am* and the mobile (hereafter *m*) group rotating about am and made up of gears *gmu, gmm, gml*. All the gears are mechanically constrained to three aluminum shafts, and secured by steel screws and bolts to avoid free spinning. The gap between *gfu* and *gfm* is different from the one existing between *gmu* and *gmm*. Shaft am is free to move axially in order to obtain two different configurations of the mechanism and let gears engage differently. Consequently, sliding up and down the m group, two different configurations can be selected: if *gfu* engages *gmu, af* and *am* perform anti-phase rotations; if *gfm* engages *gmm, af* and *am* perform in-phase rotations, thanks to the presence of *gim*. All the gears in the mechanism are 2 mm in module to guarantee a good mechanical resistance and have a double chamfer in order to ease the engagement of the gears. A selection mechanism, made up of a 3D-printed lever, is placed above the gears. If it is engaged, it works as a spacer that causes *gmu* to engage *gfu*, leading to the anti-phase rotation of *af* and *am*. If it is not engaged, *gmm* is engaged with *gfm* through *gim*, leading to an in-phase rotation of shafts. An aluminum pin p (with a 3D-printed cup) locks the position of s for safety. To track the movement performed by the patient, the mechanism is equipped with two potentiometers, measuring *af* and am rotations through two pairs of gears (*gfl, gfk and gml, gmk*) to compensate the vertical sliding of *am*: the first element of each couple is spliced and secured to the rod by a steel screw and bolt, the second one is simply spliced onto the slotted-shaft of the potentiometer. On the back-side of the device an Arduino Uno R3 board is located while two potentiometers *gfp* and *gmp* are accommodated into the base.

Figure 11. DUALarm design

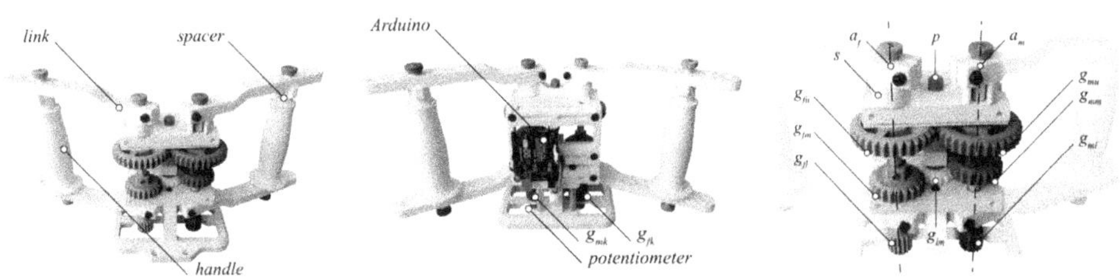

In the DUALarm configuration shown in Figure 11, two arms are connected to axis *af* and *am*. Each arm is made up of one handle, one aluminum rod (about which the handle is free to rotate) and two links connecting the shaft of the wheelgears to the handle. Because of the slight sliding movement of am to select the required configuration, every arm includes a spacer that can be placed above or below the handle in order to maintain the left and right handles at the same height. Links are constrained to the shafts of the core mechanism.

The device was manufactured thanks to the Fused Deposition Modeling (FDM) technology, an Additive Manufacturing process consisting in laying tracks of molten thermoplastic polymer onto a platform to obtain a layer of material on the XY plane of the machine (3D-printer). Once the molten layer has solidified, other subsequent layers are deposited in a bottom to top manner, in order to create the complete 3D object in the Z direction of the machine. All the 3D-printed parts of the device were made by the low-cost FDM 3D-printer Sharebot NG. The plastic used to build the entire prototype was PLA with a wire diameter of 1.75 mm. Custom parameters for the aforementioned printer have been created and main parameters used are the following: 3 layers of bottom, 3 layers of top, 0.2 mm of layer height, 4 perimeters, 30% grid infill percentage, 45_/ 45_ raster orientation for bottom and top solid layers. These parameters were used to guarantee a suitable mechanical resistance in order to test the first prototype in several conditions, from normal to heavy use, and to ensure the presence of enough material, especially around holes, to perform subsequent operations, like boring or countersinking, to remove potential imperfections coming from the 3D-printing process. In order to avoid the typical high friction between two PLA 3D-printed parts, a sheet of a very smooth polypropylene (1.5 mm in height) was positioned onto critical surfaces.

Even though DUALarm is a completely mechanical device and does not require any additional mechatronical component to perform its rehabilitative tasks, it is equipped with some basic electronical components in order to monitor the activity of patients during rehabilitation sessions. Two potentiometers (100k @ 300_) are in charge of measuring and recording the angular rotation of the DUALarm's links and, therefore, of estimating the patient's limbs positions. The data acquisition system is implemented in an Arduino UNO microcontroller, acquiring the angular position of each of the two handles in real time. Thanks to this acquisition system, it is possible to convert the analog signal coming from the potentiometers to a digital signal that can be transmitted through a serial connection to a common PC. The high-level PC program is implemented in Python, an easily customizable, open-source and cross-platform language, allowing the program to run on different operating system, even on Linux-based open-source and free distributions, contributing to lower overall use costs of the device. A graphic user interface of the device is implemented too. It is characterized by four main panels, allowing to initialize the device, displaying a tridimensional feedback for the user, a real-time updated log of measured angular positions and the actual position of the handles. Through customized python scripts it is possible to program exercises, showing to the patient different target points to be reached, tuning the required range of motion and target distances according to the actual patient capabilities.

DUALarm was tested in preliminary experimental trials to characterize the interaction with the device of healthy subjects and a patient, in three modalities:_

Monolateral (M), executed alternately with the left and the right limb), Bilateral (B) In Phase (IP) and Anti-Phase (AP). 5 right-handed healthy subjects and 1 patient (post-stroke, left hemiplegia) were enrolled to participate to the preliminary trials.

Kinematics of the upper-limbs was measured with BTS Smart Elite system to detect movement phases; EMG of Deltoid Anterior (DA) and Posterior (DP), Biceps (BIC) and Subspinatus (SS) on both limbs. Mean Muscular Activation (MMA) of each muscle was computed as:

$$MMA = \frac{1}{n} \cdot \sum_{i=1}^{n} \int_{t_b}^{t^e} EMG_i(t)dt$$

The following results were observed on Healthy Subjects (Fig. 12):

- DA is the main agonist muscle
- DA EMG is higher in (M) trials than in (B) ones.
- trials show higher activations than AP ones.
- Muscle EMGs are symmetrical between in M and IP trials, while AP trials show higher activity in the dominant limb.
- In AP trials, clear predominance of the dominant side.
- In IP trials, instead, DA of the right and left limb activate equally, and higher than in AP trials.
- The following results were observed on the Patient (Figure 12):
- Weakness of the more affected limb.
- Infact, as expected, the less affected limb strongly helps the more affected one by dragging it.
- During the bilateral IP trial, the MMAs of the less affected limb is comparable to the (M) one.
- MMAs are spotted also in the more affected limb.

LIGHTarm

Support limb weight due to gravity is a valuable strategy to enable upper limb rehabilitation in neurological patients. Raising the upper-limb is one of the more challenging tasks for neurological patients, since neurological diseases may affect the capability of performing functional movements and interacting with the environment (Krabben et al., 2012). Weight support partially discharge the muscles from the burden of supporting the limb. The effectiveness of weight-support approaches was observed in

Figure 12. EMG muscles activations during DUALarm trials

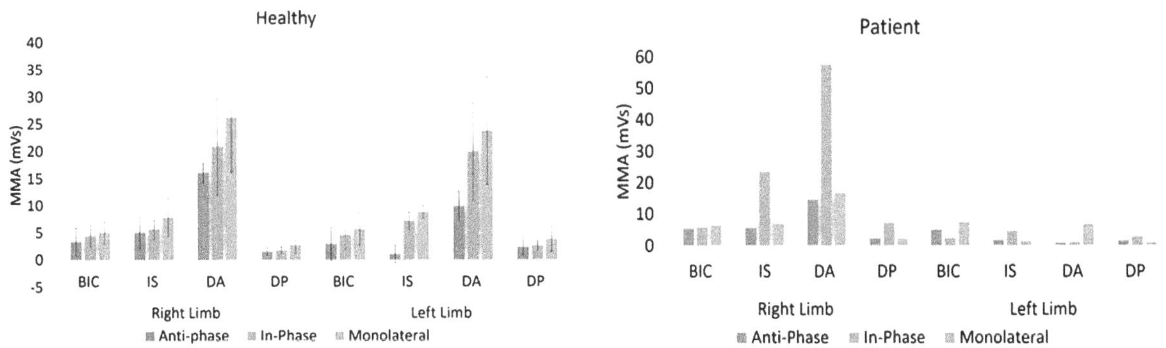

exoskeleton-assisted studies that showed benefits in terms of motor function recovery (Reinkensmeyer, 2012) and reduction of the influence of abnormal coupling (flexural-synergies) between shoulder and elbow. Starting from (Matteo Malosio, Pedrocchi, Vicentini, & Tosatti, 2011) and by a patent pending kinematic scheme (Malosio, M., Pedrocchi, N., & Tosatti, L. M. (2014)) a passively gravity-compensated exoskeleton for the upper limb neuro-rehabilitation was developed. Its peculiar kinematic structure has been specifically conceived to deal with shoulder rhythm and elbow singularity issues, achieving high backdrivability (transparency) throughout a large exploitable range of motion. An introduction describes the upper-limb and the exoskeleton kinematic models which where the basis for the development of the device (Giulio Spagnuolo et al., 2015). Then, the passively gravity-balanced exoskeleton is described. Afterwards, the results of an experimental campaign designed to test gravity support features and transparency in static and dynamic conditions, are reported (Scano, Spagnuolo, et al., 2015).

A model of the upper limb, resembling the one presented in (Malosio, M., Pedrocchi, N., & Tosatti, L. M. (2014)) is depicted in Figure 13(a) and Figure 13(c). It is a serial kinematic chain characterized by five rotational joints $J_{h,i}$ (i=1,…,5). $J_{h,1}$ denotes the horizontal add/abduction, $J_{h,2}$ the arm elevation, $J_{h,3}$ the arm internal/external rotation, $J_{h,4}$ the elbow flexo/extension and $J_{h,5}$ the forearm prono-supination. The set $\{J_{h,1}, J_{h,2}, J_{h,3}\}$ models, as a whole, the gleno-humeral joint. The three degrees of freedom translational joint $J_{h,sg}$ models the humeral head movement according to the shoulder rhythm. Moreover, let us denote by:

- $\{t\}$ and O_t a generic coordinate frame and its center, respectively; in particular, $\{hs\}$ and $\{he\}$ denote the reference frames centered in the shoulder and elbow human articulations, respectively. $\{hs\}$ translates with respect to the chest $\{w\}$ according to the shoulder girdle movement.
- $\{\mathbf{e}_{t,x}, \mathbf{e}_{t,y}, \mathbf{e}_{t,z}\}$ the unit vectors of $\{t\}$ axes;
- $\pi_{w,xy}$, $\pi_{w,yz}$ and $\pi_{w,xz}$ the coronal, transverse and sagittal planes, respectively;
- $\mathbf{u}_{h,i}$ the unit vector, respectively, of the i-th rotational axis.

The exoskeleton kinematics consists of a hybrid mechanism, composed by four distinct serial chains, denoted by S_s, S_u, S_e and S_f - nearby the shoulder, the upper arm, the elbow and the forearm, respectively, as depicted in Figure 13(c).

Figure 13. Human upper-limb modeling and LIGHTarm kinematic design. (a) Human-arm and (b) exoskeleton kinematic models.(c) Joint connection graph: joint are revolute (R), prismatic (P), 3dof translational (T), gimbal (G).

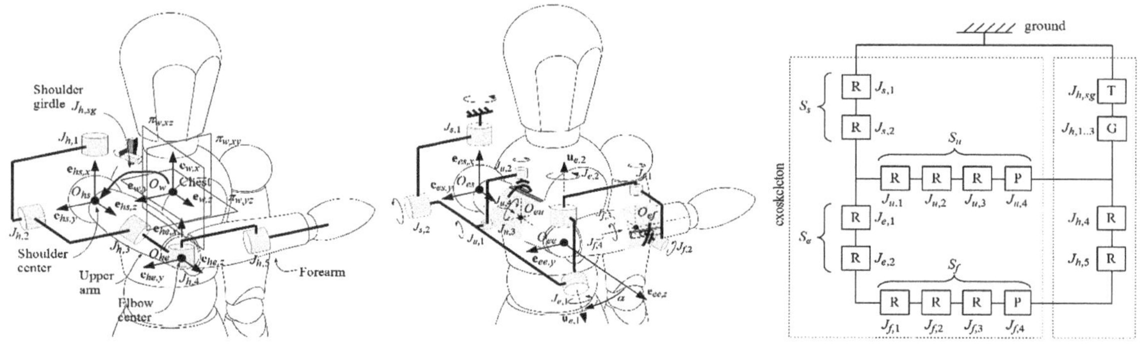

This peculiar kinematic structure a) enables human shoulder to move complying with the physiological shoulder rhythm and b) avoids the occurrence of any singularity in the whole elbow range of motion (ROM) (Matteo Malosio et al., 2011). These features, in conjunction with low inertia and low frictions, guarantee high transparency, allowing the exoskeleton to follow accurately arm movement in a large exploitable range of motion. Even though the exoskeleton kinematics allows decoupling exoskeleton and human upper limb shoulder and elbow, hereafter, without loss of generality, the analysis is performed considering the limb shoulder and elbow centres coincident with the exoskeleton shoulder and elbow centres.

An articulated mechanism is *gravity-balanced* if it is in indifferent equilibrium without requiring joint actuator forces or torques. Physically, this means that the total potential energy of the device is invariant whatever its configuration is. This condition can be achieved in different ways, using counter masses, springs and/or auxiliary parallelograms (Lu, Ortega, & Ma, 2011). The gravity-balancing system of the unactuated version of LIGHTarm is made up of a combination of these three mechanical elements, in order to partially decouple the gravitational load contributions of the upper-arm and the forearm segments (Figure 14(a)). The gravity-balancing contribution has been partially decoupled between a counterweight W_s, in charge of supporting the arm and exoskeleton masses, and a spring s in charge of supporting the elbow mass. It is important to note that, no torque is transferred from the human upper limb and the exoskeleton, through the interface cuffs (S_u, S_f, Figure 14(c)) because of the presence of the spherical joints. The counterweight W_s has to counterbalance all the arm and forearm masses. For convenience, referring to Figure 14(b) and Figure 14(c), let us denote by:

- $\mathbf{Q}_{\{t\}}$ the vector connecting the origin O_t of frame *{t}* to point Q;
- $\mathbf{Q}_{\{t\},i}$ the component of vector $\mathbf{Q}_{\{t\}}$ along the direction i (i.e. x,y,z) of frame *{t}*;
- $\mathbf{G}_a$ and m_a the center of mass and the value of the mass of a, respectively;
- $\mathbf{G}_f$ and m_f the center of mass and the value of the mass of f, respectively.

Figure 14. Gravitational load contributions of the upper-arm and the forearm segments; Auxiliary Parallelogram Counter-Balance Mechanism; Forearm Spring-Balancing Mechanism

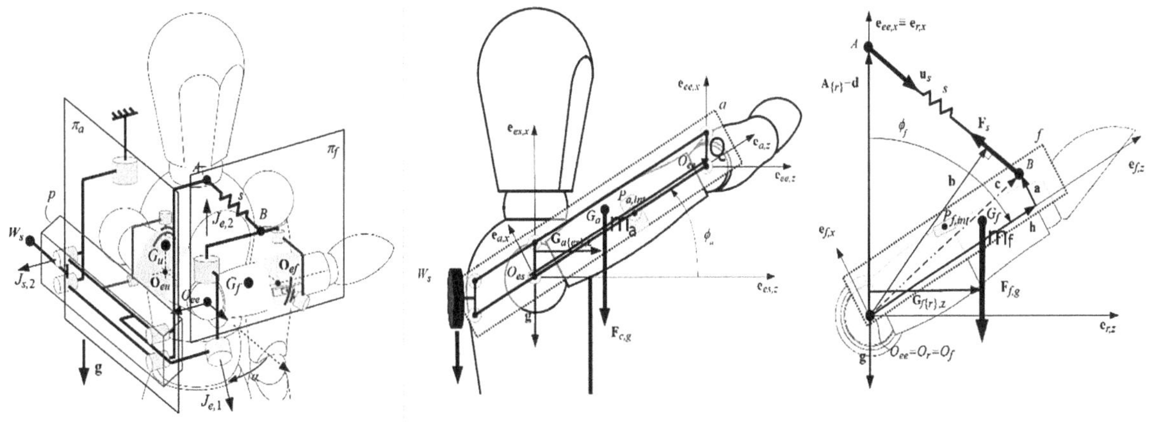

Auxiliary Parallelogram Counter-Balance Mechanism

The overall mass of the exoskeleton and of the upper limb is counterbalanced by a mass W_s, rigidly connected to the exoskeleton link parallel to the human upper arm. This link has been realized through an auxiliary-parallelogram mechanism (Figure 14(b)): referring to the plane of elevation, the planar hinged-parallelogram produces 1-DoF circular translation. It consents to the reference frame *{ee}* centered in the elbow to translate with respect to (w.r.t.) the shoulder reference frame *{es}* (Fig. 14(b)), instead of rotating about $\mathbf{e}_{es,y}$, (Figure 14(b)). This leads the point A to be above O_{ee}, having the line $\overline{OA} \,/\!/\, \boldsymbol{e}_{es,x}$ for any configuration, guaranteeing that the forces generated by the spring s supporting the forearm is unaffected by the upper-arm orientation. Referring to Figure 14(b), let us consider the fixed reference frame *{es}* w.r.t. which the body a rotates, and a reference frame *{a}* rigidly connected to a. In depth, let us analyze each contribution acting on the shoulder joints. No torque is transferred from the human upper limb and the exoskeleton, through the interface cuffs (S_u, S_f, Figure 13(c)) because of the presence of the spherical joints. The counterweight W_s has to counterbalance all the arm and forearm masses. The auxiliary parallelogram allows the mass balance in respect to O_{es} to be not influenced by the actual configuration of the masses downline the parallelogram.

The forearm mass is suspended by tension spring, generating a tension force between point A of the exoskeleton structure and point B along the forearm (Figure 14(a)). Let us refer to the model depicted in Figure 14(c), schematizing the spring balancing system, supporting a body f, modelling the forearm, hinged to a point O_f. Referring to Figure 14(a) it is a representation of the forearm kinematics on the plane π_f. Let us consider a fixed reference frame *{r}* w.r.t. which the body f rotates, and a reference frame *{f}* rigidly connected to f. A tension spring s is connected between points A and B, respectively fixed w.r.t. *{r}* and *{f}*.

Figure 15. LIGHTarm CAD render

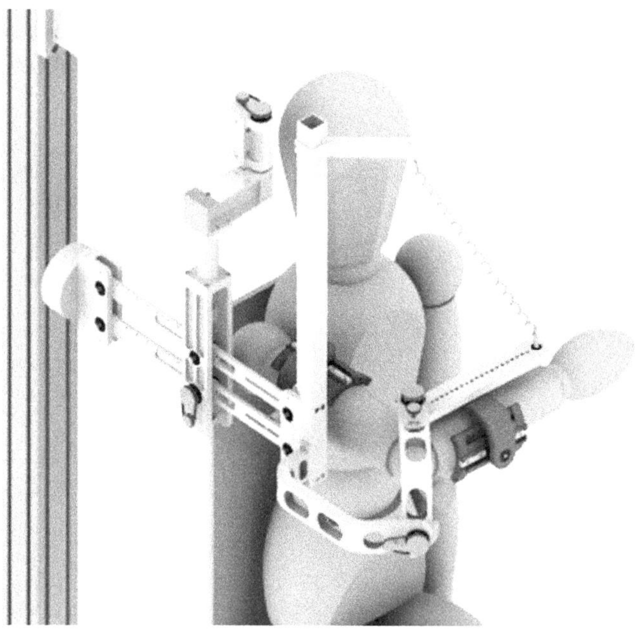

Several trials of the first LIGHTarm prototype were performed. Firstly, to test LIGHTarm weight support characteristics in different configurations; EMG activity of the Deltoid Anterior (DA), the main shoulder flexor muscle in the sagittal plane, was used as assessment parameter. Second, to investigate LIGHTarm transparency during the execution of significant functional motor tasks; an analysis of the differences of the neuromuscular activation patterns of the relevant agonist and antagonist muscles during free movements and exoskeleton-assisted ones was used for assessment.

Three healthy subjects were enrolled for this study: age 32.7±11.9, height 181±2 cm, weight 77±8 kg, 3M, all right-handed. The following equipment was used: the prototype of the LIGHTarm exoskeleton; 8-channel EMG acquisition system (FreeEMG, B|T|S, Italy); 6-TVC Marker-based Motion Capture system (Smart-D, B|T|S, Italy). s-EMG were recorded to monitor muscular activation patterns of the following muscles: Deltoids Anterior, Lateralis and Posterior, Trapezius Superior, Pectoralis Major, Triceps Lateral Head, Biceps Brachii Caput Longus and Brachioradialis. All subjects were right-handed and the recordings were performed on the right limb. Kinematics of the right limb was recorded too. Retroreflective Markers were positioned on: C7 and D5 vertebras, acromion, right epicondyle of the elbow, and styloid process of the ulna (Caimmi et al., 2015). To face the fact that the elbow marker is obscured by the exoskeleton in some configurations of the upper limb, a set of four markers was used to reconstruct elbow location. The procedure is not reported for the sake of brevity.

Two experimental conditions were investigated (Figure 16). *Static* trials were conceived to evaluate the efficacy of the gravity-support features in different configurations. *Dynamic* trials were designed to assess transparency. In both cases, the subjects' performances with LIGHTarm were compared to free movements. We denoted each trial with a 2-cyphers alphanumeric id. The letter identifies the task while the progressive number identifies different trials conditions. Static trials (Tab. II) were conducted with the exoskeleton and without exoskeleton in three different configurations (Figure 3): 1) *Begin (B)* position, where subjects held their hand on the thigh and the arm was relaxed; 2) In the *End (E)* position, where subjects held the arm raised at 90° of shoulder flexion and the elbow was extended; 3) In the *Intermediate (I)* position, subjects held the arm raised at 90° of shoulder flexion and the elbow flexed at 90° in the horizontal plane. In the Free Positioning trials, *Not Supp (0)* denoted the subjects had to keep their posture without external support, while *Supp (1)* indicated that a physiotherapist was statically supporting the upper-limb weight. In the exoskeleton trials, *Not Supp (2)* denotes that only the exoskeleton weight was supported, while *Supp (3)* indicates that both the weight of the exoskeleton and of the upper-limb were compensated with the gravity-support mechanism.

In order to investigate the perceived *transparency* during motion, three dynamic tasks were designed (Tab. III): the 3D Reaching (*R*, Figure 16), the Hand-to-Mouth (*M*, Figure 16) and the Hand-to-Nape (*N*, Figure 16). Their combination allows a daily-life purposeful interaction with the environment.

The Deltoid Anterior (DA) EMG activity, related to 8 seconds of the three static postures holding (B, I and E), in the four different trial conditions.

In the upper-left panel, the free positioning subject 1' trials are shown. The DA EMG activity is negligible in the Begin configuration (B0). This is expected because no torque is required to support the limb as the hand lies on the thigh. In the End and Intermediate configurations, the DA activity is quite high as the whole upper-limb weight is supported. The DA EMG activity is slightly superior in the End position (E0) consistently with the increase in shoulder torque due to elbow extension.

In the upper-right panel, the DA EMG activity during *exoskeleton-assisted not-supported* trials is shown for the same subject. The EMG activity in the three different configurations is highly comparable to the ones in the corresponding Free static postures. This indicates that the exoskeleton weight was

Figure 16. LIGHTarm trials: Static postures and dynamic gestures

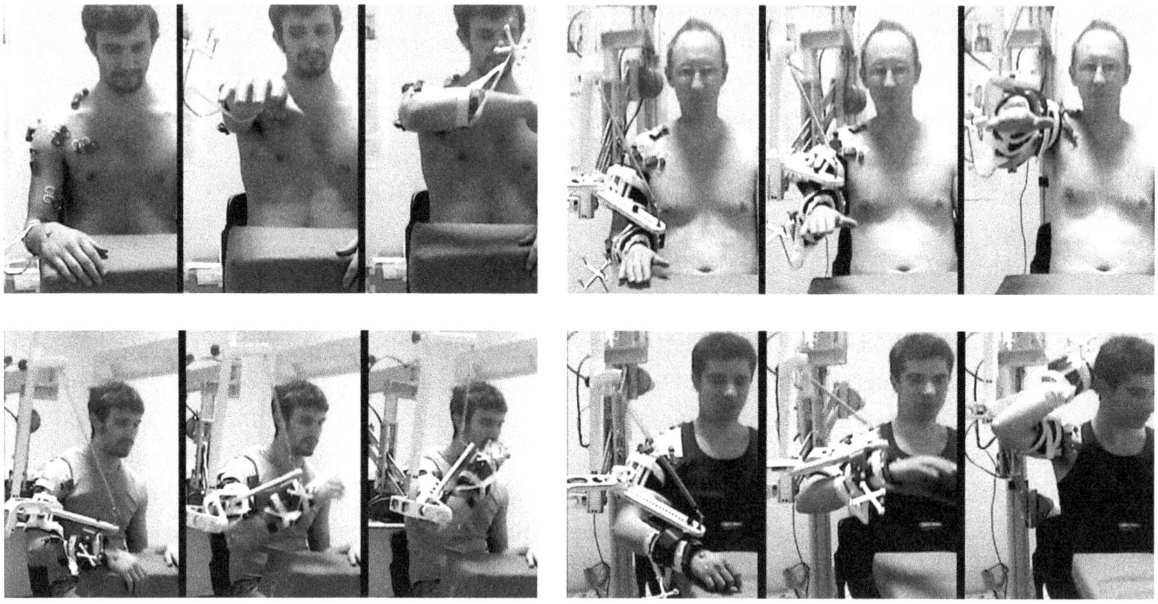

supported. A slightly lower EMG activity in trial E2 could indicate a slight, though observable, over compensation of LIGHTarm weight.

The comparison between the EMG activity of the DA when the upper-limb is supported by the therapist (trials B1, E1, I1) and when supported by the exoskeleton (trials B3, E3, I3), is reported.

Considering as gold standard the DA EMG activity when the physiotherapist was supporting the arm (left-lower panel), the following considerations follow:

- Being the weight of the upper limb fully supported by the therapist (trials 1), no difference in DA EMG activity would be expected among the End, Intermediate and Begin configurations; residual DA EMG activity in the End position is due to the inability of the subject to fully relax;
- The DA EMG activity at End position are comparable between physiotherapist (B1, I1, E1) and exoskeleton support (B3, I3, E3) trials, indicating satisfactory weight support provided by LIGHtarm.
- Having obtained a perfect support in the configuration that requires more support (End), the slightly higher DA EMG activity in the exoskeleton supported trials at the Intermediate position, could be due to the inability of the subject to completely relax. Authors cannot exclude that LIGHtarm cannot provide perfectly uniform weigh-support being the cause of the residual DA EMG activity in trial I; however, the EMG activity in supported trials is considerably lower than in free ones.

Movement phases were calculated analyzing the vertical displacement of the wrist. Dynamic trials refer to subject 1, as representative of all the three subjects. As a first result, it is remarked that, while not being trained in the use of the device, subjects were able to move at self-selected speed without incurring in any mechanical lock.

Kinematic data indicated that both free movements and exoskeleton-assisted ones are characterized by similar range of motions. The main difference lies in the fact that exoskeleton-assisted trials (1 and 3 referring to Table. III) were executed at a spontaneously slower velocity than the free ones. Such results are likely due to inertial properties of the device (new parts with lighter materials will be realized for next trials) and to motor control adaptation (fully supported trials were executed lastly).

RCH: The comparison between trials R0 and R2 shows a slightly lower activity of the DA muscle to elevate the arm. Such result can be explained by: 1) only a partial compensation of the limb weight and, 2) lower torques required by slower movements. The Deltoid Posterior (DP) contraction is strong in both trials. In R0 a slight EMG activity is shown also during shoulder flexion (DP co-contraction), indicating the physiological need of stabilizing the upper-limb. R3 instead shows a remarkable decrease of the activity both of the DA and the DP, indicating that the weight support system is reducing the amount of torque required to elevate and stabilize the arm.

HTM: The comparison between trials M0 and M2 shows similar activation patterns, for both biceps (BIC) and triceps (TRIC). The triceps activity is slightly reduced in the exoskeleton-assisted trial, again, probably indicating less need to stabilize the arm. In the M3 trial, the exoskeleton supports the biceps in flexing the elbow against gravity and the EMG activity decreases.

HTN: The comparison between trials N0 and N2 shows that the trapezius (TRAP) activates with a very similar pattern, while the DA has a lower peak value, probably due to LIGHTarm slight over weight-support action. The DP shows similar activation patterns and amplitudes.

Authors underline that EMG activity suggests that transparency features are generally achieved; the exoskeleton does not alter significantly the EMG patterns of the main agonists and antagonists muscles. The shape of the EMG activity is generally preserved, even if sometimes amplitudes are partially altered. All subjects referred their comfort when assisted by the exoskeleton. Some discomfort was noticed when moving in vertical direction in full-compensated trials. The inertial load probably accounts for the problem, as healthy subjects perform movements at high velocity. This will be not the case of the majority of pathological subjects, for whom achievable velocities are significantly lower.

VIRTUALarm

According to the World Health Organization (WHO), the essential components of successful neurorehabilitation include expert multidisciplinary assessment, goal-oriented programs and evaluation of impact on patient and goal achievement through the use of clinically appropriate, scientifically sound outcome measures incorporating the patient's perspective (World Health Organization, 2002). VIRTUALarm is a Kinect One-based platform for motor performance monitoring, including body and limb tracking and a biomechanical evaluation of the performance in relation to databases of healthy subjects. Assessments include range of motion, motion dynamics, effort, motor control indexes, body segments barycenter tracking. Thus, at the current state, VIRTUALarm does not implement feedback during the execution of games, but provides an assessment similar to the ones made with marker-based systems (Cappozzo, Catani, Leardini, Benedetti, & Della Croce, 1996; Petuskey, Bagley, Abdala, James, & Rab, 2007; Roux, Bouilland, Godillon-Maquinghen, & Bouttens, 1999).

VIRTUALarm exploits Kinect One technology that allows that tracking of human body segments as shown in Fig. 17. Preliminary tests were performed on healthy subjects and patients, for a twofold reason:

Figure 17. Kinect One graphical interface, visual feedback and graphs

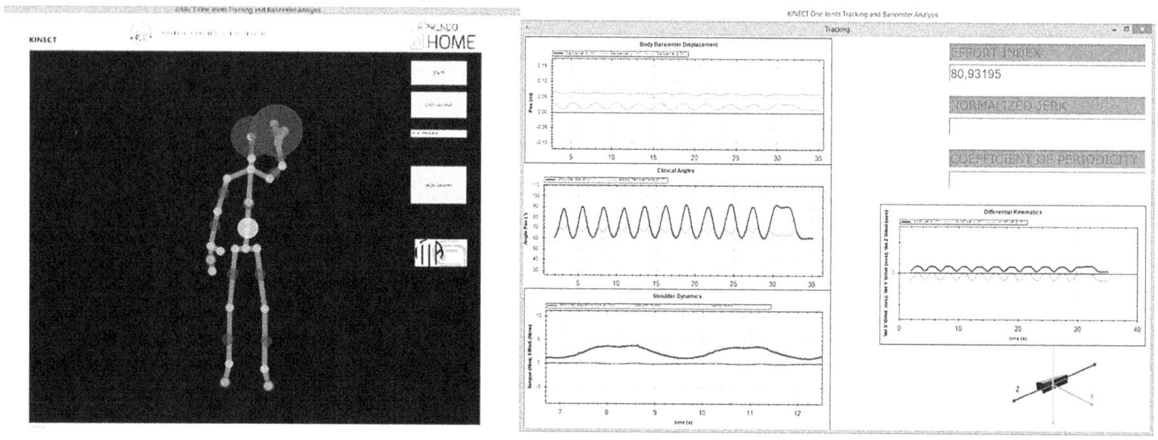

1) to build database of healthy people performances, to be used as reference when evaluating neurological patients; 2) to assess Kinect One capabilities of assessing motor performances and discriminate different levels of motion capability. In a recently published study (Scano et al., 2017), 78 healthy subjects were enrolled to perform the Reaching Against Gravity Movement (RM) and Hand-To-Mouth Movement (HtMM) according to the protocol described in (Caimmi et al., 2015). Subjects stood comfortably on a chair, with their back straight. They were asked not to move their torso during the experiment and to perform the motor tasks only by limb motion. A target for the RM was set at shoulder height, indicating the target point. The target of the HtMM was the mouth itself. Subjects were requested to perform the repetitions of the gestures at natural, comfortable, self-selected speed, with no pauses between one repetition and the following one. The starting position was with the elbow flexed of about 90°, with pronated hand leaning on the thigh. Four acquisitions was performed by each subject: RM and HtMM gestures, both performed with their dominant and non-dominant limbs. Twelve repetitions of the motor task was performed per acquisition.

Microsoft Kinect One can track human movements but does not allow defining other reference points, except for the articular joint centers. Consequently, a fine characterization of the interaction with an object, as with marker-based systems, is not possible. However, as stated by (Schaefer & Hengge, 2016) motor control analysis and knowledge acquired in tests on "standardized" reaching movements, can be considered as representative of "natural" reaching movements, involving physical interaction with real objects. Consistently, the use of the Microsoft Kinect One sensor for the analysis of reaching movements is a reasonable (and cost-effective) choice.

During the trials, the experimental set-up was composed of:

- A Microsoft Kinect One sensor version 2.0, mounted on an easel and placed at about 2.0 m distance from the torso of the subject;
- A PC with Microsoft Windows 8.1, USB 3.0, and Microsoft Kinect One S.D.K. version 2.0;
- A screen providing on-line visual feedback to the operator to visually monitor the correctness of the kinematic tracking.

Figure 18. Kinect One set-up

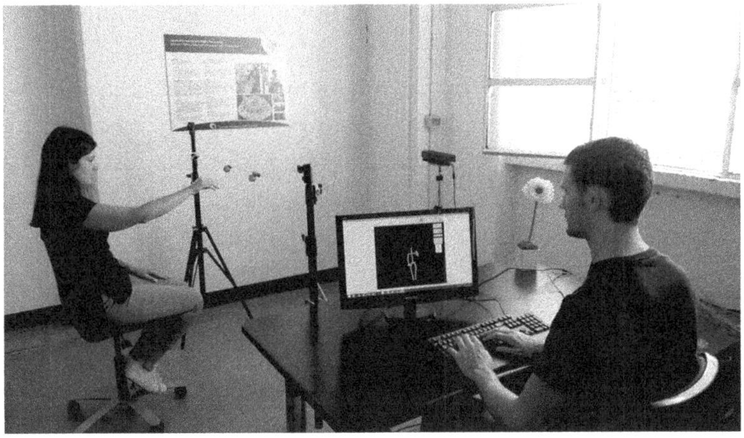

Data were recorded and logged with an in-house C# developed software for visual feedback and data acquisition. Offline analysis was performed by the means of an in-house developed Matlab software.

3D joints tracking data of shoulder, elbow, wrist and time labels were recorded and logged for off-line analysis. Tracking data were low-pass Butterworth filtered, 3rd order, cut-off frequency 6Hz. Data were used to identify the forward and backward phases of the movements, and to compute the following parameters, according to the protocol presented and described in detail in previous studies (Scano, Caimmi, Chiavenna, Malosio, & Tosatti, 2015):

- Execution Time (T, s)
- Shoulder Elevation angle (SE, °)
- Elbow Flexion and Extension angle (EF, °)

The above listed parameters account for movement kinematics (timing and range of motion) at the end of the forward phase of the movement.

- Shoulder Elevation Torque (ST, Nm)
- Shoulder Effort Index (SEI, Nms)
- Elbow Elevation Torque (ET, Nm)
- Elbow Effort Index (EEI, Nms)

The above listed parameters account for movement dynamics (torques at the end of the movement, and cumulative torque during the forward phase). They were obtained by the means of a simple biomechanical model of the upper-limb accounting gravitational and inertial contributions to articular torques.

- Normalized Jerk (NJ, a-dimensional number)
- Coefficient of Periodicity (ACP, a-dimensional number)

The above listed parameters account for movement quality of execution, representing respectively the smoothness (Normalized Jerk) and the Repeatability of the Acceleration Profile (Coefficient of Periodicity) of the forward phase.

The protocol was designed to be a simplified, synthetic version, made of a selection of the parameters computed in clinical environments during robotic therapies.

In general, Shoulder evaluations mainly characterize the RM, while Elbow evaluations, instead, are related to the HtMM. This is relevant especially for healthy subjects that execute "correctly" the motor tasks. However, patients might show impaired motor functions and compensatory motor strategies (for example, to bring the hand to the mouth, they abnormally elevate the shoulder). As reference values for the assessment of abnormal motor strategies, both Shoulder and Elbow Evaluations were computed for RM and HtMM.

In the preliminary investigation, only motor control indexes were considered (NJ and ACP) since they are more sensible to measurement because they involve derivatives of position (see Discussion for details). A previous study (Caimmi et al., 2015) performed the same analysis on healthy people using a "gold standard" marker-based system, and was used as reference to evaluate Kinect One performances. Subjects were divided into four subsets (young, middle-aged, senior, and all). Regression curves were analyzed for both RM and HtMM, NJ and ACP, and for each age subset and considering all the subjects together, for a total of 32 regression curves (2 gestures x 2 limbs x 2 motor indexes x 4 subsets each). Regression curves were compared to the above mentioned study.

In the main investigation, subjects were clustered into three groups of data subsets: data subset 1) Dominancy: dominant/non dominant limb; 2) Gender: males/females; 3) Age: young/middle-aged/senior. For each of the three group, each subset was tested in comparison to others on each variable of the biomechanical assessment. Means and standard deviations were computed along with p-values of the comparisons.

Since the study is still undergoing, results will be reported briefly in the following paragraph. Most notable remarks are:

- Regressions (t, NJ) and (t, ACP) indicate that, the higher the execution time, the lower the repeatability and the lower the smoothness. The same results were found in (Caimmi et al., 2015) this indicates that, will losing sensibility, Kinect One can be considered as a valuable tool for assessing motor control indexes. In general, senior people have higher performance decline (higher regression slopes) in comparison to young adults and middle aged-people.
- Limb dominancy does not introduce remarkable different, probably because the investigated gestures are of easy execution;
- Gender affected the execution of HtMM gestures; especially with the non-dominant limb, women proved to be more "precise" in motor execution because they elevated less the shoulder, flexed more the forearm, with best motor control and less effort. Such a result required to distinguish male and female performances when creating references for neurological evaluations;
- Age affected many parameters and was the major source of differentiation. In fact, senior people are in general slower and execute less refined gestures. Young adults and middle-aged people are usually similar in performances.

Summarizing the results, Kinect V2 can be considered as a cost-effective solution for motor evaluation. Furthermore, in order to analyze patients' performances, healthy people reference data should not be considered as a unique group, but should be clustered at least considering gender and age.

Future Research Directions

The –arm line will be exploited in future experimental works for two main objectives. First, the devices will be tested on healthy people to verify their usability and applicability and therefore planning eventual modification of the initial prototype concept. Furthermore, some clinical studies will be conducted to assess the efficacy of the devices in the clinical environment with neurological impaired patients.

In particular, some future application have already be planned:

- LINarm application on post-stroke subjects. The final version produced at the end of Echord++ project already allows the possibility of different interaction modes which can be suitable for different level of impairment. The sensors placed on the handle and the patient model assure a good monitoring of user's performances reducing risks of unwanted reactions. The possibility to orientate the device grants, in addiction, to propose different movements in the workspace like a hand-to-mouth movement, already tested on healthy subjects, and therefore customize even more the therapy.
- PLANarm application on healthy subjects. The interaction of healthy subjects with the device in different control algorithms could test the mechanical design proposed leading to eventual modification of the concept.
- Development of new handles for the DUALarm device. The positive results on healthy subjects and patients have verified the applicability of the system and the usefulness of a bimanual approach. Other configurations could possibly expand the range of possibilities and movements available for a more tailored rehabilitation protocol, maintaining the core feature of easily switch from in-phase to anti-phase movements.
- An active version of the LIGHTarm with actuators for targeted joints and force sensors to detect the intention of movement or for assist-as-needed interactions. The muscular activation profiles acquired during preliminary sessions on healthy subjects could be exploited to generate a reference for patients' usage. The passive version, with some modifications for safety and easy wearing, could undergo a clinical trial with patients to verify the actual benefit of the weight compensation.
- A Kinect-based environment for the monitoring of the patient. The human tracking with low-cost devices widens a broad list of solutions for clinical evaluation either at hospital or at home. Coupling two or more Kinect sensors placed inside the house could allow the clinician to acquire data of the patients' lifestyle and therefore program future interventions.

CONCLUSION

The devices described in this chapter share the common feature of being low-cost, while capturing many of the basic functionalities needed to promote motor recovery. The expected outcome of the arm line of device is to support the development of reliable but affordable products that can be placed in almost every clinic, even in low and middle income countries, and reach also the houses of the people, where usually a robot device is too expensive to be bought. The smart design and the exploiting of emerging technologies such as additive manufacturing may be also an example for manufacturers and developers of solutions for home rehabilitation.

REFERENCES

Caimmi, M., Guanziroli, E., Malosio, M., Pedrocchi, N., Vicentini, F., Tosatti, L. M., & Molteni, F. (2015). Normative Data for an Instrumental Assessment of the Upper-Limb Functionality. *BioMed Research International*, *2015*, 1–23. doi:10.1155/2015/484131 PMID:26539500

Cameirão, M. S., Smailagic, A., Miao, G., Siewiorek, D. P., Sander, M., Oxlund, B., ... Brooke, J. (2016). Coaching or gaming? Implications of strategy choice for home based stroke rehabilitation. *Journal of Neuroengineering and Rehabilitation*, *13*(1), 18. doi:10.118612984-016-0127-8 PMID:26921185

Cappozzo, A., Catani, F., Leardini, A., Benedetti, M. G., & Della Croce, U. (1996). Position and orientation in space of bones during movement: Experimental artefacts. *Clinical Biomechanics (Bristol, Avon)*, *11*(2), 90–100. doi:10.1016/0268-0033(95)00046-1 PMID:11415604

Cauraugh, J. H., & Summers, J. J. (2005). Neural plasticity and bilateral movements: A rehabilitation approach for chronic stroke. *Progress in Neurobiology*, *75*(5), 309–320. doi:10.1016/j.pneurobio.2005.04.001 PMID:15885874

Chin, J. H., & Vora, N. (2014). The global burden of neurologic diseases. *Neurology*, *83*(4), 349–351. doi:10.1212/WNL.0000000000000610 PMID:25049303

Di, S., Industriale, I., & Informazione, E. D. (2015). *Design and development of a planar cable-driven VSA based robot for upper-limb*. Academic Press.

Habib, M. K. (2011). Article. *Biomimetics: Innovations and Robotics, 4*(2), 113–134.

Hsieh, C.-Y., Chen, C.-H., Li, C.-Y., & Lai, M.-L. (2015). Validating the diagnosis of acute ischemic stroke in a National Health Insurance claims database. *Journal of the Formosan Medical Association, 114*(3), 254–9. doi:10.1016/j.jfma.2013.09.009

Jafari, A. (2014). Coupling between the Output Force and Stiffness in Different Variable Stiffness Actuators. *Actuators*, *3*(3), 270–284. doi:10.3390/act3030270

Krabben, T., Prange, G. B., Molier, B. I., Stienen, A. H., Jannink, M. J., Buurke, J. H., & Rietman, J. S. (2012). Influence of gravity compensation training on synergistic movement patterns of the upper extremity after stroke, a pilot study. *Journal of Neuroengineering and Rehabilitation*, *9*(1), 44. doi:10.1186/1743-0003-9-44 PMID:22824488

Krebs, H., Hogan, N., Volpe, B. T., Aisen, M. L., Edelstein, L., & Diels, C. (1999). Overview of clinical trials with MIT-MANUS: A robot-aided neuro- rehabilitation facility. *Technology and Health Care*, *7*, 419–423. Retrieved from http://www.scopus.com/inward/record.url?eid=2-s2.0-0033400738&partnerID=40&md5=1cfad3a95596bfa282229f338a7fb624 PMID:10665675

Laver, K., George, S., Thomas, S., Deutsch, J., & Crotty, M. (2012). Virtual reality for stroke rehabilitation. *Stroke*, *91*, 1–16. doi:10.1097/PHM.0b013e31826bce79.Comparison PMID:22713539

Laver, K. E., George, S., Thomas, S., Deutsch, J. E., & Crotty, M. (2015). Virtual reality for stroke rehabilitation. *Cochrane Database of Systematic Reviews*, (2): CD008349. doi:10.1002/14651858.CD008349. PMID:25927099

Legnani, G., Casalo, F., Righettini, P., & Zappa, B. (1996). A homogeneous matrix approach to 3D kinematics and dynamics - II. Applications to chains of rigid bodies and serial manipulators. *Mechanism and Machine Theory*, *31*(5), 589–605. doi:10.1016/0094-114X(95)00101-4

Legnani, G., Tosi, D., Fassi, I., Giberti, H., & Cinquemani, S. (2010). The "point of isotropy" and other properties of serial and parallel manipulators. *Mechanism and Machine Theory*, *45*(10), 1407–1423. doi:10.1016/j.mechmachtheory.2010.05.007

Lo, A. C., Guarino, P. D., Richards, L. G., Haselkorn, J. K., Wittenberg, G. F., Federman, D. G., ... Peduzzi, P. (2010). Robot-Assisted Therapy for Long-Term Upper-Limb Impairment after Stroke. *The New England Journal of Medicine*, *362*(19), 1772–1783. doi:10.1056/NEJMoa0911341 PMID:20400552

Lu, Q., Ortega, C., & Ma, O. (2011). Passive Gravity Compensation Mechanisms: Technologies and Applications. *Recent Patents on Engineering, 5*, 32–44. Retrieved from http://cat.inist.fr/?aModele=afficheN&cpsidt=24159482

Malosio, M., Caimmi, M., Cotti Cottini, M., Crema, A., Dinon, T., Mihelj, M., ... Spagnuolo, G. (2016). An affordable, adaptable, and hybrid assistive device for upper-limb neurorehabilitation. *Journal of Rehabilitation and Assistive Technologies Engineering*, *3*, 205566831668098. doi:10.1177/2055668316680980

Malosio, M., Corbetta, F., & Ram, F. (n.d.). *The kinematic architecture of a two-dof parallel and orthogonal variable stiffness actuator*. Academic Press.

Malosio, M., Pedrocchi, N., Vicentini, F., & Tosatti, L. M. (2011). Analysis of elbow-joints misalignment in upper-limb exoskeleton. *IEEE International Conference on Rehabilitation Robotics*, 1–6. 10.1109/ICORR.2011.5975393

Marchal-Crespo, L., & Reinkensmeyer, D. J. (2009). Review of control strategies for robotic movement training after neurologic injury. *Journal of Neuroengineering and Rehabilitation*, *6*(1), 20. doi:10.1186/1743-0003-6-20 PMID:19531254

Molinari-Tosatti, L., Fassi, I., & Legnani, G. (2003). Kineto-static optimisation of PKMs. *CIRP Annals - Manufacturing Technology, 52*(1), 337–341. doi:10.1016/S0007-8506(07)60596-4

Morasso, P., Casadio, M., Giannoni, P., Masia, L., Sanguineti, V., Squeri, V., & Vergaro, E. (2009). Desirable features of a "Humanoid" robot-therapist. *Proceedings of the 31st Annual International Conference of the IEEE Engineering in Medicine and Biology Society: Engineering the Future of Biomedicine, EMBC 2009*, 2418–2421. 10.1109/IEMBS.2009.5334954

Muratori, L. M., Lamberg, E. M., Quinn, L., & Duff, S. V. (2014). Article. *Extremity Rehabilitation*, *26*(2), 94–103. doi:10.1016/j.jht.2012.12.007.Applying

Petuskey, K., Bagley, A., Abdala, E., James, M. A., & Rab, G. (2007). Upper extremity kinematics during functional activities: Three-dimensional studies in a normal pediatric population. *Gait & Posture, 25*(4), 573–579. doi:10.1016/j.gaitpost.2006.06.006 PMID:16875821

Rau, G., Disselhorst-Klug, C., & Schmidt, R. (2000). Movement biomechanics goes upwards: From the leg to the arm. *Journal of Biomechanics, 33*(10), 1207–1216. doi:10.1016/S0021-9290(00)00062-2 PMID:10899329

Roux, E., Bouilland, S., Godillon-Maquinghen, A.-P., & Bouttens, D. (1999).. . *Evaluation of the GO Method Within the Upper Limb Kinematics Analysis, 35*(1), 1–14.

Scano, A., Caimmi, M., Chiavenna, A., Malosio, M., & Tosatti, L. M. (2015). Kinect One-based biomechanical assessment of upper-limb performance compared to clinical scales in post-stroke patients. *Proceedings of the Annual International Conference of the IEEE Engineering in Medicine and Biology Society, EMBS, 2015*, 5720–5723. 10.1109/EMBC.2015.7319691

Scano, A., Caimmi, M., Chiavenna, A., Malosio, M., & Tosatti, L. M. (2017). A Kinect-Based Biomechanical Assessment of Neurological Patients' Motor Performances for Domestic Rehabilitation. *Virtual Reality Enhanced Robotic Systems for Disability Rehabilitation*, 252–279. doi:10.4018/978-1-4666-9740-9.ch013

Scano, A., Caimmi, M., Malosio, M., & Tosatti, L. M. (2014). Using Kinect for upper-limb functional evaluation in home rehabilitation: A comparison with a 3D stereoscopic passive marker system. *5th IEEE RAS/EMBS International Conference on Biomedical Robotics and Biomechatronics*, 561–566. 10.1109/BIOROB.2014.6913837

Alessandro Scano, Andrea Chiavenna, Matteo Malosio, and Lorenzo Molinari Tosatti, "Kinect V2 Performance Assessment in Daily-Life Gestures: Cohort Study on Healthy Subjects for a Reference Database for Automated Instrumental Evaluations on Neurological Patients," Applied Bionics and Biomechanics, vol. 2017, Article ID 8567084, 16 pages, 2017. doi:10.1155/2017/8567084

Scano, A., Spagnuolo, G., Caimmi, M., Chiavenna, A., Malosio, M., Legnani, G., & Tosatti, L. M. (2015). Static and dynamic characterization of the LIGHTarm exoskeleton for rehabilitation. *IEEE International Conference on Rehabilitation Robotics,* (1), 428–433. 10.1109/ICORR.2015.7281237

Schaefer, S. Y., & Hengge, C. R. (2016). Testing the concurrent validity of a naturalistic upper extremity reaching task. *Experimental Brain Research, 234*(1), 229–240. doi:10.100700221-015-4454-y PMID:26438508

Schmidt, R. (1975). A Schema Theory of Discrete Motor Skill Learning. *Psychological Review, 82*(4), 225–260. doi:10.1037/h0076770

Sherwood, D. E., & Lee, T. D. (2003). Schema theory: Critical review and implications for the role of cognition in a new theory of motor learning. *Research Quarterly for Exercise and Sport, 74*(4), 376–382. doi:10.1080/02701367.2003.10609107 PMID:14768838

Spagnuolo, G., Malosio, M., Dinon, T., Tosatti, L. M., & Legnani, G. (2017). Analysis and synthesis of LinWWC-VSA, a Variable Stiffness Actuator for linear motion. *Mechanism and Machine Theory, 110*, 85–99. doi:10.1016/j.mechmachtheory.2016.11.002

Spagnuolo, G., Malosio, M., Scano, A., Caimmi, M., Legnani, G., & Tosatti, L. M. (2015). Passive and active gravity-compensation of LIGHTarm, an exoskeleton for the upper-limb rehabilitation. *IEEE International Conference on Rehabilitation Robotics, 2015*, 440–445. 10.1109/ICORR.2015.7281239

Swinnen, S. P. (2002). Intermanual Coordination: From Behavioural Principles To Neural-Network Interactions. *Nature Reviews. Neuroscience*, *3*(5), 348–359. doi:10.1038/nrn807 PMID:11988774

Van Delden, A. E. Q., Peper, C. E., Kwakkel, G., & Beek, P. J. (2012). A systematic review of bilateral upper limb training devices for poststroke rehabilitation. *Stroke Research and Treatment*, *2012*, 1–17. doi:10.1155/2012/972069 PMID:23251833

Vanderborght, B., Albu-Schaeffer, A., Bicchi, A., Burdet, E., Caldwell, D. G., Carloni, R., ... Wolf, S. (2013). Variable impedance actuators: A review. *Robotics and Autonomous Systems*, *61*(12), 1601–1614. doi:10.1016/j.robot.2013.06.009

Vitiello, N., Lenzi, T., Roccella, S., De Rossi, S. M. M., Cattin, E., Giovacchini, F., ... Carrozza, M. C. (2013). NEUROExos: A powered elbow exoskeleton for physical rehabilitation. *IEEE Transactions on Robotics*, *29*(1), 220–235. doi:10.1109/TRO.2012.2211492

World Health Organization. (2002). Towards a Common Language for Functioning, Disability and Health ICF. *International Classification, 1149*, 1–22.

Zhang, S., Guo, S., Pang, M., Gao, B., & Guo, P. (2014). *Mechanical Design and Control Method for SEA and VSA-based Exoskeleton Devices for Elbow Joint Rehabilitation*. Academic Press.

Zhou, X., Jun, S. K., & Krovi, V. (2014). Stiffness modulation exploiting configuration redundancy in mobile cable robots. *Proceedings - IEEE International Conference on Robotics and Automation*, 5934–5939. 10.1109/ICRA.2014.6907733

KEY TERMS AND DEFINITIONS

Arm: Line of devices for rehabilitation developed by ITIA.

Assistance: Paradigms of interaction that robot implement to facilitate motor recovery.

Biomechanical Evaluation: Evaluation of patients' motor capability with the techniques of biomedical and mechanical engineering.

End-Effector: Device for rehabilitation that requires the holding of a handle for interaction.

Exoskeleton: Device for rehabilitation that wraps patients' body.

Neurological Patients: Patients with injury or impairment due to neurological diseases.

Rehabilitation: Process for the recovery of motor function.

Robots: Devices for promoting the neuro-motor rehabilitation process.

Variable Stiffness Actuators: Actuators that can change their stiffness to meet patients' rehabilitative needs.

This research was previously published in the Handbook of Research on Biomimetics and Biomedical Robotics edited by Maki Habib, pages 161-190, copyright year 2018 by Engineering Science Reference (an imprint of IGI Global).

Chapter 18
Recent Trends in 3D Printing of Dental Models:
Rapid Prototyping in Dental Implants

Kayalvizhi Mohan
Agni College of Technology, India

ABSTRACT

This chapter introduces the recent trend in 3D printing (3DP) in dentistry. The advantage and disadvantages of 3DP are discussed. It elaborates on different types of 3DP techniques involved and their significance. The chapter further discuss about the biomaterial used. It also describes the complete steps involved in 3DP such as image acquisition, modeling, segmentation, and printing techniques. The merits and demerits of the different methodologies pertaining to steps involved in 3DP are illustrated. Rapid prototyping in dental implants is discussed in detail. It ends with review of a case study in implementing the technique.

INTRODUCTION

This chapter reviews the incipient role of 3D printing in dentistry, a topic of growing relevancy (Schubert, Van Langeveld, & Donoso, 2014). Further this chapter gives an insight in bioprinting that aids medical community to design structurally complex 3D constructs by comprehensive positioning and aiding complex patterns that are formed by different structures of cells, biomaterials and bioactive molecules within a single make. With the evolution of several bioprinting strategies during recent period, 3D bioprinting is applied in several research areas that include dental implant (Klein, Lu & Wang, 2013).

Bioprinting has emerged over the past decade as a prominent technology in the field of dentistry (Banks, 2013). Implementation of 3D printer helps in reducing the cost, it enhances the speed of production and is easy method for implementation in dental rehabilitation. It supports the surgeons in planning for surgery in resection and reconstruction. It represents as an exceptional aid for the academicians to train their students (Bertassoni, Cecconi, Manoharan, Nikkhah, Hjortnaes, Cristino, & Khademhosseini, 2014). Medical Rapid Prototype based pre-bended reconstruction helps in decreasing the resection dura-

DOI: 10.4018/978-1-5225-9624-0.ch018

tion, rate and anaesthesia menace bared by the subjects (Mertz, 2013). The need for reconstruction for dental rehabilitation is continuous challenge.

The very purpose of reconstruction is the repair of appropriate aesthetics and regularty of facet. Reconstruction also helps in stabilising arrangements and the strengthening the jaw (Salgueiro, & Stevens, 2010). The defects in ideal model in reconstruction is obtained by using titanium plate during resection. For mandibular reconstruction, is not easy to shape the titanium plates and the whole procedure is time consuming (Cohen, Laviv, Berman, Nashef, & Abu-Tair, 2009). The intraoperative prearrangement leads to extensive surgical operation and upsurges the budget (Yovchev, Stanimirov, & Mihaylova, 2014). The appropriate alteration of the plate to match the skeletal structure is indispensable to get a positive outcome.

BACKGROUND

The dental prosthesis fabrication includes many of manual laboratory processes and the quality of restorations depends on the dentists and dental technician's skill and expertise. The dental restoration is also a very individual and complex process. Computer aided manufacturing (CAM) is methodology is introduced to overcome the disadvantage in manual process. Introducing CAM for the processing procedure will highly rise the manufactured quality of dental restorations. The manufacturing process includes: data acquisition, data processing and model fabrication (Nayar, Bhuminathan, & Bhat, 2015).

3D Printing Intial Steps

The first step often used towards 3d printing involves, Computer Tomography scan with field of view within twenty cm, the slice thickness usually selected is slightly more than one cm (equal to scan spacing) and the gantry tilt of zero degree is set. During the whole procedure the patients are not allowed to move and the occlusal planes are set parallel to the gantry. The whole arrangement helps in reducing the artefacts in the anatomic model. The acquired images are stored in DICOM file format.

The open source 3D imaging software is used to process CT DICOM images. The commonly used software version is Vesalius 3.0.0 version. The bone mask is created using the tool for mask creation in Hounsfield unit range. The 3D surfaces are crated using the bone mask as reference. Finally, the data is transferred in stereolithography (STL) file. The process takes few minutes depending the CPU configuration.

The Autodesk Meshmixer 2.9.1 version is used for editing STL file. The selection/analysis tools are used to isolate and repair the mandible. Further the compatible model for printing is produced by transforming the complex geometry of the bone. Finally mesh of the mandible is prepared for the slicing using open source Matter Control and printing is performed using the ROBO 3D R1 with 1.75mm polylactid acid filament.

Imaging Modalities

The first step important step towards 3D printing is imaging. The imaging techniques such as CT, MRI, digital photography is commonly used.

The objectives of diagnostic imaging depend on

- The extent and nature of data required (combinations of conventional dental images)
    - Facility to envisage the implant site in the mesiodistal, buccolingual, and superioinferior dimensions
    - Facility to allow consistent, precise estimation
    - Normal anatomical structures (incisive canal, nasal floor, maxillary sinus, mandibular canal)
    - Capability to estimate trabecular bone density and cortical thickness (height, width)
- The imaging should have least radiation risk to the patient
- It should include appropriate clinical examination and patient's need.
- The reasonable access and cost-effective

Dental implant is a promptly intensifying area of dentistry. The imaging modalities range from 2D projections to complex 3D imaging. The 2D modalities are voluntarily available, less expensive with very less radiation exposure, but have restrictions like exaggerations and overlapping of planes and so clinician are not able to develop a 3D perspective of patient's anatomy with a single image. However, in complex cases, more extensive and advanced radiographic evaluation is needed. Hence, cross-sectional capturing is more and more considered crucial for optimal implant placement, important roles in pre- and postoperative evaluation of the implant patient especially in complex reconstructions and multiple implants.

The feat of endodontic method mostly depends on the detection of all root canals so that are opened, cleaned, shaped, and kept immovable (Vertucci, 1984). Conventional radiographic method is able to identify only 55% of these structures (Ramamurthy, Scheetz, Clark, & Farman, 2006). The occurrence of a second mesiobuccal canal in maxillary first molars is testified to fluctuate from 69% to 93% subject to the study method involved. This inconsistency happens in the buccolingual plane where overlapping of anatomic arrangements hampers the identification of small structural density variations (Pineada,1973, Nance, Tyndall, Levin, & Trope, 2000).

Ramamurthy et al. (2006) established that evaluators using different 2D film modalities were seldom capable of detecting more than 50% presence of MBC canals. They also established the dissimilarities in perception rates with complementary metal oxide semiconductors (CMOSs), analog film, and photo-stimulable phosphor plates (PSP) detecting 55%, 44%, and 39% of MBC canals, respectively. Matherne et al. (2008) detected the number of root canals on intraoral digital (both charged-couple device and photostimulable phosphor) plate images with CBCT in 72 extracted teeth and related the capability of three board-certified endodontists. It was established that on average the contributors did not detect even one root canal in 40% of teeth using intraoral radiographs.

CBCT assessments recognized an average of 56% root canals (RCS) per maxillary molar, 19% per mandibular premolar, and 24% per mandibular incisor (Gepreel, & Niinomi, 2005). Estimation of CCD images revealed an average number of 58% per maxillary molar, 19% RCS per mandibular incisor and 19% per mandibular first premolar. Assessment of PSP images signified an average number of 23% RCS per mandibular incisor, 18% per mandibular first premolar, and 54% per maxillary molar. In Taiwanese individuals the high occurrence of the distolingual canal was described using CBCT imaging (Alamri, Sadrameli, Alshalhoob, & Alshehri,2012; Alamri, Sadrameli, Alshalhoob, & Alshehri,2009), it also was helpful in determination of root curvature (Estreala, Bueno, Sousa-Neto, & Pécora, 2008) and acme glitches in the root canal system of mandibular premolars (Cleghorn, Christie, & & Fu-yuan,, 2008). Baratto Filho et al. (2009) studied the internal change in shapes of the extracted maxillary first molars by matching finding rates attained by means of an operating microscope and CBCT to ex vivo

Table 1. Modalities in dental implants

	Common Imaging Types	Advantage	Disadvantage
1	**Periapical radiography (Bagchi,& Joshi, 2012)**	• Long cone paralleling technique. • Best resolution (line pairs/mm). • Area of interest are inspected for types of trabecular patterns, the changes in residual root and periodontiumn and how the angulation of adjacent teeth is formed. • It is readily available, inexpensive and involves less radiation dose	• 2D perspective of 3D anatomy are inadequate for inspecting the amount of available bone (facio-lingual dimension) in the edentulous area. • The valuation is inadequate due to the limited size of mandibular, large edentulous areas and maxillary structures • The maxillary and mandibular structures face image distortion due to inadequate size and anatomic limitations in image receptor inflexibility
2	**Digital intraoral radiographic images (Lingeshwar, Dhanasekar & Aparna, 2010)**	• This gives less exposure compare to conventional periapical radiography, instant results, contrast is manipulated.	• But its limitation is, it will provide the dentist the limited area for imaging, and no facio-lingual dimension is assessed.
3	**Bitewing (Siu, Chu, Li, Chow, Deng, 2010, Kalra, Jain, Deoghare & Lambade, 2010)**	• This modality helps in the assessment of the coronal view of tooth and in relation to the adjacent edentulous space of alveolar bone	• Periapical area cannot be assessed so approximation with adjacent vital structures is a major limitation
4	**Occlusal radiograph (DelBalso, Greiner & Licata, 1994, Siu, Chu, Li, Chow, Deng, 2010, Kalra, Jain, Deoghare & Lambade, 2010)**	• Provide us with facio-lingual dimension technique is readily available, high image definition, inexpensive, less radiation exposure	• Due to distortion, it records only the widest portion of mandible, with limited reproducibility and makes it unusable in maxillary arch capturing
5	**Panoramic radiograph (Siu, Chu, Li, Chow, Deng, 2010, Kalra, Jain, Deoghare & Lambade, 2010)**	• It is commonly used for diagnosis of gross pathoses within the jaws in the implant site • It helps in assessing the anatomic structures such as foramen, sinuses, canals, fossa, and foramen • This helps in better visualization of bony anatomy clearly	• The main disadvantage is the sharpness and resolution is very much reduced resolution and does not give cross-sectional imaging • The magnification is not stable and unpredictable specifically in the vertical aspect and is extra prominent in posterior than in anterior regions. Thus, can contribute a wrong sense that extra bone subsists between the crest of the alveolar process and the inferior alveolar canal, nasal fossa or maxillary sinuses. • Incorrect subject placing can additionally contribute to image misrepresentation. Even appropriately placed and exposed panoramic radiographs are not used for direct bony measurements.
6	**Lateral cephalogram (Siu, Chu, Li, Chow, Deng, 2010, Kalra, Jain, Deoghare & Lambade, 2010)**	• Along with soft tissue profile of the patient, axial tooth inclination and dentoalveolar ridge relationship in the midline can be assessed.	• Provide dentist with cross-sectional dimension only at midline, image magnification and also contralateral side superimposition evident.
7	**Dentascan imaging (Siu, Chu, Li, Chow, Deng, 2010, Kalra, Jain, Deoghare & Lambade, 2010, Matherne, Angelopoulos, Kulild, & Tira, 2008)**	• This imaging modality provides with programmed formation, recollection, and display of the image, mandibular arch, the maxillary arch and cross sectional and at a tangent panoramic image of the alveolus that enables accurate preprosthetic treatment planning. • Diagnostic template is necessary for determining the Location and alignment of dental implants.	• Magnification of images, grayscale values, is limited, and the inclination of the patient's head all through the check-up is necessary.
8	**Interactive CT (ICT) (Tyndall & Rathore, 2008)**	• It overcomes limitations of CT, also enables radiologist to transfer the image to the computer, so clinician can view • Hence, 3D treatment is planned according to patient's anatomy and are seen before surgery thus helping in determining the number and size of implants correctly according to density of bone at suggested sites. ICT is more relevant diagnostic technique when compared to conventional techniques for implant planning.	• Determining the Exact position of implants is difficult, tedious and time consuming
9	**MRI (Manisundar, Saravanakumar, Hemalatha, Manigandan, & Amudhan, 2014)**	• In MRI, cortical bone is clearly delineated from the cancellous bone. • Vital structures such as nerves, vessels, and the floor and mucosa of the maxillary sinus are readily identified by the implant surgeon, exceptionally safe due to the absence of ionizing radiation, high image quality.	• High capital and running costs with lack of availability have a significant barrier to the use of MRI for implant assessment
10	**CBCT and multidetector CT (Alamri, Sadrameli, Alshalhoob, & Alshehri, 2012; Kumar & Satheesh, 2013, Arai, Tammisalo, Iwai, Hashimoto, & Shinoda, 1999)**	• It allows the surgeon to have an accurate information of implantsite, in edentulous patient or when multiple implants are considered thus aiding in diagnosis and provides the dentist with controlled surgical plan, is used when conventional radiograph fails to provide the needed information. • Evaluation are done in all three surfaces of space without image misrepresentation, overlapping of structures, and differential enlargement of the image based on geometry, also accurate position of anatomic markers as well as the altitude, thickness, inclination, and quality of alveolar bone for implant site. • Also, for case selection and a postsurgical evaluation to assess implant's position in the alveolus.	• Increased susceptibility to movement artefacts and metallic restoration often leads to streak artefact, in CBCT no standard grayscale system present, so universality is questionable

sections. It was stated that an ex vivo incidence of a fourth canal in 67.14% of teeth and additional root canals in 92.85% of mesiobuccal roots. CBCT showed lowest overall (37.05%) detection rate. Clinical valuation suggested slightly lower overall (53.26%) but higher (95.63%) MBC detection rates. It was also indicated that CBCT delivered a worthy method for the initial assessment of maxillary first molar internal morphology but that the practice of operating microscopes was optimal. Once the images are captured, they have to be processed.

IMAGE PROCESSING TECHNIQUES

During the past few decades numerous image-processing techniques have been proposed to delineate tooth images. Interpolation method was used to obtain an N dimensional object from a set of (N_{2-1}) dimensional objects by Bors et al (2002). They also applied their proposed method for reconstruction of the three-dimensional tooth shape with 2D CT images, but the tooth regions were segmented manually. Edge detection was proposed by Mol and van der Stelt (1991) in his work to detect the modification in bone to the morbid state, from an X-ray image. Gao et al (2006) reconstructed three-dimensional shapes using shadow speckles. Region growing method were applied for extracting the region of all the teeth by & Fu-yuan, -ri and Fu-yuan (2010).

A semiautomatic method was proposed by Omachi et al (2007) for extracting the three-dimensional shapes of teeth from dental CT images. The coutour method was really challenging because the adjacent teeth or the surrounding tissues is also extracted (Jayadevappa, Kodhandarama, Santosh, & Rashid 2010). A three-dimensional statistical model by Buchaillard et al (2007) was proposed with the objective of obtaining a good estimate of the entire tooth when only partial information of a tooth is provided. A method for segmenting and envisioning all of the teeth with CT images was recommended by Akhoondali et al. (2009). Hirogaki et al. (2001) projected a technique for rebuilding a three-dimensional dental cast model with a line laser. Carter et al. (2010) analyzed tooth reconstruction techniques. Hoy et al. (2013) suggested improved methods for extracting the contours of teeth in dental radiographs. The region growing method is used as an interactive tool for segmentation and visualization of the teeth in CT scans.

Commonly Used Image Processing Techniques

The commonly used step in reconstruction of three-dimensional shape from the CT image is region growing method (Grenier, Revol-Muller, Costes, Janier, M & Gimenez, 2005). Normalization, contrast enhancement and smoothing are introduced as pre-processing techniques for attaining precise delineation of the tooth. Normalization is used to reduce the variance of the brightness values along the elevation. It is followed by contrast enhancement that enhances the boundary between the tooth region and the surrounding tissues. Further the noise in the CT image is reduced by using a commonly used smoothing bilateral filter (Akhoondali, Zoroofi, & Shirani, 2009). The three main constraints in delineating the tooth structures are the continuity of the brightness values, the maximum brightness value and isobrightness contour information. The constraints results in over-extraction, which is a major issue when using the existing method (Bors, Kechagias, & Pitas, 2002). To overcome such constraints, region competition is accomplished by growing regions initially selecting multiple seeds. This helps in determining the boundary among the tooth to be delinated and the neighboring teeth (Bors, Kechagias, & Pitas, -Nielsan, 1996). The irregular points are eliminated by applying morphological operations (Dong & Fu-yuan, 2010).

Pre-Processing

In dental CT images, tooth regions do not have same intensity value, it mainly varies in the enamel and dental pulp (Omachi, Saito, Aso, Kasahara & Yamada Kimura, 2007). The region to be delineated contains different intensity value compared to the other regions. Normalization is the process in which the target to be extracted is processed to have uniform distribution of the intensity value. It is achieved by Normalization based on slice energy. This is achieved by taking summation of intensity values of each slice on x-y plane as normalization factor (Omachi,Saito, Aso, Kasahara & Yamada Kimura, 2008).

The precise delineation is achieved when the intensity levels of tooth to be extract is different from the surrounding regions and it is achieved by contrast enhancement (Juergens, Krol, Zeilhofer, Beinemann, Schicho, Ewers & Klug, 2009). The real challenge in extraction of tooth is the similarity in the intensity levels of regions such as alveolar bone, gum and periodontal ligament, to the desired tooth region (Stavropoulos, & Wenzel, 2007; Kalra,2010). To delineate the desired tooth with overlapping similar intensity region, the intensity value of each voxel is substituted by a square as suggested by Pratt (2007). The final step in the preprocessing is removing the noise inherent and specific in CT. The bilateral filter, which is a boundary preserving smoothing filters is preferred to remove the noise. The smoothing is achieved when the distance from one pixel to the other is minimal or these pixels have similar intensity values. This method helps in noise reduction and upholding the boundary information are accomplished instantaneously (Yau, Lin, Tsou, & Lee, 2008).

Region Extraction

On the CT image, the intensity values are distributed like the dental pulp with low intensity value, the dentine with high intensity value and the enamel with highest intensity value in a tooth (Kumar, Moni, & Rajeesh, 2011).

As suggested by R Yanagisawa et al (2011) the pixel with minimum intensity value is selected as the seed because the dental pulp has a low intensity value (Rahimov, & Farzaliyev, 2011; Roser, Ramachandra, Blair, Grist, Carlson, Christensen, & Steed, 2010). Further the pixels with a lower intensity value are explored iteratively from the adjoining region of the speculative initialized seed to determine the pixel that has minimum intensity value (Mupparapu, & Singer,, 2000). After identifying the seed, the region-growing method is applied by using this seed as the starting point (Scarfe,2006). The dental pulp, dentine and enamel of each delineated tooth are the tissues that are anticipated to be delineated. The major typical feature of these tissues is that the intensity value of a pixel increases as it approaches the surface of a tooth (Palomera, Martinez, Benitez & Ortega, 2010).

In spite of all the restrictions applied in the algorithm, adjacent teeth around the crown of the tooth are over extracted (Grenier, Revol-Muller, Costes, Janier, & Gimenez,2005; Yanagisawa, & Omachi,2011; Palomera, Martinez, Benitez & Ortega, 2010). The major disadvantage of the region growing method is the presence of disjointed points, due to the left-over noise on the image. These disjoint are often removed using morphological operation. This is achieved by opening and closing process with a structure element for n iteration (Grenier, Revol-Muller, Costes, Janier, & Gimenez, 2005; Yanagisawa, & Omachi, 2011; Palomera, Martinez, Benitez & Ortega, 2010).

Data Processing

The quality of dental or medical model is depend on data processing which is very complex step. The acquired data file format is converted into a file format that are compatible with engineering and industrial procedure. Sterolithographic file format (STL) is suitable in most situations for Rapid manufacturing.

Model Fabrication

Many types of RP technologies were used such as stereolithography, selective laser sintering, fused deposition modelling (FDM), and Ink Jet printing techniques to create dental models. The type of dental restoration will determine the chosen technology, it depended on exacting precision, finishing surface, visible manifestation of inner structures, number of colors which needed for each model, strength and properties of the materials etc. The model is ready for implementation if the team decided that there are no errors after the RP medical model is created (Klein, Lu, & Wang, 2013) .

3D MEDICAL RAPID PROTOTYPE DESIGN PROCESS

Stereolithography

Stereolithography is the most popular from the advanced RP technologies. It creates three dimensional models by computer-controlled laser beam that moves in order to build up the required objects from a liquid in a layer by layer manner. To fabricate prototypes, models, patterns, the material has to be dipped in photosensitive liquid resin, a model-construction platform, and an ultraviolet (UV) or laser for curing the resin are used in 3D printing technology. The layers sequentially bind together to form the solid that begins from the bottom to the top. A distinct visualization of anatomical structures is used for educational purposes and it preferable for surgical planning before surgery as suggested by Azariet et al (2009). It is used to produce the model for reconstructive surgical procedure and sub-periosteal dental implant surgical procedure. Currently, the main purpose for to use these models is to fabricate surgical drilling templates during the insertion of dental implant.

Advantages

- It has Good speed.
- The needed time depends on the model volume and complexity.
- The accuracy of producing the details is increased.
- It has close tolerance.
- Good finished surface and smooth.
- It is possible to make it transparent.
- The possible density is 100percent.
- High mechanical strength, its Prototypes are used as master models for injection moulding, thermoforming, blow moulding.

Disadvantages

- The equipments and materials are highly Expensive.
- It used with polymers only.
- Handling of wet materials.
- It needs Post-processing procedure.

Inkjet-Based System (3D Printing - 3DP)

3DP technique got a great importance because it is precise, cost effective and speedy. It is an additive technique that differs in duration of design and print a wax pattern of a restoration. The machine constructs wax patterns of full crowns and frameworks similar to that of manual waxed restorations.

Similar to ink-jet printing, the pattern fabricated over the surface of a powder bed by spraying binder materials on a thin distribution of powder and it spreads over the complete bed. An amount of powder is measured and added from a source compartment by movement of piston ascendant, then a roller spreads and squeezes the powder at the upper manufacturing compartment. After that a head of multi- channel jetting will deposit a fluid bonding agent in a 2D pattern on the coating of powder. After a finishing of a layer, the piston which supports the powder bed 3DP is a coating - by- coating technique which continued progressively till the sample is totally manufactured. After heat- treating, unbounded powder is flounced up and leaves the created part intact as implemented by Azari et al. (2009).

3DP technology has high tolerance and hence not useful for molding intents as SLA, this tolerance considers as a problem to make flimsy dental prototypes, to form configurations to create educational models. Due to this high tolerance, 3DP technology is not useful for impression objectives as SLA. 3DP is inaccurate technology because it breaks off easily which requires extra period for accurate fitting of medical implant throughout the surgical procedure, However, it is supportive to work-out the surgical processes.

Advantages

- They have short manufacturing time.
- The used material is inexpensive
- Their colouring ability.
- Models are used in casting purposes with low ash burnout.
- The toxicity is Low.
- Variety of materials are used.

Disadvantages

- They have High tolerance.
- The models with low strength.
- The surface finish is rough.
- The resolutions are Limited.

Selective Laser Sintering

Selective laser sintering (SLS) creates the desired three-dimensional shape solid mass by fusing small particles powdered materials like plastic, metal, ceramic or glass powders with a high-power laser (CO2 laser). A wide variety of materials are used by coating particles with thermal binders. The powders have many higher properties than resin-based technologies with higher yields and faster post-finishing. Fusing of the material is done by the laser selectively after scanning of the cross- sections that created from a 3-D digital description of the model on surface of a powder bed.

By using a roller, the powdered material will spread on the surface of a build cylinder. Then moving the piston cylinder down just one-layer thickness, to give sufficient space to the new layer of powder. The distribution system of powder is similar in action to the build cylinder, in which a piston moves upward gradually to give a measured amount of powder for each layer. Then the surface of this tightly compacted powder is exposed to laser beam. An interaction will occur between laser beam and the powder which elevates its temperature to the melting point, leading to fusing of the powder particles and solid mass forming. Laser beam energy is adjusted to melt only the powder in areas which determined by the object's form over that cross section. The temperature of manufacturing chamber is kept lower than powder's melting point and the laser heat will raised the temperature a little only to cause sintering. SLS machine keeps the bulk powder material temperature in the powder bed less than its melting point by using infrared heating to reduce thermal deformation ″curling″ and simplify binding to the prior layer. This will reduce a lot of processing time. After finishing of the first layer, extra powder layer will be added by a roller technique over the layer which scanned previously. This process is repeated until the whole object is finished, then the object is removed from the building chamber and the powder which is not scanned and fused are reused. Post-processing is needed, depending on the desired application. SLS technology is used to fabricate removable partial denture (RPD) frameworks with cobalt- chromium alloy spherical powder that has maximum particle size of 0.045mm (particle size range 0.005- 0.045mm), the mean particle size approximately 0.030mm. A complete cobalt- chrome RPD framework is fabricated with a demonstrate successful.

Advantages

- Extensive variety of materials are used especially in dentistry.
- Excellent accuracy.
- Simple post- processing.
- Un-sintered powder remains at the sites of support structure. So, support structures are not needed, after completing the model it can be recycled.

Disadvantages

- Material and equipment are costly.
- Large tolerance is required for dental procedures (0.5 ± 0.2).

Fused Deposition Modeling

FDM is the other vastly used rapid prototyping technology, following stereo lithography. It is a technique in which a thermoplastic material was threw in a layer by layer manner through a temperature-controlled head. In this system, supplied materials is released from a coil with a plastic filament in to an extrusion nozzle head. The plastic melts in to a semi liquid form by the heated nozzle, the melted plastic then permitted to flow in order to be rotated on and off (Ventola,2014).

This procedure functions in three axes, essentially, portrayal of the model in one layer at the same time. When machine nozzle is progressed over the stand table in the essential path, it drops a slim bead of extruded plastic that forms the first layer. After ejection of it from the nozzle, the plastic will solidify directly and unites with the layer below.

The procedure is composed of a compartment that is fixed at a temperature lesser than temperature that melts the plastic. This technology permit designing of a wide range of materials and colors like wax of investment casting and medical grade ABS (Kumar, Moni, & Rajeesh, SS,2011).

FDM is suitable for fabrication of bone imprints. It uses build materials which manufacture strong and tough models in one step. Furthermore, it gives excellent visualization by highlighting a chosen feature with different color. For overhanging geometries, supporting material is required which can be broken away from the object after finishing it, or can be easily washed away when a water-soluble support material is used (Kumar, Moni, & Rajeesh, 2013). The system is slow in fabricating models have wide cross sections, while in fabricating small models that have tall, thin form-factors it will be faster.

Advantages

- Wax pattern is directly fabricated.
- Constructed in multi-color.
- Relatively, fast and speed technique.

Disadvantages

- Removing the support structure after model fabrication is completed.
- Finishing surface is rough.
- Only thermoplastic material can be used.
- Its dense is less than 100 percent.

For designing frameworks of RPD, STL is the best choice. STL model is often recommended to recognize the existing undercut areas on the computer-generated CAD model which prepared for the teeth and surrounding tissues of the patient. Undesirable undercuts should be blocked out to ensure a path way for replacement and removal of the RPD into the patient mouth. Soft tissues will be relieved, then carving will be done virtually with special software tools making it similar to conventional carving by using haptic device which enabling models′ movement, rotation and translation in all axes, as well as hand moving.

3D PRINTING IN DENTISTRY

Velasco et al. (2017), in his paper discussed evaluation of right perimandibular swelling in the case study a multilocular lesion of few cms in size in the right mandible was disclosed in the study of Panorex and maxillofacial CT scan of a 40-year-old patient It also spread commencing from root of the canine to the mid-ramus. Apart from this the incisional tissue removal of the lesion resulted in solid ameloblastoma.

For the first stage a titanium reconstruction plate was planned on an mandibular resection along with safeguarding right condyle. The second stage involved delayed restoration of the flaw with iliac crest bone graft. The restoration plate is placed without modifications and hovels are pierced in healthy bone. Plate elimination and segmental mandibulectomy is accomplished, preceded by relocation of the plate and fixation to the predrilled hoveles. The whole procedure took around 4 hrs. A skull x-ray taken after the completion of the operation showed the precise placing of the restoration plate and mandibular symmetry.

Velasco et al. (2017) in his paper another a case of 34 old patient with a large anterior mandibular solid ameloblastoma formerly biopsied. A multilocular lesion around few cms were found predominantly in the symphysis region, spreading from the root of the right canine to the left first molar in the panoramic x-ray and maxillofacial CT scan. The treatment planning of mandibular resection with titanium 2.7-mm mandible reconstruction plate was suggested. The complete surgery was planned using rapid prototype Modelling.

Benefits of 3D Printing in Dentistry

The utmost benefit that 3D printers offers is the liberty to fabricate custom-made medical tools, devices and equipment. 3D printing is preferably used to tailor prosthetics and implants that bestows unlimited value for both patients and general practitioner (Ursan, Chiu, & Pierce 2013). In addition, 3D printing aids to fabricate jigs and fixtures for utility in surgical rooms. Custom-made implants, fixtures, and surgical tools have a constructive effect in terms of the time required for surgery, patient recovery time, and the success of the surgery or implant (Erkal, Lockwood, Chen, & Spence 2014).

The significant assistance offered by 3D printing is, the ability to manufacture items at minimal cost (Lipson, 2013, Li, 2014). It is realized with standard implants or prosthetics, used for craniofacial disorders. The cost to custom-print 3D object is nominal, with the first item being low-priced. This technique is beneficial for corporations that have low fabrication sizes or that assemble parts or models that are multifarious or necessitate regular changes, economical to supply to patients (Chae, Rozen, McMenamin, Findlay, Spychal, & Hunter-Smith,2015, Cui, Boland, D'Lima, & K Lotz, 2012).

The main positive attribute offered by 3D printing is the patient specific personalized design and manufacturing of medical tools and devices. An increasing group of biomaterials are accessible for use in 3D printing, and hence it decreases the cost. 3D printer allows flexibility to produce novel products for individual and marketable usage. The nature of 3D printing data files also offers an extraordinary opportunity for sharing among researchers (Cunningham, Madsen, & Peterson, 2005). Getting approval from regulatory bodies is another major obstacle in use of medical application of 3D printing. (De Vos, Casselman, J & Swennen, 2009).

BIOMATERIALS USED IN DENTAL IMPLANTS

Titanium and its alloys are the main optimal choice for implantation application. Titanium readily passives itself to form a thin oxide layer. This characteristic makes it biocompatible as well as forming a direct bond between bone and implant leading to permanent fixation of the implant. Titanium is commonly used in fracture plates, dental implants, etc.

Although 3D bioprinting schemes are laser-based, inkjet based, or extrusion-based, inkjet-based bioprinting is most common. This method deposits "bioink," or droplets of biomaterials, onto a substrate according to digital instructions to replicate dental implants. Multiple print heads are to deposit different cell types (region specific, blood vessel), a necessary feature for fabricating whole heterocellular tissues. A process for bioprinting involves to create a blueprint of a region with its vascular architecture. Then generate a bioprinting process plan. Further isolates stem cells and then differentiate the stem cells into tissue-specific cells (Tyndall DA & Brooks,2000). It is followed by preparing bioink reservoirs with tissue-specific cells, blood vessel cells, support medium and load them into the printer. Laser printers have also been employed in the cell printing process, in which laser energy is used to stimulate the cells in a particular pattern, providing spatial control of the cellular environment (Derand & Hirsch,2009).

PLANNING FOR 3D PRINTING

Dental rehabilitation planning has progressed promptly in the past two decades with the developments in medical image processing. Medical rapid prototyping (MRP) is defined as the manufacture of dimensionally accurate physical models of human anatomy derived from medical image data using a variety of rapid prototyping technology. The construction of the physical models is based on CT scans and there are many techniques for production of MRP models including stereolithography, selective laser sintering, and 3D printing (Eckardt & Swennen, 2005). There are several 3D printing processes for building 3D objects in any shape (Estrela, Bueno, Sousa-Neto & Pécora, 2008). MRP models are generally used in the perioperative period for improving the likelihood of treatment (Frederiksen,1995). Such technology is becoming increasingly acceable (Fullerton, Frodsham, G. C., & Day,2014). Recently rapid prototyping is used to produce solid object of computer models. It can form functional structures in a direct way such as metal parts, as well it is used in nano manufacturing and bio manufacturing, it is powerful manufacturing method.

Like other medical branches, dentistry also uses rapid prototyping as surgical guide, maxillofacial repair or physical model of dental implantology implemented by Yan et al (2008) and Prosthodontics implemented by Hoang et al. (2015), Sun et al. (2011) and Bibb et al. (2011). From the definition, metal-free ceramic restorations that made by CAD/CAM system falls into this group of technology.

One of the considerable other benefits of RP when used in dentistry is medical modeling fabrication, it is quite useful when used in mass production of patterns for casting purposes as analyzed by Toro (2007)). In this method, it is easy to create the difficult parts of restorations even without human involvement in a short time as experimented by Bibb (2010). Clinically, rapid prototyping technology reduces the time and possible injury which may occur during implantation process as reported by Hoang (2015).

By RP technology, dental prosthesis is constructed layer by layer directly from digital model simply and rapidly without special tooling and human interference, it makes a revolutionary progressing in the manufacturing of dental prosthesis as proposed by Tyndall (2008).

After the evolution of RP technologies there is a possibility to create many types of dental prostheses for many various implementations, Such as dental prosthesis wax pattern, dental (facial) prosthesis mold (shell), dental metallic prosthesis, and zirconia prostheses. RP thought to play an important role in prosthodontics and it will be most acceptable technique for digital manufacturing of dental prostheses as reported by Sun et al (2011). Some of RP implementations are: crowns and RPD wax pattern manufacturing, complete dentures and casting molds, maxillofacial prosthesis and dental metallic prosthesis manufacturing.

RP technologies will be an alternative method instead of usual method which depends on the skills of dentists and technicians in the manufacturing of dental.

Advantages of RP Systems

RP is fast, it needs 1to 2 days to construction and insertion of the prosthesis. The design defects are detected early and repaired with RP technique. The confidence of the design integrity is one of the properties of RP technique. It also has good management for the tissues with less trauma. It further enables accurate assessment of anatomical landmarks such as the size of the maxillary sinus in the upper jaw and location of the alveolar nerve in the lower jaw. It is well known to have accurate Osseous Topography analysis and precise location of implants. The data is got without direct patient, to avoid tissue tension. Also, can be used with patients who can't cooperate when preparing usual full face moulage impression. CAD and RP technique can be reduce lab work due to creation of positive models which invested for casting later. An object with definite accurate dimensions can be created by using RP approach of stereo lithography. If primary model is wasted or damaged during fabrication process, digital one can be used. The fabricated model has the internal detail as well as outer surface.

Disadvantages

Rp technique is unsupported in soft tissue. A two-steps impression may be required to have proper orientation of the soft tissue. Undercuts can be minimized by positioning the patient correctly, also it will be helpful to create any additional undercuts by using digital manipulation of the STL file, as proposed by Sun et al (2011).

CONCLUSION

This review chapter discusses about different technologies that were developed in past two decades in dentistry. Rapid prototyping manufacturing (RPM) is one of recent technologies, it is also known as an advanced and resilient fabrication technique as suggested by Mertz et al. (2013). There are continuous attempts to improve them by: amending its speed and accuracy, reducing the system and items cost, and developing more obtainable popular items. RP technologies plays a very important role in dental applications and it is noticeable that using of its models in dentistry will be widened in the future with continuous evolution, as mentioned by Yau et al. (2008). It will be useful in many fields of dentistry such as planning for surgery and designing for prosthodontics as reported by Winder et al, (2005).

REFERENCES

Akhoondali, H., Zoroofi, R. A., & Shirani, G. (2009). Rapid automatic segmentation and visualization of teeth in CT-scan data. *Journal of Applied Sciences (Faisalabad)*, *9*(11), 2031–2044. doi:10.3923/jas.2009.2031.2044

Alamri, H. M., Sadrameli, M., Alshalhoob, M. A., & Alshehri, M. A. (2012). Applications of CBCT in dental practice: A review of the literature. *General Dentistry*, *60*(5), 390–400. PMID:23032226

Arai, Y., Tammisalo, E., Iwai, K., Hashimoto, K., & Shinoda, K. (1999). Development of a compact computed tomographic apparatus for dental use. *Dento Maxillo Facial Radiology*, *28*(4), 245–248. doi:10.1038j.dmfr.4600448 PMID:10455389

Bagchi, P., & Joshi, N. (2012). Role of radiographic evaluation in treatment planning for dental implants: A review. *J Dent Allied Sci*, *1*(1), 21–25. doi:10.4103/2277-4696.159112

Banks, J. (2013). Adding value in additive manufacturing: Researchers in the United Kingdom and Europe look to 3D printing for customization. *IEEE Pulse*, *4*(6), 22–26. doi:10.1109/MPUL.2013.2279617 PMID:24233187

Baratto Filho, F., Zaitter, S., Haragushiku, G. A., de Campos, E. A., Abuabara, A., & Correr, G. M. (2009). Analysis of the internal anatomy of maxillary first molars by using different methods. *Journal of Endodontics*, *35*(3), 337–342. doi:10.1016/j.joen.2008.11.022 PMID:19249591

Barker, T. M., Earwaker, W. J. S., & Lisle, D. A. (1994). Accuracy of stereolithographic models of human anatomy. *Journal of Medical Imaging and Radiation Oncology*, *38*(2), 106–111. PMID:8024501

Bartlett, S. (2013). Printing organs on demand. *The Lancet. Respiratory Medicine*, *1*(9), 684. doi:10.1016/S2213-2600(13)70239-X PMID:24429271

Bertassoni, L. E., Cecconi, M., Manoharan, V., Nikkhah, M., Hjortnaes, J., Cristino, A. L., ... Khademhosseini, A. (2014). Hydrogel bioprinted microchannel networks for vascularization of tissue engineering constructs. *Lab on a Chip*, *14*(13), 2202–2211. doi:10.1039/C4LC00030G PMID:24860845

Bornstein, M. M., Scarfe, W. C., Vaughn, V. M., & Jacobs, R. (2014). Cone beam computed tomography in implant dentistry: A systematic review focusing on guidelines, indications, and radiation dose risks. *International journal of oral & maxillofacial implants*, 29. PMID:24660190

Bors, A. G., Kechagias, L., & Pitas, I. (2002). Binary morphological shape-based interpolation applied to 3-D tooth reconstruction. *IEEE Transactions on Medical Imaging*, *21*(2), 100–108. doi:10.1109/42.993129 PMID:11929098

Bro-Nielsen, M., Larsen, P., & Kreiborg, S. (1996). Virtual teeth: a 3D method for editing and visualizing small structures in CT scans. In Proc. Computer Assisted Radiology (CAR'96) (pp. 921-924). Academic Press.

Buchaillard, S. I., Ong, S. H., Payan, Y., & Foong, K. (2007). 3D statistical models for tooth surface reconstruction. *Computers in Biology and Medicine*, *37*(10), 1461–1471. doi:10.1016/j.compbiomed.2007.01.003 PMID:17336957

Carter, C. N., Pusateri, R. J., Chen, D., Ahmed, A. H., & Farag, A. A. (2010, September). Shape from shading for hybrid surfaces as applied to tooth reconstruction. In *Image Processing (ICIP), 2010 17th IEEE International Conference on* (pp. 4049-4052). IEEE. 10.1109/ICIP.2010.5653040

Chae, M. P., Rozen, W. M., McMenamin, P. G., Findlay, M. W., Spychal, R. T., & Hunter-Smith, D. J. (2015). Emerging applications of bedside 3D printing in plastic surgery. *Frontiers in Surgery, 2*, 25.

Chen, H., & Jain, A. K. (2005, January). Dental biometrics: Alignment and matching of dental radiographs. In Application of Computer Vision, 2005. WACV/MOTIONS'05 Volume 1. Seventh IEEE Workshops on (Vol. 1, pp. 316-321). IEEE.

Cleghorn, B. M., Christie, W. H., & Dong, C. C. S. (2008). Anomalous mandibular premolars: A mandibular first premolar with three roots and a mandibular second premolar with a C-shaped canal system. *International Endodontic Journal, 41*(11), 1005–1014. doi:10.1111/j.1365-2591.2008.01451.x PMID:19133090

Cohen, A., Laviv, A., Berman, P., Nashef, R., & Abu-Tair, J. (2009). Mandibular reconstruction using stereolithographic 3-dimensional printing modeling technology. *Oral Surgery, Oral Medicine, Oral Pathology, Oral Radiology, and Endodontics, 108*(5), 661–666. doi:10.1016/j.tripleo.2009.05.023 PMID:19716728

Cui, X., Boland, T., & D'Lima, D., & Lotz, M. (2012). Thermal inkjet printing in tissue engineering and regenerative medicine. *Recent Patents on Drug Delivery & Formulation, 6*(2), 149–155. doi:10.2174/187221112800672949 PMID:22436025

Cunningham, L. L. Jr, Madsen, M. J., & Peterson, G. (2005). Stereolithographic modeling technology applied to tumor resection. *Journal of Oral and Maxillofacial Surgery, 63*(6), 873–878. doi:10.1016/j.joms.2005.02.027 PMID:15944992

De Vos, W., Casselman, J., & Swennen, G. R. J. (2009). Cone-beam computerized tomography (CBCT) imaging of the oral and maxillofacial region: A systematic review of the literature. *International Journal of Oral and Maxillofacial Surgery, 38*(6), 609–625. doi:10.1016/j.ijom.2009.02.028 PMID:19464146

DelBalso, A. M., Greiner, F. G., & Licata, M. (1994). Role of diagnostic imaging in evaluation of the dental implant patient. *Radiographics, 14*(4), 699–719. doi:10.1148/radiographics.14.4.7938761 PMID:7938761

Dérand, P., & Hirsch, J. M. (2009). Virtual bending of mandibular reconstruction plates using a computer-aided design. *Journal of Oral and Maxillofacial Surgery, 67*(8), 1640–1643. doi:10.1016/j.joms.2009.03.039 PMID:19615575

Dong-ri, S., & Fu-yuan, G. (2010, April). The segmentation algorithm of dental CT images based on fuzzy maximum entropy and region growing. In *Bioinformatics and Biomedical Technology (ICBBT), 2010 International Conference on* (pp. 74-78). IEEE. 10.1109/ICBBT.2010.5479006

Eckardt, A., & Swennen, G. R. (2005). Virtual planning of composite mandibular reconstruction with free fibula bone graft. *The Journal of Craniofacial Surgery, 16*(6), 1137–1140. doi:10.1097/01.scs.0000186306.32042.96 PMID:16327572

Estrela, C., Bueno, M. R., Azevedo, B. C., Azevedo, J. R., & Pécora, J. D. (2008). A new periapical index based on cone beam computed tomography. *Journal of Endodontics*, *34*(11), 1325–1331. doi:10.1016/j.joen.2008.08.013 PMID:18928840

Estrela, C., Bueno, M. R., Leles, C. R., Azevedo, B., & Azevedo, J. R. (2008). Accuracy of cone beam computed tomography and panoramic and periapical radiography for detection of apical periodontitis. *Journal of Endodontics*, *34*(3), 273–279. doi:10.1016/j.joen.2007.11.023 PMID:18291274

Estrela, C., Bueno, M. R., Sousa-Neto, M. D., & Pécora, J. D. (2008). Method for determination of root curvature radius using cone-beam computed tomography images. *Brazilian Dental Journal*, *19*(2), 114–118. doi:10.1590/S0103-64402008000200005 PMID:18568224

Frederiksen, N. L. (1995). Diagnostic imaging in dental implantology. *Oral Surgery, Oral Medicine, Oral Pathology, Oral Radiology, and Endodontology*, *80*(5), 540–554. doi:10.1016/S1079-2104(05)80153-2 PMID:8556464

Fullerton, J. N., Frodsham, G. C., & Day, R. M. (2014). 3D printing for the many, not the few. *Nature Biotechnology*, *32*(11), 1086–1087. doi:10.1038/nbt.3056 PMID:25380438

Gahleitner, A., Watzek, G., & Imhof, H. (2003). Dental CT: Imaging technique, anatomy, and pathologic conditions of the jaws. *European Radiology*, *13*(2), 366–376. PMID:12599003

Gao, J., Xu, W., & Geng, J. (2006). 3D shape reconstruction of teeth by shadow speckle correlation method. *Optics and Lasers in Engineering*, *44*(5), 455–465. doi:10.1016/j.optlaseng.2005.04.013

Gepreel, M. A. H., & Niinomi, M. (2013). Biocompatibility of Ti-alloys for long-term implantation. *Journal of the Mechanical Behavior of Biomedical Materials*, *20*, 407–415. doi:10.1016/j.jmbbm.2012.11.014 PMID:23507261

Grenier, T., Revol-Muller, C., Costes, N., Janier, M., & Gimenez, G. (2005). Automated seeds location for whole body NaF PET segmentation. *IEEE Transactions on Nuclear Science*, *52*(5), 1401–1405. doi:10.1109/TNS.2005.859354

Gross, B. C., Erkal, J. L., Lockwood, S. Y., Chen, C., & Spence, D. M. (2014). *Evaluation of 3D printing and its potential impact on biotechnology and the chemical sciences*. Academic Press.

Hirogaki, Y., Sohmura, T., Satoh, H., Takahashi, J., & Takada, K. (2001). Complete 3-D reconstruction of dental cast shape using perceptual grouping. *IEEE Transactions on Medical Imaging*, *20*(10), 1093–1101. doi:10.1109/42.959306 PMID:11686444

Hoy, M. B. (2013). 3D printing: Making things at the library. *Medical Reference Services Quarterly*, *32*(1), 93–99. doi:10.1080/02763869.2013.749139 PMID:23394423

Jain, A. K., & Chen, H. (2004). Matching of dental X-ray images for human identification. *Pattern Recognition*, *37*(7), 1519–1532. doi:10.1016/j.patcog.2003.12.016

Jayadevappa, B. S., Kodhandarama, G. S., Santosh, S. V., & Rashid, W. T. (2010). Imaging of dental implants. *Journal of Oral Health Research*.

Juergens, P., Krol, Z., Zeilhofer, H. F., Beinemann, J., Schicho, K., Ewers, R., & Klug, C. (2009). Computer simulation and rapid prototyping for the reconstruction of the mandible. *Journal of Oral and Maxillofacial Surgery*, *67*(10), 2167–2170. doi:10.1016/j.joms.2009.04.104 PMID:19761910

Kalra, D., Jain, G., Deoghare, A., & Lambade, P. (2010). Role of imaging in dental implants. *Journal of Indian Academy of Oral Medicine and Radiology*, *22*(1), 34–38. doi:10.5005/jp-journals-10011-1007

Khaled, S. A., Burley, J. C., Alexander, M. R., & Roberts, C. J. (2014). Desktop 3D printing of controlled release pharmaceutical bilayer tablets. *International Journal of Pharmaceutics*, *461*(1-2), 105–111. doi:10.1016/j.ijpharm.2013.11.021 PMID:24280018

Klein, G. T., Lu, Y., & Wang, M. Y. (2013). 3D printing and neurosurgery—ready for prime time? *World Neurosurgery*, *80*(3), 233–235. doi:10.1016/j.wneu.2013.07.009 PMID:23871811

Krol, Z., Chlebiej, M., Zerfass, P., Sader, R., Zeilhofer, H. R., Mikołajczak, P., & Keeve, E. (2002). Surgery planning tools for the osseous grafting treatment. *Biomedizinische Technik/Biomedical Engineering, 47*(s1a), 97-100.

Kumar, S. S., Moni, R. S., & Rajeesh, J. (2011). Automatic segmentation of liver and tumor for CAD of liver. *Journal of Advances in Information Technology, 2*(1), 63-70.

Kumar, V., & Satheesh, K. (2013). Applications of cone beam computed tomography (CBCT) in implant treatment planning. *JSM, 1*(1008).

Li, Y., Yang, C., Zhao, H., Qu, S., Li, X., & Li, Y. (2014). New developments of Ti-based alloys for biomedical applications. *Materials (Basel)*, *7*(3), 1709–1800. doi:10.3390/ma7031709 PMID:28788539

Lingeshwar, D., Dhanasekar, B., & Aparna, I. N. (2010). Diagnostic imaging in implant dentistry. *International Journal of Oral Implantology & Clinical Research*, *1*(3), 147–153.

Lipson, H. (2013). New world of 3-D printing offers" completely new ways of thinking": Q&A with author, engineer, and 3-D printing expert Hod Lipson. *IEEE Pulse*, *4*(6), 12–14. doi:10.1109/MPUL.2013.2279615 PMID:24215725

Lofthag-Hansen, S., Huumonen, S., Gröndahl, K., & Gröndahl, H. G. (2007). Limited cone-beam CT and intraoral radiography for the diagnosis of periapical pathology. *Oral Surgery, Oral Medicine, Oral Pathology, Oral Radiology, and Endodontics*, *103*(1), 114–119. doi:10.1016/j.tripleo.2006.01.001 PMID:17178504

Manisundar, N., Saravanakumar, B., Hemalatha, V. T., Manigandan, T., & Amudhan, A. (2014). Implant imaging-A literature review. *Biosciences Biotechnology Research Asia*, *11*(1), 179–187. doi:10.13005/bbra/1251

Matherne, R. P., Angelopoulos, C., Kulild, J. C., & Tira, D. (2008). Use of cone-beam computed tomography to identify root canal systems in vitro. *Journal of Endodontics*, *34*(1), 87–89. doi:10.1016/j.joen.2007.10.016 PMID:18155501

Mehra, P., Miner, J., D'Innocenzo, R., & Nadershah, M. (2011). Use of 3-d stereolithographic models in oral and maxillofacial surgery. *Journal of Maxillofacial and Oral Surgery*, *10*(1), 6–13. doi:10.100712663-011-0183-3 PMID:22379314

Mertz, L. (2013). Dream it, design it, print it in 3-D: What can 3-D printing do for you? *IEEE Pulse*, *4*(6), 15–21. doi:10.1109/MPUL.2013.2279616 PMID:24233186

Mol, A., & van der Stelt, P. F. (1991). Application of digital image analysis in dental radiography for the description of periapical bone lesions: A preliminary study. *IEEE Transactions on Biomedical Engineering*, *38*(4), 357–359. doi:10.1109/10.133231 PMID:1855798

Mupparapu, M., & Singer, S. R. (2004). Implant imaging for the dentist. *Journal - Canadian Dental Association*, *70*(1), 32–32. PMID:14709253

Nance, R., Tyndall, D., Levin, L. G., & Trope, M. (2000). Identification of root canals in molars by tuned-aperture computed tomography. *International Endodontic Journal*, *33*(4), 392–396. doi:10.1046/j.1365-2591.2000.00330.x PMID:11307216

Nayar, S., Bhuminathan, S., & Bhat, W. M. (2015). Rapid prototyping and stereolithography in dentistry. *Journal of Pharmacy & Bioallied Sciences*, *7*(5Suppl 1), S216. doi:10.4103/0975-7406.155913 PMID:26015715

Omachi, S., Saito, K., Aso, H., Kasahara, S., Yamada, S., & Kimura, K. (2008). Semi-automatic reconstruction of tooth shape from CT images by contour propagation. [in Japanese.]. *IEICE Transactions on Information and Systems*, *J91-D*, 2426–2429.

Omachi, S., Saito, K., Aso, H., Kasahara, S., Yamada, S., & Kimura, K. (2007). Tooth shape reconstruction from CT images using spline curves. *Proceedings of the International Conference on Wavelet Analysis and Pattern Recognition*, 393–6.

Ozbolat, I. T., & Yu, Y. (2013). Bioprinting toward organ fabrication: Challenges and future trends. *IEEE Transactions on Biomedical Engineering*, *60*(3), 691–699. doi:10.1109/TBME.2013.2243912 PMID:23372076

Özen, T., Kamburoğlu, K., Cebeci, A. R. I., Yüksel, S. P., & Paksoy, C. S. (2009). Interpretation of chemically created periapical lesions using 2 different dental cone-beam computerized tomography units, an intraoral digital sensor, and conventional film. *Oral Surgery, Oral Medicine, Oral Pathology, Oral Radiology, and Endodontics*, *107*(3), 426–432. doi:10.1016/j.tripleo.2008.08.017 PMID:18996725

Palomera-P'erez, M. A., Martinez-Perez, M. E., Benitez-P'erez, H., & Ortega-Arjona, J. L. (2010). Parallel multiscale feature extraction and region growing: Application in retinal blood vessel detection. *IEEE Transactions on Information Technology in Biomedicine*, *14*(2), 500–506. doi:10.1109/TITB.2009.2036604 PMID:20007040

Patel, S., & Kumar, N. A. (2009). Implant radiology. *Journal of the Indian Dental Association*, 3.

Pineda, F. (1973). Roentgenographic investigation of the mesiobuccal root of the maxillary first molar. *Oral Surgery, Oral Medicine, and Oral Pathology*, *36*(2), 253–260.

Pratt, W. K. (2007). *Digital image processing* (4th ed.). Los Altos, CA: Wiley-Interscience. doi:10.1002/0470097434

Rahimov, C., & Farzaliyev, I. (2011). Virtual bending of titanium reconstructive plates for mandibular defect bridging: Review of three clinical cases. *Craniomaxillofacial Trauma & Reconstruction*, *4*(04), 223–234. doi:10.1055-0031-1293523 PMID:23205175

Ramamurthy, R., Scheetz, J. P., Clark, S. J., & Farman, A. G. (2006). Effects of imaging system and exposure on accurate detection of the second mesio-buccal canal in maxillary molar teeth. *Oral Surgery, Oral Medicine, Oral Pathology, Oral Radiology, and Endodontics*, *102*(6), 796–802. doi:10.1016/j.tripleo.2006.02.009 PMID:17138184

Roser, S. M., Ramachandra, S., Blair, H., Grist, W., Carlson, G. W., Christensen, A. M., & Steed, M. B. (2010). The accuracy of virtual surgical planning in free fibula mandibular reconstruction: Comparison of planned and final results. *Journal of Oral and Maxillofacial Surgery*, *68*(11), 2824–2832. doi:10.1016/j.joms.2010.06.177 PMID:20828910

Salgueiro, M. I., & Stevens, M. R. (2010). Experience with the use of prebent plates for the reconstruction of mandibular defects. *Craniomaxillofacial Trauma & Reconstruction*, *3*(04), 201–208. doi:10.1055-0030-1268520 PMID:22132258

Scarfe, W. C., Farman, A. G., & Sukovic, P. (2006). Clinical applications of cone-beam computed tomography in dental practice. *Journal - Canadian Dental Association*, *72*(1), 75. PMID:16480609

Schubert, C., Van Langeveld, M. C., & Donoso, L. A. (2014). Innovations in 3D printing: A 3D overview from optics to organs. *The British Journal of Ophthalmology*, *98*(2), 159–161. doi:10.1136/bjophthalmol-2013-304446 PMID:24288392

Siu, A. S. C., Chu, F. C. S., Li, T. K. L., Chow, T. W., & Deng, F. (2010). *Imaging modalities for preoperative assessment in dental implant therapy: an overview. Hong Kong Dental Journal.*

Stavropoulos, A., & Wenzel, A. (2007). Accuracy of cone beam dental CT, intraoral digital and conventional film radiography for the detection of periapical lesions. An ex vivo study in pig jaws. *Clinical Oral Investigations*, *11*(1), 101–106. doi:10.100700784-006-0078-8 PMID:17048029

Tomasi, C., & Manduchi, R. (1998, January). Bilateral filtering for gray and color images. In *Computer Vision, 1998. Sixth International Conference on* (pp. 839-846). IEEE. 10.1109/ICCV.1998.710815

Toro, C., Robiony, M., Costa, F., Zerman, N., & Politi, M. (2007). Feasibility of preoperative planning using anatomical facsimile models for mandibular reconstruction. *Head & Face Medicine*, *3*(1), 5. doi:10.1186/1746-160X-3-5 PMID:17224060

Tyndall, D. A., & Brooks, S. L. (2000). Selection criteria for dental implant site imaging: A position paper of the American Academy of Oral and Maxillofacial Radiology. *Oral Surgery, Oral Medicine, Oral Pathology, Oral Radiology, and Endodontics*, *89*(5), 630–637. doi:10.1067/moe.2000.106336 PMID:10807723

Tyndall, D. A., & Rathore, S. (2008). Cone-beam CT diagnostic applications: Caries, periodontal bone assessment, and endodontic applications. *Dental Clinics*, *52*(4), 825–841. doi:10.1016/j.cden.2008.05.002 PMID:18805231

Ursan, I. D., Chiu, L., & Pierce, A. (2013). Three-dimensional drug printing: A structured review. *Journal of the American Pharmacists Association*, *53*(2), 136–144. doi:10.1331/JAPhA.2013.12217 PMID:23571620

Ventola, C. L. (2014). Medical applications for 3D printing: Current and projected uses. *P&T*, *39*(10), 704. PMID:25336867

Velasco, I., Vahdani, S., & Ramos, H. (2017). Low-cost method for obtaining medical rapid prototyping using desktop 3d printing: A novel technique for mandibular reconstruction planning. *Journal of Clinical and Experimental Dentistry*, *9*(9), e1103. PMID:29075412

Vertucci, F. J. (1984). Root canal anatomy of the human permanent teeth. *Oral Surgery, Oral Medicine, and Oral Pathology*, *58*(5), 589–599. doi:10.1016/0030-4220(84)90085-9 PMID:6595621

Winder, J., & Bibb, R. (2005). Medical rapid prototyping technologies: State of the art and current limitations for application in oral and maxillofacial surgery. *Journal of Oral and Maxillofacial Surgery*, *63*(7), 1006–1015. doi:10.1016/j.joms.2005.03.016 PMID:16003630

Yanagisawa, R., & Omachi, S. (2010, November). Extraction of 3D shape of a tooth from dental CT images with region growing method. In *International Workshop on Computational Forensics* (pp. 68-77). Springer.

Yau, H. T., Lin, Y. K., Tsou, L. S., & Lee, C. Y. (2008). An adaptive region growing method to segment inferior alveolar nerve canal from 3D medical images for dental implant surgery. *Computer-Aided Design and Applications*, *5*(5), 743–752. doi:10.3722/cadaps.2008.743-752

Yovchev, D., Stanimirov, P., & Mihaylova, H. (2014). Lower jaw reconstruction using prototype from cone-beam computed tomography data. *Int Journal of Sciences and Research*, *3*, 57–59.

KEY TERMS AND DEFINITIONS

Fused Deposition Modeling (FDM): It is an additive manufacturing (AM) technology commonly used for modeling, prototyping, and production. It is one of the techniques used for 3D printing. Thus, FDM is also known as a solid-based AM technology.

Implantology: A dental implant (also known as an endosseous implant or fixture) is a surgical component that interfaces with the bone of the jaw or skull to support a dental prosthesis such as a crown, bridge, denture, facial prosthesis or to act as an orthodontic anchor.

Prosthesis: An artificial body part, such as a limb, a heart, or a dental implant.

Rapid Prototyping: Rapid prototyping is the idea of quickly assembling a physical part, piece or model of a product. This is often done using sophisticated computer-aided design or other assembly software, and physically implemented using 3-D printers.

Selective Laser Sintering: It is an additive manufacturing (AM) technique that uses a laser as the power source to sinter powdered material (typically nylon/polyamide), aiming the laser automatically at points in space defined by a 3D model, binding the material together to create a solid structure.

Stereolithography: Stereolithography is a process for making models and parts in which a solid is built up a layer at a time by hardening parts of a container of resin using light. In stereolithography, parts are produced by successive solidification of thin resin layers using a UV laser beam.

This research was previously published in Computational Techniques for Dental Image Analysis edited by K. Kamalanand, B. Thayumanavan, and P. Mannar Jawahar, pages 217-237, copyright year 2019 by Medical Information Science Reference (an imprint of IGI Global).

Section 5

Trends and Future Directions

Chapter 19
The Evolution of Manufacturing Systems:
From Craftsmanship to the Era of Customisation

Dimitris Mourtzis
University of Patras, Greece

Michael Doukas
University of Patras, Greece

ABSTRACT

The transition from local economies to the global competitive landscape and the current fluctuating customer demands drives a never-ending transformation of manufacturing systems. In combination with prominent technological advances and other socio-political reasons of each era, such changes led to the evolution of manufacturing paradigms systems from craftsmanship to automation and decentralisation. This transition created the necessity for new means of designing, planning, and controlling manufacturing companies and networks. In production early stages, manufacturing was viewed as a functional area. On the other hand, current production paradigms are based on the customer focus and view manufacturing as an enterprise. This major diversification led to the gradual reorganisation of manufacturing systems in order to achieve mass customisation of typically mass-produced products. The purpose of this chapter is to specify the evolution of manufacturing paradigms, their basic principles, and the links among them. A discussion on future trends of manufacturing paradigms derives from the review and is finally presented.

DOI: 10.4018/978-1-5225-9624-0.ch019

INTRODUCTION

Manufacturing is among the key driving forces of any modern economy. Specifically, in Europe, in 2010, 34 million people were employed in the EU-27 manufacturing sector, representing 15.9% of the total employment. Indirectly (with related sectors and activities), manufacturing ac-counts for close to 50% of the European economy (Eurostat, 2011).

During the last decades, local economies have evolved into global and highly competitive players. Industries started to operate on an intercontinental basis, expanding the limits of their business. The export of finished goods to foreign markets has been the dominant theme in the international trade up to the 1990s, and gained even more attention the last decade. Moreover, location-specific factors such as low-cost labour and highly-skilled personnel enabled the rapid globalisation. Enterprises started to seek fertile production environments into developing or developed countries (Abele et al., 2006). In addition, the advent of the Internet and the increasing computational power led to the development of virtual entities (Chryssolouris, 2005) and transformed competition and collaboration practices (Lazonick, 1991 and Mourtzis, 2011).

Historically, manufacturing paradigms, driven by the change of the environment in which they operate, change in character and evolve in patterns over time. The various patterns witnessed up to now can be roughly correlated to movements between three stages: (i) craft shops that employ skilled artisans, (ii) long-linked industrial systems using rigid automation and (iii) post-industrial enterprises characterised by flexible resources and information intensive intellectual work (Gunasekaran and Ngai, 2012).

METHODOLOGY OF THIS REVIEW CHAPTER

For the preparation of this review chapter, a search was performed between for academic journals, articles and books primarily in the Scopus and Google Scholar databases using as main keywords the fields of interest. These keywords include: production paradigm, evolution, craft production, mass production, lean production, just in time, Kanban, mass customization/customisation, additive manufacturing, sustainable production, rapid manufacturing, personalized/personalised production, personalization/personalization, global manufacturing networks, cluster concepts and agile manufacturing.

This literature review considered a number of constraining criteria. The first one was the time span of the journals or articles, which was set from 1985 to 2013, with only minor notable exceptions. This constraint was not considered for books. In Scopus, additional limitations were imposed. Journals from the scientific areas of: materials of science, chemistry, chemical engineering science, physics, astronomy, agricultural sciences, earth and planet sciences, medicine, toxicology, biological sciences and other referring to health professions were excluded from the results. Sciences that were included were: engineering, management, business, economic, mathematic and environmental.

STRUCTURE OF THE CHAPTER

The rest of the chapter is structured in the following way. Section 2 contains the evolution of manufacturing systems from the era prior to the Industrial Revolution up to currently dominant paradigms of mass customisation and personalisation. Section 3 includes a comparison of the different manufacturing

Table 1. Evolution of manufacturing paradigms (Adapted from Mourtzis & Doukas, 2012)

Production Paradigms	Time Span	Manufacturing System	Business Model	Workforce
Production until Industrial Revolution (Koren, 2009)	600 – 1800	General Purpose Tools	Pull	Artisans
Craft Production (Hu et al., 2011).	1850 – 1900	General Purpose Machine Tools	Pull	Specialists & Non-specialists
American Production (Hounshell, 1984 and Gilmore and Pine, 1997)	1900 - 1920	Dedicated Manufacturing Line (DML)	Push	Highly skilled craftsmen & unskilled operators
Mass Production (Ford, 2012)	1913 – 1955+	Dedicated Manufacturing Line (DML)	Push	Relatively unskilled workers
Lean Production (Womack et al., 1990)	1955 – 1980+	Dedicated Manufacturing Line (DML)	Push	Highly motivated, loyal and skilled workers
Mass Customisation (Pine, 1992)	1980 – 2000+	Flexible Manufacturing System (FMS)	Push – Pull	Actively engaged loyal workforce to handle CNC, FMS, CAD, CAM, CIM, EDI technologies (Ulrich et al., 2003)
Global Manufacturing (Gunasekaran and Ngai, 2012)	2000 – 2008+	Reconfigurable Man. System (RMS) & On Demand Manufacturing Systems (DMS)	Pull	

paradigms based on the technologies used in each one, economic perspectives and criteria of social nature, such as the consideration of the human factor. Section 4 concludes the chapter and outlines future directions for manufacturing paradigms.

EVOLUTION OF MANUFACTURING PARADIGMS

The manufacturing industry, since its birth two centuries ago, has evolved through a number of paradigms. The most prevailing manufacturing paradigms that characterised significant periods of time are chronologically: Craft Production, American Production, Mass Production, Lean Production, Mass Customisation and Global Manufacturing. Each paradigm had its time-span, was supported by a specific type of manufacturing system and operated based on specific business model (Table 1).

The establishment of the aforementioned manufacturing paradigms took place to better satisfy particular demands and requests of their respective era. In the beginning, craft production focused on creating high quality products in order to satisfy customer demands but at very high prices and without any standardisation (Koren, 2009 and Ford, 2012). As a result, in the early 1910's "Mass Production" offered the capability for low cost manufacturing of large quantities of goods with limited variety, which was enabled by dedicated assembly systems (Ford, 2012). The evolution in technology systems offered the possibility to create more and more products and led to the birth of new manufacturing paradigms. A little later in Japan, in order to refine mass production practices and make them applicable to the Japanese mentality, Toyota focused on reducing wasteful activities, with utmost goal the reduction of product cost, improvement of product quality and improvement of the customer perceived product value. This led to the birth of a new manufacturing system, namely the Lean Production (Womack et al., 1990). A few years later, in the 1980's a new paradigm arose in response to consumer demands for low cost and high quality products that are available in high variety. Mass Customisation (Pine, 1992) was an innovative paradigm because it offered to the customers the capability to select among a variety of features and accessories, in order to shape a final customised assembly combination of a basic product (Hu et al., 2011). As a result, new technology systems and assembly lines were required to achieve such an exploding product variety. The evolution in technology (Chryssolouris et al., 2008), the rapid development of the Internet services and globalisation created very demanding customers that requested unique, high quality, low cost products, fitted to their personal needs and preferences (Hu et al., 2011). As a result, novel decentralised manufacturing approaches emerged and partially replaced the traditional centralised structures (Leitao, 2009). Consequently, at the dawn of the 21st century, a new manufacturing paradigm arose, namely the globalised manufacturing paradigm. It is obvious from Table 1 that changes in the market landscape drives the changes in manufacturing systems, which must adapt in order to cope with the demands and maximize their efficiency.

Craft Production

A phrase that best characterizes this production paradigm could be "Skilled workers, using general-purpose machines, making exactly what the customer paid for; one product at a time." (Koren, 2009). In this production paradigm every unique product was constructed separately in a small machine shop, supported by limited technological aids; the shops had only general purpose machine tools. However, even if with these simple machine tools (lathes, drills and milling machines) the highly skilled work force was able to produce high quality products. It must be mentioned that the same work force was able to do the bodywork and to assembly the products. The organisations were extremely decentralised, although concentrated in a single city. Most of the parts for a product were coming from small machine shops. The system was coordinated by an entrepreneur in direct contact with everyone involved (customers, employees and suppliers). The production volume was obviously very low, while the production costs were too high and even though the volume could be increased, the costs remained the same. The customer had to pay in advance. Moreover, each product was a prototype so reliability and consistency were elusive traits. Finally, one of the main characteristics of craft production is the pull-type business model that can be described by the sequence: Sell – Design – Make. Each part of a product was produced separately in a small machine shop that had general-purpose machine tools. Highly skilled workers that knew how to operate machine tools (such as lathes, drills, and milling machines) took great pains to produce precision parts, especially the complex mechanism like engines and drive trains. These workers also did all the bodywork and assembly. Figure 1 depicts the structure and operation of a craft production system. It must be noted here that the method used for the construction of every production paradigm structure is the IDEF0 method (IDEF0, 2010).

Application

The Craft Production paradigm is still applied today to a very limited and exclusive market, in production of sports cars, really unique and traditional product, with the exception that advanced tools and software infrastructure used are present. In the 1880s craft methods for car industry were used especially in France, but by the end of nineteenth century craft car industry production flourished in the whole Europe. Some years later there were hundreds of craft modelled companies in Western Europe and in the United States.

Figure 1. Structure of Craft Production

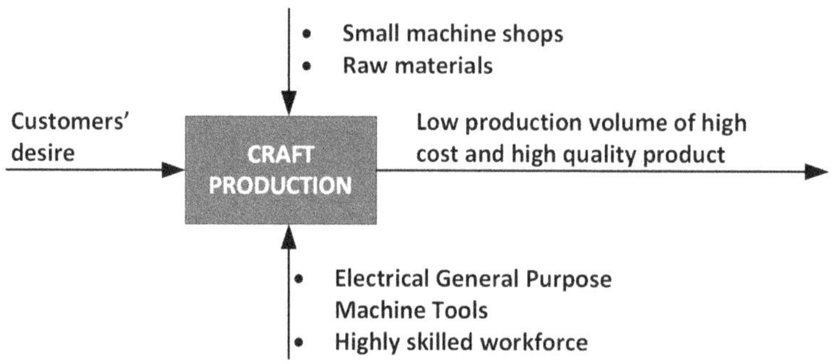

Benefits and Drawbacks

The major benefits and drawbacks that characterise Craft Production are summarised in Table 2.

American Production

The American system had its roots in Craft Production. The main focus was given on the process of production rather than on individual products. The dependence on worker skill, know-how, and flexibility extended to machine tool industry. The successes of companies due to the skill of their craftsmen were based on their skills, flexibility and ability to apply what they learned to a far-ranging set of problems. The interchangeable parts simplified the production process and saved large amounts of labour. Furthermore, they facilitated the effective repair and maintenance of products. It was necessary to use machines that would process parts of products with tight tolerances in order to eliminate hand fitting. To achieve exact interchangeability, systems of gauges and fixtures were developed as key components of the new machinery. This helped in the quick detection of any discrepancy of the monitor machines (Hounshell, 1984).

The first years of the American system, the first practitioners built their machines in-house. However, they soon joined the machine tool industry because they could focus more in their own production problems, avoid the costs of machine development and share the technological advancements made by machine tool suppliers. Company owners focused on the entire organisation rather than only providing the isolated craftsmen what they needed to manufacture. As a result, a functional organisation of the

Table 2. Benefits and drawbacks of the Craft Production paradigm

Benefits	Drawbacks
High product variety	Very low production volume
High quality products	Very high production costs
Assembly of the final product in the small machine shop by the same work force	Low reliability and consistency due to prototype products
Highly skilled work force	No supply network

factory was developed where materials and semi-finished products were moved in batches. Along with this development, supervision and worker accountability appeared in the working environment.

This development would quite naturally lead to greater efficiency. Even if the craftsmen did not want to lose their skills and their responsibility, this approach led to increased efficiency and productivity. Machines, even if they were being made to automate work, skilled craftsmen were required to control them. That was an effort to improve their skills and not to be replaced. Two kinds of workers were developed: (i) those who built, maintained, set up and improved machines and (ii) unskilled operators who operated machines (Doll & Vonderembse, 1991). The American businessmen were innovators, inventors and imitators. They were motivated by the idea that there is the expectation an improvement to be found in everything, as an anonymous American stated in 1832.

For the first time, an extensive use of interchangeable parts and mechanised means to produce them was observed, which resulted in more efficient use of labour compared to hand methods Figure 2. The American System was based on semi-skilled human labour utilising machine tools and jigs to produce standardized interchangeable parts. Tolerances were incorporated so that products could be assembled with a minimum of time and skill, requiring little to no fitting.

Application

This system was applied in the firearms industry and was supported by the U.S. government. Probably the most indicative example of purely American production was the Springfield armoury in Massachusetts (Uselding, 1971).

Benefits and Drawbacks

The major benefits and drawbacks that characterise American Production are summarised in Table 3.

Figure 2. Structure of American production

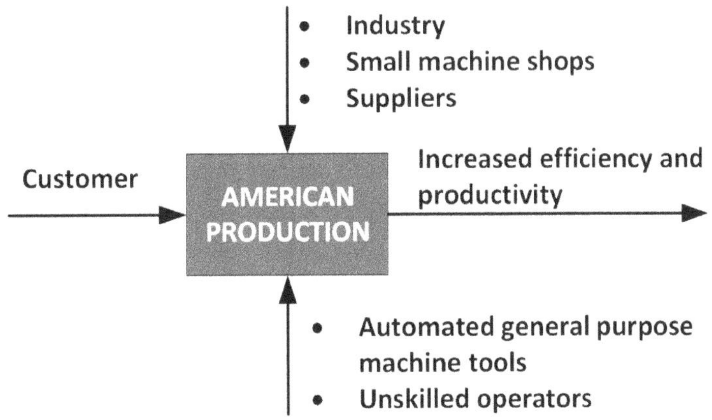

Table 3. Benefits and drawbacks of the American Production paradigm

Benefits	Drawbacks
Highly skilled craftsmen	High costs of machine developments
Interchangeable parts	Without supply network, no sharing of the technological developments made by suppliers
Effective repair and maintenance of products	Automated machines led to unskilled operators
Greater efficiency and productivity	Not focus on lowering costs but in higher output

Mass Production

This production paradigm is generally linked with the invention of the automobile moving assembly line in Detroit, Michigan, by Henry Ford in 1913. The inability of the small independent shops to develop new technologies and the lack in resources to pursue fundamental innovations led Henry Ford to find a way to overcome these problems of Craft production. Henry Ford in the first years of Highland Park in Detroit was an assembler. By 1915 he had stopped buying the components for a product from others and he gathered these functions in-house. He was well on his way to achieve vertical integration. He managed to achieve his target by 1931. He had perfected mass production techniques before his suppliers and managed to achieve substantial cost savings by doing everything in-house due to his difficulty to trust anyone else. With that new approach, Ford could lower significantly the costs and increase product quality.

Mass production began in the early 1910s and managed to prosper for a major part of the 20th century. It must be mentioned that the main characteristics of mass production are exactly the opposite of those of craft production; mass production targets high volume per product and at the same time low product variety. A significant factor for the success of mass production is the availability of high quality interchangeable parts. It resulted in the reduction of production costs and offered the possibility to low skilled workers to be a significant part of the assembly of the car industry. In Mass production, the products are initially designed by an Original Equipment Manufacturer (OEM) and then they are constructed with the hope that there always are enough customers to buy them (Figure 3). This kind of business model is a pure push type and is characterised by the sequence: Design – Make – Sell (Koren, 2009). The one and significant penalty with this system was inflexibility, because changing the dedicated machines to do a new task was time-consuming and costly.

Application

The basis of mass production was the Henry Ford's factory in Detroit specialised in automobile industry. During the Second World War, US industry was remarkably successful in converting its production system to military applications. Industries also managed to join forces with scientific researchers to perfect rapidly relatively new products such as airplanes and radars and to produce them in large volumes (Hayes et al., 1988). Although Mass production is an old paradigm it is still used today. More specifically, with new compact, robust and flexible CNC controllers nowadays all the requirements from job shop manufacturing to mass production could be met.

Figure 3. Structure of Mass Production

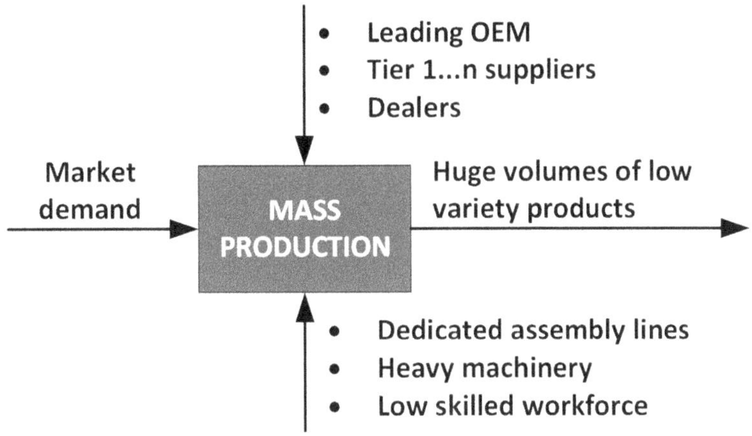

Benefits and Drawbacks

The major benefits and drawbacks that characterise Mass Production in Table 4.

Lean Production

After a visit to Ford's Rouge Plant in Detroit, Eiji Toyota and his production genius, Taijchi Ohno, concluded that mass production could never work in Japan (Hines et al., 2004). This was the beginning of the Toyota Production System and ultimately its generalization through the inclusion of additional techniques and methods, under the umbrella of Lean Production (Figure 4). The term "lean" was coined by Krafcik (Krafcik, 1988) and has only been prevalent since the 1990s (Thinking Lean, 2010). Figure 5 highlights the key events of Lean Production from its birth till today. This system is regarded as a route to the world class manufacturing (Womack et al., 1990). One of the most important publications on the field that also tries to define a common definition of lean is (Shah and Ward, 2007). According to that research work, "Lean production is an integrated socio-technical system, whose main objective is to eliminate waste by concurrently reducing or minimizing supplier, customer, and internal variability."

Many studies of lean production, sociotechnical and tayloristic production systems have been reported in literature (Womack and Jones, 2003, Niepcel and Molleman, 1998 and Lin et al., 2010). Most case studies compare the most recent methodologies, LP and STS, and comments are made in relation to

Table 4. Benefits and drawbacks of the Mass Production paradigm

Benefits	Drawbacks
High volume per product	Very limited product variety
High quality interchangeable parts	Inflexibility
Low-skilled workers become a significant part of the assembly of the product	Many constructed products without knowledge if there will be possible customers
Standardisation	No concern about customer's preferences

Taylorism. A comparison among these three production systems has been done in (Rask and Johansson, 2008). The analysis is depicting the linkages between lean and other systems. However, the study was limited to basic descriptions and principles for its methodologies.

There exist two frequent approaches towards interpreting and implementing lean manufacturing. The focal principles of the first approach are to concentrate on value creation by identification and steady elimination of waste and the perfection by continuous improvement. Standardised operations are been performed with minimal variance due to the focus on steady flow in the production (Rask and Johansson, 2008). Moreover, in this production approach, the workers, even if they are in the lowest level, are encouraged to contribute to the production and the entire assembly line can be stopped by one of them in case they observe a quality or throughput issue. The second approach focuses on improving the material flow through improved system layout and structuring by using exact methods and techniques such as Group Technology, Production Flow Analysis, cellular manufacturing, Single Minute Exchange of Dies (SMED), quick changeover, Total Preventive Maintenance, Value Stream Mapping and other. The two approaches are fundamentally different and applicable in different manufacturing contexts. However, the second approach entails the advantage that it inherently considers a system-wide perspective. The second approach is mainly promoted throughout this book.

A main component of Lean Production can be regarded the Just In Time (JIT) system, which refers to the timing of production flow. The goods are delivered to the manufacturing lines just in time to be used, in the immediately needed quantities and just to the production processes that need them. Goods need to arrive within minutes of their use in the production line. In this way the elimination of waste can be achieved. In a first approach, the JIT system seems simple, but in fact it is really complicated and many of the things that characterize it cannot be understood until they have been tried out in the factory (Hirano, 2009).

Some of the basic principles and techniques of Lean Production are discussed next:

Flow Manufacturing: It requires the whenever possible elimination of pile – ups. The target is every good to move through the chain of processes in order to be correctly processed within the cycle time.

Multi–Process Handling: When a single worker can move with the work pieces from process to process until the work piece processing is finished. It can be achieved when machines that make up an entire sequence of processes are grouped together.

Figure 4. Structure of Lean Production

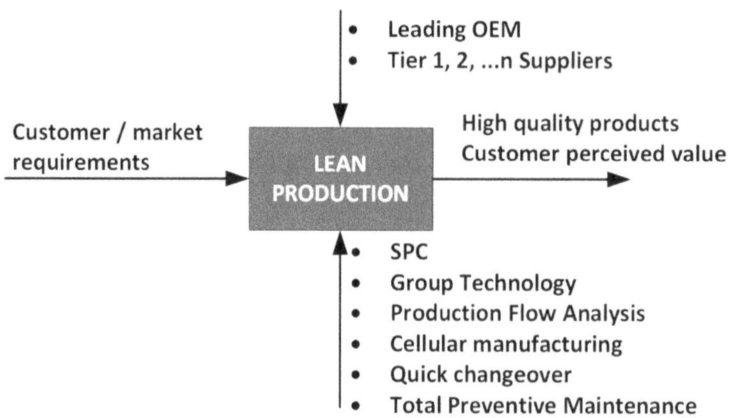

Kanban: The JIT philosophy began to attract significant attention from the West in the late 1970s under the label of the Kanban system. It was rather misleading because Kanban is only one part of the total JIT system (Lai and Cheng, 2009). The Kanban system is one major reason why the JIT production system gained such an attention. In Kanban system signs are attached to goods and replace vouchers as the medium for giving operating instructions and production orders. This type is useless for factories that use the conventional type of production method. The factory has to target firstly to switch over to flow manufacturing. The Kanban system is a method of control designed to maximize the potential of the Toyota production system as well as a system with its own independent functions (Shigeo and Andrew, 1989).

Manpower Reduction: The JIT production system organizes production using the minimum number of workers required to meet the demand of the next process.

Visual Control: A method that aims and makes the failures or the factory floor problems visible when no one can easily spot them. There is a variety of such devices that can make the production line problems more visible.

Levelling: In mass production era the lots were processed and then warehoused. The idea of levelling calls for product types and volumes to be spread out to produce as level a production flow as possible. As a result, levelling is fundamental both in JIT production and in flow manufacturing.

Changeover: The goal of changeover is to shorten the time needed for operations like the revision of standards and the replacement of assembly parts and other materials. More reductions have to be done in labour-hour requirements in order to build a strong, flexible and adaptive manufacturing line.

Quality Assurance: Quality is not something that just happens when there exists a good production equipment. Having equipment operators work more cautiously, does not necessarily reduce the number of detective products. Quality assurance requires a comprehensive approach that addresses all production factors (e.g. people, goods, and production equipment and production methods).

Standard Operations: Operations that have been developed to preserve an effective combination of people, goods and machines to produce high quality products in an economic, quick and safe way.

Maintenance and Safety: In JIT production system high production capacity is achieved and in the same time emphasis is placed on safety in order to prevent breakdowns and accidents.

Jidoka: To bring humans into the automation process to ensure reliability, flexibility and precision.

Specific types of industries may benefit from a combination of agile and lean practices in their organisation. This have been stressed in (Prince and Kay, 2003) through an extensive case study in a cable manufacturing company. A decoupling point in the production procedure between the application of lean and agile production should be easily adaptable according to the demand. The complementarity between the two practices will lead to improved overall performance and profitability of the factory. Similarly, companies that apply delayed differentiation in their products, such as PC suppliers can benefit by combining lean and agile manufacturing paradigms. This is based on the fact that agility is required for supporting a highly fluctuating demand (Naylor et al., 1999).

Application

The basis of Lean Production was in Nagoya, Japan. Toyota had fully worked out the principles of Lean Production by 1960s. That period Japanese firms had a huge advantage over the mass-producers. In the 1980s the world's economy was at the same point in the diffusion of LP that it was with the mass production in the 1920s. Gradually, other Japanese firms adapted their paradigm. Until the oil crisis in 1973,

Mazda had not fully adapted Ohno's ideas for the organisation of the factory and the supplier system. The model Toyota followed was that it held an equity stake of its suppliers, and they had a small equity stake in Toyota. Soon, the Toyota industrial group exhibited some of the circular equity structure seen in the keiretsu. Toyota held a strong position at the centre (Lewis, 2000).

The implementation of Lean Production should be structured according to the organisational context of the company and specifically based on plant size, unionisation and plant age. Friction is generated when lean practices are imposed on a unionised work force. Similarly, older plants are often employing older workforce that are resistant to change their work practice. Finally, large plants are more likely to possess the resources required for implementing lean practices (Shah and Ward, 2003).

Moreover, the linkage between environmentally friendly manufacturing and lean practices is obvious. Reducing waste, as one of the main notions behind lean manufacturing, perfectly suits the goal towards energy consumption reduction and greener strategies' development (McKinsey, 2007). In addition, a correlation between quality standards adoption and environmental management standards was identified from a large scale investigation, as reported by King and Lenox (2001).

Benefits and Drawbacks

The major benefits and drawbacks that characterise Lean Production are presented in Table 5.

There is an important drawback in the way that LP is designed. The workers, in order to be more productive must have opportunities for career advancements. In LP system there is a lack in variety of challenges. In this way they feel that they are honing their skills and are valued for many kinds of expertise they have attained. As a result they feel that they have reached a dead end at an early point in their career. Also, the main advantage of lean production disappears since the disappointed workers hold back their know-how and their commitment.

Total Quality Control

As Hirano says "JIT improvement is not a theory or a technique". It is a discipline, or more sophisticated a form of art. The principles that characterize it are based on faith, on accumulated experience, wisdom and gut feelings. JIT improvement is a process that actually is difficult to be learned from books or to be taught. This movement was spread not only in the manufacturing but in hospitals and government offices too. It seemed to have boundless limits and applications. The basic principle of TQC was that

Table 5. Benefits and drawbacks of the Lean Production paradigm

Benefits	Drawbacks
Concentration of value	Lack of variety of challenges
Perfection by continuous improvement	Disappointed workers hold back their know-how
Workers encouragement to contribute	Slow ramp-up until lean practices are perfected
Considerable decrease in Work In Progress (WIP)	
Significant reduction in Lead Times	
Improved quality	
Elimination of waste	

there is nothing at all wrong with asking everyone in the company to brainstorm solutions to problems and make improvements. Moreover, the results were not very satisfactory because people were aiming their improvement efforts towards impressive announcement meetings rather than substantial changes in workplace. On the other hand, JIT improvement eliminates all emphasis on formalities and announcement meetings. As result, factories that adapted TQC became full of posters and nothing got improved (Hirano, 2009).

Seven Basic Tools of Quality

The Seven Basic Tools of Quality is a designation given to a fixed set of graphical techniques identified as being most helpful in troubleshooting issues related to quality (Montgomery, 2012). They are called basic because they are suitable for people with little formal training in statistics and because they can be used to solve the vast majority of quality-related issues (Kaoru, 1985). The Seven Basic Tools of Quality include: Ishikawa (fishbone) diagram, Check sheet, Control chart, Histogram, Pareto chart, Scatter diagram and Flowchart.

Human Resource Management

HRM has been defined "as a set of politics designed to maximize organisational integration, employee commitment, flexibility and the quality of work" and hard and soft versions have been identified (Guest, 1987). Soft HRM is a method of releasing untapped reserves of human resourcefulness any increasing employee commitment, participation and involvement and has greater emphasis on human relations. Hard HRM is designed to maximize the economic return from labour resources by integrating HRM into business strategy. Hard HRM has a broader engineering base and it is strongly oriented towards meeting market requirements by means of greater production flexibility and product improvement (Appelbaum and Batt, 1994). The HRM has a clear affinity with Taylor's vision of scientific management. That vision includes continuous improvement in quality and performance, JIT inventory systems and statistical process control design. The purpose of HRM is to foster a pre-emptive rather than reactive approach to operational efficiency, quality control and innovation by shifting responsibility and accountability for decision making towards the shop floor.

High Performance Manufacturing (HPM)

Total quality management, just-in-time, world class manufacturing, lean production and concurrent engineering, broadly labelled as high performance manufacturing (HPM) provide companies with a wide range of practices and approaches aimed at achieving higher performance levels (Flynn et al., 1997; Filippini et al., 1998). Moreover the availability of instruments for analysing and mapping the communication requirements induced by HPM practices is a prerequisite for taking advantage of the tremendous power of the developments of IT, such as enterprise resource planning (ERP) and product data management (PDM) systems (Jablin et al., 1987).

Figure 5. Historical evolution of Lean Production

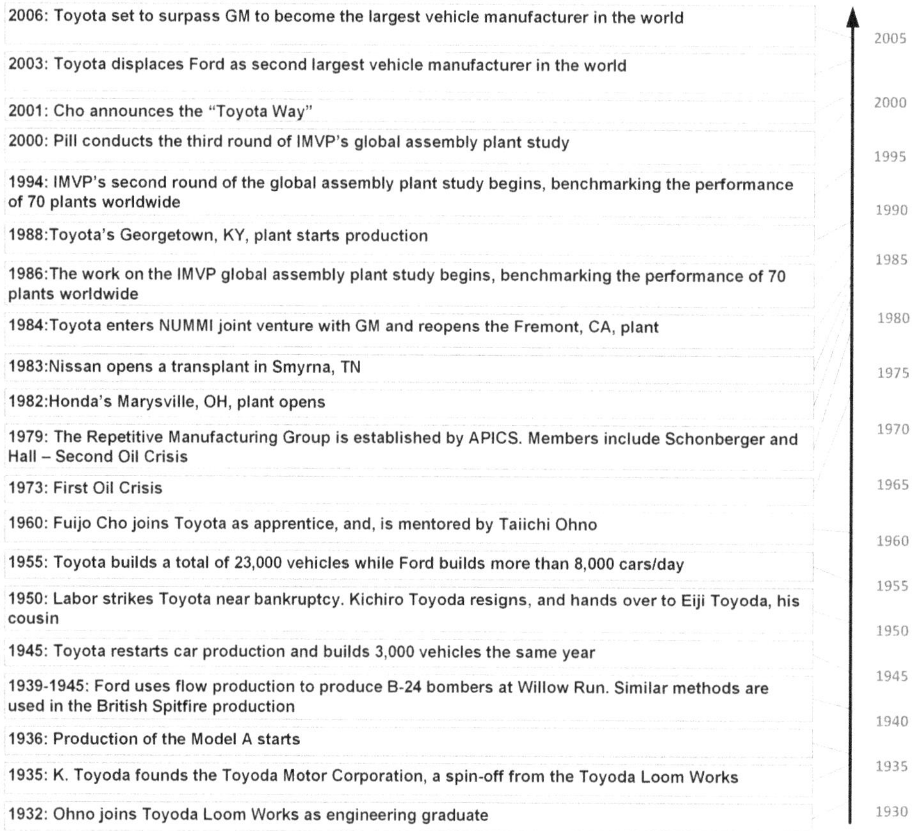

Mass Customisation

Mass Customisation emerged as a new production paradigm in the late 1980s in order to respond to consumer demands for product variety and appears as an alternative to differentiate companies in a highly competitive and segmented market (Pine, 1992; Da Silveira et al., 2001). Manufactures began to offer larger variations of their standard product (Mourtzis et al., 2008). The designing procedure was cost effective because the basic product is re-designed and consumers are allowed to select the assembly combination they preferred most. In this way, economy of scale could be achieved at the component level, whereas economy of scope of the final product could be achieved by reconfigurable assembly systems.

This production paradigm aims at producing goods with services catering to individual customer's selections with near mass production volume-efficiency. The goal of mass customisation is to create more selective consumers. Today's customer is a far cry from the compliant customer of the past, who would be satisfied with a single priority. He/she wants it all: a product with the highest quality, fastest delivery, and highest level of product customisation, and yet is unwilling to pay the commensurate price (Kumar, 2007). Based on the definition of mass customisation, Koren supports that "A wider product variety increases demand and sales" (Koren, 2009). It seeks to enable business to exploit market trends for greater product variety and individualisation (McCarthy, 2004). The tools of mass customisation can substantially enable product personalisation (Montreuil and Poulin, 2005). Finally, faced with an objective for efficient, flexible, multi-product manufacture, a need for rapid and high quality changeovers becomes paramount (McIntosh et al., 2010). In any case, MC can be considered as a systemic idea involving all aspects of product sale, development, production and delivery, spanning from the customer option up to receiving the finished product (Jiao et al., 1998). In the literature, four business strategies for mass customisation can be identified as presented in the following sections..

Strategy 1: Off-the-Shelf Product Variety

In this strategy, customers are allowed to select from a variety of products whatever best fits their needs (Figure 6). This is a push implementation of the mass customisation business model. The two major advantages of this approach are: (i) the price is lower than more highly customised products and (ii) the delivery lag time is zero since the whole range is offered at the same time and place.

Strategy 2: Intense Options of Products

Intense product options is the basic strategy implemented in the mass customisation approach. In this strategy, the customer orders a product with personalised characteristics and the product is constructed afterwards. This is also a push business model however; every product is produced after the order so it is an obvious pull model too (Figure 7).

Strategy 3: Point-of-Delivery Personalisation

This strategy pulls the type of accessories that are offered from the customer's desires. Then, the manufacturer pushes the product with these accessory offers. It is a pull – push business model. Products of this strategy are the second more expensive among the four strategies (Figure 8).

Figure 6. Off-the-shelf product variety

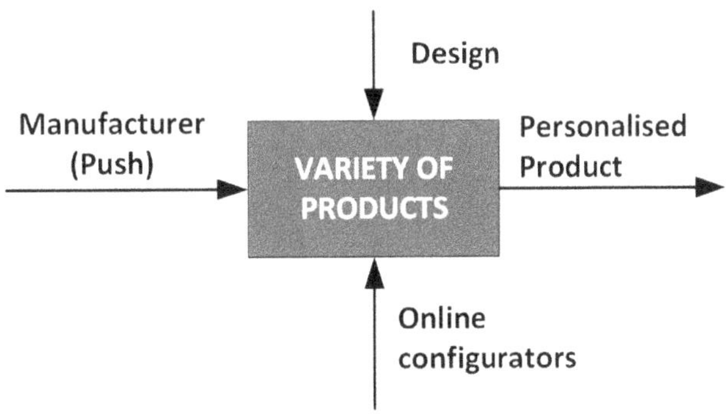

Figure 7. Intense options of products

Figure 8. Point-of-delivery personalisation

Strategy 4: Personalised Products

In this strategy the buyer designs the product option set. The products in this strategy are differentiated during their manufacturing process. Flexible and reconfigurable manufacturing systems are required for these products. Finally, these products are the most expensive than any of the previous strategies. It is the pinnacle of the pull business model (Figure 9).

According to the following production model, demand can always grow if new variants of the product family are offered. As a result there will be more customers to whom new variants will offer a better fit to their needs. On the other hand the production model for mass production constitutes a volume-based loop. Contrarily, mass customisation loop is a variety-based loop. In other words it could be said that MC loop is based on economies of scope. The general structure of a mass customisation paradigm can be created as depicted in Figure 10.

Figure 9. Personalised products

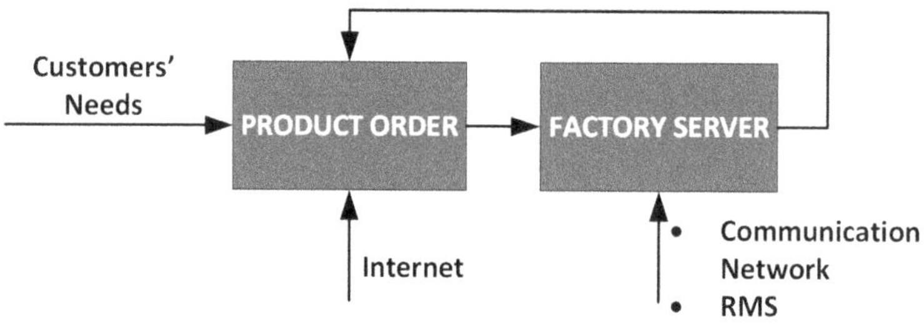

Figure 10. Structure of Mass Customisation paradigm

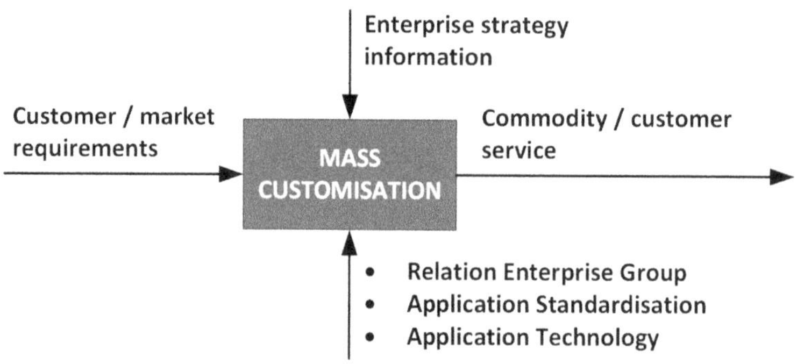

Application

Koren et al., (1996) depicted that from the early 70s to the late 90s there is an impressive increase in the variety of products that are used in the daily life. As a result, consumers demand a variety of products that fit their needs and preferences. The first implementation of the mass customisation (MC) paradigm started in the 1980s. In contrast with the mass production, mass customization focuses on relatively low volume per product [6]. The basic idea behind this approach is not revolutionary, since customers have always been able to customize certain mass – produced products by choosing from predefined modules. Nowadays, the blossom of the MC paradigm is obvious, since there is a large variety of products from heterogeneous industrial/market domains that are made with MC techniques. Successful applications of MC have been reported in literature. High – visibility studies have covered sectors including the food industry (McIntosh et al., 2010), electronics (Partnaten and Haapasalo, 2004), large engineered products (Lu et al., 2009), mobile phones (Comstock et al., 2004), personalised nutrition (Boland. 2008), automotive industry (Mourtzis and Doukas, 2012 and Mourtzis et al., 2012) and CNC machine building industry (Mourtzis et al., 2013).

Benefits and Drawbacks

The major benefits and drawbacks that characterise Mass Customisation are depicted in Table 6.

Global Manufacturing and Personalisation

The last decade of the 20th century can be characterised by many significant changes in terms of enterprise strategy and manufacturing paradigms in manufacturing community. Customers demand high quality products that can also respond to their specific preferences and requirements (Vastag & Kasarda, 1994). In order to cope with the ever – changing customer requirements, the manufacturing enterprises should have the capability to produce a wide variety of essentially the same product (Eastwood, 1994). Customer demands and market situations are crucial factors that characterize the manufacturing strategies. The answer to meet these demands is the global manufacturing (GM) network structure. These networks link partnering companies from different parts of the world. Their united goal in a collaborative way can be achieved throughout research, development, design, marketing and line production (Bidanda et al., 1993). A number of global networks (of companies) have been developed and it can be observed that most of them are mainly designed to provide services to the customers such as product selection, electronic payment and customer service feedback (Tso et al., 2000). Shi et al. (1997) suggested that the

Table 6. Benefits and drawbacks of the Lean Production paradigm

Benefits	Drawbacks
Product mix is high and changes very often	Variety leads to increased product development cost
Convertibility	Final assembly workers make more mistakes if there is a variation
Scalability	Not dedicated machining but flexible equipment needed
Smaller volume compared to mass production	More skilled workers are required
	Complex disassembly of the product at the end of lifetime

framework of international manufacturing network configurations can be classified into four maturity levels with progressive levels of "configuration complexity" and proposed a practical self – assessment tool to support GM strategy formulation. A number of conceptual frameworks and formulation strategies related to GM systems have been proposed (Chung et al., 1997 and Link et al., 1997). In this case study a technique is suggested for companies to assess their maturity levels terms of global configuration and subsequently formulate their GM strategy. Moreover, a network configuration for the Global Coordination Network has been examined in many examples (Flaherty, 1986; Maruca, 1994). A Virtual Global manufacturing architecture is been proposed by Zhang and Chen (Zhang & Chen, 1997). This architecture is designed to be technically responsive and proactive in accepting, validating and implementing change. They also suggest a new decentralised manufacturing system, named "AMI", which is claimed to be particularly suitable for developing VGM networks. Karlsson and Ahlstrom (Womack and Jones, 1994) propose a framework for building a "lean" global network. The concept that this network was based on, had previously been researched by Womack and Jones (Bose, 1996). Apart from research work related to GM systems there are also studies in the adoption of intelligent agents for enhancing operations in general areas. The research in the agent – based systems has been reported in a number of publications (Kasanoff, 2009). To make personalized production a cost effective reality, several enabling technologies must be developed. These technologies include: Methods and tools for understanding and capturing consumers' needs and preferences, designing by non-designers, cyber systems for collaboration (Mourtzis & Doukas, 2012), on-demand manufacturing systems, assembly system. Moreover, Tam and Ho (2006) defined three types of personalisation: (i) User-driven personalisation: when the user specifies in advance the desired web layout and content that matches her interests and preferences with the tools and options provided, (ii) Transaction-driven personalisation, an online merchant generates the personalised layout and content, and thus personalisation is driven by previous transactions, and (iii) Context-driven personalisation employs an adaptive mechanism to personalize content and layout for each individual user based on the context and inference of user's processing objectives in real time (e.g. product inspection versus random browsing).

A global manufacturing network involves a complex number of participants that include enterprises (Figure 11), organisations and institutions covering several countries or continents. A simple structure example of the structure of this type of network is presented (Monroy and Arto, 2010).

Figure 11. Structure of personalisation paradigm

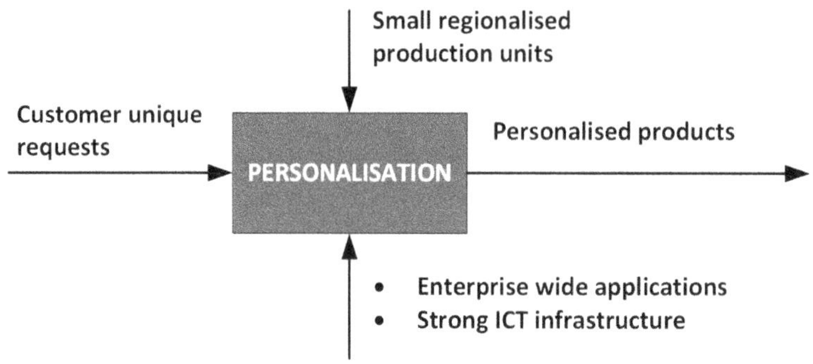

Application

True product personalisation can be described as producing according to one individual's requests. Therefore, apart from very specialised applications of high cost products, the massive application of personalisation is for now very restricted to high cost product, due to process, production and manufacturing network constraints.

Benefits and Drawbacks

Table 7 is a comparison of the various manufacturing paradigms that are carried out. The comparison focuses on the differences between Lean, Mass production and Mass customisation.

Comparison between Manufacturing Paradigms

When comparing the two major plants of the mass and lean production paradigms, namely the General Motors Assembly Plant versus Toyota Takaota Assembly Plant, the differences in the paradigms are pointed out (Table 8), although the two plants performed exactly the same operations (Krafcik, 1988).

Table 7. Benefits and drawbacks of the Personalised Production paradigm

Benefits	Drawbacks
Unlimited product variety	Technological constraints (process planning, CAD, CAM, ERP, etc.)
Design by non-designers	Highly-skilled knowledge workers are required
	Increased production complexity

Table 8. Differences of Mass, Lean Production and Mass Customisation (Womack et al., 1990, Koren, 2009, Ford, 2012, Pine, 1992, Womack and Jones, 2003, Kee and Cheng, 2009 and Cusumano, 1991)

	Mass Production	Lean Production	Mass Customization
Focus	Efficiency through stability and control	The reduction of waste and improvement of the services	Variety and customisation through flexibility and quick responsiveness
Goal	Developing, producing, marketing and delivering goods and services at prices low enough that nearly everyone can afford them	Elimination of waste, value flow and perfection by continuous improvement	Supply of varied products that fit the specific customer's needs in order to increase his interests with maintaining low prices
Key features	Stable demand Large, homogeneous markets, Long product development time, Long product lifecycle	Avoid high cost, teams of multi skilled workers, flexible automated machines to produce volumes of products in enormous variety	Fragmented demand, Heterogeneous niches, Short product development time, Short product lifecycles
Market	Demand > Supply	Demand > Supply	Demand < Supply
Conditions	Homogenous markets	Homogenous markets	Fragmented markets
Products	A few products – Long product lifespan	A few products – Long product lifespan	Variety of products – Short product lifespan
Business strategy	Ignore niche markets (Economies of scale)	Economies of scale	Sell to niche markets (Economies of scope)

The result of this survey showed that Takaoka Plant (LP) was almost twice as productive and three times as accurate as Framingham in performing the same tasks with respect in some criteria of the study such as gross and assembly hours per car, assembly defects per 100 cars and average inventories of parts. It is thought that Takaoka is in some ways a more impressive achievement than Ford's Plant (at Highland) because it represents an advance on so many dimensions. More specifically, in Takaoka plant the effort is cut down in half, defects are reduced by a factor of three, inventories and manufacturing space is slashed. Moreover, Takaoka plant is able to change over in a few days from one type of vehicle to the next generation of product. In contrast, mass production plants have to close for months while switching new products.

To conclude, the reason why lean production is unlikely to prove more oppressive than mass production is that lean production is fragile. On the other hand mass production is designed with buffers everywhere in order to make it function. In situations that parts don't arrive on time or some workers cannot respond to their duties (e.g. fail to detect a problem before the system is mass produced), the system still runs. Once, a lean production manager stated: "Mass Production is Lean run by the rule book. No one takes initiative and responsibility to continually improve the system". See Table 9 for the design differences between Lean and Mass Production.

The paradigms of Toyota Production System (TPS), Lean Production (LP), Mass Customisation (MC) and other instantiations such as Agile Manufacturing (AgM), Holonic Manufacturing Systems (HMS) and Quick Response Manufacturing (QRM), are all focused towards the general goal of increased profitability and improved customer satisfaction, focus on achieving objectives.

Paradigms like TPS and Lean Manufacturing are focused on elimination of waste and the improvement of the production (Table 10). QRM and AM help making the organisation respond quickly to changing customer demands. MC is focused on improving the company's ability to manufacture products especially through modularity. HMS helps with the reconfiguration and scaling manufacturing systems in response to the changes in product volume and variety. As a conclusion, all these paradigms contribute towards making the company more agile, reducing operating costs, improving market-share and increasing profitability (Nambiar, 2010).

Table 9. Differences in the design methods between Lean and Mass Production paradigms

Criteria	Lean Production	Mass Production	Differences
Leadership	Shusa system (the leader, the boss)	Team leader	In mass production there is less power and career paths for the leader
Teamwork	Small teams for each project	Individuals on each project (including the team leader)	In Lean firms there are fewer people during the life of a project due to efficient organisation and little turnover in Japanese teams
Communication	Team members sign formal pledges to do exactly what everyone has agreed upon as a group. The design process is sequential	Development efforts fail to resolve critical design trade – offs until very late in the project due to great reluctance to confront conflicts directly	The mass producer keeps the line moving at all costs and massive amounts of rework are required. The Lean producer spends more effort up front correcting problems before they multiply and ends with less total effort and higher quality.
Simultaneous development	Lean die makers are much better at scheduling production in die-cutting shop	Time consuming die – making approach. One step after the other in expensive equipment.	In LP can be produced a complete set of production – ready dies for a new product in a year, exactly half the time needed in typical mass production die – making. As a result in LP there are fewer tolls, lower inventories and less human effort.

Table 10. Areas of focus

Feature/Trait	TPS	Lean	AgM	MC	HMS	QRM
Customisation			+	++	++	++
Flexibility			++	+	++	+
Inventory Reduction	++	++				+
Lead Time		+	+	+		++
Organisation-focus		+	+			+
Quality	+	+	+			+
Re-Configurability			++		++	
Responsiveness			++	+	+	++
Waste elimination	++	++	+			

Comparison Based on Career Opportunities

Craft Production

The highly skilled workers in design, machine operations and fitting progressed through an apprenticeship to a full set of craft skills. Many could also hope to run their own machine shops, becoming self-employed contractors to assembler firms.

Mass Production

The shop-floor worker had no career path, except probably to become a foreman. The newly emerging professional engineers had a direct climb up the career ladder. Unlike the skilled craftsman their career paths did not lead toward ownership of a business. They would only advance within their profession form young engineer – trainee to senior engineer. Then they would be in charge of coordinating engineers at lower levels.

Lead Production

The employees in a company environment in order to be productive must be offered new and continuous challenges. In this way they feel skilful and think they can implement their knowledge. This system calls for more professional skills and applying them in a team setting rather than in a rigid hierarchy. The better a worker is at team work the less he knows about a specific specialty. Many employees find the lack of a step career ladder with even more elaborating titles and job descriptions both disappointing and disconcerting. In LP system, thus, workers feel they have reached an early death in their careers.

Table 11. Career opportunities in Craft, Mass and Lean production (Womack et al., 1990, Kee and Cheng, 2009 and Benko and Weisberg, 2007)

Production Paradigms	Early Career Stage	Top Career Level
Craft Production	Trainees	Highly skilled workers
Mass Production	No career path for operators with the exception of Foreman/engineer-trainee	Competitive advantage through increased employee job satisfaction and loyalty
Lean Production	Skilled workers	No career step
Mass Career Customisation	Talented employees	Senior Engineer

Mass Career Customisation (MCC)

MCC provides a structure, systematic approach and corporate lexicon that allows organizations to correlate employees' talents, career aspirations and evolving life circumstances over time in ways that match up with the enterprise's evolving marketplace strategies and commensurate need for talent. Moreover, MCC recognizes, validates and embraces the changing tempos of today's knowledge workers, offering a scalable solution to the enervating dilemmas in their search for career-life integration. In this way, MCC delivers distinct competitive advantage through increased employee job satisfaction and loyalty. Greater potential for the kinds of continuous, long-term relationships with higher-performing employees that improves productivity and lower costs related to less employee churn. The career opportunities for these three production paradigms can be summarized in Table 11.

Morale of Work Force

A significant point for the workers' general attitude in the company is the morale of the work force. The work space was harder in Japan, because there was a sense of purposefulness. In the U.S., the workers simple execute their tasks under the supervision of the foreman. In Japan, the workers were lifetime employees of the company. The security of their jobs returned a full commitment to their work. Another benefit of LP production is that the employees tend to remain in the same company for their entire career span. In this way they can improve their skills at a constant rate. In the same time the company benefits by this situation. In this way the employee loyalty can be achieved, together with low turnover costs and the fulfilment of company goals (Benko and Weisberg, 2007).

Timeline of Production Paradigms

An overview of the major characteristics, methods and tools used in each one of the aforementioned prominent manufacturing paradigms throughout the period of time is depicted in Figure 12. The timeline includes also some events prior to the Craft production paradigm. These events, such as the invention of the windmill are significant technological evolutions that predate any organised form of production as we know it today.

CONCLUSION AND FUTURE TRENDS

Summary of Findings

Manufacturing experienced notable changes and transitions in nature, mainly due to the social and political situations and advances in technology. Interpreting these transitions, one can claim that the factory systems and technologies have been evolving in two directions. At first, they became more versatile in the variety of products that they can produce, resulting in innovations in production, in design technology and in management techniques. Second, companies have extended factories like tools and techniques. Factories emerged from firms that introduced a series of product and process innovations that made possible the efficient replication of a limited number of designs in massive quantities. This tactic has been called economies of scale (Pindyck and Rubinfeld, 1989 and Chandler, 1990). Factory systems replaced craft modes of production as firms learned how to rationalise product designs as well as production work itself (Hounshell, 1984). Although factory organisations provided higher worker and capital productivity, their structure made it difficult to introduce new products or processes quickly and economically, or to meet the demands of customers with distinctive tastes; factory-oriented design and production systems have never completely replaced craftsman or job shops even if the new technologies continue to appear. The result, in both manufacturing and design has been a shift from simple economies of scale, as in the conventional mass production of a limited number of products, to economies of scope and integration (Piller, 2002). It is clear that mass-production factories or their analogies, are not appropriate for all types of products or competitive strategies. Moreover, they have traditionally worked best for limited numbers of designs suited to mass replication and mass consumption. The craft approach offers a less efficient process at least for commodity products but remains necessary for technologies that are still new or emerging and continues to serve specific market niches, such as for tailoring products for individual needs and luxury or traditional items.

In several industries, new managerial and engineering techniques along with advanced computerised technologies have been reducing this trade-off of flexibility for efficiency (Alexopoulos et al., 2010). This is achieved not through simple economies of scale based on the mass replication of a small number of designs, but through economies of scope based on efficiencies in the design and production of a variety of products in individual production or engineering facilities (Baumol et al., 1982). Advances in computer processing capacity and in programming have facilitated the development of the coding and retrieval systems needed for group technology to work well for large numbers of items, leading to extensive monetary savings in firms. The basic concept, however, is as old as interchangeable parts (Hyer & Wemmerloc, 1984) (Figure 13).

Apart from summarising the findings of these chapters, Table 12 also comprises a comparison of the particularities of the production paradigms. The compared characteristics are: the societal need that led to the creation of each paradigm, the specific market demand that shaped the paradigms, the goal that each paradigm strives to achieve, the underlying technology that fuels the operation of the system, the optimal form of manufacturing system that can effectively respond to the market needs and achieve its goal, the prevalent architecture of the factory, the business model that it pursuits and finally, the type of work force that is required by the system.

Needs evolved in parallel with societal and political situation of their respective era. At first, a standardisation of product structure and quality was pursuit driven by the establishment of living standards. A great enabler for that was harnessing electricity. Afterwards, the need for interchangeable parts, mainly

Figure 12. The evolution of manufacturing systems

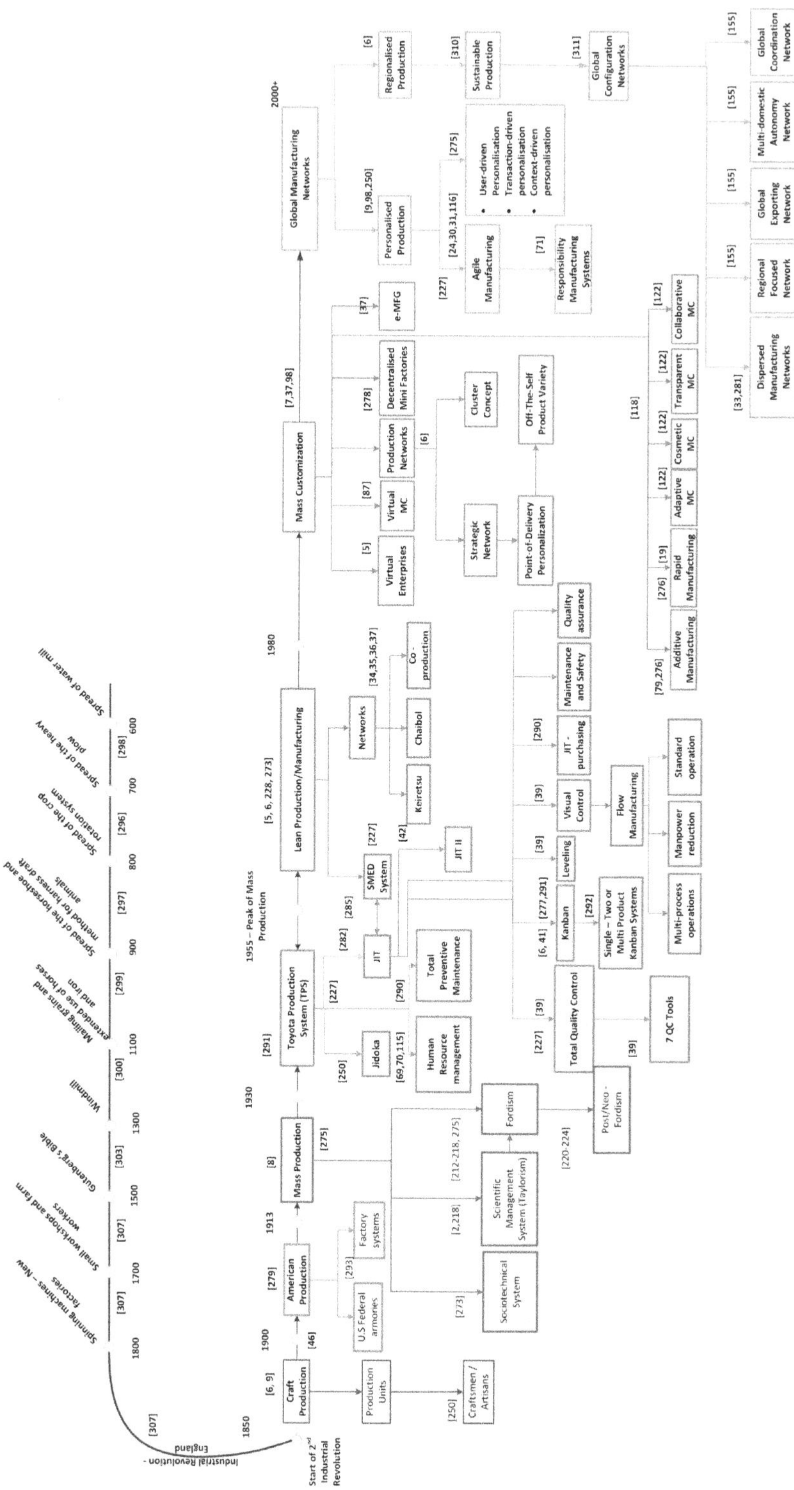

Figure 13. Phases of factory structuring over the last five decades

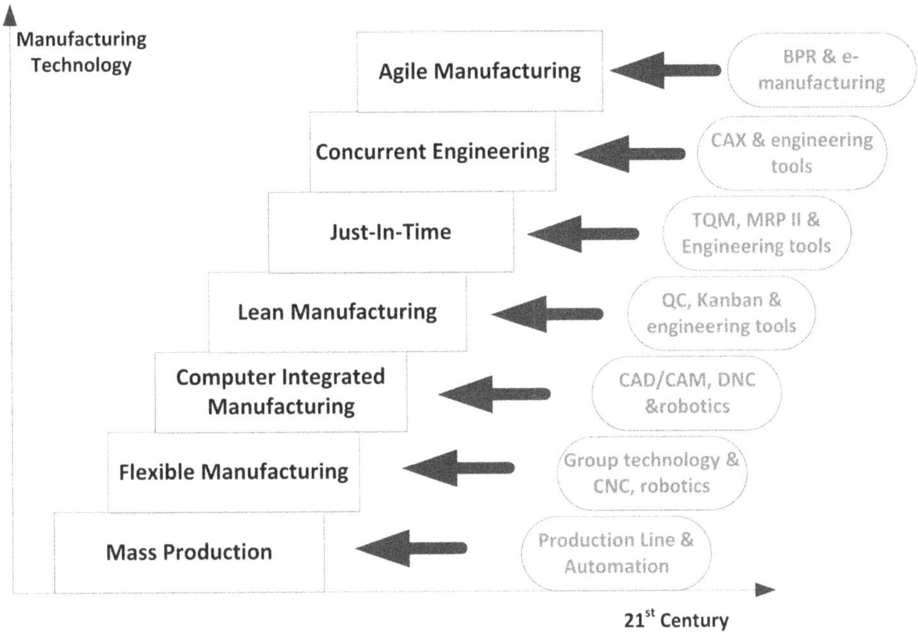

due to warfare, and simultaneously the increase in quality of life, led to the establishment of the up to now longest living form of production, mass production, which thrived in America and later on in the rest of the world. The Japanese counterpart of mass production was then implemented in the factories of Toyota. It was Lean Production that continues to evolve today, both in methods and tools, as well as in theory, as demonstrated by the subsequent chapters. The saturation of the customers towards mass produced products, generated the needs for increased product variety. The volatile market generated the needs for flexibility and changeability in the production systems. The result was the creation of mass customisation and personalisation production paradigms/concepts. The customer is allowed to select design combinations that he prefers the most in the first and contribute to the design by incorporate unique individual preferences in the second.

Future Trends

Mass Customisation (MC) and Personalisation are in the centre of research and industrial interest. These paradigm is facing a rapid increase in demand and technology. Focusing on MC as the current strategy in the manufacturing real world and in academia with a huge number of implementations there are some questions that are not fully discussed in the literature. For example which is the concept that MC will mainly be based on in the next years? Which of the MC available tools can affect immediately the real word conditions? What really the customers want from MC? How can all these aspects of MC can be controlled or organised? Is a Lean Customisation paradigm an emerging paradigm and can it be viable? Based on such sample questions, some promising areas for future research are mentioned:

Additive Manufacturing: Additive manufacturing, also known as rapid manufacturing (RM) has evolved from rapid prototyping technologies (Bateman and Cheng, 2006 and Tuck et al., 2008). These technologies,

Table 12. Manufacturing systems comparison

Production Systems	Need	Market Demand	Goal	Required Technology	Optimal MFG Systems	Production Architecture	Business Model	Work Force
Craft Production	Unique products and commodities	Periods of seasonal demand for the wares (Bidanda et al., 1993)	Customer's satisfaction	Electricity	Electrical General -Purpose Machine Tools	Workshop	Pull	Artisans
American System	Low cost, quality products	Seasonal	Low cost	Interchangeable parts	Specialised machines	Modular	Push	Highly Skilled Work Force
Mass Production (Ford, 2012)	Low cost products	Steady	Low cost	Interchangeable parts, general purpose machine tools	Serial/ Moving Assembly Line	Unified	Push	Relatively Unskilled Work Force
Lean Production (Womack et al., 1990)	Explicity on perfection, High-perceived quality products	Stable (Hall, 1983)	Elimination of waste, continuous improvement (Kaneko and Nojiri, 2008)	Interchangeable parts, automated machines	Moving Assembly Line	Cellular	Push	Highly motivated skilled workers (Kee and Cheng, 2009)
Mass Customisation	Huge product variety / High quality	Unstable	Wide variety	Information technology / Internet	Assembly Line, FMS, RMS	Modular	PushPull	Job Rotation (Michalos et al., 2011)
Global Manufacturing - Personalisation	Regional personalised products	Highly volatile	Rapid esponsiveness	IT, Internet, CNCs, RMTs, Virtual Software	Reconfigurable, Non – serial Assembly Line systems	Highly Modular	Pull	Knowledge workers

although existed for a long period of time, only recently, have been investigated in literature and practice in MC (Atzeni et al., 2010 and Pallari et al., 2010). Moreover, to our knowledge, only in (Bateman and Cheng, 2006 and Atzeni et al., 2010), the research is focused on business management implications of RM. Although, these technologies are in an early stage, the centralised and decentralised RM supply chains could be a short lived phenomenon as home based additive technologies become a commercial reality. RM of MC internet designs could become as common as the home printing of photographs or music downloading (Reeves et al., 2011). The broader organisation and economic implications of RM technology will receive greater attention in the near future.

Quality Control: Quality issues became increasingly important in MC environments because product variety has increased as well as the uncertainty of the demand. Therefore, the literature has focused on this direction in recent years (Tu et al., 2007). In (Zhao, 2008) the way that QC tools can adapt to such situations is presented. Moreover, research efforts on the subject concentrate on transferring customer's demands into product design (Xinggang et al., 2006 and Mourtzis and Doukas, 2012), or on incorporating quality indices in the design of a product platform (Jiao & Tseng, 1999; Rai & Allada, 2003). Research has followed this direction the last few years, however, there is still a huge research and development potential in the field.

Although there is still enormous potential for traditional and electronic MC, there might be a new trend in the horizon – virtual and lean MC. These computer-based simulated environments are intended for its users to inhabit and interact in the form of personalised avatars, while on the background lean practices are executed. However, until now, the focus of these analyses has mainly been on their function as sales agents in business – to – consumer relationships. One very important fact that both companies and people have to realise is that virtual worlds are not video games but an extension of real life (Holtzwarth et al., 2006).

Moreover, the incorporation of lean thinking in training through delivery mechanisms such as serious games and simulation can support the training of the engineers of tomorrow. Using industrial engineering techniques only to achieve lean is not the answer. Lean is also about human development, thus workers and managers should be able to understand lean practices deeply and this can be achieved only through appropriate training.

The establishment of robust repeatable lean processes that are continuously improved is another topic of consideration for the future of lean manufacturing. These processes will be tightly integrated and supported by the future Information and Communication Technologies. Considering that the computer savvy generation of 1985 – 2004 will shape the future of manufacturing, their relation with ICT is a definitely positive factor towards improving lean.

Regarding core manufacturing processes, the evolution of 3D printing is ideal for lean manufacturing, as this technology generated virtually no waste. Moreover, handheld devices and industrial touchscreens are very promising solutions for asset tracking, material localisation and optimisation of kanban systems. Nevertheless, for these to happen, a steady improvement in lean practices must be assured, through the diffusion and application of the core ideas of lean in more enterprises.

ACKNOWLEDGMENT

This work has been supported by the EC funded project "A Web-based Collaboration System for Mass Customization – e-CUSTOM" (260067).

REFERENCES

Abele, E., Elzenheimer, J., Liebeck, T., & Meyer, T. (2006). *Reconfigurable manufacturing systems and formable factories - Globalization and decentralization of manufacturing*. New York, NY: Springer-Verlag.

Alexopoulos, K., Papakostas, N., Mourtzis, D., & Chryssolouris, G. (2010). A method for comparing flexibility performance for the lifecycle of manufacturing systems under capacity planning constraints. *International Journal of Production Research*, *49*(11), 3307–3317. doi:10.1080/00207543.2010.482566

Appelbaum, E., & Batt, R. (1994). *The new American workplace: Transforming work systems in the United States*. New York, NY: ILR Press.

Atzeni, E., Iuliano, L., Minetola, P., & Salmi, A. (2010). Redesign and cost estimation of rapid manufactured plastic parts. *Rapid Prototyping Journal*, *16*(5), 308–317. doi:10.1108/13552541011065704

Bateman, R. J., & Cheng, K. (2006). Extending the product portfolio with 'devolved manufacturing': Methodology and case studies. *International Journal of Production Research*, *44*(16), 3325–3343. doi:10.1080/00207540500536947

Baumol, W. J., Panzer, J. C., & Willig, R. D. (1982). *Contestable markets and the theory of industry structure*. Fort Worth, TX: Saunders College Publishing/Harcourt Brace.

Bidanda, B., Cleland, D. I., & Dharwadkar, S. R. (1993). *Manufacturing: Shared a global perspective*. Fort Worth, TX: McGraw-Hill.

Boland, M. (2008). Innovation in the food industry: Personalised nutrition and mass customisation, innovation: Management. *Policy & Practice*, *10*(1), 53–60. doi:10.5172/impp.453.10.1.53

Bose, R. (1996). Intelligent agents framework for developing knowledge-based decision support systems for collaborative organisational processes. *Expert Systems with Applications*, *11*(3), 247–261. doi:10.1016/S0957-4174(96)00042-5

Chandler, A. D. (1990). *Scale and scope*. Cambridge, MA: Harvard University Press.

Cheng, K., & Bateman, R. J. (2008). e-Manufacturing: Characteristics, applications and potentials. *Progress in Natural Science*, *18*(11), 1323–1328. doi:10.1016/j.pnsc.2008.03.027

Chryssolouris, G. (2005). *Manufacturing systems: Theory and practice* (2nd ed.). New York, NY: Springer Verlag.

Chryssolouris, G., Mavrikios, D., Papakostas, N., Mourtzis, D., Michalos, G., & Georgoulias, K. (2008). Digital manufacturing: History, perspectives and outlook. *Proceedings of the I MECH E Part B Journal of Engineering Manufacture, 222*, 451-462.

Chung, W., Li, K. C., & Chik, S. (1997). Towards the formulation of a practical global manufacturing strategy: A theory-driven empirical research approach. In *Proceedings of CIRP International Symposium - Advanced Design and Manufacture in the Global Manufacturing Era* (pp. 397–404). Hong Kong, China: CIRP.

Comstock, M., Johansen, K., & Winroth, M. (2004). From mass production to mass customization: Enabling perspectives from the Swedish mobile telephone industry. *Production Planning and Control*, *15*(4), 362–372. doi:10.1080/0953728042000238836

Cusumano, M. A. (1991). *Shifting economies: From craft production to flexible systems and software factories*. Cambridge, MA: Sloan School of Management.

Da Silveira, G., Borenstein, D., & Fogliatto, F. S. (2001). Mass customisation: Literature review and research directions. *International Journal of Production Economics*, *72*(1), 1–13. doi:10.1016/S0925-5273(00)00079-7

Doll, W. J., & Vonderembse, M. A. (1990). The evolution of manufacturing systems: Towards the post – industrial enterprise. *Omega*, *19*(5), 401–411. doi:10.1016/0305-0483(91)90057-Z

Eastwood, M. A. (1996). Implementing mass customizations. *Computers in Industry*, *30*, 171–174. doi:10.1016/0166-3615(96)00010-3

Eurostat. (2011). *High-tech statistics, employment in high-tech*. Retrieved October 25, 2012 from http://epp.eurostat.ec.europa.eu/statistics_explained/index.php/High-tech_statistics

Filippini, R., Forza, C., & Vinelli, A. (1998). Sequences of operational improvements: Some empirical evidence. *International Journal of Operations & Production Management*, *18*(2), 195–207. doi:10.1108/01443579810177067

Flaherty, M. T. (1986). Coordinating international manufacturing and technology. In M. E. Porter (Ed.), *Competition in global industries*. Boston, MA: Harvard Business School Press.

Flynn, B. B., Schroeder, R. G., Flynn, E. J., Sakakibara, S., & Bates, K. A. (1997). World-class manufacturing project: Overview and selected results. *International Journal of Operations & Production Management*, *17*(7), 671–685. doi:10.1108/01443579710175592

Fogliatto, F. S., Da Silveira, G. J. C., & Borenstein, D. (2012). The mass customization decade: An updated review of the literature. *International Journal of Production Economics*, *138*(1), 14–25. doi:10.1016/j.ijpe.2012.03.002

Ford, H. (2012). *My life and work – An autobiography of Henry Ford*. Charleston, SC: Forgotten Books.

Gilmore, J. H., & Pine, J. (1997). The four faces of mass customization. *Harvard Business Review*, *75*(1), 91–101. PMID:10174455

Guest, D. (1987). Human resource management and industrial relations. *Journal of Management Studies*, *24*(5), 503–521. doi:10.1111/j.1467-6486.1987.tb00460.x

Gunasekaran, A., & Ngai, E. W. T. (2012). The future of operations management: An outlook and analysis. *International Journal of Production Economics*, *135*(2), 687–701.

Hall, R. W. (1983). *Zero inventories*. New York, NY: McGraw-Hill Publications.

Hayes, R. H., Wheelwright, S. C., & Clark, K. B. (1988). *Dynamic manufacturing: Creating the learning organization*. New York, NY: The Free Press.

Hines, P., Holweg, M., & Rich, N. (2004). Learning to evolve: A review of contemporary lean thinking. *International Journal of Operations & Production Management*, *24*(10), 994–1011. doi:10.1108/01443570410558049

Hirano, H. (2009). *JIT implementation manual: The complete guide to just-in-time manufacturing* (2nd ed.). New York, NY: Productivity Press. doi:10.1201/b10234

Holzwarth, M., Janiszewski, C., & Neumann, M. M. (2006). The influence of avatars on online consumer shopping behaviour. *Journal of Marketing*, *70*(4), 19–36. doi:10.1509/jmkg.70.4.19

Hounshell, A. (1984). *From the American system to mass production, 1800-1932: The development of manufacturing technology in the United States*. Baltimore, MD: Johns Hopkins University Press.

Hyer, N. L., & Wemmerloc, U. (1984). *Group technology and productivity*. Boston, MA: Harvard University Press.

IDEF0 Function Modelling Method. (2010). Retrieved from http://www.idef.com/IDEF0.htm

Jablin, F. M., Putnam, L. L., Roberts, K. H., & Porter, L. (1987). *Handbook of organizational communication: An interdisciplinary perspective*. London, UK: Sage Publications.

Jiao, J., Tseng, M., Duffy, M. M., & Lin, F. (1998). Product family modelling for mass customisation. *Computers & Industrial Engineering*, *35*(3-4), 495–498. doi:10.1016/S0360-8352(98)00142-9

Jiao, J., & Tseng, M. M. (1999). A methodology of developing product family architecture for mass customization. *Journal of Intelligent Manufacturing*, *10*(1), 3–20. doi:10.1023/A:1008926428533

Kaneko, J., & Nojiri, W. (2008). The logistics of Just-in-Time between parts suppliers and car assemblers in Japan. *Journal of Transport Geography*, *16*(3), 155–173. doi:10.1016/j.jtrangeo.2007.06.001

Kaplan, A. M., Schoder, D., & Haenlein, M. (2007). Factors influencing the adoption of mass customization: The impact of base category consumption frequency and need satisfaction. *Journal of Product Innovation Management*, *24*(2), 101–106. doi:10.1111/j.1540-5885.2007.00237.x

King, A. A., & Lenox, M. J. (2001). Lean and green? An empirical examination of the relationship between lean production and environmental performance. *Journal of Production and Operations Management*, *10*(3), 244–256. doi:10.1111/j.1937-5956.2001.tb00373.x

Koren, Y. (2009). *The global manufacturing evolution*. New York, NY: Wiley.

Koren, Y., Pasek, Z., Ulsoy, G., & Benchetrit, U. (1996). Real-time open-control architectures and system performance. *CIRP Annals*, *45*(1), 377–380. doi:10.1016/S0007-8506(07)63084-4

Krafcik, J. F. (1988). *A methodology for assembly plant performance determination. International Motor Vehicle Program*. Boston, MA: Massachusetts Institute of Technology.

Kumar, A. (2007). From mass customization to mass personalization: A strategic transformation. *International Journal of Flexible Manufacturing Systems*, *19*(4b), 533–547. doi:10.100710696-008-9048-6

Lai, K., & Cheng, T. C. E. (2009). *Just-in-time logistics*. Farnham, UK: Gower.

Lazonick, W. (1991). *Business organisation and competitive advantage: Capitalist transformations in the twentieth century, technological enterprise in a historical perspective*. Oxford, UK: Oxford University Press.

Lean Thinking Pty Ltd. (2010). *About lean*. Retrieved from http://www.leanthinking.info/aboutlean.html#lean

Lewis, M. A. (2000). Lean production and sustainable competitive advantage. *International Journal of Operations & Production Management*, *20*(8), 959–978. doi:10.1108/01443570010332971

Lin, C. T., Hong, W. C., Chen, Y. F., & Dong, Y. (2010). Application of salesman-like recommendation system in 3G mobile phone online shopping decision support. *Expert Systems with Applications*, *37*(12), 8065–8078. doi:10.1016/j.eswa.2010.05.081

Link, N. R., & Darlow, T. S. (1997). A framework for the role of manufacturing strategy in global manufacturing. In *Proceedings of CIRP International Symposium - Advanced Design and Manufacture in the Global Manufacturing Era* (pp. 417-422). Hong Kong: CIRP.

Lu, R. F., Petersen, T. D., & Storch, R. L. (2009). Asynchronous stochastic learning curve effects in engineering-to-order customisation processes. *International Journal of Production Research*, *47*(5), 1309–1329. doi:10.1080/00207540701484921

Maruca, P. F. (1994). *The right way to be global: an interview with Whirlpool CEO David Whitwam.* Boston, MA: Harvard Business Review.

McCarthy, I. P. (2004). Special issue editorial: The what, why and how of mass customization. *Production Planning and Control, 15*(4), 347–351. doi:10.1080/0953728042000238854

McIntosh, R. I., Matthews, J., Mullineux, G., & Medland, A. J. (2010). Late customisation: Issues of mass customisation in the food industry. *International Journal of Production Research, 48*(6), 1557–1574. doi:10.1080/00207540802577938

McKinsey. (2007). *Climate change special initiative.* Author.

Michalos, G., Makris, S., & Mourtzis, D. (2011). A web based tool for dynamic job rotation scheduling using multiple criteria. *CIRP Annals-Manufacturing Technology, 60*(1), 453–456. doi:10.1016/j.cirp.2011.03.037

Monroy, C. R., & Arto, J. R. V. (2010). Analysis of global manufacturing virtual networks in the aeronautical industry. *International Journal of Production Economics, 126*(2), 314–323. doi:10.1016/j.ijpe.2010.04.008

Montgomery, D. (2012). *Introduction to statistical quality control* (6th ed.). Hoboken, NJ: Wiley.

Montreuil, B., & Poulin, M. (2005). Demand and supply network design scope for personalized manufacturing. *Production Planning and Control, 16*(15), 454–469. doi:10.1080/09537280500088068

Mourtzis, D. (2011). Internet based collaboration in the manufacturing supply chain. *CIRP Journal of Manufacturing Science and Technology, 4*(3), 296–304. doi:10.1016/j.cirpj.2011.06.005

Mourtzis, D., & Doukas, M. (2012). A web-based platform for customer integration in the decentralised manufacturing of personalised products. In *Proceedings of Procedia CIRP, 45th CIRP Conference on Manufacturing Systems* (pp. 209-214). CIRP.

Mourtzis, D., & Doukas, M. (2012). *Decentralized manufacturing systems review: Challenges and outlook.* Springer. doi:10.1007/978-3-642-30749-2_26

Mourtzis, D., Doukas, M., & Psarommatis, F. (2012). A multi-criteria evaluation of centralized and decentralized production networks in a highly customer-driven environment. *CIRP Annals - Manufacturing Technology, 61*(1), 427-430.

Mourtzis, D., Doukas, M., & Psarommatis, F. (2013). Design and operation of manufacturing networks for mass customisation. *CIRP Annals - Manufacturing Technology, 63*(1), 467-470.

Mourtzis, D., Papakostas, N., Makris, S., Xanthakis, V., & Chryssolouris, G. (2008). Supply chain modeling and control for producing highly customized products. *CIRP Annals – Manufacturing Technology, 57*(1), 451-454.

Nambiar, A. N. (2010). Modern manufacturing paradigms - A comparison. In *Proceedings of the International Multi-Conference of Engineers and Computer Scientists 2010.* IMECS.

Naylor, J. B., Naim, M. M., & Berry, D. (1999). Leagility: Integrating the lean and agile manufacturing paradigms in the total supply chain. *International Journal of Production Economics*, *62*(1–2), 107–118. doi:10.1016/S0925-5273(98)00223-0

Niepcel, W., & Molleman, E. (1998). Work design issues in lean production from a sociotechnical systems perspective: Neo-Taylorism or the next step in sociotechnical design? *Human Relations*, *51*(3), 259–287. doi:10.1177/001872679805100304

Pallari, J. H. P., Dalgarno, K. W., & Woodburn, J. (2010). Mass customization of foot orthoses for rheumatoid arthritis using selective laser sintering. *IEEE Transactions on Bio-Medical Engineering*, *57*(7), 1750–1756. doi:10.1109/TBME.2010.2044178 PMID:20211798

Partanen, J., & Haapasalo, H. (2004). Fast production for order fulfilment: Implementing mass customization in electronics industry. *International Journal of Production Economics*, *90*(2), 213–222. doi:10.1016/S0925-5273(03)00214-7

Piller, F. T., & Möslein, K. M. (2002). From economies of scale towards economies of customer interaction: Value creation in mass customisation based electronic commerce. In *Proceedings of 15th Bled Electronic Commerce Conference* (pp. 214-228). München, Germany: Bled.

Pindyck, R. S., & Rubinfeld, D. L. (1989). *Microeconomics*. New York, NY: MacMillan.

Pine, B. J. (1992). *Mass customization: The new frontier in business competition*. Boston, MA: Harvard Business School Press.

Prince, J., & Kay, J. M. (2003). Combining lean and agile characteristics: Creation of virtual groups by enhanced production flow analysis. *International Journal of Production Economics*, *85*(3), 305–318. doi:10.1016/S0925-5273(03)00118-X

Rai, R., & Allada, V. (2003). Modular product family design: Agent-based Pareto- optimization and quality loss function-based post-optimal analysis. *International Journal of Production Research*, *41*(17), 4075–4098. doi:10.1080/0020754031000149248

Rask, K., & Johansson, J. (2008). Similarities and differences between lean production, Tayloristic and socio-technical systems revealed in the methodology characteristics map. In *Proceedings of the 18th International Conference on Flexible Automation and Intelligent Manufacturing (FAIM2008)* (pp. 995-1001). Skövde, Sweden: FAIM.

Reeves, P., Tuck, C., & Hague, R. (2011). *Additive manufacturing for mass customization*. New York, NY: Springer.

Shah, R., & Ward, P. T. (2003). Lean manufacturing: Context, practice bundles, and performance. *Journal of Operations Management*, *21*(2), 129–149. doi:10.1016/S0272-6963(02)00108-0

Shah, R., & Ward, P. T. (2007). Defining and developing measures of lean production. *Journal of Operations Management*, *25*(4), 785–805. doi:10.1016/j.jom.2007.01.019

Shi, Y., Gregory, M. J., & Naylor, M. (1997). International manufacturing configurations map: A self-assessment tool of international manufacturing capabilities. *Integrated Manufacturing Systems*, *8*(5), 273–282. doi:10.1108/09576069710179724

Shigeo, S., & Andrew, D. (1989). *A study of the Toyota production system from an industrial engineering viewpoint, produce what is needed, when it's needed*. New York, NY: OR Productivity Press.

Sowinski, L., & Orton, C. (2001). Move over, JIT. *World Trade*, *14*(4), 54–56.

Tam, K. Y., & Ho, S. Y. (2006). Understanding the impact of web personalization on user information processing and decision outcomes. *Management Information Systems Quarterly*, *30*(4), 865–890.

Tso, S. K., Lau, H., & Ho, J. K. L. (2000). Coordination and monitoring in an intelligent global manufacturing service system. *Computers in Industry*, *43*(1), 83–95. doi:10.1016/S0166-3615(00)00053-1

Tu, Y. L., Xie, S. Q., & Fung, R. Y. K. (2007). Product development cost estimation in mass customization. *IEEE Transactions on Engineering Management*, *54*(1), 29–40. doi:10.1109/TEM.2006.889065

Tuck, C. J., Hague, R. J. M., Ruffo, M., Ransley, M., & Adams, P. (2008). Rapid manufacturing facilitated customization. *International Journal of Computer Integrated Manufacturing*, *21*(3), 245–258. doi:10.1080/09511920701216238

Ulrich, P. V., Anderson-Connel, L. J., & Wu, W. (2003). Consumer co-design of apparel for mass customization. *Journal of Fashion Marketing and Management*, *7*(4), 398–412. doi:10.1108/13612020310496985

Uselding, P. J. (1971). Technical progress at the Springfield armoury: 1820-1850. *Explorations in Economic History*, *9*, 291–316. doi:10.1016/0014-4983(71)90062-3

Vastag, G., Kasarda, J. D., & Bonne, T. (1994). Logistical support for manufacturing agility in global markets. *International Journal of Operations & Production Management*, *14*(11), 73–85. doi:10.1108/01443579410068657

Womack, J. P., & Jones, D. T. (2003). *Lean thinking: Banish waste and create wealth in your corporation*. New York, NY: Productivity Press.

Womack, J. P., Jones, D. T., & Roos, D. (1990). *The machine that changed the world*. New York, NY: Rawson and Associates.

Womack, L. P., & Jones, D. T. (1994). *From lean production to the lean enterprise*. Boston, MA: Harvard Business Review.

Xinggang, L., Dingwei, W., & Jiafu, T. (2006). Model for product configuration based on quality deployment. *Control and Decision*, *21*(12), 1360–1364.

Zhang, S., & Chen, B. (1997). A concept of virtual global manufacturing based on autonomous manufacturing islands. In *Proceedings of CIRP International Symposium - Advanced Design and Manufacture in the Global Manufacturing Era* (pp. 346-352). Hong Kong: CIRP.

Zhao, F., He, Z., & Wu, D. (2008). Quality assurance of mass customization: A state-of-the-art review. In *Proceedings of IEEE Symposium on Advanced Management of Information for Globalized Enterprises*. (pp. 1-3). Tianjin, China: IEEE.

This research was previously published in the Handbook of Research on Design and Management of Lean Production Systems edited by Vladimír Modrák and Pavol Semančo, pages 1-29, copyright year 2014 by Business Science Reference (an imprint of IGI Global).

Chapter 20
3D Printing Technology Diffusion:
A Revolution or an Illusion?

Kemal Yayla
Ege University, Turkey

Basak Ozdemir
Independent Researcher, Turkey

Serhat Burmaoglu
Izmir Katip Celebi University, Turkey

Haydar Yalcin
Izmir Katip Celebi University, Turkey

ABSTRACT

3D printing technology has been considered one of the most potentially groundbreaking technologies for the future, as the customer expectations, market requirements, and the competition grows in a global scale. In order to understand the potential effect of 3D printing technology and if it is a disruptive innovation that will change the traditional manufacturing paradigm, it is essential to examine the diffusion of knowledge in this area. In this study, 3D printing technology has been reviewed and patent analysis regarding 3D printing technology has been conducted in order to understand the diffusion of 3D printing technology. The results of the patent analysis indicate that the diffusion of 3D printing technology which is represented by the patents of four key methods not expected to fit with Bass diffusion model. According to the findings, it can be concluded that 3D printing technologies are in a situation where a state of maturity has not been reached, yet the growth still continues.

DOI: 10.4018/978-1-5225-9624-0.ch020

INTRODUCTION

In advanced economies, the range of economic variables, such as firm performance and technological specialization, play important role in explaining economic growth. The essence of this argument is that innovation, which involves resembles aspects of evolutionary process of accumulating and consolidating existing technologies into new technologies, is a natural process. However, the diffusion of knowledge plays an active role in shaping the direction of innovation, technological and economic evolution. It is only through the dissemination of information that the knowledge that a product or service can be created in the medium in which it could generate an additional opportunity for the economic units in every society (Hur, 2017). The diffusion of new ideas to different environments through recombination of existing technologies depends on the flow of technology in between different sectors (Fleming, 2001; Fleming & Sorenson, 2001). At this point, digital transformation creates opportunities and advantages for the global dissemination of any kind of information that has been produced (Kiesling, Günther, Stummer, & Wakolbinger, 2012; Rifkin, 2012; Samuelson, 2000).

Digitalization is rapidly transforming all sectors that it touches. Although it is not yet clear how digitalization finds a precise response in the society, it seems to affect the certain layers of society more rapidly. Production technology is at the forefront of this change. Cloud Computing, Internet of Things and Robotics are regarded as major technologies, which have emerged as the benchmarks of digitization (Manyika et al., 2013). All of these technologies are considered as technical innovations which generally create changes in business processes, product functions or customer-oriented services, rather than creating value directly.

Among the varied emerging technologies, 3D printing technology stands out as a phenomenon within the concept of technological innovation. 3D printing applications are used in a wide range of areas, from direct production of digital goods to hobby products or biomedical applications (Forrest & Cao, 2013; Gibson, Rosen, & Stucker, 2015; Pierrakakis, Gkritzali, Kandias, & Gritzalis, 2015). 3D printing promises a new wave of innovation that will enable the production of wide variety of objects for a wide user range, from amateurs, to new business ventures, and large-scale companies (Desai & Magliocca, 2013).

3D printing technology is predicted the one of driving forces of technological revolution, which will have a substantially profound economic impact on the manufacturing sector. In addition, 3D printing will spark a new industrial revolution and, technology will have transformational effects in the coming years. 3D printing is generally the nomenclature of mass media, marketers of the technology, and decision makers. The similarity between Additive Manufacturing and 2D printing has given rise to the alternate common name of 3D printing. However, AM (Additive Manufacturing) is broader term that embraces all the technologies within the name of a single one (Bourell, 2016). There has been an accelerating amount of interest in AM technologies since the early 1990s. At the same time, scholars in the field of technology of management have been increasingly attracted the business model and technological development perspectives.

In particular, there is clearly articulated perception by both scholars and practitioners that AM technologies have the potential to change the traditional manufacturing paradigm. However, there are only a small number of early global adopters in the new business model of AM systems, which have seen predominantly in sectors that produce small batches of products and/or require customization (Jiang, Kleer, & Piller, 2017; Rogers, Baricz, & Pawar, 2016; Steenhuis & Pretorius, 2016; Weller, Kleer, & Piller, 2015). It appears to be significantly unrealistic to expect AM technologies will disrupt the existing manufacturing and innovation practices (Rayna & Striukova, 2016). To be specific, this paper aims to answer one research question as "What is the diffusion characteristic of AM technologies?"

In order to answer the research question, the outline of this chapter is as follows: a brief literature on disruptive technology and concept of disruptive innovation is discussed in the first part of background. Moreover, to demonstrate how disruptive technologies penetrate in real life problems, AM technologies are investigated. Patent analysis, diffusion models and growth curves are reviewed in the methodology section. Analysis is performed and findings are demonstrated. The chapter concludes with findings and possible future research directions regarding to AM technologies.

BACKGROUND

Technology refers to the theoretical and practical knowledge and skills that can be used to develop products or services for production and distribution systems. In this sense, technological change has the power to transform the way we live. Many technologies have potential to bring about broad economic value to be affected, and have high-tech change potential for disruptive economic impact. Technological change can affect lifestyles, the way people work, and it can also provide new business opportunities for economic growth (Abernathy & Clark, 1985; Bower & Christensen, 1996; Cheng, Huang, Ramlogan, & Li, 2016; Danneels, 2004; Kostoff, Boylan, & Simons, 2004; Manyika et al., 2013; Walsh, 2004; Walsh & Linton, 2000).

The concept of disruptive technology, first introduced by Bower and Christensen (1996) has become a new feature in continuous or accelerated change rates for customer demands in the mainstream market, providing new technological features in terms of price / performance ratio, and rate of change for the products in the product portfolios (Cheng et al., 2016; Manyika et al., 2013). Regarding disruptive technologies Christensen argues that "disruptive technologies are typically simpler, cheaper, and more reliable and convenient than established technologies'' (Danneels, 2004). In other words, disruptive technologies can be regarded as scientific discoveries that transcend ordinary product / technology capabilities, and form the basis of a new competitive paradigm (Kostoff et al., 2004). Disruptive technologies attract the attention of niche market customers because they have penetrated the market due to their new features. When disruptive technology enters mainstream markets, the performance of features that disruptive technology meets the expectations of the main market (Vecchiato, 2016)

Although the terminology used to define the concept is varied such as hurricanes, typhoons or emerging technologies have been made the notion of disruptive technology has been accepted as general term (Walsh & Linton, 2000). Disruptive technologies create new significant growth opportunities in industries where those with different skill sets and firms penetrate existing sectors by adding value through stepping, or by creating innovations from the value they provide (Kostoff et al., 2004). Research on successful disruptive technology cases indicate that they create new values for customers who find existing production paradigm less satisfactory.

Disruptive technology first coined by Christensen, described the technologies that were rather simple or cheap when introduced, but have a potential to replace the existing competitors, however the term itself was variously defined, and this situation caused debate among the academics, as it is became unclear what constitutes a disruptive technology (Danneels, 2004; Schmidt & Druehl, 2008; Weeks, 2015; Yu & Hang, 2010). The term "Disruptive Innovation" took place of "disruptive technologies", as the concept grew and the cases described exceeds the mere example of technological changes, and started to includes other forms of disruption cases (Yu & Hang, 2010). In this sense, as Yu and Hang suggests, disruptive innovation implies a process (Yu & Hang, 2010).

Disruptive innovation frequently results from changing social practices, reframing the problems from their very roots. In order to create a common understanding, Schmidt and Druehl proposed a framework that would clarify the patterns and degrees of diffusion of disruptive innovations, as well as managerial guidelines, how to assess whether an innovation is sustaining or disruptive (Schmidt & Druehl, 2008). The framework indicates that each of the 4 types of innovation can be equivalent to one of 5 types of "encroachment" (Schmidt & Druehl, 2008) which have been used to define a cannibalization of an older product or service by according to whether diffusion starts the low-end market or the high-end market (Schmidt & Druehl, 2008). Danneels suggested that the term 'disruptive technology' is somewhat vaguely defined and therefore is not able to help managers to predict which technologies will be disruptive earlier on (Danneels, 2004). As disruptive innovation has been described as a process (Yu & Hang, 2010), identifying potentially disruptive technologies is essential for companies to be able to survive and thrive in their respective areas. After reviewing the disruptive technology and innovation concept, which is the focus of the chapter and the most appropriated AM case in literature, is reviewed from various perspectives from historical roots to methods and application areas.

MAIN FOCUS OF THE CHAPTER

Historical Roots of Additive Manufacturing Technologies

AM has come to the fore as a popular field, as well as scientific research conducted in many different areas and its history can be traced to three technological developments dating back around 90–150 years: photo sculpture, topography, and material deposition. In view of the historical development of additive manufacturing, the roots of technologies originated with photo sculpture (1860) and topography (1890) in contrast to the general belief that the methods emerged only with the widespread use of modern production techniques (Gao et al., 2015; Küchler, 2014). One of first example of additive manufacturing is a method of producing three-dimensional landscape images from wax plates patented by J.E Blanther in 1862 (Bertling & Rommel, 2016; Bourell, Beaman, Leu, & Rosen, 2009). The method consisted of smoothing the topographic contour lines across a series of wax plates, cutting the wax along the contour lines, and then smoothing out the wax sections. It was possible to create a three-dimensional map of the shape in accordance with the negative and positive gradients of the surface. The method is regarded as having a similar approach to creating objects with current three-dimensional printing technologies (Bourell, 2016).

Although methods for 3D printing were developed from the early 19th century, the commercialization of methods has increased considerably since late 1950s. The first patent was filed by Munz (1956), which possessed current features of stereolithography (SLA) technique (Bourell et al., 2009). The method proposed by Munz, was designed to produce three-dimensional objects with photochemical substance that displays the image on the squeeze cylinder in the transparent photo emulsion of the layered structure of the cross-section of the scanned object. Another method proposed by Swainson (1977), suggests creating a three-dimensional figure by means of three-dimensional polymerization at the intersection of a laser-sensitive polymer and two laser beams, had similar properties. Laser, electron beam or plasma particles were used to create a continuous layer in a laboratory environment, but not for a commercialized production apparatus (Bourell, 2016; Bourell et al., 2009).

The commercialization of 3D printing methods started in mid 1980s. It started with the first well documented additive manufacturing patent, stereolithography (Hull, 1986), recognized as the most influential patent in additive manufacturing technologies. Further patents included laminated object manufacturing (Feygin, 1988), selective sintering (Deckard, 1991), fused deposition modeling (Crump, 1992), three-dimensional printing (Sachs, Haggerty, Cima, & Williams, 1993) and Laser engineered net shaping (Jeantette, Keicher, Romero, & Schanwald, 2000). More recently, the AM systems has been slowly gaining attraction, specifically within design, however, expiration of patented methods have amplified potential growth and there have become accessible to a range of small businesses and supportive communities also to individual designers.

Modern Additive Manufacturing Technologies

ASTM F42 Technical Committee defines additive manufacturing (AM) as "layered manufacturing process for forming objects with three-dimensional (3D) model data" (Simon Ford & Despeisse, 2016; Gross, Erkal, Lockwood, Chen, & Spence, 2014; Guo & Leu, 2013). Unlike mechanical machining processes such as grinding or milling, additive manufacturing is a group of emerging technologies that make up objects by adding layers of material in cross-sections (Guo & Leu, 2013; Pierrakakis, Kandias, Gritzali, & Gritzalis, 2014).

Different additive manufacturing technologies have been applied for decades in a wide variety of sectors such as aviation, biomedicine, automotive and digital arts. Additive manufacturing technology entered rapid period of change at the end of the 1980s, with the widespread use of selective laser sintering printers. However, the definition of additive manufacturing has evolved since then as consequences of developing new methods and business models. So far, there are a range of AM techniques ranging from well-established methods, employed in industrial settings to those still under development in research laboratories designed for more specific applications. In the literature, alternative terms used for AM techniques include rapid prototyping (Bak, 2003), free form fabrication (Hutmacher, Sittinger, & Risbud, 2004), rapid manufacturing (Levy, Schindel, & Kruth, 2003), additive fabrication (Giannatsis & Dedoussis, 2009) and direct digital manufacturing (Chen et al., 2015). Although AM is described as a synonym for such terms, it can be considered as an umbrella term for the concept of layered production (Coon, Pretzel, Lomax, & Strlič, 2016; Jin, Ji, Li, & Yu, 2017). AM processes are divided into three main categories, liquid, powder and solid states, which are related with the initial state of materials, landmark AM methods include Stereolitograhpy (SLA), selective laser sintering (SLS) fused deposition modeling (FDM) and laminated object manufacturing (LOM). In order to understand diffusion characteristics and technological development of AM, these major methods are investigated in detail (Coon et al., 2016; Han & Sohn, 2015; Huang, Liu, Mokasdar, & Hou, 2013; Levy, 2010).

SLA is the oldest liquid-based manufacturing process and uses a light-sensitive monomer resin and a UV laser to form layers in the production of the base part (Bourell, 2016). Supports are needed to connect the component to the building platform. On each layer, the laser beam follows the cross section of the particle on the surface of the liquid resin to solidify the pattern. Subsequently, the building platform is lowered to completely cover the part then a blade is used to wipe off the resin and resulting in a level that leaves a layer of resin on top of the part. The part is then lowered with a single layer and left until a flat surface is obtained before the next layer is formed. When the part is complete, support can be removed manually. SLA is particularly appropriate for the manufacturing industry because it reduces the time required to produce a prototype part and achieves good surface coverage. The SLA-generated part was constructed from cross-sectional data obtained from the stereolithography file (.STL), which is the standard format for all AM technologies (Coon et al., 2016; Gross et al., 2014; Huang et al., 2013). SLA has advantages and disadvantages in terms of application procedure. The advantage in terms of dimensionally correct parts production is the limited risk of shrinkage dimensional change after thermal degradation of the thermo set photopolymers used in the production process, and the main limitation is that the product size is relatively small, generally less than a 2-meter cube.

Another approach is powder-based methods, which are based on the principle that the material is selectively formed by a local heat source to form each layer of powdered material. In first place, powdered materials is used to spread in a layer and then selectively scanned by a laser called as Selective Laser Sintering (SLS) method. Three dimensional parts can be created with SLS having relatively wide powder form materials, including wax, polymers, polymer / glass composites, polymer / metal powders, metals and ceramics (Coon et al., 2016; Levy, 2010). SLS creates a three-dimensional CAD model by cross-sectioning on a powder bed and layering with the laser. The laser is used to heat the powder particles to the melting temperature rather than to the complete melting state, as this should cause deformation. The powder material bed used during production reduces thermal degradation and melting temperature to ensure coalescence between the layers (Gibson et al., 2015). SLS does not require support holder during the production of parts, since the manufactured part is surrounded by undiluted powder (Gao et al., 2015). A number of factors may inhibit the production of quality prototypes, such as the different thermal properties of different materials, the presence of powdered polymers with appropriate particle size and morphology, large thermal processing windows, and constraints in laser sintering systems (Simon Ford & Despeisse, 2016). The density and porosity of the sintered part will depend on the size of the particles and how often they are packed in the structural part. Other parameters, such as laser wavelength, laser energy and temperature distribution in the cartridge, can also affect the part's mechanical properties, such as porosity or anisotropy (Caulfield, McHugh, & Lohfeld, 2007).

Materials extrusion (Fused Deposition Modeling (FDM)) and sheet lamination (Laminated Object Manufacturing (LOM)) are two others methods used for solid state materials. An AM process, trademarked as Fused Deposition Modeling(FDM), creates layers mechanically on the principle of phase change of a thermoplastic material, this requires high operating temperatures, and the finished prints typically exhibit high porosities (Coon et al., 2016; Gao et al., 2015). A thermoplastic filament (mostly an ABS filament) is heated to a semi-molten state, and extruded by a nozzle moving around on the x–y plane, depositing traces on the working platform. Since the material is extruded in a semi-molten state, the newly deposited material fuses with adjacent material that has already been deposited (Ahn, Montero, Odell, Roundy, & Wright, 2002; Gao et al., 2015). Lowering the platform in the z direction to a specified distance allows for deposition of subsequent layers. A second nozzle extrudes support material where required. The molten material fuses with the layer underneath. This layering in different orientations gives FDM parts anisotropic properties (Coon et al., 2016). LOM is another AM process designed to build a part sequentially from layers of paper, plastic, metal, ceramics or composite sheets, all coated with a thermally activated adhesive (Mueller & Kochan, 1999; Park, Tari, & Hahn, 2000). Generally, the LOM technique was developed for processing paper. The process begins with the lamination of 20 layers of paper onto the retractable platform. The slices of material sheets are cut in the required shape from a roll of material using a laser beam. The cutting material sheet is laid on machine platform and bonded to the previous sheet using a hot roller, which activates a heat sensitive adhesive. After addition of all sheets, the solid part of the material is removed from the platform. Surrounding material and material in regions of the part that are hollow must be removed in a post processing step (Gao et al., 2015). Finally, it is apparent that these methods were developed for solving practical real-life problems. To understand the AM concept from the output side, the AM applications in different sectors are reviewed and briefly explained in the following section.

Applications of Additive Manufacturing Technologies

In the late 1980s and early 1990s a plethora of AM methods appeared. In the emerging stage of AM technology, the research community generally applied the processes and process variations in novel ways in order to address to a wide variety of research problems in diverse technical areas, whether the application is technically possible or economically feasible largely depends on its production volume, part size, complexity, and material cost (Berman, 2012; Rayna & Striukova, 2016). The impacts of AM technologies continue to grow, in terms of both commercial and scholarly activities. Double digit annual growth is expected in the global additive manufacturing market for both rapid prototypes and functional end-use parts in 2016 (Gordon, 2015). In addition, giant leading manufacturers all over the world have applied AM technologies in many fields, including the aerospace & defense, automotive and medical industries.

The aerospace and defense (A&D) industry was an early adopter of AM technology. There are two main reasons why the aerospace industry is a pioneer of adaptation of AM technologies. First, the industry requires parts that are lightweight, strong, and geometrically complex. Second, the lot sizes in aviation are typically produced in small quantities (Uriondo, Esperon-Miguez, & Perinpanayagam, 2015). Many of these parts are costly to produce in high quantities and AM can minimize the amount of materials used in stringent operating conditions. In addition, AM has significant impact for the parts in complex geometries relative to other conventional manufacturing processes (Lyons, 2014; Paritala, Manchikatla, & Yarlagadda, 2017). The advantages of AM in A&D industry include reducing the expenses of physical warehousing, the capability design to complex tools, the low cost of parts, waste reduction and part simplification. Furthermore, AM machines produce less scrap than traditional machines and increase the availability of parts at the point of use.

Aerospace firms are increasingly adapting to the AM technologies to reduce the costs of developing models and of creating components. Several aerospace companies have started to adjust supply chain operations to accommodate AM parts (Cotteleer & Joyce, 2014; Sharon Ford, 2014; Kellner, 2014; Lyons, 2014). Current AM applications in the A&D industry have ranged from arm rests to more complex parts such as engine components. For instance, GE has been using additively produced parts inside commercial jet engines since approval of FAA in 2015 (Steenhuis & Pretorius, 2017). Airbus has also updated certain cabin brackets for the A380, making this the company's first commercial plane to use AM components (Bogue, 2013). In addition, NASA has already built its prototypes of tools and instrument parts for current missions using AM methods (Kietzmann, Pitt, & Berthon, 2015).

The modern automotive industry was another early adopter, and has used AM technology to make tool prototypes and small custom parts with short production runs for decades. Since production volumes in the automotive industry are often high, companies in this industry use AM technology to offers an increasing variety of different car models that are customizable due to mass customization, vast amounts of components create considerable overheads during production. AM could provide a novel approach to creating new products and reach the market faster than traditional subtracting manufacturing (Gibson et al., 2015; Jiang et al., 2017). Despite many advantages, however, most manufacturers have not incorporated to AM parts on their mass-produced car models. In fact, AM methods can be cost effective for specialty cars or low-volume manufacturers. The American company KOR Ecologic has applied AM technologies to mainly produced their car Urbee for only 50 components (Bechthold et al., 2015). There also have been some niche applications of AM technologies, some racing teams started to use AM parts for their cars. The racing organization Formula 1 has been using AM for rapid prototyping since the early to mid-2000s (Bechthold et al., 2015; Bogue, 2013; Gibson et al., 2015). Additionally, AM could make spare part inventories become obsolete, it means manufacturers need only to keep the CAD files of the parts, and then manufacture the spare parts demand.

Another application area is medical sector. AM methods have been used for medical applications when this technology was first commercialized. In recent years, AM technologies have gained momentum in the healthcare sector by providing new solutions in the areas of surgical guides, medical manufacturing and bioengineering (Giannatsis & Dedoussis, 2009; Gibson et al., 2015). The use of AM technologies for diagnostic purposes was presumably the first medical application. AM has provided opportunities to meet the demand for individualized health care products based on the data obtained from various well-developed techniques since the digitalization of patient data. For instance, AM methods are effective in transforming medical imaging data into bio models' due to the organic nature of the complex forms (Bechthold et al., 2015). AM methods have found widespread use as a tool to bioengineered tissue, varying in composition from bone and teeth to vascular and organ scaffolding. It currently uses a variety of biomaterials with different physical and biochemical properties to construct organic scaffolds for the tissue repair process (Hutmacher et al., 2004). Thus, customized AM manufactured scaffold allows for individualized treatment for tissue regeneration while avoiding the need to sample from the patient's own tissue for scaffold formation (Gross et al., 2014). It also allows for the design of customized prosthetic implants suited to individuals need, including size, shape and mechanical properties. AM methods also eliminate the need for time consuming steps of prosthetic fabrication (Jardini et al., 2014).

It is clear that AM has disruptive effect on different sectors from these application areas. It can be seen that AM has diffused into many sectors with different motivations and potential advantages. For instance, AM technologies are expected to present substantial positive effects on the structure of value chains, although this will be a gradual change, as it first depends on companies engaging in component and product redesign. Product and component redesign will affect the production system, as companies simplify complex multi-component products into single component products. Another benefit is that the traditional model of disruptive innovation underlines that disruptive technologies typically have a lower unit price, despite having a higher usage cost. Moreover, industries such are Aerospace or Medical have higher unit costs, in particular incurred due to the nature of materials. The cost of parts will likely to be higher due to the greater complexity in the architectural design of these product systems, in comparison to predecessors. In contrast, disruptive innovations do not succeed in isolated environments. New markets are needed to increase the value of the potential disruptive innovation likely be caused by simpler, more affordable innovations that create a need in the customers, motivating them to transfer from the traditional mainstream market to the new one generated by innovators. Put differently, a disruptive innovation processes are based on long-term characteristics. These characteristics will only be visible after the existing products and existing firms have been replaced by new products and new entrants. Even though, literature reviews give insight into AM applications; this review cannot show the overall views of diffusion, therefore a vital role of AM technologies bring out with patent analysis.

SOLUTIONS AND RECOMMEDATIONS

Innovation Diffusion Models and Growth Curves

Diffusion is defined as a factor stimulates the adoption of technologies by product which an innovation is adopted. According to E. M. Rogers (1962) Diffusion theory, innovation is defined as an idea, object or practice that is perceived as new by individuals or organizations. Thus, diffusion is defined as the process which ''is communicated through some certain channels over time among members of a social system'' (Soffer, Nachmias, & Ram, 2010). In the technological diffusion view, technology diffusion is defined as the process by which innovations spread within and across economies. From a theoretical perspective, Epidemic, Rank and Game Theoretic models are major approaches that have been characterized diffusion process (Bocquet, Brossard, & Sabatier, 2007). The use of epidemic models which consider diffusion as resulting from the spread of information. An early model proposed by Mansfield (1961) suggests that diffusion follows a simple logistic penetration of the market, and depends on speed of diffusion based on the frequency of contacts between potential users and actual users. Consequently, the potential determinants of the speed of adoption depend on the process by which people become informed about technology (Kiesling et al., 2012).

However, different models have been proposed to take into account different situations. The "Diffusion Model" proposed by Bass (1969) is one of the well-known models in the field. The theoretical foundation of the Bass diffusion model (Equation 1) is combination of models of Fourt and Woodlock (1960) and Mansfield (1961) (Chang, Lai, & Chang, 2009). Bass's model estimates how a new idea or new behavior spreads across the market over time. It analyzes the diffusion process through consumers' behavior, under the basic assumption that the probability of initial purchase is related linearly to the number of previous buyers. Buyers or adopters may be innovators or imitators. The original Bass model is shown in Equation (1).

$$n\left(t\right)=\left(p+q\frac{N\left(t-1\right)}{m}\right)\left(m-N\left(t-1\right)\right) \tag{1}$$

In Equation (1) n(t) is the sales of a new product at time t; N(t) is the cumulative sales of a new product at time t; m is the market potential; p is the innovation coefficient and q is the imitation coefficient. In a pure innovation scenario ($p > 0$, $q = 0$), diffusion follows a modified exponential model; ia pure imitation scenario ($p = 0$, $q > 0$), it follows a logistic curve model (Meade & Islam, 2006). In patent related analysis, Bass model has been modified and the new form is given close approximation compared to the original (Cunningham & Kwakkel, 2011; You et al., 2017). The Bass-equivalent model, which is validated by its prediction for patent citation growth, is defined in Equation (2) (You et al., 2017).

$$n\left(t\right)=a+b\sum_{i=1}^{t-1}n\left(i\right)+c\left[\sum_{i=1}^{t-1}n\left(i\right)\right]^{2} \tag{2}$$

In Equation (2), n(t) denotes the net patent amount at the time t while the symbols a, b and c are all fitting coefficients in model.

Another quantitative approach for understanding innovation is forecasting the diffusion of innovation. It has become a major research topic in the fields of innovation and technology research. Growth curve is an empirical model, originally developed in analogy to the growth of a living organism, designed to estimate the evolution of a quantity over time based on historical data (Meade & Islam, 2006; Young, 1993). Thus, they have been widely used to forecast analysis of technology lifecycles. Various studies have been conducted to identify the estimation of current development stage of technology domains with different growth curves (Meade & Islam, 2006; Yoon, Park, Kim, Lee, & Lee, 2014). For the patent data, the most frequently used models are The Gompertz model, the general Logistic model, the Pearl model and the Fisher-Pry model. The Pearl and Fisher–Pry models are generally used to forecast the cumulative growth of new adoptions (Cunningham & Kwakkel, 2011). Among those growth curves, Gompertz is considered appropriate for analyzing the life cycles of rapidly growing technologies (Bengisu & Nekhili, 2006; Martino, 2003; Yoon et al., 2014). The Gompertz model is shown in Equation (3)

$$Y_t = Le^{-ae^{-bt}} \quad (3)$$

In Equation (3) has three primary coefficients a, b, and L. The coefficients a and b, respectively, determine the location and shape of the curve, while L is its asymptotic maximum value (Meade & Islam, 2006; Yoon et al., 2014).

Patent Analysis

Patents, as a vital data source, are main outputs of research and development, and represent the characteristics of technology (Harhoff, Narin, Scherer, & Vopel, 1999; Trajtenberg, 1990). A vast amount of recent technical knowledge is available in patent documents, it has become increasingly important to exploit this knowledge since electronic patent database become accessible worldwide (Ernst, 2003). Hence, patent data have been considered as a key source in technology diffusion research (Cao & Zhao, 2013; Chang et al., 2009; Choi, Kim, & Park, 2007; Jun, Park, & Jang, 2015; Yan & Luo, 2016).

Patent analyses have been used to assess the knowledge diffusion and transfer processes in research and development which extract useful knowledge from databases. There are, of course, classification on using patent data as indicators of technology analysis (Karki, 1997). Patent analysis can be divided into three types of research level: Macro-level research of national or industrial analysis (Curran & Leker, 2011), micro-level research of particular technological diffusion analysis (Lee, Kim, & Cho, 2010) and forecasting (Chang et al., 2009). At the macro level, the major topics are technological innovation and evaluation of technological competitiveness of nations. At the micro-level, there are research activities such as identifying technological advantages and disadvantages of competitors (Choi et al., 2007; Cotropia, Lemley, & Sampat, 2013; Harhoff et al., 1999; Trajtenberg, 1990).

A patent contains "prior art'', which is to clearly explain the path of invention through citing linkages, such as patents, journals or other literature. In other words, the more a patent is cited, the wider the technology is diffused, and it implies a great number of applications and higher values (Lai & Wu, 2005). Patent citation analysis is widely used method for advanced analysis on technological diffusion through the examination of citation links among various patents. It is able to evaluate both the spill-over effects between industries and the importance of the technology-based transformation.

In this study, forward citation, a method for patent analysis, is applied. Forward citation in the patents is used to indicate the extent to which patents are closely related to technological breakthroughs and are considered technologically strong (Harhoff et al., 1999; Trajtenberg, 1990). In other words, a patent which is cited by other patents indicates its potential to create technological and economic value by making important contributions to the future path of technological diffusion (Karvonen & Kässi, 2013; Nemet & Johnson, 2012). In patents, there are key terms referred to as basic technologies which indicate the direction of information flow of the new technologies developed. The effectiveness of the technology is determined by the strength of the direction of information flow from key patents to newly filed ones. In typical patent evaluations, forward citation is used as "technological value" (Hu, Rousseau, & Chen, 2012). In order to compare each technology by in terms of technological strength, two performance indicators, Impact Index and Marketability index, are defined.

Impact Index (II)

The impact index is defined as the extent to which technology penetrate into other areas. In this case, number of citations to key patent can be determining as its impact. As each patent has different technological life time, these should be normalization for comparison. Impact index is shown in Equation (4).

$$II = \frac{C_i}{P_i} \tag{4}$$

e C_i denotes cumulative number of times i was cited and P_i. the number of patentin selected patent family. In this case, key patent has been already defined due to design of research. In order to compare patents citation periods are fixed.

Marketability Index (MI)

Marketability index (MI) is defined as the applicability of the products that utilizes a technology, and is evaluated using the number of citation over 4- digit CPC (Cooperative Patent Code) ratio in citing patents. This index can signal how technology diffuses into new technological territories and also perceived as qualifying technologies market size. MI is calculated is using Equation (5).

$$MI = \frac{F_i}{C_i} \tag{5}$$

where F_i denotes number of CPC codes in citing patents of key patents and C_i is number of times i was cited.

Analysis and Findings

Forward citation measure is frequently used for determining technological value of patents (Wang, Chiang, & Lin, 2010; Yasukawa & Kano, 2015). The definition of AM technologies covers a wide area of application. To understand the strength of technological breakthrough and growth based on time, forward citation analysis is mostly applied (Fallah, Fishman, & Reilly, 2011; Nair & Mathew, 2015). In this study, first commercial patents developed for liquid, powder, and solid-state materials, consisting SLA, SLS, FDM and LOM were selected as key AM methods (Coon et al., 2016; Han & Sohn, 2015; Huang et al., 2013). It is assumed that selected key AM methods are applied for different type of materials and application methods. Descriptive data for key methods are demonstrated in Table 1.

As can be seen in Table 1 the greatest number of patent citations was found for SLA method. Even though, FDM method is younger than others, it has more citation counts therefore it can be interpreted that solid-state materials have more impact than liquid and powder state materials. The change in the number of patent citations is a proxy measure of that technology's diffusion. In order to the understand diffusion status of AM technologies, Bass equivalent model was applied using the patent data between 1993 and 2016. Bass model evaluates technological change based on innovation and imitation ratios, unlike other diffusion models (Daim, Iskin, & Ho, 2011; Radojičić & Marković, 2009). The diffusion of key patents based on Bass model is demonstrated in Figure 1.

To compare diffusion and growth of selected key patents, the 1993-2016 periods was chosen because citation periods are equal for each patent. In this period, diffusion of AM technologies was evaluated based on all citation counts. For the evaluation of diffusion, Bass equivalent model, proposed by Cunningham and Kwakkel (2011) was used. As can be seen in Figure 1, fit ratio of model (R^2=0.60134) can be interpreted as low and suggesting that AM technologies have not developed as much as expected. In previous innovation diffusion studies, fit ratios of Gompertz model were higher than our findings (Lim et al., 2012; Nam, Kim, & Lee, 2008; Turk & Trkman, 2012). In addition, growth curves were used to test the technologies to determining in which state of product life cycle. From the reviewed literature, it is clear that Gompertz curve is more appropriate for determining growth stage in technological forecasting (Bengisu & Nekhili, 2006; Chang et al., 2009; Martino, 2003; Suominen & Seppänen, 2014; Yoon et al., 2014). The findings based on Gompertz curve analysis are demonstrated in Table 2.

Table 1. Key AM patents and forward citation counts

Method	Patent No.	Patent Title	Patent Registration Date	Forward Citation Counts*[1]
SLA	4575330	Apparatus for production of three-dimensional objects by stereolithography	03/ 11 / 1986	555
SLS	4863538	Method and apparatus for producing parts by selective sintering	09/ 05 /1989	355
FDM	5121329	Apparatus and method for creating three-dimensional objects	06 /09 /1992	351
LOM	4752352	Apparatus and method for forming an integral object from laminations	06 / 21 / 1988	192

Figure 1. Diffusion trends of key patents

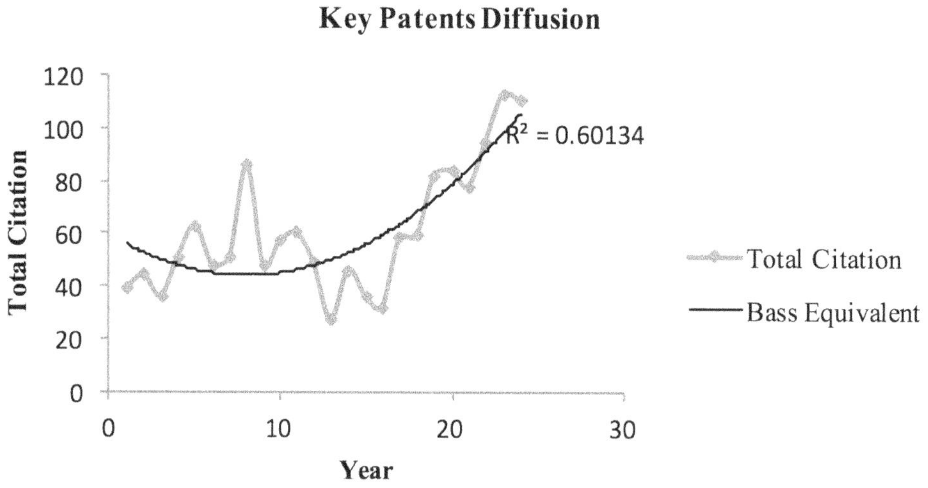

Table 2. Growth curve fit with Gompertz Model

Patent No	Coeff Det. (R^2)	Stand. Error
4575330	0,96	0,08
4752352	0,93	0,07
4863538	0,96	0,08
5121329	0,94	0,1

Based on the findings in Table 2, according to the Gompertz model, the growth curve of cumulative citations seems to be highly compatible with previous studies (Sood, James, Tellis, & Zhu, 2012; Suominen & Seppänen, 2014). As mentioned before, the Gompertz model is a growth curve used to describe the period in which the growth that has not yet reached maturity continues in the technological life cycle. Although selected key patents have lost their preemptive right based on the year limitation for intellectual property rights, the four key patent citation curves reveal that for the technological strength that the patents represent, is still retained and has not yet reached the maturity. Individual citation performances of each key patent are shown in Figure 2, Figure 3, Figure 4 and Figure 5 respectively.

In Figure 2, there is only slightly fluctuation between 1989 and 2005. After this period, there is rapid growth in citation of patent 4575330. It can be interpreted that those patents approximately same period but attention to methods differs from each other.

Figure 3 should has same citation behavior as for the years 1990-2005. The main differences between Figure 2 and Figure 3 are ceiling slopes of peak years. In Figure 3 there is a non-continuous growth trend. In Figure 4, patent is rather new than compared to other key patents. Its citation trend has been increasing since 2008. The time lag of inflection point is same as Figure 2.

In comparison with Figure 3, Figure 4 has only slightly amount of forward citations. By contrast to filling dates of each patent, it is expected more differences than current status.

Figure 2. Citation distribution of patent 4575330

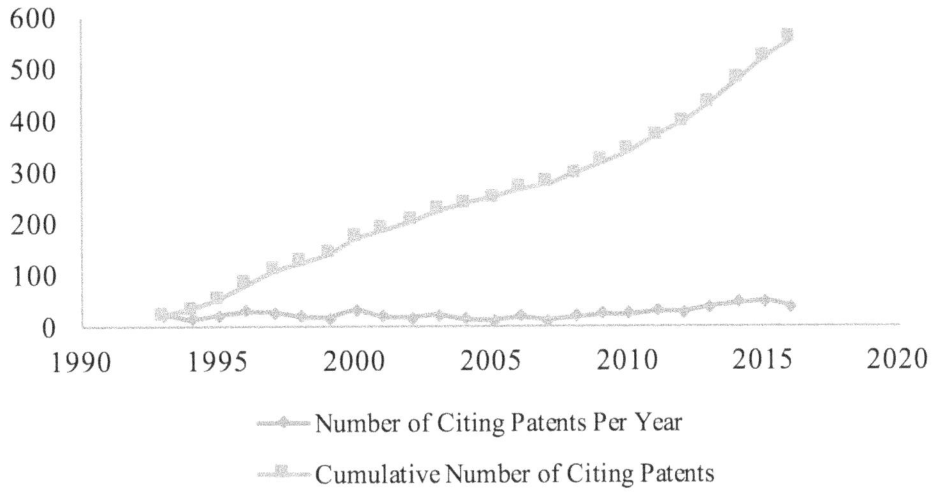

Figure 3. Citation distribution of patent 4863538

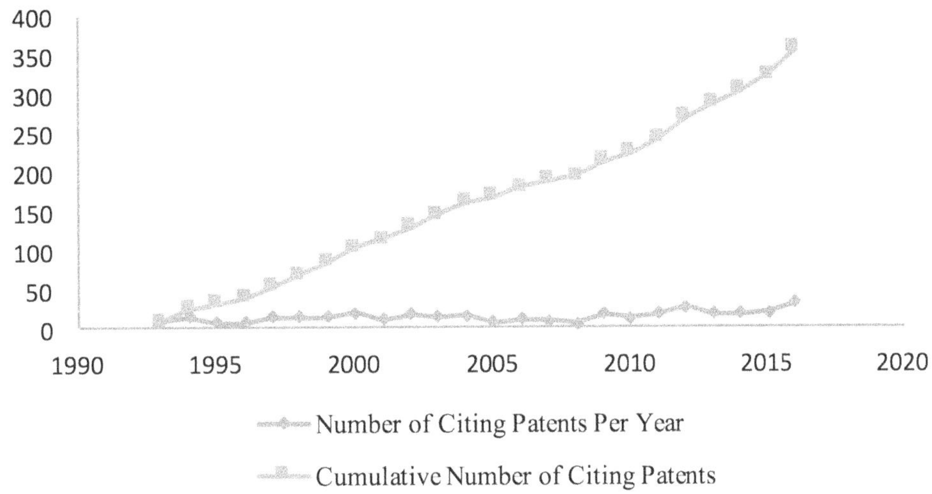

In Figure 5, it can be concluded that the yearly trend in patent citation is the most fluctuating patent among key patents, although, this patent was filed earlier than Figure 3, and Figure 4 citation count is lower than those patents. This implies that this patent has smaller impact than others. Annually growing trends per patents indicate more attention to specific technology in terms of time activity pattern. However, this method of explanation suffers from limitations to provide how technology dependence paths evolve. To do so, performance indicator will help to make pair wise comparison between patents. The impact index values of each patent are shown in Table 3.

Patent 4575330 has the highest impact index. Patent 5121329 is relatively more recent than orthers, its impact indexed as high as patent 4863538. Those patents represent different AM methodology in terms of initial material state. Patent 4752352 has lowest impact index. Thus, it is in the same material

Figure 4. Citation distribution of patent 5121329

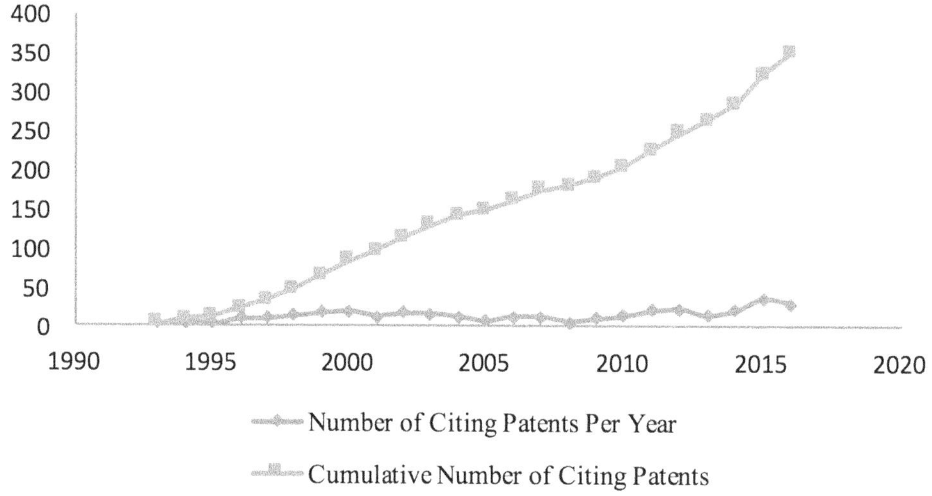

Figure 5. Citation distribution of patent 4752352

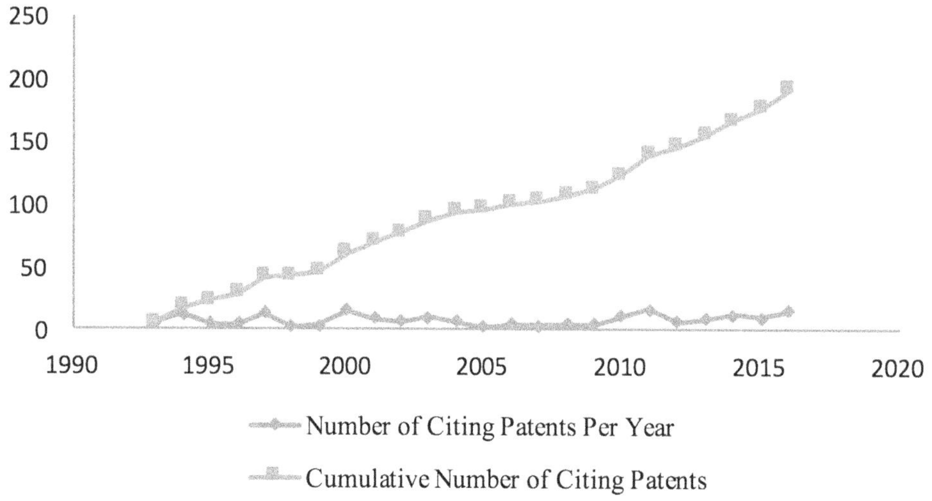

Table 3. Impact index of key patents

Patent No	Impact Index
4575330	0,38
4863538	0,25
5121329	0,24
4752352	0,13

family with Patent 4752352. This suggesting that time lag is not discriminative factor of technological impact and it may depend on other factors such as material, size, shape accuracy etc. Patent 4575330 is frequently applied, because it is liquid-based and often used for hobby activities (Bai, Liu, Wang, & Wen, 2017). In addition to technological impact, marketability impact is also analyzed based on the ratio of variety of CPC codes and citation count. Greater varied CPC codes showing that the method is found different application areas, and diffused into them. Marketability index values can be seen at Table 4.

As seen in Table 4, the smaller ratio of the MI indicates that technological effect more widely spread. Values reveal that commercial application of technology changes according to technology adaptation. 4752352 patent has the highest index value because, showing by the lower CPC code number compared to others.

FUTURE RESEARCH DIRECTIONS

The direction of further research is to study how researchers and industrial managers combine the selection of different types of AM methods or choose between them. For instance, to get more quantitative data, academic papers and patents can use as a proxy to measuring diffusion performance of each country or region. As a result of such research, both R&D managers and national and international policy makers get information on the ways of technological diffusion of 3D printing technologies which will be responding to the existing challenges of socio-technical issues at acceptable levels of risks and costs.

CONCLUSION

AM technologies have been considered as one of the most innovative technologies by both scholars and practitioners. Its potential to change the dynamics in many industries has made 3D Printing technologies a topic of great interest. In this study, 3D printing technologies have been examined from different standpoints.

In the literature reviewed, it can be seen that 3D printing technologies penetrate markets in which highly customized products are needed such as medicine and the automotive industry. Also, our review of literature has provided historical background of AM technologies and the four cores AM methods investigated in the analysis, Stereolithography (Patent no: 4575330), Selective Laser Sintering (Patent no: 4863538), Fused Deposition Modelling (Patent no: 5121329) and Laminated Object Manufacturing (Patent no: 4752352). In the analysis, forward citations have been used in order to understand the diffusion of the AM methods of choice.

Table 4. Marketability index of key patents

Patent No	MI	4-Digit CPC Number
4575330	0,22	120
4863538	0,26	98
5121329	0,28	98
4752352	0,44	84

The result of the conducted analysis indicates that AM technology has not yet reached the state of maturity. Result of the Bass Model suggests that the diffusion rate of the core AM technologies are not in accordance with the general fit of Bass model, indicating although there has been diffusion, it is not the expected level. The analysis on Impact Indices point out that despite emerging in slightly different timelines, the methods are very close in terms of extent of market penetration. This shows that the time between patents have limited effects compared to factors such as materials.

From the disruptive technologies and innovation points of view, this study points out that the current state of AM technologies is addressing to a niche customer. As it is known that disruptive technologies start with addressing the needs of niche customers, and later diffuse to a mainstream market, it can state that AM technologies have been slowly growing and evolving, and it can be considered disruptive technology with potential to creating major changes in the future. But it is also important to keep in mind that according to the results of the study, this is a gradual diffusion and it may take longer than scholars and practitioners predicted. Disruptive innovation is a process rather than an instantaneous event (Yu & Hang, 2010). Accordingly, in the case of 3D printing, this process is in motion, but is not as rapid as predicted. For this case, it is possible that 3D printing technologies are not yet in a position to be regarded as a revolution, but as we see its effects and growth in different sectors. We cannot say that it is an illusion either.

REFERENCES

Abernathy, W. J., & Clark, K. B. (1985). Innovation: Mapping the winds of creative destruction. *Research Policy*, *14*(1), 3–22. doi:10.1016/0048-7333(85)90021-6

Ahn, S.-H., Montero, M., Odell, D., Roundy, S., & Wright, P. K. (2002). Anisotropic material properties of fused deposition modeling ABS. *Rapid Prototyping Journal*, *8*(4), 248–257. doi:10.1108/13552540210441166

Bai, X., Liu, Y., Wang, G., & Wen, C. (2017). The pattern of technological accumulation: The comparative advantage and relative impact of 3D printing technology. *Journal of Manufacturing Technology Management*, *28*(1), 39–55. doi:10.1108/JMTM-10-2016-0136

Bak, D. (2003). Rapid prototyping or rapid production? 3D printing processes move industry towards the latter. *Assembly Automation*, *23*(4), 340–345. doi:10.1108/01445150310501190

Bass, F. M. (1969). A new product growth for model consumer durables. *Management Science*, *15*(5), 215–227. doi:10.1287/mnsc.15.5.215

Bechthold, L., Fischer, V., Hainzlmaier, A., Hugenroth, D., Ivanova, L., Kroth, K., . . . Sitzmann, V. (2015). *3D printing: A qualitative assessment of applications, recent trends and the technology's future potential.* Retrieved from https://www.econstor.eu/bitstream/10419/156627/1/StuDIS_2015-17.pdf

Bengisu, M., & Nekhili, R. (2006). Forecasting emerging technologies with the aid of science and technology databases. *Technological Forecasting and Social Change*, *73*(7), 835–844. doi:10.1016/j.techfore.2005.09.001

Berman, B. (2012). 3-D printing: The new industrial revolution. *Business Horizons*, *55*(2), 155–162. doi:10.1016/j.bushor.2011.11.003

Bertling, J., & Rommel, S. (2016). A Critical View of 3D Printing Regarding Industrial Mass Customization Versus Individual Desktop Fabrication. In S. Dickel, J.-P. Ferdinand, & U. Petschow (Eds.), The Decentralized and Networked Future of Value Creation (pp. 75-105). Springer.

Bocquet, R., Brossard, O., & Sabatier, M. (2007). Complementarities in organizational design and the diffusion of information technologies: An empirical analysis. *Research Policy*, *36*(3), 367–386. doi:10.1016/j.respol.2006.12.005

Bogue, R. (2013). 3D printing: The dawn of a new era in manufacturing? *Assembly Automation*, *33*(4), 307–311. doi:10.1108/AA-06-2013-055

Bourell, D. L. (2016). Perspectives on Additive Manufacturing. *Annual Review of Materials Research*, *46*(1), 1–18. doi:10.1146/annurev-matsci-070115-031606

Bourell, D. L., Beaman, J. J., Leu, M. C., & Rosen, D. W. (2009). A brief history of additive manufacturing and the 2009 roadmap for additive manufacturing: looking back and looking ahead. *Proceedings of RapidTech*, 24-25.

Bower, J. L., & Christensen, C. M. (1996). Disruptive technologies: Catching the wave. *Journal of Product Innovation Management*, *1*(13), 75–76.

Cao, Y., & Zhao, L. (2013). Analysis of patent management effects on technological innovation performance. *Baltic Journal of Management*, *8*(3), 286–305. doi:10.1108/BJOM-May-2012-0033

Caulfield, B., McHugh, P., & Lohfeld, S. (2007). Dependence of mechanical properties of polyamide components on build parameters in the SLS process. *Journal of Materials Processing Technology*, *182*(1), 477–488. doi:10.1016/j.jmatprotec.2006.09.007

Chang, S.-B., Lai, K.-K., & Chang, S.-M. (2009). Exploring technology diffusion and classification of business methods: Using the patent citation network. *Technological Forecasting and Social Change*, *76*(1), 107–117. doi:10.1016/j.techfore.2008.03.014

Chen, D., Heyer, S., Ibbotson, S., Salonitis, K., Steingrímsson, J. G., & Thiede, S. (2015). Direct digital manufacturing: Definition, evolution, and sustainability implications. *Journal of Cleaner Production*, *107*, 615–625. doi:10.1016/j.jclepro.2015.05.009

Cheng, Y., Huang, L., Ramlogan, R., & Li, X. (2016). Forecasting of potential impacts of disruptive technology in promising technological areas: Elaborating the SIRS epidemic model in RFID technology. *Technological Forecasting and Social Change*.

Choi, C., Kim, S., & Park, Y. (2007). A patent-based cross impact analysis for quantitative estimation of technological impact: The case of information and communication technology. *Technological Forecasting and Social Change*, *74*(8), 1296–1314. doi:10.1016/j.techfore.2006.10.008

Coon, C., Pretzel, B., Lomax, T., & Strlič, M. (2016). Preserving rapid prototypes: A review. *Heritage Science*, *4*(1), 40. doi:10.118640494-016-0097-y

Cotropia, C. A., Lemley, M. A., & Sampat, B. (2013). Do applicant patent citations matter? *Research Policy*, *42*(4), 844–854. doi:10.1016/j.respol.2013.01.003

Cotteleer, M., & Joyce, J. (2014). 3D opportunity: Additive manufacturing paths to performance, innovation, and growth. *Deloitte Review*, *14*, 5–19.

Crump, S. S. (1992). *Apparatus and method for creating three-dimensional objects*. Patent No. US5121329 A.

Cunningham, S. W., & Kwakkel, J. (2011). Innovation forecasting: A case study of the management of engineering and technology literature. *Technological Forecasting and Social Change*, *78*(2), 346–357. doi:10.1016/j.techfore.2010.11.001

Curran, C.-S., & Leker, J. (2011). Patent indicators for monitoring convergence–examples from NFF and ICT. *Technological Forecasting and Social Change*, *78*(2), 256–273. doi:10.1016/j.techfore.2010.06.021

Daim, T. U., Iskin, I., & Ho, D. (2011). Technology forecasting for residential energy management devices. *Foresight*, *13*(6), 70–87. doi:10.1108/14636681111179609

Danneels, E. (2004). Disruptive technology reconsidered: A critique and research agenda. *Journal of Product Innovation Management*, *21*(4), 246–258. doi:10.1111/j.0737-6782.2004.00076.x

Deckard, C. R. (1991). *Method and apparatus for producing parts by selective sintering*. Patent No. US5017753 A.

Desai, D. R., & Magliocca, G. N. (2013). Patents, meet Napster: 3D printing and the digitization of things. *Geological Journal*, *102*, 1691.

Ernst, H. (2003). Patent information for strategic technology management. *World Patent Information*, *25*(3), 233–242. doi:10.1016/S0172-2190(03)00077-2

Fallah, M. H., Fishman, E., & Reilly, R. R. (2011). Forward patent citations as predictors for patent valuation. *International Journal of Intellectual Property Management*, *4*(3), 165–178. doi:10.1504/IJIPM.2011.041081

Feygin, M. (1988). *Apparatus and method for forming an integral object from laminations*. Patent No. US4752352 A.

Fleming, L. (2001). Recombinant uncertainty in technological search. *Management Science*, *47*(1), 117–132. doi:10.1287/mnsc.47.1.117.10671

Fleming, L., & Sorenson, O. (2001). Technology as a complex adaptive system: Evidence from patent data. *Research Policy*, *30*(7), 1019–1039. doi:10.1016/S0048-7333(00)00135-9

Ford, S. (2014). Additive Manufacturing Technology: Potential Implications for U.S. Manufacturing Competitiveness. *Journal of International Commerce & Economics*.

Ford, S., & Despeisse, M. (2016). Additive manufacturing and sustainability: An exploratory study of the advantages and challenges. *Journal of Cleaner Production*, *137*, 1573–1587. doi:10.1016/j.jclepro.2016.04.150

Forrest, E., & Cao, Y. (2013). Digital Additive Manufacturing: A Paradigm Shift in the Production Process and Its Socio-economic Impacts. *Engineering Management Research*, *2*(2), 66. doi:10.5539/emr.v2n2p66

Fourt, L. A., & Woodlock, J. W. (1960). Early prediction of market success for new grocery products. *Journal of Marketing*, *25*(2), 31–38. doi:10.2307/1248608

Gao, W., Zhang, Y., Ramanujan, D., Ramani, K., Chen, Y., Williams, C. B., ... Zavattieri, P. D. (2015). The status, challenges, and future of additive manufacturing in engineering. *Computer Aided Design*, *69*, 65–89. doi:10.1016/j.cad.2015.04.001

Giannatsis, J., & Dedoussis, V. (2009). Additive fabrication technologies applied to medicine and health care: A review. *International Journal of Advanced Manufacturing Technology*, *40*(1), 116–127. doi:10.100700170-007-1308-1

Gibson, I., Rosen, D., & Stucker, B. (2015). *Additive manufacturing technologies: 3D printing, rapid prototyping, and direct digital manufacturing*. Academic Press.

Gordon, R. (2015). Trends in commercial 3D printing and additive manufacturing. *3D Printing and Additive Manufacturing, 2*(2), 89-90.

Gross, B. C., Erkal, J. L., Lockwood, S. Y., Chen, C., & Spence, D. M. (2014). Evaluation of 3D printing and its potential impact on biotechnology and the chemical sciences. *Analytical Chemistry*, *86*(7), 3240–3253. doi:10.1021/ac403397r PMID:24432804

Guo, N., & Leu, M. C. (2013). Additive manufacturing: Technology, applications and research needs. *Frontiers of Mechanical Engineering*, *8*(3), 215–243. doi:10.100711465-013-0248-8

Han, E. J., & Sohn, S. Y. (2015). Patent valuation based on text mining and survival analysis. *The Journal of Technology Transfer*, *40*(5), 821–839. doi:10.100710961-014-9367-6

Harhoff, D., Narin, F., Scherer, F. M., & Vopel, K. (1999). Citation frequency and the value of patented inventions. *The Review of Economics and Statistics*, *81*(3), 511–515. doi:10.1162/003465399558265

Hu, X., Rousseau, R., & Chen, J. (2012). A new approach for measuring the value of patents based on structural indicators for ego patent citation networks. *Journal of the American Society for Information Science and Technology*, *63*(9), 1834–1842. doi:10.1002/asi.22632

Huang, S. H., Liu, P., Mokasdar, A., & Hou, L. (2013). Additive manufacturing and its societal impact: A literature review. *International Journal of Advanced Manufacturing Technology*, 1–13.

Hull, C. W. (1986). *Apparatus for production of three-dimensional objects by stereolithography*. Patent No. US4575330 A.

Hur, W. (2017). The patterns of knowledge spillovers across technology sectors evidenced in patent citation networks. *Scientometrics*, *111*(2), 595–619. doi:10.100711192-017-2329-7

Hutmacher, D. W., Sittinger, M., & Risbud, M. V. (2004). Scaffold-based tissue engineering: Rationale for computer-aided design and solid free-form fabrication systems. *Trends in Biotechnology*, *22*(7), 354–362. doi:10.1016/j.tibtech.2004.05.005 PMID:15245908

Jardini, A. L., Larosa, M. A., Maciel Filho, R., de Carvalho Zavaglia, C. A., Bernardes, L. F., Lambert, C. S., ... Kharmandayan, P. (2014). Cranial reconstruction: 3D biomodel and custom-built implant created using additive manufacturing. *Journal of Cranio-Maxillo-Facial Surgery*, *42*(8), 1877–1884. doi:10.1016/j.jcms.2014.07.006 PMID:25175080

Jeantette, F. P., Keicher, D. M., Romero, J. A., & Schanwald, L. P. (2000). *Method and system for producing complex-shape objects*. Patent No. US6046426 A.

Jiang, R., Kleer, R., & Piller, F. T. (2017). Predicting the future of additive manufacturing: A Delphi study on economic and societal implications of 3D printing for 2030. *Technological Forecasting and Social Change*, *117*, 84–97. doi:10.1016/j.techfore.2017.01.006

Jin, Y., Ji, S., Li, X., & Yu, J. (2017). A scientometric review of hotspots and emerging trends in additive manufacturing. *Journal of Manufacturing Technology Management*, *28*(1), 18–38. doi:10.1108/JMTM-12-2015-0114

Jun, S., Park, S., & Jang, D. (2015). A technology valuation model using quantitative patent analysis: A case study of technology transfer in big data marketing. *Emerging Markets Finance & Trade*, *51*(5), 963–974. doi:10.1080/1540496X.2015.1061387

Karki, M. (1997). Patent citation analysis: A policy analysis tool. *World Patent Information*, *19*(4), 269–272. doi:10.1016/S0172-2190(97)00033-1

Karvonen, M., & Kässi, T. (2013). Patent citations as a tool for analysing the early stages of convergence. *Technological Forecasting and Social Change*, *80*(6), 1094–1107. doi:10.1016/j.techfore.2012.05.006

Kellner, T. (2014). World's First Plant to Print Jet Engine Noozles in Mass Production. *General Electric Reports*. Retrieved from http://www.gereports.com/post/91763815095/worlds-first-plant-to-print-jet-engine-nozzles-in/

Kiesling, E., Günther, M., Stummer, C., & Wakolbinger, L. M. (2012). Agent-based simulation of innovation diffusion: A review. *Central European Journal of Operations Research*, *20*(2), 183–230. doi:10.100710100-011-0210-y

Kietzmann, J., Pitt, L., & Berthon, P. (2015). Disruptions, decisions, and destinations: Enter the age of 3-D printing and additive manufacturing. *Business Horizons*, *58*(2), 209–215. doi:10.1016/j.bushor.2014.11.005

Kostoff, R. N., Boylan, R., & Simons, G. R. (2004). Disruptive technology roadmaps. *Technological Forecasting and Social Change*, *71*(1), 141–159. doi:10.1016/S0040-1625(03)00048-9

Küchler, S. (2014). Additive technology and material cognition: A view from anthropology. *Journal of Cognition and Culture*, *14*(5), 385–399. doi:10.1163/15685373-12342133

Lai, K.-K., & Wu, S.-J. (2005). Using the patent co-citation approach to establish a new patent classification system. *Information Processing & Management*, *41*(2), 313–330. doi:10.1016/j.ipm.2003.11.004

Lee, M., Kim, K., & Cho, Y. (2010). A study on the relationship between technology diffusion and new product diffusion. *Technological Forecasting and Social Change*, *77*(5), 796–802. doi:10.1016/j.techfore.2010.01.003

Levy, G. N. (2010). The role and future of the laser technology in the additive manufacturing environment. *Physics Procedia*, *5*, 65–80. doi:10.1016/j.phpro.2010.08.123

Levy, G. N., Schindel, R., & Kruth, J.-P. (2003). Rapid manufacturing and rapid tooling with layer manufacturing (LM) technologies, state of the art and future perspectives. *CIRP Annals-Manufacturing Technology*, *52*(2), 589–609. doi:10.1016/S0007-8506(07)60206-6

Lim, J., Nam, C., Kim, S., Rhee, H., Lee, E., & Lee, H. (2012). Forecasting 3G mobile subscription in China: A study based on stochastic frontier analysis and a Bass diffusion model. *Telecommunications Policy*, *36*(10), 858–871. doi:10.1016/j.telpol.2012.07.016

Lyons, B. (2014). Additive manufacturing in aerospace: Examples and research outlook. *The Bridge, 44*(3).

Mansfield, E. (1961). Technical change and the rate of imitation. *Econometrica*, *29*(4), 741–766. doi:10.2307/1911817

Manyika, J., Chui, M., Bughin, J., Dobbs, R., Bisson, P., & Marrs, A. (2013). *Disruptive technologies: Advances that will transform life, business, and the global economy*. Retrieved from http://www.mckinsey.com/~/media/McKinsey/Business Functions/McKinsey Digital/Our Insights/Disruptive technologies/MGI_Disruptive_technologies_Full_report_May2013.ashx

Martino, J. P. (2003). A review of selected recent advances in technological forecasting. *Technological Forecasting and Social Change*, *70*(8), 719–733. doi:10.1016/S0040-1625(02)00375-X

Meade, N., & Islam, T. (2006). Modelling and forecasting the diffusion of innovation–A 25-year review. *International Journal of Forecasting*, *22*(3), 519–545. doi:10.1016/j.ijforecast.2006.01.005

Mueller, B., & Kochan, D. (1999). Laminated object manufacturing for rapid tooling and patternmaking in foundry industry. *Computers in Industry*, *39*(1), 47–53. doi:10.1016/S0166-3615(98)00127-4

Munz, J. O. (1956). *Photo-glyph recording*. Patent No. US2775758 A.

Nair, S. S., & Mathew, M. (2015). The dynamics between forward citations and price of singleton patents. *International Journal of Innovation and Technology Management*, *12*(03), 1540003. doi:10.1142/S0219877015400039

Nam, C., Kim, S., & Lee, H. (2008). The role of WiBro: Filling the gaps in mobile broadband technologies. *Technological Forecasting and Social Change*, *75*(3), 438–448. doi:10.1016/j.techfore.2007.04.008

Nemet, G. F., & Johnson, E. (2012). Do important inventions benefit from knowledge originating in other technological domains? *Research Policy*, *41*(1), 190–200. doi:10.1016/j.respol.2011.08.009

Paritala, P. K., Manchikatla, S., & Yarlagadda, P. K. (2017). Digital Manufacturing-Applications Past, Current, and Future Trends. *Procedia Engineering*, *174*, 982–991. doi:10.1016/j.proeng.2017.01.250

Park, J., Tari, M. J., & Hahn, H. T. (2000). Characterization of the laminated object manufacturing (LOM) process. *Rapid Prototyping Journal*, *6*(1), 36–50. doi:10.1108/13552540010309868

Pierrakakis, K., Gkritzali, C. D., Kandias, M., & Gritzalis, D. (2015). *3D Printing: A Paradigm Shift in Political Economy?* Paper presented at the 65th International Studies Association's Annual Convention, New Orleans, LA.

Pierrakakis, K., Kandias, M., Gritzali, C., & Gritzalis, D. (2014). *3D Printing and its regulation dynamics: The world in front of a paradigm shift.* Paper presented at the 6th International Conference on Information Law and Ethics, Thessaloniki, Greece.

Radojičić, V. D., & Marković, G. Z. (2009). *New technology forecasting using the bass model.* Paper presented at the Telecommunication in Modern Satellite, Cable, and Broadcasting Services, 2009. TELSIKS'09. 9th International Conference on. 10.1109/TELSKS.2009.5339534

Rayna, T., & Striukova, L. (2016). From rapid prototyping to home fabrication: How 3D printing is changing business model innovation. *Technological Forecasting and Social Change*, *102*, 214–224. doi:10.1016/j.techfore.2015.07.023

Rifkin, J. (2012). The third industrial revolution: How the internet, green electricity, and 3-d printing are ushering in a sustainable era of distributed capitalism. *World Financial Review, 1.*

Rogers, E. M. (1962). *Diffusion of innovations.* New York: Free Press.

Rogers, H., Baricz, N., & Pawar, K. S. (2016). 3D printing services: Classification, supply chain implications and research agenda. *International Journal of Physical Distribution & Logistics Management*, *46*(10), 886–907. doi:10.1108/IJPDLM-07-2016-0210

Sachs, E. M., Haggerty, J. S., Cima, M. J., & Williams, P. A. (1993). *Three-dimensional printing techniques.* Patent No. US5204055 A.

Samuelson, R. J. (2000). The internet and gutenberg. *Newsweek (Atlantic Edition), 135,* 2-2.

Schmidt, G. M., & Druehl, C. T. (2008). When is a disruptive innovation disruptive? *Journal of Product Innovation Management*, *25*(4), 347–369. doi:10.1111/j.1540-5885.2008.00306.x

Soffer, T., Nachmias, R., & Ram, J. (2010). Diffusion of Web Supported Instruction in Higher Education-The Case of Tel-Aviv University. *Journal of Educational Technology & Society*, *13*(3), 212–223.

Sood, A., James, G. M., Tellis, G. J., & Zhu, J. (2012). Predicting the path of technological innovation: SAW vs. Moore, Bass, Gompertz, and Kryder. *Marketing Science*, *31*(6), 964–979. doi:10.1287/mksc.1120.0739

Steenhuis, H.-J., & Pretorius, L. (2016). Consumer additive manufacturing or 3D printing adoption: An exploratory study. *Journal of Manufacturing Technology Management*, *27*(7), 990–1012. doi:10.1108/JMTM-01-2016-0002

Steenhuis, H.-J., & Pretorius, L. (2017). The additive manufacturing innovation: A range of implications. *Journal of Manufacturing Technology Management*, *28*(1), 122–143. doi:10.1108/JMTM-06-2016-0081

Suominen, A., & Seppänen, M. (2014). Bibliometric data and actual development in technology life cycles: Flaws in assumptions. *Foresight*, *16*(1), 37–53. doi:10.1108/FS-03-2013-0007

Swainson, W. K. (1977). *Method, medium and apparatus for producing three-dimensional figure product.* Patent No. US4041476 A.

Trajtenberg, M. (1990). A penny for your quotes: Patent citations and the value of innovations. *The Rand Journal of Economics*, *21*(1), 172–187. doi:10.2307/2555502

Turk, T., & Trkman, P. (2012). Bass model estimates for broadband diffusion in European countries. *Technological Forecasting and Social Change*, *79*(1), 85–96. doi:10.1016/j.techfore.2011.06.010

Uriondo, A., Esperon-Miguez, M., & Perinpanayagam, S. (2015). The present and future of additive manufacturing in the aerospace sector: A review of important aspects. *Proceedings of the Institution of Mechanical Engineers. Part G, Journal of Aerospace Engineering*, *229*(11), 2132–2147. doi:10.1177/0954410014568797

Vecchiato, R. (2016). Disruptive innovation, managerial cognition, and technology competition outcomes. *Technological Forecasting and Social Change*.

Walsh, S. T. (2004). Roadmapping a disruptive technology: A case study: The emerging microsystems and top-down nanosystems industry. *Technological Forecasting and Social Change*, *71*(1), 161–185. doi:10.1016/j.techfore.2003.10.003

Walsh, S. T., & Linton, J. D. (2000). Infrastructure for emergent industries based on discontinuous innovations. *Engineering Management Journal*, *12*(2), 23–32. doi:10.1080/10429247.2000.11415071

Wang, J.-C., Chiang, C., & Lin, S.-W. (2010). Network structure of innovation: Can brokerage or closure predict patent quality? *Scientometrics*, *84*(3), 735–748. doi:10.100711192-010-0211-y

Weeks, M. R. (2015). Is disruption theory wearing new clothes or just naked? Analyzing recent critiques of disruptive innovation theory. *Innovation*, *17*(4), 417–428. doi:10.1080/14479338.2015.1061896

Weller, C., Kleer, R., & Piller, F. T. (2015). Economic implications of 3D printing: Market structure models in light of additive manufacturing revisited. *International Journal of Production Economics*, *164*, 43–56. doi:10.1016/j.ijpe.2015.02.020

Yan, B., & Luo, J. (2016). Measuring technological distance for patent mapping. *Journal of the Association for Information Science and Technology*.

Yasukawa, S., & Kano, S. (2015). Comparison of examiners' forward citations in the United States and Japan with pairs of equivalent patent applications. *Scientometrics*, *102*(2), 1189–1205. doi:10.100711192-014-1466-5

Yoon, J., Park, Y., Kim, M., Lee, J., & Lee, D. (2014). Tracing evolving trends in printed electronics using patent information. *Journal of Nanoparticle Research*, *16*(7), 2471. doi:10.100711051-014-2471-6

You, H., Li, M., Hipel, K. W., Jiang, J., Ge, B., & Duan, H. (2017). Development trend forecasting for coherent light generator technology based on patent citation network analysis. *Scientometrics*, *111*(1), 297–315. doi:10.100711192-017-2252-y

Young, P. (1993). Technological growth curves: A competition of forecasting models. *Technological Forecasting and Social Change*, *44*(4), 375–389. doi:10.1016/0040-1625(93)90042-6

Yu, D., & Hang, C. C. (2010). A reflective review of disruptive innovation theory. *International Journal of Management Reviews*, *12*(4), 435–452. doi:10.1111/j.1468-2370.2009.00272.x

KEY TERMS AND DEFINITIONS

3D Printing: An innovative technology in the production and services industries that changes, substitutes, or supports the parts of/the whole chain of the traditional production paradigm through offering easier and cost-efficient solutions to businesses value or consumers.

Additive Manufacturing: Nontraditional production concept that is building objects up to one layer at a time from digital design without any mold or manufacturing background.

Disruptive Innovation: Is conceived as results from a combination of the emergent qualities of ideas based on observing the world differently, thinking the unthinkable and taking pre-emptive action to influence the environment.

Disruptive Technology: A bunch of scientific discoveries that changed ordinary product or technology capabilities into recent form in the basis of new competitive paradigm.

Forward Citation: The type of citation that receives citations from later papers or patents.

Growth Curve: Is an S-shaped curve that can frequently use in technology diffusion literature to describe performance of technology and diffusion over the life of technology.

Innovation Diffusion: Seeks to understand how new ideas, social learning styles, organized information dissemination and products spread throughout a society over time.

Patent Analysis: Is a method based on structured information of patent data such as filing dates, assignees, citations to analyze by bibliometric methods or data mining techniques.

ENDNOTE

1 In order to compare each patent citation period is fixed into 1993-2016.

This research was previously published in Disruptive Technologies for Business Development and Strategic Advantage edited by Anatoly V. Zhuplev, pages 213-244, copyright year 2018 by Business Science Reference (an imprint of IGI Global).

Chapter 21
Big Data, 3D Printing Technology, and Industry of the Future

Micheal Omotayo Alabi
University of the Witwatersrand, South Africa

ABSTRACT

This article describes how 3D printing technology, also referred to as additive manufacturing (AM), is a process of creating a physical object from 3-dimensional digital model layers upon layers. 3D printing technologies have been identified as an emerging technology of the 21st century and are becoming popular around the world with a wide variety of potential application areas such as healthcare, automotive, aerospace, manufacturing, etc. Big Data is a large amount of imprecise data in a variety of formats which is generated from different sources with high-speed. Recently, Big Data and 3D printing technologies is a new research area and have been identified as types of technologies that will launch the fourth industrial revolution (Industry 4.0). As Big Data and 3D printing technology is wide spreading across different sectors in the era of industry 4.0, the healthcare sector is not left out of the vast development in this field; for instance, the Big Data and 3D printing technologies providing needed tools to support healthcare systems to accumulate, manage, analyse large volume of data, early disease detection, 3D printed medical implant, 3D printed customized titanium prosthetic, etc. Therefore, this article presents the recent trends in 3D printing technologies, Big Data and Industry 4.0; including the benefits and the application areas of these technologies. Emerging and near future application areas of 3D printing, and possible future research areas in 3D printing and Big Data technologies as relating to industry 4.0.

1. INTRODUCTION

Wohlers (2010) based on ASTM F2792 standard, additive manufacturing is officially defined as the process of joining materials to make objects from 3D model data, usually layer upon layer, as opposed to subtractive manufacturing methodologies. Additive manufacturing is also referred to as 3D printing technology. In this paper, 3D printing technology will frequently be used instead of additive manufac-

DOI: 10.4018/978-1-5225-9624-0.ch021

turing. The 3D printing technology uses .STL files which specifies the information about the objects to be printed, as shown in Figure 1 (Schubert et al., 2014). The 3D printing technology has already been applied in several sectors like automotive, aeronautic, medical and dental, manufacturing, academic, tooling, construction, electronic, moulding, etc. (Gross et al., 2014).

Industry 4.0 is one of the major discussed topics in the industry and academia, because of its great potential for the manufacturing industries and new opportunities for various applications (Foidl and Felerer, 2016). Several many years, organizations and businesses have been analysing data, but the advancements in computing have brought new opportunities and thereby unleashing the Big Data potentials (Eastwood, 2017). The IBM 2014 innovation report identifies 3D printing technology and Big Data as part of the emerging industries and IBM has been able to use Big Data technologies to detect flaws in their operations or gauge user's opinions on the products they buy (Parmar et al., 2014; Eastwood, 2017). Furthermore, from healthcare perspective, "Big data in health' encompasses high volume, high diversity biological, clinical, environmental, and lifestyle information collected from single individuals to large cohorts, in relation to their health and wellness status, at one or several time points" (Heinrich et al., 2016).

This paper is arranged as follows: the next section introduces the industry of the future known as industry 4.0 and its benefits; section 3 presents the key important areas of 3D printing in industry 4.0; section 4 briefly describes the emerging and near future applications of the 3D printing technology; Section 5 introduces an overview of the term 'Big Data', the V's of Big Data, the technologies behind Big Data and the Big Data challenges; Section 6 introduces possible Future Research for 3D Printing, Big Data and Industry 4.0 and the final section is conclusions.

2. THE INDUSTRY OF THE FUTURE - INDUSTRY 4.0

Industry of the future is referred to as the industry 4.0 which stands for the fourth industrial revolution and other terms used for Industry 4.0 are 'Industrial Internet', 'Smart Factory' or 'Digital Factory'. In 2011, the term industry 4.0 was originated from Germany at Hannover Messe; Hannover Messe is one of the world's largest trade fairs. Geissbauer et al. (2016) defined "Industry 4.0 as an industry that focuses on the end-to-end digitization of all physical assets and integration into digital ecosystems with value chain partners". Industry 4.0 has the potential to analyse, generate and communicate data smoothly and continuously thereby supporting the aim and purposes of Industry 4.0, which networks a wide range of novel technologies to create value".

Industry 4.0 is also referred to as "the digital transformation of manufacturing, leveraging third platform technologies, such as Big Data/Analytics and innovation accelerators, such as the Industrial Internet of Things; and requiring the convergence of IT (Information Technology) and OT (Operational

Figure 1. Generalized 3D Printing Processes (Chamberlin, 2014)

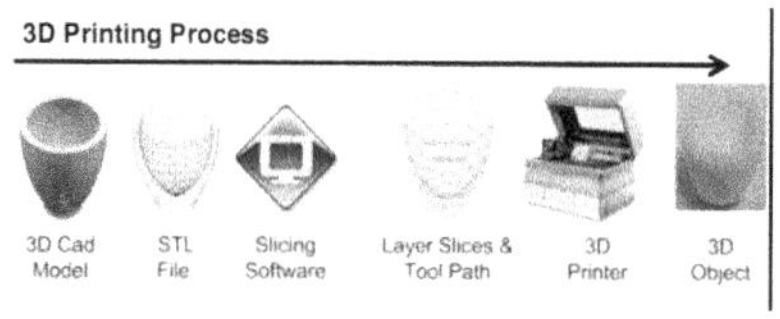

Technology), robotics, data and manufacturing processes to realize connected factories, smart decentralized manufacturing, self-optimizing systems and the digital supply chain in the information-driven cyber-physical environment of the fourth industrial revolution" (I-Scoop, 2016).

Industry 4.0 is a system-based platform, real-time end-to-end planning and horizontal collaboration that uses cloud-based planning platforms (Geissbauer et al., 2016). Industry 4.0 framework comprises of nine digital technologies and each technology have different application areas. These nine digital technologies include (augmented reality, cloud computing, Internet of things (IOT), Big Data analytics, cyber security, 3D printing, smart sensors, system integration, autonomous robots, and simulation) as shown in Figure 2.

In 2016, Boston Consulting Group (BCG) report presents the nine existing application areas of industries 4.0 as shown in Table 1 and U.S manufacturing industries was used as the case study. In the past few years, it is obvious that some companies have invested in a few of these nine technologies; most especially in cloud computing, Big Data/Analytics, 3D printing and Industrial Internet of Things (I-Scoop, 2017).

Figure 2. The Nine Digital Technologies of Industry 4.0 framework (I-Scoop, 2017)

Table 1. Identification of the nine industries 4.0 digital technology and application areas

S/N	The Nine Digital Technologies of Industry 4.0	Emerging Application Areas of Industry 4.0
1.	Advanced Robotics	• Autonomous, cooperating industrial robots • Numerous integrated sensors and standardized interfaces
2.	Additive Manufacturing/3D Printing Technologies	• 3D printing, particularly for spare parts and prototypes • Decentralized 3D facilities to reduce transport distances and inventory
3.	Augmented Reality	• Augmented reality for maintenance, logistics, and all kinds of standard operating procedures (SOP). • Display of supporting information, for instance, through the use of glasses
4.	Simulation	• Industry 4.0 allows value networks for simulation of purposes • Optimization based on real-time data from intelligent system.
5.	System Integration	• Cross-company data integration based on data transfer standards • Precondition for a fully automated value chain (from supplier to customer, from management to shop floor)
6.	Cyber-Security	• Uses of operation in networks and open systems • High level of networking between intelligent machines and products and systems.
7.	Internet of Things (IoT)	• IoT enables networking of machines and products. • Multi-directional communication between networked objects
8.	Cloud Computing	• Management of huge data volumes in open systems • In production system real-time communication is needed for production systems through industry 4.0
9.	Big Data and Analytics	• Available data can be fully evaluated (e.g. from ERP, SCM, MES, CRM, and machine data) • Industry 4.0 allows real-time decision-making support and optimization

Source: (I-Scoop, 2017; BCG, 2016)

3. THE BENEFITS OF INDUSTRY 4.0

Industry 4.0 is providing important benefits for industry and assisting them to change the conventional way they work. Industry 4.0 goals are to make manufacturing and related industries such as supply chain, manufacturing, logistics, etc.; to be customer-centric and efficient in their day-to-day operations; Adoption of Industry 4.0 helps industries in automation, optimization and achieving new business models. The following are the key benefits of industry 4.0 as highlighted below (I-Scoop, 2017):

1. Industry 4.0 uses optimization and automation to enhance productivity
2. In a real-time economy, industry 4.0 allows a real-time data collection for a real-time supply chain
3. Higher business continuity through advanced maintenance and monitoring possibilities
4. Industry 4.0 encourages good working condition and sustainability.
5. Industry 4.0 increases real-time monitoring and uses Internet of Things to improve better products quality.
6. Industry 4.0 uses Internet of Things to enable quality improvement
7. Personalization and customization for the ‘new’ customer and improvement in agility
8. Industry 4.0 brings about innovative capabilities and new revenue models.

3.1. The 3D Printing Technologies

3D printing is an integral component of industrial 4.0 and industrial 4.0 has provided an enabling environment where processes, software and technology work together perfectly. For industrial 4.0 to attain its full potential, a complete workflow has to be established which would allow other technologies such as Big Data, Smart Sensor, Cloud Computing, and 3D Printing to work alongside digital technologies to form a part of global manufacturing operations (RP-PLATFORM, 2017). In the past five years, some software has been developed by industry partner in additive manufacturing such as SAP, Siemen, and General Electric to keep up with the growth the manufacturing of sophisticated 3D printers and discovery of new materials, which allows integration with manufacturers' CAD platform, slicing program and Enterprise Resources Planning (ERP) system. This software can also provide real-time quality control of 3D printing products.

In the past three decades, many additive manufacturing and 3D printing technologies have been designed and these technologies are based on the way layers are being deposited to create physical components and the type of materials used (HS3DP, 2017). Additive manufacturing and 3D printing technologies are divided into two basic methods (can also be seen in Table 2) as highlighted below:

- **Powder based methods**: This is a type of technologies that uses a soften material to produce physical 3D printed components. Such technologies under powder-based method includes Fused Deposition Modelling and Electron Beam Melting.
- **Liquid based methods**: This method uses a cure liquid material with the assistance of some other sophisticated technologies to produce physical 3D printed parts, such liquid-based technologies includes: Stereolithography and Laminated Object Manufacturing.

Table 2. Six 3D Printing Technologies

S/N	3D Printing Technologies	Brief Description of the Technologies
1.	Stereolithography (SLA)	Stereolithography technology uses an ultraviolet beam to harden liquid resin, bonding each successive layer to produce physical objects
2.	Fused Deposition Modeling (FDM)	This technology uses a stream of melted thermoplastic material which are extruded through the nozzle to create physical layers of 3D components, and each layer bonding to the previous layer. The common materials include Acrylonitrile Butadiene Styrene (ABS) and polyactic acid polymers
3.	Selective Laser Sintering (SLS)	This technology uses powdered based materials (for example, nylon, titanium, aluminium, polystyrene, and glass). Powder is jetted from many nozzles onto the print surface. Laser is used to sinter the powdered materials layer by layer to produce physical parts.
4.	Selective Laser melting (SLM)	This technology uses powdered based material and melted at a very high temperature. Selective Laser Melting and Selective Laser Sintering are similar technology, but the differences is SLM melted material and SLS fused material.
5.	Electron Beam Melting (EBM)	This technology uses an electron beam as the power source and EBM is like SLS.
6.	Laminated Object Manufacturing (LOM)	LOM are bonded in successive layers and cut into shapes. This technology uses metals, plastics or paper as materials. In some cases, the parts produced through LOM are worked on further using drilling and machine to complete the final product.

Source: (Curran, 2016)

4. EMERGING AND NEAR FUTURE APPLICATION AREAS OF 3D PRINTING

Various advancement in research and innovation has brought so many discoveries of 3D printing technology. The use of 3D printing technology has rapidly increased and several future uses of 3D printing have emerged. Table 3 shows some of the industrial sectors and among others; some emerging industrial application areas of 3D printing has been identified.

5. AN OVERVIEW OF BIG DATA

Big Data is referred to as a large volume of both structured and unstructured data that are very large and difficult to be analysed through the conventional databases and software technologies. Data are collected through different sources. For instance, sensors are used to gather climate information, social media posts, digital pictures and videos through mobile devices, purchase transaction records, GPS signals through cell phone, etc. (Johnsen, 2017). Big Data is also defined as "the collection of complex and large datasets that are very difficult to capture, process, store, search and analyse using traditional data systems" (Kaur and Shah, 2017; Walker, 2015).

Despite of the huge benefits of Big Data technologies, there are some certain challenges facing Big Data which includes data capturing, data storage, data analysis, data curation, data searching, data sharing and transfer, data visualization, querying and updating of data and information privacy. It was noted that as the data increases exponentially, the current data processing techniques are becoming outdated. Big Data definition can be simplified using 3V's, i.e. data volume, data velocity and data variety. The three 'V''s are the driving forces for Big Data quantification as stated by Gartner analyst (Beyer and Laney, 2012).

Table 3. Emerging Application Areas of 3D Printing Technology in Different Industrial Sectors

S/N	Industry Sector	Some Emerging and Near Future Applications of 3D Printing
1.	Automotive and Industrial Manufacturing	• Integrating of several components into a one complex part • 3D printing creates capacities for production tooling • It allows the production of space parts • It speed-up product development cycle
2.	Aerospace Sector	• Produce complex geometry parts that cannot be manufactured using traditional manufacturing • 3D printing controls the stiffness, density, and other material properties of an aerospace part. • 3D printing assisting the aerospace industry to design and produce lighter parts
3.	Pharma/Healthcare	• Used to precise anatomical models based on ICT scan or MRI in surgery planning. • Orthopaedic implants and prosthetics can be customized for medical purposes. • For medical training, 3D printed cadavers can be used as an alternative.
4.	Retail	• 3D printing is being used to produce custom toys, jewellery, games, home decoration products. • For auto and home repair, it can be used to print spare or replacement parts
5.	Sports	• Produce complex geometry parts that cannot be manufactured using traditional manufacturing • Used to create perfect custom protective gear for better fit and safety • To manufacture custom spike plates for soccer shoe using biomechanical data.

Source: (Curran, 2016)

5.1. The V's of Big Data

Gartner defines Big Data using the 3V's as follows "*Big data is high-volume, high-velocity and/or high-variety information assets that demand cost-effective, innovative forms of information processing that enable enhanced insight, decision making, and process automation*"

Recently, it was argued that 3V's of Big Data designed by Gartner Inc in 2001, are not enough to illustrate the term "Big Data" technologies. Therefore, IBM data scientists break Big Data into four specific categories as data volume, data variety, data velocity, and data veracity as shown in Figure 3.

The 4V's of Big Data as described by IBM data scientist below:

1. **Data Volume:** This is the scale of data size or the amount of data from datasets with sizes of terabytes to zettabyte, the volume of data stored could be through but not limited to transaction, live streaming, and sensors (Kaur and Shah, 2017; Su, 2017).
2. **Data Variety:** This describes the category of data collected and the type of Big Data it belongs; that is the Data can be classified into different data categories, for instance historical data, structured, unstructured, semi-structured data.
3. **Data Velocity:** This is the speed of the data. Huge amounts of data from transactions with high speed rate will result in data streams entering at a high speed. This led to a shift from batch processing to real time streaming of data (Su, 2017).
4. **Data Veracity:** There are degree of uncertainty from the data collected and incorrect data can cause a lots problem for the organization. Accurate data is important especially in an organization where automated decision-making that is human-centric. (van Rijmenam, 2014).

Van Rijmenam (2013) in his book mentioned additional three V's namely, Variability, Visualization, and Value as briefly described below:

1. **Data Variability:** Big data is extremely variable and data variability is very important when performing sentiment analyses. For instance, "in the same online 'tweets', a word can have a totally several meanings. Therefore, to carry out a proper sentiment analyses, algorithms need to be able

Figure 3. The 4V's of Big Data as Characterized by IBM. (Allen, 2016)

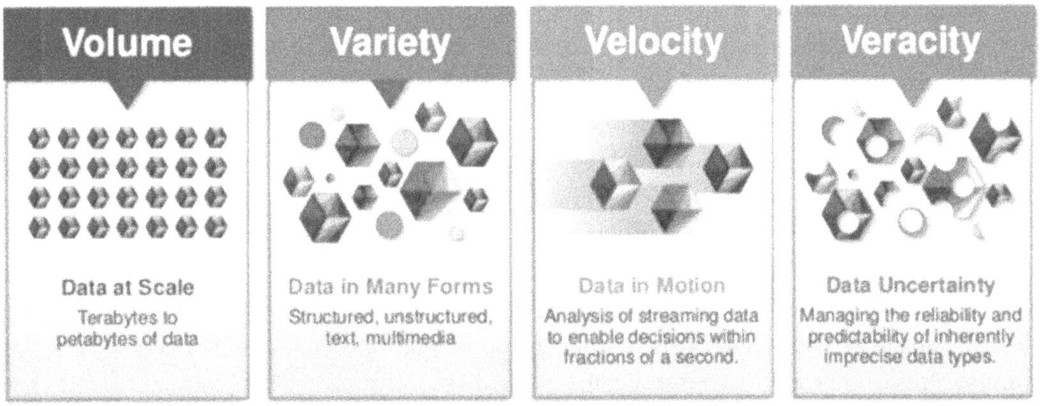

to decode or interpret both the context and decipher of the exact meaning of a word in that context, which is very difficult" (Van Rijmenam, 2013).

2. **Data Visualization:** This is the most difficult aspect of Big Data, the large amount of data collected must be comprehensible in such a manner that it is easy to understand and read. Visualization does not mean ordinary graphs or pie charts, but complex graphs that includes different variables of data while remains understandable and readable (Van Rijmenam, 2013).
3. **Data Value:** All the data collected produces lots of value for the enterprises, societies, and consumers. The value of the data lies in the way organization uses the data and turn the organization into an information-centric enterprise which relies on the insights derived from data analyses for rightful decision-making for the organization (Van Rijmenam, 2013).

5.2. Key Technologies Behind Big Data

The Big Data is an emerging technology being driven by some technologies, namely, Hadoop, MapReduce and NoSQL. Enterprises are using these technologies to analyse, visualize and rationalize Big Data in a way that it provides them the opportunity to quickly separate the actionable insight from huge waste raw input of data. These technologies are open source licensing which brings the cost of Big Data implementation under control. The four technologies Hadoop, MapReduce, NoSQL, and MongoDB are briefly explained as follows:

- **Hadoop:** It is a software framework used for distributed processing of large datasets across large clusters of computers; which can be installed on a commodity Linux cluster which allows large scale distributed data to be analysed (Eltabakh, 2012). Hadoop is an open-source implementation for Google MapReduce which is based on simple data model which allows any data to fit in. Hadoop requires no hardware modification, except the changes to meet recommended disk space, RAM, etc (Taylor, 2010). This technology comprises of two main layers framework, i.e. Execution Engine (MapReduce) and Hadoop Distributed File System, HDFS (Eltabakh, 2012). Most importantly, Hadoop is a widely used tool for historical Big Data analytics and it is not designed to handle real-time data or streaming data, this is one of the disadvantage of Hadoop (Scott, 2017). Despite this disadvantage, there are other technologies options for handling streaming and real-time data such as MongoDB, Spark Streaming, Apache Samza/Apache Flink, etc. (Scott, 2017).
- **MapReduce:** This is a programming model and an associated implementation for processing and generating large datasets. It allows the users to specify a map function which processes a value or key pair that generates a set of intermediate value or key pairs, and a reduce function that merges all intermediate values associated with the same intermediate key (Dean and Ghemawat, 2004). MacLean (2011) explains that the within MapReduce technology framework users can achieve the following, namely, a web-link graph traversal, distributed search, distributed sort, and machine learning.
- **NoSQL:** NoSQL databases take a different approach in solving Big Data problems and has been identified as the next generation databases which is not relational database. NoSQL is horizontally scalable to hundreds of nodes with high availability and transparent load balancing (Mousavi, 2011). Many inherent architecture and design considerations such as schema-less design, compromise an ACID property, predominantly RAM-based processing for large set of data, key-val-

ue/document-based storage have been considered in most of the NoSQL databases (AspireSys, 2017). The term "ACID" property means Atomicity, Consistency, Isolation, Durability, and ACID is a concept commonly used when referring to a database system of four transaction properties.

- **MongoDB:** It is an open source cross-platform document-oriented database program and it is being classified as a NoSQL database program. MongoDB can "underpin many Big Data systems, not only as a real-time, operational data store but in offline capacities as well". MongoDB allows enterprises to provide more data, extend users capabilities and more insight with greater ease (MongoDB, 2017a). MongoDB easily incorporates any type of data, irrespective of the structure, format or source of the data. MongoDB has the capabilities to analyse data of any structure such as sensor data, market data, click streams, social data, log data; directly within the database, and the results are provided in real-time without expensive data warehouse loads (MongoDB, 2017b).

5.3. Type of Big Data Analytics

The advent of Big Data has brought about better performance which allows manufacturing industries and enterprises to make significant and strategic decisions based on thorough analysis of evidence rather than mere intuition. Smith (2017) and Shah (2016) explain that "a rightful Big Data analytics can assist the organization to deliver richer insight due to multiple sources of data which helps to uncover some hidden patterns and relationship". There are four types of Big Data analytics as described below:

- **Prescriptive Analytics:** Prescriptive analytics uses optimization and simulation algorithm to give possible outcomes of a situation (Smith, 2017). A prescriptive data can provide impressive results when used correctly (Nida, 2015).
- **Predictive Analytics:** Predictive analytics identifies the past data patterns and gives some likely outcomes for a given situation. When the current and historical data is studied, predictive analysis will present a forecast of what may happen in the future. Predictive analytic method is one of the most commonly used analytical techniques. This method can be used for Big Data to predict other data that is not yet available (Nida, 2015).
- **Diagnostic Analytics:** This is used for discovery or to determine why something happened (Shah, 2016). The diagnostic analysis looks at past performance to understand what happened and why it happened (Nida, 2015).
- **Descriptive Analytics:** Descriptive analytic examines what is happening in real-time based on the in-coming data. Big Data can be converted into useful bite-sized nuggets using descriptive analytic technique. Major enterprises and organization uses descriptive analytics technique because it provides a platform to monitor the results in real-time through email reports or from a dashboard (Nida, 2015). Descriptive analytics is referred to as data mining.

5.4. Recent Trends and Activities of Big Data

Big Data is being adopted across various sectors which includes but not limited to: health and wellbeing, banking, urban transport, waste management, health care, economic development, retail, predicting and understanding crime, fire service, and 3D printing technology. The recent trends and activities of Big Data in some sectors are explained in Table 4:

Table 4. Some of the Trends and Activities of Big Data in Different Areas

S/N	Application Sectors	Recent Trends in Big Data Technologies
1.	Health Care	• In the health care sectors, Big Data has been used to improve the ability to detect the prevalence of flu, faster than official source through Google Flu. • Big Data has been applied across specific geographical and demographics areas to predict the cost of managing specific illness. • Big Data has assisted the United States to reduce patient mortality by 20% through stream analysis of patient data.
2.	Waste Management	• Waste management made easier using Big Data technologies, because it helps to deliver a more efficient and environmentally friendly waste collection and recycling services using historical data and insertion of sensor in waste container to identify the filling level.
3.	Economic Development	• Big Data technology was used by UN initiative, called Global Pulse, to strength global development. A sentiment analysis (SA) was conducted using social network messages and this allow them to predict job losses, outbreak of diseases, and reduction in spending within a region.
4.	Urban Transport	• Urban transport industry uses Big Data technologies to capture data real-time by installing magnetic sensors across the road network; GPS system can also be used to track public transport and monitoring system can be used to capture social media data. This has assisted various transport agencies to manage potential road traffic by passing accurate information to drivers through mobile applications providing them with alternative driving routes during traffic and indicating approximate driving times.
5.	Retail	• Big Data is used by organisation to tailor product selection for individual stores. The timing of price markdowns is determined using Big Data analytic. It gives access to real-time accurate information as relating purchase delivery.
6.	Health and Wellbeing	• Big Data is used for self-tracking of phenomenon by actively entering data or passively through the sensors, gadgets, and apps. It allows individual to track his/her own sleeping and eating habits, sex, moods, fertility patterns, physical and emotional health, blood and glucose levels.
7.	Predicting and Understanding of Crime	• In the past few decades, it was recorded that Los Angeles Police Department of the United States, collected data on more than 130 million crimes using a software, and update the software continually as crimes being committed. Because of these, the analysis has resulted in 33% reduction in burglaries, 21% in violent crime, 12% in property crimes and they were able to understand why crime occur in some regions. Some countries like UK, Canada, Australia, Italy now using same system to collect data of different crimes across their big cities.
8.	Fire Services	• With the use of Big Data, fire services collect data from different sources such as government open-source data, emergency vehicle GPS tracking, surveillance systems, smoke and fire detectors. The data collected by fire services are used to calculate risks emanating from the building, streets, transport link and waterways.

Source: (Walker, 2015)

5.5. The Challenges of Big Data

Gantz and Remsel (2012) and Yin and Kaynak, (2015) divides Big Data technologies into four sequential standards, i.e. data generation, data acquisition, data storage, and data analytics. In each of this data sequential standards, there are diverse challenges that comes with each. Some of the challenges of Big Data technologies are highlighted as follows (Spears & Gold, 2016, Paoletti & Dearden, 2017):

1. Performing real-time analysis on multiple high-velocity data streams using Big Data Analytics.
2. Difficulties in data management, i.e. processing of the data in real-time and archival storage of data.
3. Identifying the amount of data to be processed (>1TB/day) in real time is one of the challenges. The type of data coming from different sensors have different sampling rates and noise characteristics

and for effective Big Data analytic relevant features must be extracted from the data to identify process and part defects.

4. Data usability in Big Data, that is, does the data good enough to be used for further data analysis, because of the increase in dirty and imperfect data collected from different sources.
5. Inability to test distributed formulations of data mining algorithm that are suitable for Big Data technologies like MapReduce and Hadoop paradigm.

6. POSSIBLE FUTURE RESEARCH FOR 3D PRINTING, BIG DATA AND INDUSTRY 4.0.

Big Data and 3D printing technology are an integral part of the nine digital factories of industry 4.0 and widely discussed in public media, industry and academia. Currently, theory is being explored rigorously with various engineering disciplines, by data scientist and computer scientist. The application of Big Data and 3D printing technologies in the manufacturing industry is increasingly competitive and little research has been conducted in these areas.

Recently, within the industry and the academics, there are different novel research springing up as relating to 3D printing, Big Data and Industry 4.0 because of the interdisciplinary attribute of this research topic. The topic has the potential to attract the attention of various industries across the world. Some possible future research areas in 3D printing, Big Data and Industry 4.0 are highlighted, but are not limited to:

- *Intelligent Big Data Analytics for 3D Printing Technology/Manufacturing/Healthcare Applications*
  The nature of this research will explore new Big Data analytics using machine learning, deep learning, big data, Internet of Things or other techniques to enhance 3D printing processes/ manufacturing industry/supply chain management/production lines/healthcare. Application of intelligent Big Data Analytics would increase productivity, enhance performance and provide optimum output rates. Manufacturing industry can monitor the data gathered from the production process in real-time using intelligent systems. This could provide information needed to make a rightful decision on time and to maintain consistency in production (Cheng et al., 2013; Manogaran et al., 2017). Siriweera et al. (2015) recognised the conventional way of collecting, processing and storing data as time and resources consuming and suggests an automated real-time and online analytical approach as the greatest cognitive solution.
- *Knowledge-Based Process Planning for Big Data:*
  Increasingly numerous data are being generated within the manufacturing and medical sectors daily and the data gathered are good to support effective decision making. It has been recorded that manufacturing industries store lots of data than many other sectors and a future research can be carried out to investigate the appropriate techniques on knowledge of capturing bio-medical data, 3D printing and additive manufacturing data, Computer Numerical Controller (CNC) machines data, and how the data can be re-used to improve the operations of new product development. For instance, additive manufacturing service bureaus have few available techniques or tools for users to optimize the process and production plan from a multiple part production context, Zhang & Bernard (2014) dealt with this issue by introducing an AM feature and knowledge based systematic process planning strategy and a developing CAPP

system based on a systematic process planning framework for AM was presented for multiple parts production.

- *Big Data Analytics and Mining for Additive Manufacturing*:

Large and complex data set are very difficult to process using traditional, relational and sequential data processing applications. It is not about the size of the data collected but what is the impact and insight of the data analysis/mining that can be generated from this set of complex data. Basically, providing effective and efficient algorithms and tools for Big Data analytics and Data Mining is very vital. In this case, distributed approaches like Hadoop and MapReduce programming techniques can be used to investigate and test distributed formulations of Big Data mining algorithm that are suitable for data-driven and distributed paradigm of Big Data technologies or distributed computing such as MapReduce (Zhang, 2017).

- *Big Data Analysis from Real Time Streaming*:

Real-time streaming is the process of using real-time analytics to deliver information on business operations as they occur meaning "near-zero latency, i.e. the information is available anywhere from milliseconds to five second after the event, the data can be stored later or in parallel" (Scott, 2017). For data streaming model, data come at a very high speed and algorithm that will process such data must do so under very strict constraints of space and time. In a data stream mining, three things must be considered, namely; accuracy, amount of space needed, and the time required to predict (Bifet, 2015). In this type of research, a platform for mining Big Data streams can be developed and could assist to adjust the time and space used, influence accuracy, and stored precomputed information. Although, in 2015, De Francisci Morales and Bifet (2015) developed a platform for mini Big Data streams called SAMOA, (Scalable Advanced Massive Online Analysis), written in Java and is an open source, but still currently undergoing incubation at the Apache software foundation.

- *Monitoring Sensor of Big Data Analysis for 3D Printing Technologies, manufacturing environment and Other related industries such as healthcare*:

Most of the data collected on shop floor are not used to monitor or control the building process in real-time. The need for robust big data platforms to store, secure, archive and otherwise manage data; and monitoring of sensor data coming from printers and other connected devices within complex fabrication processes. Therefore, monitoring sensor need to be placed in the systems to monitor and control the process in real-time. An algorithm can be developed to analyse data coming from the different sensor. Relevant data can be extracted, and the data can be visualized using different big data analytics methods to handle these issues. According to Konyha and Banyai (2017) as relating to Industry 4.0, a smart manufacturing perspective, it is very important to support the processes with intelligent sensor networks; sensors are often used in manufacturing processes. The sensor network generates a large and increasing amount of data that needs to be processed. In such cases, computationally intensive algorithms are used for the data processing, such as (image, voice and signal processing, different classification functions, numeric optimization routines) (Konyha and Banyai, 2017). Manogaran et al. (2017) in a recent paper titled "Big Data Security Intelligence for healthcare industry 4.0"; proposes a secure Industrial Internet of Things (IoT) architecture for storing and processing scalable sensor data i.e. Big Data, for effective healthcare applications. Manogaran et al. (2017) in a paper proposed a Metal Cloud Redirection (MC-R) architecture using Big Data system. The sensor medical machines fixed into human body (patient) in order to collect clinical measures such as (blood pressure, body temperature, blood sugar level, heart rate and etc.) and through the sensor devices, an alert message would be sent to the Doctor containing the clinical value through a wireless network (Manogaran et al., 2017).

- *Big Data Visualization*

The goal of Big Data visualisation is to recognize interesting patterns and the correlations, and the dimensions of the data to be visualized need to be carefully selected. For instance, interesting patterns can be lost if low dimensions were chosen or if all the dimensions were used and this could result in low data visualization (Ali et al., 2016). Ali et al. (2016) concludes that interactivity of data visualization is highly important in the era of industry 4.0 and choosing the appropriate data visualization tools would produce effective interactive visualization. Data visualization is very important, and it allows people to understand data in a graphical manner. Most especially in the 'Big Data' era, data grows daily bigger and bigger. Currently, the traditional data visualization is inadequate to handle Big Data and it is becoming more demanding. Big Data visualization comes with challenges and opportunities, and the challenges include perceptual scalability, real-time scalability, and interactive scalability. While the opportunities of Big Data visualization present better ways to visualize Big Data such as data reduction, reducing latency, etc. In this case, data visualization tools and techniques can be developed to help the users to identify missing, erroneous or duplicate values (Kadadi et al., 2015).

7. CONCLUSION

Industry 4.0 comprises of nine digital technologies; and for industry 4.0 to reach its full potential it cannot work in isolation of other systems. The recent trend and activities of 3D printing and Big Data have shown that they have significant roles to play in the next industrial revolution. Research in this area are still at the infant stage, and more research needs to be explored in collaboration with academia and manufacturing industry across the world. Within the healthcare sector, Big Data and Analytics has great potential to unlock an immense productivity bottlenecks and radically improving the quality and accessibility of the healthcare system globally (Heinrich et al., 2016). Likewise, application of 3D printing technology in the healthcare industry is very exceptional and such applications includes customization and personalization of medical products, drugs and equipment, tissue and organ fabrication, creation of customized prosthetics, implants and anatomical models (Ventola, 2014).

Finally, lot of data are being generated or collected on the shop floor within the manufacturing industry environment daily, such as large and small scale industrial 3D printing lab, Computer Numerical Controller (CNC) machines, or a biomedical environment, etc. Most of these data are not useful because they are not being analysed; to assist manufacturing industries to become a data-driven organization and to facilitate rightful decision making in businesses. This paper has been able to present recent trends in 3D printing and Big Data technologies; the challenges, and possible future research areas.

REFERENCES

Agrawal, D., Bernstein, P., Bertino, E., Davidson, S., Dayal, U., Franklin, M., & Widom, J. (2012). Challenges and Opportunities with Big Data (white paper). Computing Research Association. Retrieved from http://cra.org/ccc/resources/ccc-led-whitepapers/

Ali, S. M., Gupta, N., Nayak, G. K., & Lenka, R. k. (2016). Big Data Visualization: Tools and Challenges. In *2nd International Conference on Contemporary Computing and Informatics* (IC3I) (pp. 656-660).

Allen, M. (2016). The Challenge of Big Data – It's more than just big files. Retrieved 30 Oct. 2017 from https://pro2col.com/challenge-big-data-more-than-just-big-files/

AspireSys. (2017). Big Data with NoSQL (White Paper). Retrieved from www.aspiresys.com/White-Papers/BigData_with_NoSQL_Whitepaper.pdf?pdf=nosql-whitepaper

BCG. (2016). The Boston Consulting Group. Sprinting to Value in Industry 4.0: Perspective from and Implications for U.S Manufacturers. Retrieved 30 Oct. 2017 from https://www.slideshare.net/TheBostonConsultingGroup/sprinting-to-value-in-industry-40

Beyer, M. A., & Laney, D. (2012). *The Importance of 'Big Data': A Definition*. Gartner Inc.

Bifet, A. (2015). Real-Time Big Data Stream Analytics. In *Proceedings of the 2nd Annual International Symposium on Information Management and Big Data (SIMBig 2015)*, Cusco, Peru, September 2-4.

Bull, G., Maddox, C., Marks, G., McAnear, A., Schmidt, D. & Schrum, L. (2010). Educational implications of the Digital Fabrication Revolution. *Journal of Research on Technology in Education*, *42*(4).

Chamberlin, B. (2014). 3D Printing – A 2014 HorizonWatching Trend Summary Report. Retrieved 06 Nov. 2017 from https://www.slideshare.net/HorizonWatching/s12-3-d-printing-2014-horizonwatching-trend-summary-report-17apr2014

Cheng, S., Shi, Y., Qin, Q., & Bai, R. (2013). *Swarm Intelligence in Big Data Analytics. In Intelligent Data Engineering and Automated Learning (IDEAL), LNCS* (Vol. 8206, pp. 417–426). Berlin, Heidelberg: Springer.

Curran, C. (2016). The road ahead for 3-D printing. Retrieved 22 Oct. 2017 from http://usblogs.pwc.com/emerging-technology/the-road-ahead-for-3d-printing/

De Francisci Morales, G., & Bifet, A. (2015). SAMOA: Scalable Advanced Massive Online *Analysis. Journal of Machine Learning Research*, *16*, 149–153.

Dean, J. (2014). *Big Data, Data Mining, and Machine Learning: Value Creation for Business Leaders and Practitioners*. SAS Institute Inc. doi:10.1002/9781118691786

Dean, J., & Ghemawat, S. (2004). *MapReduce: Simplified Data Processing on Large Clusters*. Google, Inc.

DiFatta, G. (2017). Big Data Analytics and Mining: investigating and testing distributed formulations of data mining algorithms that are suitable for the MapReduce paradigm and for other distributed computing approaches. Retrieved 07 Nov. 2017 from https://www.findaphd.com/search/projectdetails.aspx?PJID=81074&LID=4021

Eastwood, G. (2017). How big data Is driving technological innovation. *Networkworld*. Retrieved 30 Oct. 2017 from https://www.networkworld.com/article/3208764/big-data/how-big-data-is-driving-technological-innovation.html

Eltabakh, M. Y. (2012). Hadoop: A Framework for Data-Intensive Distributed Computing CS561-Spring. Retrieved from https://web.cs.wpi.edu/~cs561/s12/Lectures/6/Hadoop.pdf

Foidl, H., & Felderer, M. (2016). *Research Challenges of Industry 4.0 for Quality Management. In Innovations in Enterprise Information Systems Management and Engineering* (Vol. 245, pp. 121–137). Cham: Springer.

Gantz, J., & Remsel, D. (2012). The digital universe in 2020: Big data, bigger digital shadows and biggest growth in the far east. In *Proc. IDC iView, IDC Anal.*

Geissbauer, R., Vedso, J., & Schrauf, S. (2016). Industry 4.0: Building the digital enterprise. In *2016 Pwc Global Industry 4.0 Survey*. Retrieved from https://www.pwc.com/gx/en/industries/industries-4.0/landing-page/industry-4.0-building-your-digital-enterprise-april-2016.pdf

Gross, B. C., Erkal, J. L., Lockwood, S. Y., Chen, C., & Spence, M. D. (2014). Evaluation of 3D printing and its potential impact on biotechnology and the chemical sciences. *Journal of Analytical Chemistry*, *86*(7), 3240–3253. doi:10.1021/ac403397r PMID:24432804

Heinrich, A., Menasalvas, E., Lojo, A., & González, R. A. (2016). Big Data Technologies in Healthcare: Needs, opportunities and challenges. *Big Data Value Association*. Retrieved from http://www.bdva.eu/sites/default/files/Big%20Data%20Technologies%20in%20Healthcare.pdf

Home Shop 3D Printing (HS3DP). 2016. 3D Printing Process and Technologies. Retrieved 07 Nov. 2017 from http://homeshop3dprinting.com/3d-printing-qa/3d-printing-process-and-technologies/

Hughes, A. (2017). Industry 4.0 is About More Than Data: 3D Printing in Manufacturing. *LNSResearch*. Retrieved 21 Oct. 2017 from http://blog.lnsresearch.com/industry-4.0-is-more-than-data-3d-printing-in-manufacturing

Hughes, R. (2015). Big data is pushing 3D printing to its tipping point. IBM Bigdatahub. Retrieved 09 Nov. 2017 from http://www.ibmbigdatahub.com/blog/big-data-pushing-3d-printing-its-tipping-point

I-Scoop. (2017). Industry 4.0: the fourth industrial revolution – guide to Industrie 4.0. Retrieved 26 Oct. 2017 from https://www.i-scoop.eu/industry-4-0/

IBM. (2014). IBM Big Data & Analytics Hub: The Four V's of Big Data. Retrieved 26 Oct. 2017 from http://www.ibmbigdatahub.com/infographic/four-vs-big-data

Johnsen, M. (2017). The Future of Artificial Intelligence in Digital Marketing: The next big technological break.

Kadadi, A., Agrawal, R., Dai, X., & Andres, F. (2015). Challenges and Opportunities with Big Data Visualization. In *MEDES '15*, October 25-29, Caraguatatuba, Brazil (pp.169-173).

Kaur, M., & Shah, A. (2017). A Novel Framework for Big Data Security Infrastructure Components. *International Journal on Recent and Innovation Trends in Computing and Communication*, *5*(6), 1083–1088.

Konur, S. (2017). Intelligent Big Data Analytics for Manufacturing and Engineering Applications. Retrieved 6 Nov. 2017 from https://www.findaphd.com/search/projectdetails.aspx?PJID=87033&LID=3247

Konyha, J., & Bányai, T. (2017). Sensor Networks for Smart Manufacturing Processes. *Diffusion and Defect Data, Solid State Data. Part B, Solid State Phenomena*, *261*, 456–462. doi:10.4028/www.scientific.net/SSP.261.456

Laney, D. (2012). Others Claiming Gartner's Construct for Big Data. *Gartner*. Retrieved 26 Oct. 2017 from https://blogs.gartner.com/doug-laney/deja-vvvue-others-claiming-gartners-volume-velocity-variety-construct-for-big-data/

MacLean, D. (2011). A Very Brief Introduction to MapReduce for CS448G. Retrieved from http://hci.stanford.edu/courses/cs448g/a2/files/map_reduce_tutorial.pdf

Manogaran, G., Thota, C., Lopez, D., & Sundarasekar, R. (2017). *Big Data Security Intelligence for Healthcare Industry 4.0. Cybersecurity for Industry 4.0.* Springer Series in Advanced Manufacturing. doi:10.1007/978-3-319-50660-9_5

MongoDB. (2017a). Big Data Explained. Retrieved 13 Nov. 2017 from https://www.mongodb.com/big-data-explained

MongoDB. (2017b). Real-Time Analytics. Retrieved 13 Nov. 2017 from https://www.mongodb.com/use-cases/real-time-analytics

Mousavi, C. (2011). What is Big Data? What are NoSQL databases? What is Hadoop, Pig, Hive? Retrieved 07 Dec 2017 from https://catmousavi.wordpress.com/what-is-big-data-what-are-nosql-databases-what-is-hadoop-pig-hive/

Mukherjee, S., & Shaw, R. (2016). Big Data – Concepts, Applications, Challenges and Future Scope. *International Journal of Advanced Research in Computer and Communication Engineering*, *5*(2).

Nida, R. (2015). Big data and analytics: Which type of analytics does your business need? *Smartshifttech*. Retrieved 07 Nov. 2017 from http://www.smartshifttech.com/big-data-and-analytics-which-type-analytics-does-your-business-need

Paoletti, P., & Dearden, G. G. (2017). Sensor data analysis for additive manufacturing process monitoring. Retrieved 06 Nov. 2017 from https://www.findaphd.com/search/projectdetails.aspx?PJID=86052

Parmar, R., Cohn, D. L., & Marshall, A. (2014). *Driving innovation through data. IBM Global Business Services*. IBM Institute for Business Value.

Ross, A. (2016). *The Industries of the Future: Summary & Analysis*. Instaread.

RP-PLATFORM. (2017). How 3D printing technology will be a key part of Industry 4.0. Retrieved 09 Nov. 2017 from https://rpplatform.com/2017/08/21/3d-printing-technology-industry-4-0/

Schubert, C., van Langeveld, M. C., & Donoso, L. A. (2014). Innovations in 3D printing: A 3D overview from optics to organs. *The British Journal of Ophthalmology*, *98*(2), 159–161. doi:10.1136/bjophthalmol-2013-304446 PMID:24288392

Scott, S. (2017). Real-Time Analytics: Streaming Big Data for Business Intelligence. *Dzone*. Retrieved 09 Nov. 2017 from https://dzone.com/articles/real-time-analytics-streaming-big-data-for-busines

Shah, P. (2016). Four Types of Big Data Analytics and examples of their use. *LinkedIn*. Retrieved 07 Nov. 2017 from https://www.linkedin.com/pulse/four-types-big-data-analytics-examples-use-parvez-shah/

Singh, S., Singh, P., Garg, R., & Mishra, P K, P. K. (2015). Big Data: Technologies, Trends and Applications. *International Journal of Computer Science and Information Technologies*, *6*(5), 4633–4639.

Siriweera, T. H. A. S., Paik, I., Kumara, B. T. G. S., & Koswatta, K. R. C. (2015). Intelligent Big Data Analysis Architecture based on Automatic Service Composition. In *IEEE International Congress on Big Data*. 10.1109/BigDataCongress.2015.46

Smith, A. (2017). Big Data Analytics: Challenges and Implementation. Retrieved 07 Nov. 2017 from http://www.hiddenbrains.com/blog/big-data-analytics-challenges-implementation.html

Spears, T. G., & Gold, S. A. (2016). In-process sensing in selective laser melting (SLM) additive manufacturing. Integrated Materials and Manufacturing Innovation, 5(2).

Su, X. (2017). Introduction to Big Data (Department of Informatics and e-learning at NTNU Learning material is developed for course IINI3012 Big Data).

Taylor, R. C. (2010). An overview of the Hadoop/MapReduce/HBase framework and its current applications in bioinformatics. *Taylor BMC Bioinformatics*, *11*(Suppl. 12).

Van Rijmenam, M. (2013). Why The 3V's Are Not Sufficient To Describe Big Data. *Datafloq*. Retrieved 26 Oct. 2017 from https://datafloq.com/read/3vs-sufficient-describe-big-data/166

Van Rijmenam, M. (2014). *Think Bigger: Developing a Successful Big Data Strategy for Your Business. Publisher*. AMACOM.

Vassakis, K., Petrakis, E., & Kopanakis, I. (2018). Big Data Analytics: Applications, Prospects and Challenges. In *Mobile Big Data* (pp. 3–20). Cham: Springer. doi:10.1007/978-3-319-67925-9_1

Ventola, C. L. (2014). Medical Applications for 3D Printing: Current and Projected Uses. *P&T*, *39*(10), 704–711. PMID:25336867

Walker, B. (2015). Every day Big Data Statistics – 2.5 Quintillion Bytes of Data Created Daily. *Vcloudnews*. Retrieved 26 Oct. 2017 from http://www.vcloudnews.com/every-day-big-data-statistics-2-5-quintillion-bytes-of-data-created-daily/

Wang, L., & Alexander, C. A. (2015). Big Data in Medical Applications and Health Care. *Current Research in Medicine*, *6*(1). doi:10.3844/amjsp.2015.1.8

Wang, L., & Alexander, C. A. (2016). b). Additive Manufacturing and Big Data. *International Journal of Mathematical. Engineering and Management Sciences*, *1*(3), 107–121.

Wanga, L., & Wang, G. (2016a). Big Data in Cyber-Physical Systems, Digital Manufacturing and Industry 4.0. *International Journal of Engineering and Manufacturing*, *4*.

Wikipedia. (2017). MongoDB. Retrieved 13 Nov. 2017 from https://en.wikipedia.org/wiki/MongoDB

Wohlers, T. (2010). *State of the Industry, Wohlers Annual Worldwide Progress Report*. Wohlers Associates.

Wohlers Associates, Inc. (2010). what is Additive Manufacturing? Retrieved 06 Nov. 2017 from https://wohlersassociates.com/additive-manufacturing.html

Yin, S., & Kaynak, O. (2015). Big Data for Modern Industry: Challenges and Trends. *Proceedings of the IEEE*, *103*(2), 143–146. doi:10.1109/JPROC.2015.2388958

Zhang, X. (2017). Knowledge based process planning using manufacturing Big Data. Retrieved 06 Nov. 2017 from https://www.findaphd.com/search/ProjectDetails.aspx?PJID=88039&LID=1921

Zhang, Y., & Bernard, A. (2014). AM Feature and Knowledge Based Process Planning for Additive Manufacturing in Multiple Parts Production Context. In *Proceedings of 25th Annual. International Solid Freeform Fabrication Symposium* (pp.1259–1276).

This research was previously published in the International Journal of Big Data and Analytics in Healthcare (IJBDAH), 2(2); edited by Mu-Yen Chen, pages 1-20, copyright year 2017 by IGI Publishing (an imprint of IGI Global).

Index

3D Printer 15, 74, 116-118, 120-121, 123, 126, 128, 131, 137, 144, 181, 196, 335, 339-341, 343, 346, 352, 424, 434
3D Printing 4, 9, 15, 17, 19, 24, 74, 86-88, 91-93, 96, 99, 102, 107, 111-112, 116-124, 126, 128-129, 131, 136-138, 142, 164-165, 180, 199, 220, 244, 259, 270, 323, 332-341, 344, 346-348, 352, 368, 424-425, 430-431, 434-435, 443, 471, 478-479, 481-482, 493-494, 502-505, 507-508, 511, 513, 515
3D Scanners 163, 333
3DCloud 126-130

A

Adaptive Façades 37
Additive Manufacturing 2, 4, 9-10, 24, 44-46, 52-53, 55, 66, 73-74, 76-77, 86-88, 91-93, 96, 114, 116-117, 123, 129, 137-138, 165, 180, 198-199, 207-208, 220-221, 223-224, 226-228, 232, 240, 244, 258, 272, 278, 296-298, 300, 321-322, 333-334, 352-353, 355, 368, 370, 373, 389, 394-395, 399, 408, 420, 443-444, 447, 469, 479, 481-482, 484, 502-503, 507
Architectural Heritage 163-165, 168, 199
Arm 112, 259, 276, 281, 301, 353, 394-396, 399, 403, 408, 410-414, 419, 423, 484
Assistance 296, 397, 402-403, 423, 434

B

Big Data 503-505, 507-513, 515
Big Data Analytics 505, 511
Bimodal Structure 270
Biofouling 246, 270
Biomechanical Evaluation 415, 423
Business Modeling 116, 123-124

C

CAD 4, 9-10, 32, 40-41, 53, 65-67, 93-94, 100, 103-104, 117-118, 121, 127, 198, 220, 225, 244, 259, 299, 302, 322, 326-329, 333, 353, 355-356, 368, 412, 433, 435-436, 483, 485, 507
CAE 53, 65-67, 328-329
Cloud based 3D Printing 116, 123-124, 131
Cloud Computing 117, 121-123, 126, 131, 479, 505, 507
Coefficient of Friction (COF) 244
Collaboration 6, 86-88, 93-94, 96, 108, 110, 112, 120, 126, 128, 131, 165, 198-199, 296, 303, 336, 447, 462, 471, 505, 515
complexity measure 372, 389
Composite 38, 90, 147, 220, 224, 237-241, 244, 249, 255, 259-260, 273, 278, 280, 283, 309, 326, 483
Computational Simulation 271, 273, 284
Computer Aided Design 327, 333
Computer Aided Engineering 328
Core 4, 20, 40-41, 111, 121, 313, 321-322, 326-329, 342, 346, 399, 406-408, 471, 494

D

Dental implants 424, 435
Design for Additive Manufacturing 52, 73, 76-77
Design for Six Sigma 52, 55
Diffusion of Innovation 487
Digital Fabrication 25-26, 30, 37-38, 44
Disruptive creativity 86
Disruptive Innovation 88, 92, 111, 478, 480-481, 485, 494, 502

Disruptive Technology 111, 117, 480-481, 494, 502
Dynamic Mechanical Analysis 353, 357-358, 365, 368

E

Early Stage 89, 136-142, 144, 150-154, 310, 327, 332, 470
Electronics 5, 10, 13, 16, 19, 118, 120, 232, 300, 335, 340, 342-343, 345, 405, 461
End-Effector 31-33, 397, 404, 423
Exoskeleton 410-415, 423

F

FDM 16, 74, 120, 180, 298, 300-301, 306-307, 309, 313, 323, 329, 333, 354-357, 359-363, 365-366, 371, 408, 430, 433, 443, 482-483, 489
FEA 46, 66, 70, 72
FGM 239-240, 244
Forward Citation 488-489, 502
Fused Deposition Modeling 16, 99, 180, 298, 300, 323-324, 333, 336, 353-355, 357, 359, 368, 371, 408, 430, 433, 443, 482-483, 493

G

Generation Y (Gen Y) 24
Green Technology 321
Growth Curve 487, 490, 502

H

Healthcare 339, 342, 485, 503-504, 515
Human-Centered 1-2, 4-5, 8, 11, 14, 19-20, 24

I

Implantology 435, 443
Industrial Design 9-10, 24, 86-87, 91, 94, 97, 111-112
Industry 4.0 42, 74, 503-506, 513, 515
Inkjet-based system 431
Innovation Diffusion 139, 486, 489, 502
Internet Of Things 7-10, 24, 122, 336, 340, 348, 352, 479, 504-505
investment casting 296-297, 306-307, 310-312, 433
IoT 7, 24, 122, 332-348, 352, 505

L

Laminated Object Manufacturing 298, 302, 326-327, 354, 356-357, 368, 482-483, 493
Laser Beam Diameter 226, 244
Laser Beam Profile 274
Laser Metal Deposition 220-221, 223-229, 232-233, 240-241, 244-245, 259-260, 270, 279
Laser Surface Modification Techniques 271, 275
Laser-Material Interaction 273-275
Layer Thickness 207-209, 213, 260, 300-301, 306, 323, 355, 362, 364-365, 368
Linear Coefficient of Expansion 251, 270
LOM 298, 301-302, 311, 326, 329, 354, 356-360, 365-366, 482-483, 489

M

Mass Customization 26, 37, 76, 116, 120, 122, 131, 447, 461, 471, 484
Mechanical properties 46, 74, 76, 118, 208, 224, 228, 246, 249-250, 252-255, 260, 266, 271, 282, 299, 323, 326, 354, 358, 374, 389, 483, 485
Microhardness 220, 233, 236-239, 260, 263, 265
Microstructure 220, 222, 227, 233-238, 246, 252, 254, 260-261, 263, 272, 276-283, 292, 370
Model Interior 362, 364, 368
Modern 5, 10, 52-54, 116-117, 131, 163-164, 338, 348, 394, 447, 481-482, 484
MTBF 347, 352
Multi Jet Fusion 53, 73-76, 81, 83
Multi-Objective Optimization 213-214, 217, 373

N

Neurological Patients 396, 409, 416, 423
New product 53, 321, 332-335, 337, 341-342, 345-348, 486
Nontraditional Manufacturing Process 244

O

Ontogenetic 2, 4, 19, 24
Orientation 33, 42, 172-173, 186, 191, 208, 306, 353, 356-357, 361-362, 365-366, 368, 408, 412, 436

P

Part Orientation 208, 306, 353, 368
Patent Analysis 478, 480, 485, 487-488, 502
Patterned Microstructure 276
Powder Metallurgy 246, 258, 270
product complexity 373
Product Development 8, 11, 52-54, 56, 208, 313, 321-322, 327, 332, 334-335, 337, 339, 341-342, 345, 347

product life cycle management 335
Prosthesis 425, 435-436, 443
Prototype 11, 13-14, 31-33, 94, 117, 297, 299-300, 310, 322, 327-329, 335, 342, 354, 356, 361, 404-406, 408, 413, 419, 424, 430, 434, 449, 483
Prototyping 52, 73-74, 86, 88, 91, 93, 116-118, 122-123, 128, 131, 138, 180, 208, 277, 298, 302, 309, 321-323, 327-329, 332, 335-338, 342, 346, 353-356, 358, 365, 368, 370, 424, 433, 435-436, 443, 469, 482, 485
PSO 207-208, 214-217

Q

QFD 52-53, 55-56, 61
quality characteristics 296, 306

R

Rapid Prototyping 52, 91, 93, 118, 122-123, 131, 138, 180, 208, 277, 298, 321-323, 332, 353-354, 356, 365, 368, 424, 433, 435-436, 443, 469, 482, 485
Rehabilitation 394-399, 401-404, 406-409, 420, 423-425, 435
Relationships 1-2, 10, 14, 17-18, 26, 61, 63, 112-113, 128, 150-151, 153, 280, 346, 466, 470
Research teams 86, 88, 90, 110, 112-113
Risk 9, 11, 53, 88, 90-91, 110, 152, 396, 483
Robotic Fabrication 30-32, 37
Robots 3, 18, 25, 30-31, 340, 395-397, 423, 505

S

Scanning Speed 226-227, 233, 237, 244, 260, 262-263, 265, 275, 279, 283
Selective Laser Sintering 75, 98-100, 278, 298, 301, 323, 325, 333, 354-355, 368, 371, 430, 432, 435, 444, 482-483, 493
Smart Mirror 352
Stasis 2, 24
Stereolithography 99, 180-181, 208-209, 322, 354-355, 357, 368, 425, 430, 435, 444, 481-483, 493
Sustainability 4, 6-10, 16-17, 19-20, 208, 217
Systems 1-4, 7-8, 10-11, 14-15, 17-20, 25-28, 37-40, 45, 53, 55, 65-66, 70, 90, 121-122, 136, 164-166, 168, 170, 175, 186, 191, 196-199, 208, 248, 272, 277, 280, 296-299, 302, 305, 311, 322, 327-329, 337-340, 372, 397, 399, 415-416, 436, 446-450, 453-454, 457-458, 460, 462, 464, 467-469, 471, 479-480, 482-483, 485, 503, 505, 508, 515

T

Tensile Modulus 365, 369
Textile 16, 39, 94, 101-102
Time to Market 52, 91, 328-329, 336
tissue and organ printing 371
tissue engineering 370, 373, 375, 389
Topographic 163-164, 169, 171, 481
Topology Optimization 10, 44-46, 76, 81, 83

U

Ubiquitous Computing 1-2, 7-9, 11-14, 18-19, 24
Ultimate Tensile Strain 365, 369
Ultimate Tensile Stress 364, 369
User Acceptance 136-140, 142, 153
UTAUT2 136-143, 148, 150-155

V

Variable Stiffness Actuators 399, 423
Virtual Communities 117

www.ingramcontent.com/pod-product-compliance
Lightning Source LLC
LaVergne TN
LVHW081249011026
849162LV00006BA/1230